Advances in fire retardant materials

Related titles:

Fire retardant materials
(ISBN 978-1-85573-419-7)
This authoritative reference work provides a comprehensive source of information for readers concerned with all aspects of fire retardancy. The emphasis is on the burning behaviour and flame retarding properties of polymeric materials. It covers combustion, flame retardants, smoke and toxic products, and then concentrates on some more material-specific aspects of combustion in relation to textiles, composites and bulk polymers. Developments in all areas of fire retardant materials are covered, including research into nanocomposites.

Flammability testing of materials used in construction, transport and mining
(ISBN 978-1-85573-935-2)
Flammability testing of materials used in construction, transport and mining provides an authoritative guide to current best practice in ensuring safe design of testing procedures. The book uses practical examples via the careful selection of case studies with an overall emphasis on specific types of tests that are widely accepted and internationally practised in a variety of applications. This book simplifies the difficult and often confusing area of national regulations and fire test procedures.

Flame retardants for plastic
(ISBN 978-1-85957-385-3)
This technical market report highlights the recent work on flame retardants by different companies and for different resins; it describes the situation of flux in the marketplace with the new changes to legislation and gives data on the market size and possible future changes.

Details of these and other Woodhead Publishing materials books, as well as materials books from Maney Publishing, can be obtained by:

- visiting our web site at www.woodheadpublishing.com
- contacting Customer Services (e-mail: sales@woodhead-publishing.com; fax: +44 (0) 1223 893694; tel.: +44 (0) 1223 891358 ext. 130; address: Woodhead Publishing Limited, Abington Hall, Granta Park, Great Abington, Cambridge CB21 6AH, England)

If you would like to receive information on forthcoming titles, please send your address details to: Francis Dodds (address, tel. and fax as above; e-mail: francisd@woodhead-publishing.com). Please confirm which subject areas you are interested in.

Maney currently publishes 16 peer-reviewed materials science and engineering journals. For further information visit www.maney.co.uk/journals.

Advances in fire retardant materials

Edited by
A. R. Horrocks
and
D. Price

Woodhead Publishing and Maney Publishing
on behalf of
The Institute of Materials, Minerals & Mining

CRC Press
Boca Raton Boston New York Washington, DC

WOODHEAD PUBLISHING LIMITED
Cambridge England

Woodhead Publishing Limited and Maney Publishing Limited on behalf of
The Institute of Materials, Minerals & Mining

Published by Woodhead Publishing Limited, Abington Hall, Granta Park,
Great Abington, Cambridge CB21 6AH, England
www.woodheadpublishing.com

Published in North America by CRC Press LLC, 6000 Broken Sound Parkway, NW,
Suite 300, Boca Raton, FL 33487, USA

First published 2008, Woodhead Publishing Limited and CRC Press LLC
© 2008, Woodhead Publishing Limited
The authors have asserted their moral rights.

This book contains information obtained from authentic and highly regarded sources. Reprinted material is quoted with permission, and sources are indicated. Reasonable efforts have been made to publish reliable data and information, but the authors and the publishers cannot assume responsibility for the validity of all materials. Neither the authors nor the publishers, nor anyone else associated with this publication, shall be liable for any loss, damage or liability directly or indirectly caused or alleged to be caused by this book.

Neither this book nor any part may be reproduced or transmitted in any form or by any means, electronic or mechanical, including photocopying, microfilming and recording, or by any information storage or retrieval system, without permission in writing from Woodhead Publishing Limited.

The consent of Woodhead Publishing Limited does not extend to copying for general distribution, for promotion, for creating new works, or for resale. Specific permission must be obtained in writing from Woodhead Publishing Limited for such copying.

Trademark notice: Product or corporate names may be trademarks or registered trademarks, and are used only for identification and explanation, without intent to infringe.

British Library Cataloguing in Publication Data
A catalogue record for this book is available from the British Library.

Library of Congress Cataloging in Publication Data
A catalog record for this book is available from the Library of Congress.

Woodhead Publishing Limited ISBN 978-1-84569-262-9 (book)
Woodhead Publishing Limited ISBN 978-1-84569-507-1 (e-book)
CRC Press ISBN 978-1-4200-7961-6
CRC Press order number WP7961

The publishers' policy is to use permanent paper from mills that operate a sustainable forestry policy, and which has been manufactured from pulp which is processed using acid-free and elementary chlorine-free practices. Furthermore, the publishers ensure that the text paper and cover board used have met acceptable environmental accreditation standards.

Project managed by Macfarlane Book Production Services, Dunstable, Bedfordshire, England (e-mail: macfarl@aol.com)
Typeset by Godiva Publishing Services Limited, Coventry, West Midlands, England
Printed by TJ International Limited, Padstow, Cornwall, England

Contents

	Contributor contact details	xiii
1	**Introduction**	1
	A R HORROCKS and D PRICE, University of Bolton, UK	
Part I	**Advances in fire retardant materials**	
2	**Flame retardancy of textiles: new approaches**	9
	S BOURBIGOT, Ecole Nationale Supérieure de Chimie de Lille (ENSCL), France	
2.1	Introduction	9
2.2	Developments in assessing levels of flame retardancy of textiles	10
2.3	Flame retardant natural fibres	14
2.4	Flame retardant synthetic fibres	23
2.5	Flame retardant inorganic man-made fibres	34
2.6	Future trends	36
2.7	References	37
3	**Developments in phosphorus flame retardants**	41
	S V LEVCHIK, Supresta LLC, USA and E D WEIL, Polytechnic University of Brooklyn, USA	
3.1	Introduction	41
3.2	Main types of phosphorus flame retardants	42
3.3	Polycarbonate and its blends	43
3.4	Polyesters and nylons	45
3.5	Epoxy resins	48
3.6	Polyurethane foams	50
3.7	Polyolefins	51
3.8	Future trends	55
3.9	Sources of further information and advice	57

Contents

3.10	Conclusions	60
3.11	References	60

4 Halogen-free flame retardants — 67
Y-Z WANG, Sichuan University, China

4.1	Introduction	67
4.2	Challenges posed by replacing halogen-containing flame retardants	67
4.3	Attempts at successful halogen replacement in special applications	69
4.4	Flame retardancy and anti-dripping properties	73
4.5	Flame retardancy and mechanical properties	76
4.6	Future trends	86
4.7	Acknowledgements	88
4.8	Sources of further information and advice	88
4.9	References	88

5 Nanocomposites I: Current developments in nanocomposites as novel flame retardants — 95
C A WILKIE, Marquette University, USA and A B MORGAN, University of Dayton Research Institute, USA

5.1	Introduction	95
5.2	Brief history of polymer nanocomposites in flame retardancy applications	97
5.3	Nanoparticles for polymer nanocomposite use	98
5.4	Polymer nanocomposite formation	103
5.5	Polymer nanocomposite flammability – measured effects without conventional flame retardants	106
5.6	Mechanism of flammability reduction	108
5.7	Polymer nanocomposites combined with conventional flame retardants	112
5.8	Applications for flame retardant polymer nanocomposites	114
5.9	Future trends and conclusions	115
5.10	Acknowledgements	117
5.11	Sources of further information and advice	117
5.12	References	117

6 Nanocomposites II: Potential applications for nanocomposite-based flame retardant systems — 124
A R HORROCKS, University of Bolton, UK

6.1	Introduction	124
6.2	Applications and processing challenges for nanocomposite polymers	125

6.3	Potential application areas	133
6.4	Future trends	153
6.5	References	153

| 7 | **Flame retardant/resistant textile coatings and laminates** | **159** |

A R HORROCKS, University of Bolton, UK

7.1	Introduction	160
7.2	Main types of fire retardant/resistant coatings and laminates	161
7.3	Halogen replacement in back-coated textile formulations	171
7.4	Role of nanoparticles	174
7.5	Introduction of volatile phosphorus-containing species	175
7.6	Novel or smart ways of introducing flame retardant coatings to textiles and laminates	179
7.7	Plasma-initiated coatings	181
7.8	References	184

| 8 | **Environmentally friendly flame resistant textiles** | **188** |

P J WAKELYN, National Cotton Council of America (retired), USA

8.1	Introduction	188
8.2	Key environmental/ecotoxicological issues	189
8.3	Desirable properties of environmentally friendly flame retardant chemicals	190
8.4	Examples of potentially environmentally friendly flame retardant applications for cotton-based textiles	193
8.5	Potentially environmentally friendly flame retardant applications for mattresses, bedclothes and upholstered furniture	197
8.6	Inherently flame retardant fibers	201
8.7	Risk management, risk mediation and risk reduction	204
8.8	Future trends	206
8.9	Acknowledgements	208
8.10	References	208

| 9 | **Recycling and disposal of flame retarded materials** | **213** |

A CASTROVINCI, University of Applied Sciences of Southern Switzerland, Switzerland, M LAVASELLI, Centro di Cultura per l'Ingegneria delle Materie Plastiche, Italy and G CAMINO, Politecnico di Torino, Italy

9.1	Introduction on flame retardants	213
9.2	Recycling	214
9.3	Directives on recycling of flame retarded materials	220

9.4	Recyclability of flame retarded polymers	221
9.5	Conclusions	223
9.6	References	224

Part II Testing, regulation and assessing the benefits of fire retardant materials

10 Challenges in fire testing: a tester's viewpoint 233
M L JANSSENS, Southwest Research Institute, USA

10.1	Introduction	233
10.2	Type I flammability tests	234
10.3	Type II.A flammability tests	240
10.4	Type II.B flammability tests	246
10.5	Challenges in assessing material flammability	249
10.6	Use of material flammability tests	249
10.7	Limitations of material flammability tests	251
10.8	Future trends in material flammability assessment	252
10.9	Sources of further information and advice	253
10.10	References	253

11 Challenges in fire testing: reaction to fire tests and assessment of fire toxicity 255
T R HULL, University of Central Lancashire, UK

11.1	Introduction	255
11.2	Fire testing – principles and problems	262
11.3	Fire resistance tests	263
11.4	Reaction to fire tests	263
11.5	Measurement of heat release	269
11.6	Fire toxicity	275
11.7	Future trends: prediction of fire behaviour from material properties and models of fire growth	286
11.8	Sources of further information and advice	287
11.9	References	287

12 New and potential flammability regulations 291
J TROITZSCH, Fire and Environmental Protection Service, Germany

12.1	Introduction: overview of present fire safety regulations in Europe	291
12.2	Building	293
12.3	Fire testing of construction products in the European Union	301

12.4	Fire safety requirements and tests for electrical engineering and electronics equipment	309
12.5	Fire safety requirements and tests in transportation for rail vehicles and ships	320
12.6	Fire safety requirements and tests in furniture	324
12.7	Future trends	326
12.8	Sources of further information and advice	329
12.9	References	329

13 Life cycle assessment of consumer products with a focus on fire performance 331

M Simonson and P Anderson, SP Technical Research Institute of Sweden, Sweden

13.1	Introduction	331
13.2	Life Cycle Assessment	332
13.3	Television case study	337
13.4	Cable case study	343
13.5	Furniture case study	349
13.6	Conclusions	355
13.7	Future trends	356
13.8	References	357

Part III Applications of fire retardant materials

14 The risks and benefits of flame retardants in consumer products 363

A M Emsley and G C Stevens, University of Surrey, UK

14.1	Background	363
14.2	The importance of understanding risks and benefits	364
14.3	Primary risks of flame retardants in consumer products	369
14.4	Balance of risk	373
14.5	Benefits of flame retardants in consumer products	377
14.6	Differentiating the effects of smoking and smoke alarm trends	381
14.7	Future trends	391
14.8	Sources of information and advice	394
14.9	References	396

15 Composites having improved fire resistance 398

B K Kandola and E Kandare, University of Bolton, UK

15.1	Introduction	398
15.2	Composites and their constituents	400
15.3	Key issues and performance requirements for different sectors	404

15.4	Flammability of composites and their constituents	407
15.5	Fire retardant solutions	416
15.6	Mechanical property degradation during and after fire	430
15.7	Post-fire mechanical performance of fire retardant composites	433
15.8	Future trends	435
15.9	References	436

16 Improving the fire safety of road vehicles 443
M M HIRSCHLER, GBH International, USA

16.1	Introduction	443
16.2	Regulatory requirements	445
16.3	United States fire loss statistics	447
16.4	State of the art of regulation and guidelines	448
16.5	Activities of US NFPA Hazard and Risk of Contents and Furnishings Technical Committee	450
16.6	2007 Draft Guide NFPA 556	452
16.7	Alternate approach	463
16.8	Conclusions	464
16.9	References	464

17 Firefighters' protective clothing 467
H MÄKINEN, Finnish Institute of Occupational Health, Finland

17.1	Introduction	467
17.2	Different tasks and requirements for fire and heat protection	468
17.3	Types and design of clothing required for protection	471
17.4	Materials used in firefighters' clothing	472
17.5	Measuring flame and thermal performance	475
17.6	Future trends	483
17.7	References	486
17.8	Appendix: Standards	489

18 Fire protection in military fabrics 492
S NAZARÉ, University of Bolton, UK

18.1	Introduction	492
18.2	Fire hazards	494
18.3	Military burn injuries	496
18.4	Clothing protection for military personnel	499
18.5	Existing flame retardant solutions for military uniforms	500
18.6	Flame retardant textiles: military applications	507
18.7	Types of military clothing	510
18.8	Testing standards and methodologies	514

18.9	Performance standards and durability requirements	521
18.10	Future trends	522
18.11	Sources of further information and advice	523
18.12	References	524

19 Flame retardant materials for maritime and naval applications 527
U SORATHIA, Naval Surface Warfare Center, USA

19.1	Introduction	527
19.2	Fire threat	528
19.3	Fire safety regulations	529
19.4	Methods of reducing flammability	531
19.5	Fibre reinforced plastics	538
19.6	Fire performance and test methods for fibre reinforced plastics	540
19.7	Summary	564
19.8	Sources of further information and advice	566
19.9	References	567

20 Materials with reduced flammability in aerospace and aviation 573
R E LYON, Federal Aviation Administration, USA

20.1	Introduction	573
20.2	History of aircraft fire regulations	574
20.3	Materials used in commercial aircraft	577
20.4	Fire test methods for aircraft materials	585
20.5	Future trends	596
20.6	Acknowledgements	597
20.7	References	597

Index 599

Contributor contact details

(* = main contact)

Chapter 1
Professor A. Richard Horrocks and
 Professor Dennis Price
Centre for Materials Research and
 Innovation
University of Bolton
Deane Road
Bolton BL3 5AB
UK
E-mail: A.R.Horrocks@bolton.ac.uk;
 D.Price@bolton.ac.uk

Chapter 2
Professor Serge Bourbigot
Laboratoire Procédés d'Elaboration
 des Revêtements Fonctionnels
 (PERF)
LSPES – UMR/CNRS 8008
Ecole Nationale Supérieure de
 Chimie de Lille (ENSCL)
Avenue Dimitri Mendeleïev – Bât.
 C7a
BP 90108, 59652 Villeneuve d'Ascq
 Cedex
France
E-mail: serge.bourbigot@ensc-lille.fr

Chapter 3
Sergei V. Levchik*
Supresta LLC
430 Saw River Mill Rd
Ardsley, NY 10502
USA
E-mail: sergei.levchik@icl-ip.com

Professor Edward D. Weil
Polytechnic University of Brooklyn
Six Metrotech Center
Brooklyn, NY 11201
USA
E-mail: eweil@poly.edu

Chapter 4
Professor Yu-Zhong Wang
Center for Degradable and Flame-
 Retardant Polymeric Materials
College of Chemistry
Sichuan University
Chengdu 610064
China
E-mail: yzwang@scu.edu.cn

Chapter 5

Professor Charles A. Wilkie*
Department of Chemistry
Marquette University
PO Box 1881
Milwaukee, WI 53201
USA
E-mail:
   charles.wilkie@marquette.edu

Dr Alexander B. Morgan
Advanced Polymers Group
Multiscale Composites and Polymers
   Division
University of Dayton Research
   Institute
300 College Park
Dayton, OH 45469-0160
USA
E-mail: alexander.morgan@
   udri.udayton.edu

Chapters 6 and 7

Professor A. Richard Horrocks
Centre for Materials Research and
   Innovation
University of Bolton
Deane Road
Bolton BL3 5AB
UK
E-mail: A.R.Horrocks@bolton.ac.uk

Chapter 8

Dr Phillip J. Wakelyn
National Cotton Council of America
   (retired) Consultant
1521 New Hampshire Ave
Washington, DC 20036
USA
E-mail: pwakelyn@cotton.org

Chapter 9

Andrea Castrovinci*
University of Applied Sciences of
   Southern Switzerland
SUPSI
Galleria 2
CH-6928 Manno
Switzerland
E-mail: andrea.castrovinci@supsi.ch

Matteo Lavaselli
Centro di Cultura per l'Ingegneria
   delle Materie Plastiche
Viale T. Michel 5
15100 Alessandria
Italy

Professor G. Camino
Politecnico di Torino
DISMIC – Dipartimento di Scienza
   dei Materiali e Ingegneria
   Chimica
V.le Duca degli Abruzzi 24
Torino 10129
Italy

Chapter 10

Dr Marc L. Janssens
Southwest Research Institute
Department of Fire Technology
P.O. Drawer 28510
San Antonio, TX 78228-0510
USA
E-mail: marc.janssens@swri.org

Chapter 11
Professor T. Richard Hull
Professor of Chemistry and Fire
 Science
Centre for Fire and Hazards Science
School of Forensic and Investigative
 Sciences
University of Central Lancashire
Preston PR1 2HE
UK
E-mail: trhull@uclan.ac.uk

Chapter 12
Dr Jürgen Troitzsch
Fire and Environmental Protection
 Service
Adolfsallee 30
65185 Wiesbaden
Germany
E-mail: jtroitzsch@troitzsch.com

Chapter 13
Dr Margaret Simonson* and Dr Petra
 Andersson
SP Technical Research Institute of
 Sweden
Box 857
SE-501 15 Borås
Sweden
E-mail: margaret.simonson@sp.se
 petra.andersson@sp.se

Chapter 14
Dr Alan M. Emsley and Professor
 Gary C. Stevens*
GnoSys UK Ltd
University of Surrey
Guildford GU2 7XH
UK
E-mail: g.stevens@gnosysgroup.com

Chapter 15
Professor Baljinder K. Kandola* and
 Dr Everson Kandare
Centre for Materials Research and
 Innovation
University of Bolton
Deane Road
Bolton BL3 5AB
UK
E-mail: B.Kandola@bolton.ac.uk

Chapter 16
Dr Marcelo M. Hirschler
GBH International
2 Friar's Lane
Mill Valley, CA 94941
USA
E-mail: GBHint@aol.com

Chapter 17
Dr Helena Mäkinen
Finnish Institute of Occupational
 Health
Topeliuksenkatu 41 aA
FIN-00250 Helsinki
Finland
E-mail: Helena.makinen@ttl.fi

Chapter 18
Dr Shonali Nazare
Centre for Materials Research and
 Innovation
University of Bolton
Deane Road
Bolton BL3 5AB
UK
E-mail: S.Nazare@bolton.ac.uk

Chapter 19
Usman Sorathia
Naval Surface Warfare Center
Carderock Division
West Bethesda, MD 20817-5700
USA
E-mail: usman.sorathia@navy.mil

Chapter 20
Dr Richard E. Lyon
Aircraft and Aircraft Safety R&D
  Division
Federal Aviation Administration
William J. Hughes Technical Center
Atlantic City International Airport
NJ 08405
USA
E-mail: richard.e.lyon@faa.gov

1
Introduction

A R HORROCKS and D PRICE, University of Bolton, UK

Because of the many advances in flame retardant materials made over the past few years, this new book has been assembled to update our previously successful text, *Fire Retardant Materials*, published in 2001 by Woodhead Publishing and CRC Press LLC. The two books complement each other.

The increasing use of flammable, polymeric materials in domestic, industrial and other situations necessitates the development of means to render them resistant to both the initiation and sustaining of fires. Driving these developments are the annual losses of life and property resulting from fires involving polymeric materials. In the 31 countries covered by CTIF statistics, fires result in the deaths of some 37 000 persons per annum with at least 10 times that number of associated injuries with a total cost of 1% GDP estimated in terms of property loss and replacement, cost of medical services, etc. These data refer to the 2.3 billion inhabitants living in these countries (www.ctif.org, www.cefic-efra.org). Approximating the world's population to over 6 billion, it can be estimated that roughly 6–24 million fires occur in the world annually. These would cause some 100 000 deaths per annum with a cost of about £500 billion. Such data presents an overwhelming case for further development of improved technologies for flame retarding inherently flammable materials. Such developments are often driven by new legislation motivated by either tragic events, e.g. the English Channel tunnel fire in 1996, domestic furniture fires, and skyscraper hotel fires, or environmental concerns, e.g. the dioxan problem. The UK provides an excellent example of this. Following a sequence of domestic fire tragedies over the Christmas period in 1988, the UK government's introduction of legislation requiring domestic furniture to be flame retarded has resulted in about 140 fewer deaths per year compared to what was previously the case.

This book is structured into three parts. In Part I we have combined those chapters which consider the advances made in fire retardant materials during recent times, whereas in Part II the whole associated area of testing, regulation and assessing the benefits of fire retardant materials is considered. Part III focuses on the major applications of flame retardant materials where the most stringent levels of fire retardant behaviour are demanded and these are often driven by legislation, regulation and/or the specific level of fire hazard associated with the user of these materials.

However, within these three parts there are a number of cross-cutting themes of which the reader should be aware and these fall principally within the areas of the environment, the increasing impact of nanotechnology and the continual development in regulations and legislation.

The major risk regarding the use of flame retardants has been their potential environmental impact which was first raised as an issue in the late 1980s with claims of dioxin formation associated with the incineration of consumer plastic items containing brominated flame retardants. Since this time, there has been a drive to curtail the use of such flame retardants, culminating in bans on the production and usage of all brominated diphenyl species and, more specifically, of penta- and octabromo diphenyl ethers during the 2005/06 period. In parallel and especially in the USA and EU, risk analyses have been and continue to be undertaken on brominated flame retardants generally, with attention especially being given to those in very common use such as decabromodiphenyl ether (decaBDE), hexabromocyclododecane (HBCD) and tetrabromobisphenol A (TBBPA). This has had the consequence of pressure being brought to bear on the manufacturers of consumer goods to remove brominated flame retardants from their products, even if no health or environmental risk has been associated with their use and in spite of the benefits their presence confers in terms of improved fire safety. Chapter 14 by Elmsley and Stevens fully analyses the risks and benefits of using flame retardants in consumer products both qualitatively and quantitatively.

Not surprisingly, these environmental concerns over the use of halogen flame retardant systems have enhanced the need for successful alternatives. The strongest competitors are the systems based on phosphorus. Sergei Levchik and Edward Weil have a wealth of knowledge in this area. Chapter 3 is their up-to-date account of the status of phosphorus flame retardant technology. Particular applications include polycarbonate and blends; polyesters and nylons; epoxy resins; polyurethane foams and polyolefins. Chapter 4 by Wang follows on with a review of the challenges posed by replacing halogen-containing flame retardants and the recent attempts to achieve successful alternatives, which have similarly high efficiencies, without loss in mechanical properties, because of the high concentrations possibly required, as well as associated risks of matrix polymer degradation. He also discusses the important area of attempting to reduce melt-dripping while increasing flame retardancy in thermoplastic polymers.

Within certain applications, such as textiles, the proximity of the product to the consumer whether as a clothing item in intimate contact with the skin or as furnishing fabrics located in closed domestic environments, has posed considerable environmental concerns which has have received considerable attention as noted by Elmsley and Stevens (Chapter 14) and this area is addressed in detail by Wakelyn in Chapter 8.

End-of-life disposal strategies also pose environmental challenges and with the ever expanding use of polymeric materials generally, the need for suitable, environmentally friendly technologies for their recycling and disposal becomes paramount. The presence of flame retardant species both as additives and as components of polymeric chains within such a material adds further complication to this problem. Chapter 9 by Casrovinci, Lavaselli and Camino provides a full account of the problem and potential solutions. It is an excellent starting point from which to instigate an investigation of this area.

The second major cross-cutting interest of the flame retardant science community during the last ten years has been the observation that inclusion of nanoparticulate species, when introduced into most polymer matrices, can significantly improve their fire performance, often noted as a reduction in the cone calorimetrically determined values of peak heat release rate (PHRR). However, when present alone these do not confer flame retardant properties in terms of increasing ignition times and/or reducing burning rates and burning times. In the presence of conventional flame retardants, however, they may show remarkable synergies, thereby enabling overall reduced levels of flame retardant to be used in a given polymer in order to achieve a desired level of flame retardancy. Recent developments in this whole area are intensively reviewed by Wilkie in Chapter 5 in terms of the history and nature of nanocomposite polymers but also with regard to the effectiveness of different nanoparticle species, especially in combination with conventional flame retardants, as novel flame retardant components. The proposed mechanisms of action are also reviewed and discussed. While commercial applications of nanocomposite-based flame retardant polymers have been few during the last 10 years, Horrocks, in Chapter 6, reviews their potential in bulk, film, fibre, composite and foam applications and signals increasing interest and an expectation that their commercial exploitation will expand in the near future. It is in this emerging nanotechnological area where benefits of the combined improvements in mechanical properties and fire performance at relatively low cost cannot be ignored.

Thirdly, public concern over fire safety has resulted in legislation to control the flammability specification/requirements for polymeric materials for domestic and industrial use. This in turn has necessitated Standard Fire Tests to ascertain the suitability for use of the many polymeric materials in the market place. Part II contains four chapters which deal with subject matter in great detail. Janssens provides a tester's viewpoint. Type I and II flammability tests are described in detail whilst the challenges in assessing material flammability are discussed in detail as are the uses and limitations of fire testing. This is complemented by Chapter 11 where Hull gives an alternative but complementary view with an analysis of reaction to fire tests and the assessment of fire toxicity. He completes his review with a discussion of the prediction of fire behaviour from material properties and models of fire growth. In Chapter 12,

Troitzsch discusses the status and trends in fire safety regulations and testing. Finally, Margaret Simonson in Chapter 13 provides an account of life cycle analysis, i.e. production to disposal, of flame retarded consumer products. This is illustrated with reference to case studies concerning TV sets, cables and furniture.

The above themes are continually addressed in those chapters dealing with applications. Advances at the research level in improving the flame retardancy of natural and synthetic fibres and textiles as a whole are considered in detail by Bourbigot in Chapter 2. Within this, the considerable research into the application of nanoparticulates both within fibres and fabric coating formulations as well as the development of fabric structures having the highest levels of fire performance are also detailed. While environmental challenges are introduced here, they are more fully explored in Chapter 8 by Wakelyn, as mentioned above. The particular importance of textile coatings and laminates has been addressed by Horrocks in Chapter 7, where once again the challenges posed by both environment and nanotechnology are included.

Those textiles requiring the highest levels of flame retardancy and fire performance are usually those required by the civil emergency and defence organisations. Mäkinen, in Chapter 17, considers firefighters' protective clothing in terms of the different tasks undertaken by different systems, the requirements for fire and heat protection, the associated clothing design and materials factors and the methods used to assess performance. In a similar fashion, in Chapter 18 Nazare considers the fire hazards present and fire protection levels required in military fabrics, the clothing requirements for personnel, existing flame retardant solutions, the types of military clothing and, finally, the testing, performance standards and durability requirements.

The not-unrelated application areas of textile and fibre-reinforced composites are described by Kandola and Kandare. Here the key features required of composites having improved fire resistance, the different types of polymer, metal and ceramic matrix composites and the principal issues regarding the fire performance requirements for the different sectors of aerospace, automotive, rail and marine transport are explained. This is complemented by a discussion of the flammability of composites and their constituents in general, with a focus on both current and potential flame retardant solutions. Degradation of mechanical properties of composites during a fire is becoming of increasing importance and so receives attention here.

Within the transport sectors, regulations are often of paramount importance and also international in character. However, in the automotive sector, while there are no specific international regulations in force regarding aspects of component fire retardancy, the global nature of the industry has caused most major manufacturers to adopt various versions of the US FMVSS 302 standard, which is a rather mild horizontal burn test method for all textile components within the passenger compartment. In addition to textiles, such as seating fabrics and padding and interior décor, linings and carpets, manufacturers of modern cars and road transporters make extensive use of polymer materials, e.g. in panelling, together with the current development of suitable light but strong plastics and composites for use in engines and engine compartments. In the USA, some 350 000 or so road vehicle fires cause some 470 deaths, 1850 injuries and over $1.3 billion loss in property (see News 6-2002 on www.cefic-efra.org) per annum. Development of suitable flame retarded materials is essential to improve the fire safety of automotive vehicles. Hirschler's chapter (Chapter 16) provides an excellent account of this area dealing with the regulation requirements 2007 draft of guide NFPA 556 and possible alternative approaches.

However, the maritime and aerospace transport sectors are far more regulated with regard to fire performance requirements. For instance, the construction and fittings of naval ships and maritime vessels make extensive use of combustible materials and components. Such vessels are vulnerable to exposure to fire, particularly in a war situation. Thus there is a need for suitable flame retarded polymeric materials for use in such vessels. In Chapter 19 Sorathia gives a full account of the regulations identifying the requirements of suitable materials, methods for reducing flammability, test methods and fire performance.

In a similar manner to the maritime industry, metallic structures in aircraft are increasingly being replaced by polymeric materials and especially composites in order to lose weight and thus reduce fuel requirements. Plastics and textiles are also extensively used throughout the cabin areas. Fire safety is designed into aircraft to prevent in-flight fires and mitigate post-crash fires, which account for about 20% of the fatalities resulting from airplane accidents. In the final chapter (Chapter 20) Richard Lyon provides a full account of this developing area of flame retardant materials as well as considering ways in which it will develop in the future.

Part I
Advances in fire retardant materials

2
Flame retardancy of textiles: new approaches

S BOURBIGOT, Ecole Nationale Supérieure de Chimie de Lille (ENSCL), France

Abstract: Three approaches can be considered to reduce the flammability of textiles: (i) to use inherently flame retarded textiles comprising the so-called high performance fibres; (ii) to chemically modify existing textiles; (iii) to incorporate flame retardants in synthetic fibres and/or to make specific surface treatment. Those three aspects will be considered in this chapter focusing our discussion on the new directions and concepts emerging in this field. The chapter is organized in five sections presenting the recent standards developed to assess the flame retardancy of the textiles, the flame retardant fibres (natural, synthetic and man-made fibres) and future trends.

Key words: flame retardancy, flame retardant, textile, synthetic fibres, natural fibres.

2.1 Introduction

Fire is undoubtedly an emotive subject, especially when it comes to scenarios that we can imagine in a 'closed system' (e.g., ship or aircraft) and the possibilities to escape are restricted. The high fire hazards posed by textiles, both in historical times and to the present day, are a consequence of the high surface area of the fibres present and the ease of access to atmospheric oxygen. As comfort and safety are the two relevant factors when furnishing modern living environments, the latter should be designed in such a way that the effect it is as aesthetically pleasing as possible. This may have the effect of being frequently contrary to the desire to achieve the requisite safety requirements such as flame protection. As textiles are used in a wide range of applications and in many sectors (home textiles, transportation, clothing etc.), it becomes obvious that flame retardancy of textiles is (or should be) required dependent upon risk and regulatory demand.

Statistical studies of the flammability of soft furnishings and contents have shown that this class of items is typically among the items first ignited in residential fires most heavily associated with fire fatalities. US statistics by the National Fire Protection Association (NFPA) report the items first ignited and the associated fire fatalities for all items causing over 500 fires per year during the period between 1999 and 2002 (Table 2.1).[1,2] It shows that 44% of textile-related materials are involved in fire fatalities and also that worn clothing fires, which are considered to be individual by nature, may not be neglected.

Table 2.1 Statistical data on items first ignited in home fires (from ref. 1)

Furnishing	% of total home fire fatalities	Fatalities per fire
Upholstered furniture	19	1 in 18
Mattress of bedding	14	1 in 42
Floor covering	4	1 in 70
Apparel (not worn)	3	1 in 161
Apparel (worn)	4	1 in 4

UK statistics follow the same trend and show that textiles can also be the material mainly responsible for the development of the fire in dwellings and other buildings. For example, in 2004, 18% of the fires in dwellings were caused by textiles leading to 51% of the fatal casualties in the UK (Table 2.2).[3]

The flammability behaviour of polymers and textiles is defined on the basis of several processes and/or parameters, such as burning rates (e.g., solid degradation rate and heat release rate), flame spread rates (e.g., flame, pyrolysis, burn-out, smoulder), ignition characteristics (e.g., delay time, ignition temperature, critical heat flux for ignition), product distribution (in particular, toxic species emissions), smoke production, etc. Our goal is then to inhibit or even suppress the combustion process acting chemically and/or physically in the solid, liquid or gas phases. We can interfere with combustion during a particular stage of this process, e.g. during heating, decomposition, ignition or flame spread. Three approaches can then be considered to reduce the flammability of textiles: (i) to use inherently flame retarded textiles comprising the so-called high performance fibres (e.g., polyoxazoles, poly(ether-etherketone) or polyimides[4]); (ii) to chemically modify existing textiles (e.g., copolymerization of flame-retardant monomer into PET chains[5]); (iii) to incorporate flame retardants in synthetic fibres and/or to make specific surface treatment (this last approach is applicable for both synthetic and natural fibres). Those three aspects will be considered in this chapter, which is not intended to be an encyclopaedic review of the literature associated with textile-based flame retardants, but instead, will discuss the new directions and concepts emerging in this field of science. The chapter is organized in five sections presenting the recent standards developed to assess the flame retardancy of the textiles, the flame retardant fibres (natural, synthetic and man-made fibres) and future trends.

2.2 Developments in assessing levels of flame retardancy of textiles

To design flame retardant textiles with the required levels of performance, fire scenarios need to be specified to achieve acceptable levels of safety. For example, textiles used for aircraft seats require higher levels of safety than those used for domestic sofas. Therefore, testing and performance standards are used

Table 2.2 Fires and casualties from fires in dwellings and other buildings by textile mainly responsible for the development of the fire in the UK (2004)

Material or item mainly responsible	Dwellings			Other buildings		
	Fire	Casualties		Fire	Casualties	
		Fatal	Non-fatal		Fatal	Non-fatal
Total of materials	*59743*	*375*	*11977*	*37582*	*55*	*1519*
Total of textiles, upholstery and furnishings	*10991*	*193*	*3420*	*3823*	*14*	*352*
Clothing on person (not nightwear)	96	21	58	38	3	32
Clothing nightwear on person	17	3	12	12	2	7
Other textiles and clothing	2931	15	678	1170	7	111
Bedding (on bed or mattress)	1070	24	398	332	1	73
Bed or mattress used as bed	1119	28	473	185	–	44
Bed or mattress not used as bed	243	2	66	117	–	3
Furniture, not upholstered	493	6	165	361	1	12
Combustion-modified foam upholstery	86	7	45	15	–	2
Other foam upholstery	830	31	316	302	–	16
Other upholstery, covers	1549	43	499	631	–	14
Curtains, blinds	1000	2	273	163	–	8
Floor coverings	1117	1	250	300	–	14
Furniture and furnishings – other	440	10	187	197	–	16

to determine whether the objectives will be met by a specified textile material and/or design for a specified fire scenario under the specified assumptions. The flame retardancy of an individual textile is usually evaluated at the small-scale level by measurement of its 'reaction to fire' in terms of the ignition, flame spread and heat release usually compared with an unretarded material. These last parameters are cumulative because in the case of flame spread, it means that ignition has already taken place and that a high rate of heat release will lead to a high rate of flame spread. In this section, we will only discuss the fire behaviour of the individual textile because it is out of the scope of this article to take into account the role of textile in the whole system or potential fire scenario (see, e.g., ref. 6 for more details).

Flammability tests exist for apparel textiles, many household textiles, including carpets and soft furnishings, as well as for textiles used in public places such as theatres and on transport systems.[7] Many factors, such as fibre content or fibre blend, fabric weight and structure, finishes, and garment design, may affect the flammability of clothing and textile products. Most flammability tests have a similar basis: the textile is held in a specified manner and a flame of carefully specified nature and size is applied for a set time. The flammability performance may be assessed in terms of the time taken for a certain degree of burning to take place in the case of the less demanding test specifications, through to minimal effects of burning for the highest standards (e.g., in applications such as aerospace). Reference will often be made to the amount of smoke and glowing char or other by-products of burning that may be produced, and all these factors can be taken into account during the test. As a typical example, the vertical fabric strip test EN 15025 may be used to quantify time to ignition, burning rate and extent of damage. However, a convenient method of assessing and ranking the flammability of textile fibres is to determine the limiting oxygen index (LOI), which is the concentration of oxygen/nitrogen mixture to sustain burning, although this test is rarely used commercially.

Our purpose here is to survey the latest work undertaken in the area of the assessment of the flame retardancy of textiles and a part of the recent literature has focussed on apparel on the person. Apparel can ignite or melt when exposed to an open flame or to any other intensive heat source for a sufficient length of time followed by flame spread and possible burning of the wearer. European standard EN 1103 requires that all types of apparel exhibit a flame propagation rate (FPR) of at most 90 mm/s.[8] Rossi *et al.*[9] pointed out that little was known about the relationship between flame propagation rate and actual skin burns. The hazard of becoming injured by burning garments depends on different factors such as ignitability, heat released and heat transferred by burning and/or melting materials which are not measured by EN 1103. These researchers assessed the flame propagation rate of 94 different natural and synthetic fabrics and commercially available garments according to the EN 1103 bench-scale test apparatus. To further evaluate the potential burn hazard, the fabrics were then formed into upper garments, put on a full-scale mannequin equipped with 122 heat flux sensors and ignited with a small flame. They concluded that the flame propagation rate is a good measure of predicting the potential hazard of fabrics made of natural fibres but for synthetics and blends of natural and synthetic fibres, the heat transfer to the skin has to be considered as well. Indeed, garments using fabrics made of blends with high percentages (>50%) of synthetic fibres could present an increased burn risk due to the combined flame spread and melting effects from which determinations of heat transfer are necessary, in addition to the flame propagation rates, to assess potential injury hazards. This type of study was also performed by the US army to evaluate the protection of soldiers against fire hazards.[10] The protective performance of these clothing systems against burn injuries was investigated with manikin tests compared to bench-scale tests. It was found that if the air gap distribution of a clothing system is known, bench-scale tests (usually made with zero air gap measurement) could provide useful information for full-scale performance taking into account additional heat transfer effects via convection. This study provides an acceptable correlation between the mannequin test and the usual bench-scale tests, which is consistent with the study of Rossi *et al.* and which also pointed out the effect of the apparel construction on the human body. Finally, the approaches described above were also used to address fire safety concerns associated with the use of flammable fabrics during space travel.[11] It was found that flame spread rates and heat fluxes were considerably lower in microgravity than under Earth's gravity, which resulted in longer than predicted times to produce skin burns.

Very recently Hirschler *et al.*[1] conducted a study evaluating the flammability of 50 fabrics (both cellulosic and thermoplastic for apparel and others) using the NFPA 701 small-scale test.[12] The fabrics covered a broad range of area densities (weights) and many of them were not intended for apparel. To make reliable analysis, the fabrics were separated into three groups: melting fabrics, fabrics that do not melt and do not char to the top of the sample and fabrics that melt and char to the top. It was noticed that the latter two groups could be characterized broadly, with regard to their performance in this test. The conclusions of the study are: (i) charring fabrics that weigh more than $200 \, g/m^2$ will probably spread flame at a rate of *ca.* 4 mm/s, but will not char completely, and (ii) charring fabrics weighing less than that threshold of $200 \, g/m^2$ will produce full char length, and the flame spread rate will be inversely proportional to the fabric weight (or area density).

Textiles are widely used in transportation and so contain a certain amount of combustible materials. Examining the test methods and regulations for the determination of the flammability properties of interior transportation material in general and seating materials of automobiles in particular, it was clear that there is essentially only one test method used around the world, which is the Federal Motor Vehicle Safety Standard (FMVSS) 302.[13] Basically, it is a test measuring the flame spread on a particular material impinged upon by a small ignition source. It is not very severe and it gives only a method for excluding particularly unsafe materials from being used. Searpoint *et al.*[14] selected ten materials as representative of those used as seat coverings of private and commercial passenger vehicles. They compared data from ignition tests conducted in the cone calorimeter and the FIST (Forced Flow Ignition and Flame Spread Test) apparatus[15] with tests conducted using the FMVSS 302 horizontal flame spread apparatus. The time-to-ignition of new and used materials subjected to exposure heat fluxes between $20 \, kW/m^2$ and $40 \, kW/m^2$ was measured. The flame spread rates in the FMVSS 302 apparatus were determined and a comparison was made between the performance of the materials in the flame spread apparatus, the cone calorimeter and the FIST. According to the results, the authors suggested that a critical heat flux criterion could be used to provide an equivalent pass/fail performance requirement to that specified by the horizontal flame spread test. They also pointed out that the time-to-ignition data measured in the cone calorimeter and the FIST could be used to determine the ignition temperature of the materials.

Other than flame spread, heat release is another important parameter for evaluating the fire behaviour of textiles. Kandola et al.[16] evaluated textiles (exotic animal hair fibres including mohair, polyester and the blends of them) used for aircraft interiors by cone calorimetry by oxygen consumption and using an Ohio State University (OSU) calorimeter. These two methods permits measuring heat release rate of the fabric undergoing a given external heat flux. At $35\,kW/m^2$ and using standard procedures, they observed no correlation between the two tests. They suggested the two explanations: (i) in the cone calorimeter, the sample is mounted horizontally whereas the OSU calorimetric method requires vertical sampling with exposure to a vertical radiant panel, and (ii) the ignition source in the cone is spark ignition, whereas in the OSU it is flame ignition; hence, samples in the OSU calorimeter ignite more easily compared to those in the cone at the same incident heat fluxes. However, they demonstrated that cone calorimetric exposure at $50\,kW/m^2$ heat flux gives similar peak heat release results as the $35\,kW/m^2$ heat flux of OSU calorimeter. They concluded when designing patterned fabrics comprising exotic animal hair fibre (e.g., mohair), if a silk warp is used, then design variations will have little or no effect on fire performance. However, with the more flammable polyester warps present, the design should maximize the animal hair-containing weft face percentage if the fabric is to pass the FAA (Federal Aviation Administration) requirements for OSU tests at $35\,kW/m^2$ with confidence.

In this section, we have shown that depending on the test method, the nature and magnitude of the fire parameters being assessed may be different. In general, the tests mentioned cannot simulate a complete fire scenario but they only compare the flammability of certain materials under fixed conditions. These tests are extremely simplistic, and are, most likely, not really representative of the actual burning behaviour that can occur in real fires. Such predictability would require the use of a test based on heat release, or on flame spread of a very large sample. However, they are severe enough to select materials able to provide fire safety in a given environment.

2.3 Flame retardant natural fibres

Most natural fibres are highly combustible and the flammability of derived fabrics also largely depends on the construction and density of the fabric (see previous section). Most flame retardant treatments, formulations, and additives were derived from chemistry developed in the 1950–1980 period,[17] and those having current commercial interest, have been very recently reviewed.[18,19] Basically, several approaches can be used to enhance the fire behaviour of natural fibre-based fabrics used either alone or in blends with synthetic fibres:[20]

- Coatings and/or finishing treatments may be applied to shield fabrics from heat sources and prevent volatilization of flammable materials. These may take the form of simple protective coatings or, more commonly, the treatment of fabrics with inorganic salts that melt and form a glassy coating when exposed to ignition sources. In more advanced forms, intumescent coatings produce a char that has sufficient plasticity to expand under the pressure of the gases to yield a thick, insulating layer.[21-23]
- Thermally unstable chemicals, usually inorganic carbonates or hydrates, are incorporated in the material, often as a back-coating so as to preserve the surface characteristics of the carpet or fabric. Upon exposure to an ignition source, these chemicals release CO_2 and/or H_2O, which in a first step, dilute and cool the flame to the point that it is extinguished, and in a second step form a protective ceramic around the charred fibres.
- Materials that are capable of dissipating significant amounts of heat are layered with the fabric or otherwise incorporated in a composite structure. These may be as simple as metal foils or other heat conductors or as complicated as a variety of phase-change materials that absorb large quantities of heat as they decompose or volatilize. If sufficient heat is removed from the point of exposure, the conditions for ignition are not reached.
- Char-promoting chemical treatments which may be fibre-reactive or -unreactive to yield launderable or non-durable flame retardancy respectively.[17,19]
- Chemicals capable of releasing free radical trapping agents, frequently organobromine or organochlorine compounds, may be incorporated into the fabric. These release species such as Br$^\bullet$ and Cl$^\bullet$ which can intervene in the oxidation reaction of the flame and break the chain reaction necessary for continued flame propagation.

2.3.1 Flame retardant cotton

Cellulosics, such as cotton and rayon, as well as other non-thermoplastics that are char formers, are not inherently ignition resistant and usually must be chemically treated to prevent ignition by small flames.[24] The most common approach to reduce the flammability of cotton is to apply a flame retardant finishing system which can be durable or not. A well known durable flame retardant chemical for cotton is based on organophosphorus compounds, such tetrakis (hydroxymethyl) phosphonium chloride (THPC).[25] The THPC system is effective in imparting durable flame resistance to cotton fabrics, but it requires the use of special equipment in its application. Based on this concept, Yang *et al.*[26,27] recently used a combination of a hydroxyl functional organophosphorus oligomer (HFPO) and dimethyloldihydroxyethyleneurea (DMDHEU) as new flame-retardant finishing system. DMDHEU functions as the binder to form covalent bonding with both cotton cellulose and HFPO (Fig. 2.1). At certain HFPO/DMDHEU ratios, those two compounds are also able to form a cross-

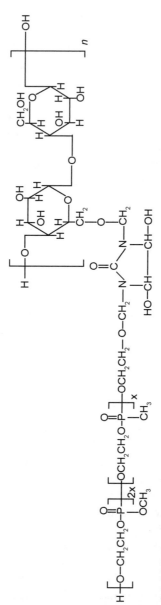

2.1 DMDHEU functions as a binder forming covalent bond with cotton cellulose (from ref. 27).

linked network on cotton. The authors claimed that the cotton fabric treated with this flame retardant finishing system exhibits high levels of durable flame retardant performance without significant change in fabric hand and whiteness.

More recently, the same authors used the well-established[17] N-methylol dimethylphosphonopropionamide (MDPA) as durable flame retardant for cotton,[28] to enable comparison of the flame resistant performance of the previous system with the new one. The cotton fabric treated with MDPA/trimethylolmelamine (TMM) has a higher initial LOI than that of the fabric treated with HFPO/TMM due to higher nitrogen content in the former system. The LOI of the cotton fabric treated with the HFPO and MDPA systems becomes identical when the treated fabric contains equal amount of phosphorus and nitrogen. The MDPA/TMM shows higher laundering durability on cotton than HFPO/TMM system, however. In their studies, no real investigation on the mechanism of action was made but according to the results, we suspect a mechanism of action involving char enhancement.

Concurrent with work of Yang, Lecoeur et al.[29,30] investigated flame retarding finishing (padding) of cotton using different formulations based on monoguanidine dihydrogenphosphate (MGHP) and 3-aminopropylthoxysilane (APS) combined or not with phosphoric acid. Typically for similarly treated cotton containing phosphorus- and nitrogen-containing retardants, FR fabrics start to decompose at lower temperature than the untreated cotton but they exhibit a transient char much higher than the latter (50 wt.-% compared to 20 wt.-% at 400 °C) but the final residues are similar 1–2 wt.-% at 1000 °C. The formulations are ranked M1 to the electrical burner test[31] but only those applied using phosphoric acid keep this ranking after water-soaking. In addition to this, one of them is still rated as M1 after washing, probably because of greater extent of phosphorylation reactions between the cellulose and the application chemicals. Cone calorimetry confirms those results. RHR (rate of heat release) curves exhibit extremely low values for the formulations containing phosphoric acid (Fig. 2.2). This also confirms visual observations showing very small flames combined with smouldering at the surface of the samples. Charred residues are obtained at the end of the experiment suggesting a mechanism involving char enhancement.

A novel approach has been developed by Mostashari et al.,[32] namely synthesizing zinc carbonate hydroxide by means of the multiple bath method and padding it on cotton. Prolonged burning was observed of treated specimens, however. The existence of zinc oxide was detected in the ashes, but no traces of the metallic zinc were detected. They explain the enhancement of the flame retardancy by reduction–oxidation reactions occurring during the smouldering process. It is noteworthy they did not observe any char enhancement.

As shown above, char enhancement provides low flammability to cotton. Intumescence is also a charring phenomenon combined with the swelling of the carbonaceous structure. Horrocks et al. (ref. 19 and references therein) has

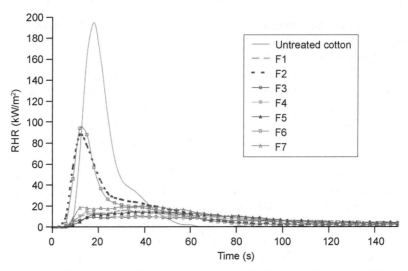

2.2 RHR curves of different padded fabrics. F1: 150 g/l MGHP, 10% APS, pH 4; F2: 75 g/l MGHP, 10% APS, pH 9.5; F3: 150 g/l MGHP 10% APS, pH 4, melamine, catalyst; F4: 150 g/l MGHP 10% APS, pH 4, melamine, catalyst; F5: 150 g/l MGHP, melamine, catalyst, 10% APS, pH 4; F6: 75 g/l MGHP 10% APS, pH 9.5, melamine, catalyst; F7: 150 g/l MGHP, melamine, 10% APS, pH 9.5 (from ref. 30).

undertaken considerable work on intumescents applied to textile structures, in particular substantive fibre treatments for cellulose. Based on the work of Halpern[33] on cyclic organophosphorus molecules, they developed a phosphorylation process for cotton fibres achieving intumescent cotton fabric with considerable durability. Char enhancement is as high as 60 wt.-% compared to the pure cotton, which is associated with very low flammability.

Coating (or back-coating) on fabric is another way to provide flame retardancy to cotton. Horrocks' group[34,35] used intumescent back-coatings based on ammonium polyphosphate (APP) as the main FR combined with metal ions as synergist. Metal ions promote thermal degradation of APP at lower temperatures than in their absence, and this enables flame retardant activity to commence at lower temperatures in the polymer matrix thereby enhancing flame retardant efficiency. Giraud et al.[36–39] developed the concept of microcapsules of ammonium phosphate embedded in polyurethane and polyurea shells (Fig. 2.3) to make an intrinsic intumescent system[40] compatible in normal polyurethane (PU) coating for textiles. The advantage of this concept is to reduce the water solubility of the phosphate and to produce textile back-coatings with good durability. The flame retarding behaviour of these coated cotton fabrics was evaluated with the cone calorimeter and exhibits a significant FR effect. Development of intumescent char at the surface of the fabric was observed confirming the expected mechanism.

Flame retardancy of textiles: new approaches

2.3 Microspheres of ammonium phosphate encapsulated by PU membrane synthesized by evaporation of solvent (median value = 0.8 μm and mean size = 100 μm) (from ref. 40).

The pioneering work of Gilman et al.[41,42] has reported that the presence of nanodispersed montmorillonite (MMT) clay in polymeric matrices produces a substantial improvement in fire performance and more recently other groups have shown similar effects with different nanoparticles including carbon nanotubes (CNT), polyhedral silsesquioxanes (POSS), nanoparticles of silica, etc.[43] Very little reported research has been undertaken regarding development of nanocomposite structures in textiles, however. The first published work devoted to cotton was the use of PU nanocomposite containing MMT and POSS as coating on cotton fabrics.[44] The incorporation of POSS (molecular silica) does not enhance the fire behaviour of the fabric but the addition of MMT provides a substantial benefit in terms of reduced heat release. Some char enhancement without swelling was observed during the cone calorimetric experiments suggesting a mechanism of action in the condensed phase. More recently, White[45] produced cotton with MMT clay in a 50% solution of 4-methylmorpholine N-oxide (MMNO). Large plaques of cotton-clay nanocomposites were made but no fibre or textile structures can be obtained from this approach. Strong char enhancement was measured and it was claimed by the author without additional proof that the material is flame retardant.

Cold plasma technologies are surface modification and/or surface coating processes. Cold plasma reactive species can promote surface functionalization reactions or generate organic or inorganic thin layers as a result of recombination of radicals or molecular fragments species on the surfaces. This second approach

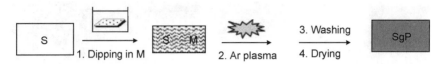

S: Substrate
M: Monomer
SgP: Substrate grafted polymer

2.4 Experimental procedure for the plasma-induced graft-polymerization of monomers (from ref. 46).

has been used to develop fire retardant protective coatings for textile materials. In our laboratory, we have used the plasma induced graft-polymerization (PIGP),[47] which consists of the simultaneous grafting and polymerization of functionalized monomers, such as acrylate monomers, on the surface of a material. This technology requires the experimental procedure reported in Fig. 2.4. The monomer (mixed with or without photoinitiators and solvent) is in a first part put on the substrate by dipping, padding, back-coating or other techniques. The coated substrates are then introduced in the plasma reactor. The reactor is then evacuated, the gas flow rate is adjusted and the discharge is initiated. The coated materials are then washed to remove non-grafted monomers or polymers and dried. The flame retardancy of cotton treated with polyethylene methacrylate phosphate grafted on the fabric is significantly improved. LOI values rise from 21 vol.-% for the pure cotton to 32 vol.-% for the FR cotton and an M2 rating (compared to M3 rating) can also be achieved with this plasma treatment. It is also noteworthy that the durability is very good as well. Results are very promising and work is still in progress to elucidate the mechanism of action.

2.3.2 Flame retardant wool

Within the area of natural fibres, wool has the highest inherent non-flammability. It exhibits a relatively high LOI of about 25 vol.-% and low flame temperature of about 680 °C.[20] The inherent flame retardant activity of the fibre can be associated with char-forming reactions which may be enhanced by a number of flame retardants. Very little was published recently on this topic and mechanisms on FR wool have neither been well researched nor reviewed apart from the work published by Horrocks *et al.* since 2000.[20,48] In this section, we will review briefly the traditional finishes used for flame retard wool associated with the postulated as well as the latest developments.

Various flame retardants are based on the combination of boric acid/borax mixtures, sulphamic acid and ammonium salts such ammonium phosphates and sulphates which promote char formation in wool. These treatments are not durable and the most commonly used as durable finish is the so-called Zirpro process.[17] The advantage of this latter process is the absence of any discoloration or other effect on wool aesthetics. The Zirpro process is based upon the exhaustion of negatively charged complexes of zirconium or titanium on to positively charged wool fibres under acidic conditions at 60 °C. The effectiveness of the Zirpro process is not fully understood from the mechanistic point of view but the flame retardancy effect is attributed to enhanced intumescent char formation or at least, to the ability to create effective flame and heat barrier properties. It is also reported that Zirpro leads to antagonism with other flame retardant treatments.[49]

As an alternative to the Zirpro process and based on the fundamental work on cellulosic materials to enhance char formation, Horrocks *et al.* offers intumescent formulations based on melamine phosphate to flame retard wool.[50] From TGA and SEM characterization, they proposed a comprehensive model on the mechanism of protection via an intumescent process which is described in Fig. 2.5. More recently, they used spirocyclic pentaerythritol phosphoryl chloride (SPDPC) phosphorylated wool to achieve intumescent wool which exhibits large char expansion and good flame retardancy.[51]

Recently launched on the market with the target of contract wool seating (particularly airline seating) is a new phosphorus-based FR called Noflan (Firestop Chems. Ltd., UK).[52] It is an organophosphorus flame retardant based on complex alkyl phosphonates (a complex of the amide of alkylphosphonic acid ammonium salt with ammonium chloride).[53] It is claimed to be a modern alternative of Zirpro. As far as we know, no paper describes the mechanism of protection and the fire behaviour of this new product, but according to the chemical structure of the flame retardant, we may reasonably assume it should be a mechanism via char enhancement.

Region A: 200–350 °C
Wool ⟶ Partial liquefaction + volatile formation ⟶
Intumescence ⟶ Partial liquefaction + phosphoric acid formation ⟶

Enhanced volatile formation
and relative residual char decrease ⟵

Region B: 350–450 °C
> 350 °C Peptide chain catalyzed by phosphoric acids
⟶ increased volatiles

> 400 °C Cross-linking and aromatization reactions dominate

Region C: > 450 °C Char oxydation rates decreased because of enhanced cross-linking including possible formation of P-N and P-O bonds

2.5 Intumescence char formation of FR wool according to Horrocks *et al.* (from ref. 50).

2.3.3 Others

Natural fibres other than cotton and wool already widely used, give the opportunity to the textile industry to develop newer materials that have both economic and environmental benefits. The potential of using other natural fibres based on cellulose such as, wood, silk, jute, kenaf, hemp, coir, sisal, pineapple, etc., has received considerable attention among scientists and remains incompletely explored. Nevertheless, some work has been published on the flame retardancy of natural fibres and in this section, we will report the work done in this area.

Silk fibre is a natural protein fibre and it is viewed as an environmentally favourable fibre because it contains only natural amino acid units. It is widely used to produce nightwear and so, it is vital for silk fabrics to be finished with flame retardants. Silk has low natural flammability, which can be attributed to its high nitrogen content (about 15–18%).[54] Its LOI value is about 23 vol.-%, but it still needs a flame retardant finish to fulfil some commercial requirements. A recent flame retardant finishing process for silk fabrics used immersion of the fabric in the mixture of borax and boric acid in solution[55] although in the late 1960s, inorganic salts and quaternary ammonium salts were suggested in Japan and America. The best-known commercial flame retardant additive products were non-durable.[56] However, in the mid 1980s, Achwal[57] reported that silk fabrics treated with a urea phosphoric acid salt, U4P, by a pad-dry process had high flame retardancy with an LOI value higher than 28% and that the flame retardancy finish was fast to dry cleaning while not fast to washing.[6] In the 1990s, Kako et al.[58] treated silk fabrics with an organophosphorus flame retardant and trimethylolmelamine (TMM) and gained good flame retardancy. More recently, there have been other reports on the subject. For example in 2006, Guan et al.[59] reported the application to silk of MDPA (Pyrovatex CP, Ciba), and it was found to produce excellent flame retardancy. The LOI value of the treated sample prepared via the pad-dry-cure-wash method is above 30 vol.-% and after 50 laundry cycles, it still has some residual flame retardancy. The reaction between flame retardant and silk only occurs in the amorphous region of silk fibre and thermogravimetric analysis (TGA) and differential scanning calorimetry (DSC) analyses show that the retardant causes silk fabrics to decompose below its ignition temperature (600 °C), and forming carbonaceous residue or char when exposed to fire. The char behaves as a thermal barrier to fire providing silk fabrics with good flame retardancy.

Nonwoven technologies offer a great opportunity in using natural fibres to make cost-effective, environmentally friendly fabrics. Of the vegetable fibres finding wide applications are jute, linen, flax, coir and sisal while among the animal fibres are cut wool, horsehair, horse and camel hair fibres. Because of their high flammability, flame retarding treatments must be applied. For example, Kozlowski et al.[60] have prepared nonwovens of linen treated with

2.6 Flax fibres and flax nonwoven used as interlayer fire blocker for seats (from ref. 61).

flame retardants and of blend of wool/hemp. Flame retardants were based upon a mixture of urea polyborates and the sodium salt of an alkyl-aryl sulphonic acid (trade name FOBOS). They found a large improvement of the flammability performance for the FR linen (LOI jumps from 23 to 29 vol.-% and the peak of RHR is decreased by 80%) and acceptable performance for the wool/hemp blend (LOI = 28 vol.-%). No details about the fire behaviour are described and no tentative of mechanism of action is proposed, but char enhancement of the material in the fire conditions may be suspected. They concluded that non-wovens made of natural raw materials with increased resistance to fire may be used as effective barrier fabrics in upholstery systems for furniture and car seats.

Using the same flame retarding treatment as above, Flambard et al.[61] suggested the use of FR nonwoven materials (linen and flax) as interlayer fire blockers in seats for application in public transportation (Fig. 2.6). They found greater effectiveness of this approach in upholstery materials (flax or linen combined with polyester) in terms of flammability and heat release.

2.4 Flame retardant synthetic fibres

The common synthetic fibres include rayon (1910), acetate (1924), aliphatic polyamide (1931), modacrylic (1949), olefin (1949), acrylic (1950), polyester (1953), polylactide or PLA (2002), etc. These fibres melt upon heating and are highly flammable. To render these fibres flame retardant, several approaches can be considered: (i) the incorporation of flame retardant additive(s) in the polymer melt or in solution prior to extrusion, (ii) the copolymerization or the grafting of FR molecules to the main polymeric chain, and (iii) the use of semi-durable or durable finishing. Note that the mechanism of action described in the previous sections is also applicable here. Other fibres generally called high performance fibres are very often inherently flame retardant. This group includes aramids (1961), Dyneema (note that this fibre is not inherently flame retardant because it is based on polyethylene) (1979), polybenzimidazole or PBI (1983), Sulfar (1983), Zylon (1991), M5 (1998), etc. All these examples will be discussed in the following.

2.4.1 Biopolymers

There is no single, universally agreed definition for the biopolymers. The widest possible definition would define a biopolymer as: 'Any polymer derived directly or indirectly from biomass'. This includes: (i) natural polymeric materials directly extracted from biomass for example; cellulose and (ii) synthetic polymers prepared from naturally occurring monomers extracted directly or produced from biomass, for example; poly (lactic acid) or PLA. Under this definition, we will comment the flame retardancy of the following fibres made from biopolymers.

Cellulose acetate or acetate rayon fibre is one of the earliest synthetic fibres and it is based on cotton or tree pulp cellulose. It is highly flammable with a rapid flame and melts when burning. No reports of any comprehensive flame retardancy study exist for this fibre but according to its chemical structure (similar to that of a cellulosic fibre), flame retarding treatments developed for cotton should be applicable to acetate fibres. Lyocell is also a fibre made from wood pulp cellulose but it only appeared on the market in 1992. Its flammability has been studied in detail by only one group.[62] From this work, the authors showed that FR Lyocell could be produced using the commercial flame retardant treatment Pyrovatex CP (Ciba). FR Lyocell exhibits a LOI value of 30 vol.-% while maintaining tensile properties comparable to the untreated Lyocell as well as having excellent whiteness and handle. As a bonus, they also showed that only half the normal amount of this FR treatment applied to cotton cellulose produces an equivalent degree of flame resistance.

PLA fibres were the first man-made (synthetic) fibres made from 100% annually renewable resources and were publicly launched by Cargill Dow in early 2003.[63] It is commercialized under the trade name of Ingeo fibres. The main applications of this new fibre are in bedding – pillows, mattresses and duvets, apparel, and household textiles – floor, wall, and furniture textiles. Hence, flame retardancy properties are required to meet the relevant regulations and legislative demands. Upon heating PLA melts and drips and reaches its temperature of ignition very quickly. Very little work has been carried out on the flame retardancy of PLA textiles, but in our laboratories,[64] we have developed PLA-clay nanocomposites able to be melt spun to make multifilament yarn. Various quantities of organomodified (OM)-MMT (from 1 to 4 wt.-%) have been added to PLA by melt blending via usual procedures to produce PLA nanocomposites and then into yarn by melt spinning. It was found

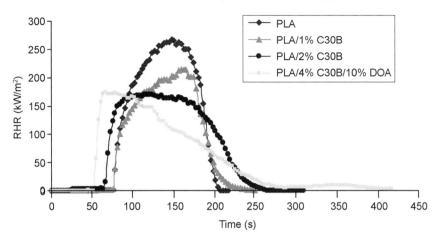

2.7 RHR values as a function of time of PLA and PLA nanocomposites evaluated as knitted fabrics by cone calorimetry at 35 kW/m² (dioctyl adipate or DOA is the used plasticizer) (from ref. 64).

that it is necessary to use a plasticizer to melt-spin a blend with 4 wt.-% of OM-MMT and the dispersion of the clay in the yarns is quite good. The multifilaments were knitted and the flammability studied using cone calorimetry at an external heat flux of 35 kW/m². Depending on the clay loading, the peak value of RHR is decreased by up to 38% demonstrating improved fire performance of these PLA fibres (Fig. 2.7). Formation of char is observed in the case of the nanocomposites suggesting a mechanism of condensed-phase, char enhancement.

2.4.2 Polyester

Polyester fibres are the main synthetic fibres used in the industrial manufacturing sector and can be found in several areas of application. Polyester fibres are used in apparel for overcoats, jackets, leisure and sportswear, protective clothing, and so forth. In home furnishings, their uses range from drapery and curtain fabrics to furniture coverings, pillows and pillow stuffing, and table and bed linen to wall and floor coverings. As polyester fibres are easily flammable, flame retardancy is a significant issue.

One of the classical solutions to flame retard polyester is to incorporate a comonomeric phosphinic acid unit into the PET polymeric chain (trade name Trevira CS).[20] Trevira CS polyester does not promote any char formation and there is evidence that the phosphorus compound acts in the gas phase. Several flame retardants have also been designed for polyester extrusion (bisphenol-S-oligomer derivatives from Toyobo, cyclic phosphonates (Antiblaze CU and 1010) from Rhodia or phosphinate salts from Clariant). Note that the cyclic phosphonate may also be applied as textile finish as well as a melt additive. All

2.8 Synthesis scheme of poly(2-hydroxy propylene spirocyclic pentaerythritol bisphosphonate) or PPPBP (from ref. 65).

those flame retardants were developed in the 1980s (except the phosphinate salt) and their modes of action have been described in the literature[20] but very little on this topic has been published recently.

In the recent published literature, Chen et al.[65] proposed the use of a novel anti-dripping flame retardant, poly(2-hydroxy propylene spirocyclic pentaerythritol bisphosphonate) (PPPBP) (Fig. 2.8) to impart flame retardancy and dripping resistance to PET fabrics. Flammability of PET fabrics treated with PPPBP was investigated by the vertical burning test which showed a significant enhancement of the flame retardancy (producing a non-ignitable fabric) and either a significant reduction of melt dripping at low levels or absence of dripping at higher levels of PPPBP.

The same authors investigated in detail the mechanism of flame retardancy of their FR PET fabrics.[66] They showed that it is a condensed-phase mechanism via char promotion (Fig. 2.9) in which PPPBP produces phosphoric or polyphosphoric acid during thermo-decomposition leading to the formation of phosphorus-containing complexes at higher temperatures. They suggest that the high yields of char are protected from thermo-oxidation by the presence of phosphoric acid contained in the charred residue and because of the high thermal stability of C=C groups in the char.

Concurrently, a new halogen-free FR master batch for polyester has been developed in our laboratories which at only 5 wt-% incorporation enables PET to obtain classification according to several standards such as the NF P 92 501 or NF P 92 503 (M classification), FMVSS 302 or BS 5852 (Crib 5).[67] Further studies are still in progress to investigate the mode of action.

2.4.3 Polyamides

Like polyester, polyamides are synthetic fibres made from semicrystalline polymers which find use in a variety of applications in textiles almost similar to those of polyester. Polyamides, however, have proved difficult to render durably flame retardant by incorporation of additives because of their melt reactivities.

2.9 Flame retardancy mechanism of FR PET fabrics treated with PPPBP (from ref. 66).

Semi-durable finishes based on thiourea derivatives are used (chemicals developed in the 1980s) but usually only on industrial polyamide textile where launderability is not an issue.[20] The recent developments for the flame retardancy of polyamides concern mainly the inclusion of nanoparticles.

In our laboratories, nylon 6 or PA-6/clay hybrid fibres have been made by melt blending and by melt spinning.[68] PA-6 nanocomposite exhibits an exfoliated structure and no degradation of the clay is observed after processing. Nevertheless, we observed a decrease in the degree of homogeneity determined by solid state NMR in PA-6 nanocomposite multifilaments after melt spinning (unpublished results). This is ascribed to a reagglomeration of the MMT platelets in the polymer and it might partially explain the poor mechanical properties of the yarns. RHR curves recorded by cone calorimetry of knitted PA-6 and PA-6/clay nanocomposite fabrics at an external heat flux of 35 kW/m^2 showed that the peak of RHR of the nanocomposite is decreased by 40% compared to that of the pure PA-6. Visually, while a char layer can be seen on each of the two textiles, that in PA-6 alone seems to be crumbly and contains holes, whereas the char produced by PA-6nano appears as uniform without holes. This structure could explain the better performance of PA-6nano against PA-6.

Recent work of Horrocks et al.[69] has investigated the effect of adding selected flame retardants based on ammonium polyphosphate, melamine phosphate, pentaerythritol phosphate, cyclic phosphonate, and similar formulations into nylon 6 and 6.6 in the presence and absence of nanoclay. They found that in nylon 6.6 all of the effective systems comprising the nanoclay demonstrated significant synergistic behaviour except for melamine phosphate because of the agglomeration of the clay. They then report in the case of nylon 6 that the presence of nanoclay acts in an antagonistic manner (in terms of LOI). To explain why the nanoclay lowered the LOI value of the FR-free nylon 6 film but not that of the nylon 6.6, they proposed that the nanoclay reinforces the fibre structure both in solid and molten phases, thereby reducing its dripping capacity. Such an effect would be likely to reduce the LOI value as the melting polymer has greater difficulty receding from the igniting flame.

2.4.4 Polypropylene

Polypropylene (PP) is presently one of the fastest growing fibres for technical end-uses where high tensile strength coupled with low-cost are essential features. PP fibres have been widely used in apparel, upholstery, floor coverings, hygiene medical, geotextiles, car industry, automotive textiles, various home textiles, wall-coverings and so on. Because of its wholly aliphatic hydrocarbon structure, polypropylene by itself burns very rapidly with a relatively smoke-free flame and without leaving a char residue. The excellent review of Zhang and Horrocks[70] describes the different approaches for making FR PP fibres and it will be very briefly commented in the following before focusing on the recent developments.

While polypropylene fibres may be treated with flame retardant finishes and back-coatings in textile form with varying and limited success,[71] the ideal flame retardant solution for achieving fibres with good overall performance demands that the property is inherent within the fibre. An acceptable flame retardant for polypropylene and especially fibre-forming grades, should have (i) thermal stability up to the normal PP processing temperature (<260 °C), (ii) compatibility with PP and no migration of the additives, (iii) flame retardancy properties when present in the fibre and (iv) efficiency at a relatively low level (typically less than 10 wt.-%) to minimize its effect on fibre/textile properties as well as cost. With such an approach, Zhang and Horrocks identified five principal types of generic flame retardant systems for inclusion in polypropylene fibres as phosphorus-containing, halogen-containing, silicon-containing, metal hydrate and oxide and the more recently developed nanocomposite flame retardant formulations (this last aspect will be extensively commented upon here because of the recent developments in this area for fibres). They concluded that apart from antimony–halogen or in some cases, tin–halogen formulations only one single flame retardant system, tris(tribomoneopentyl) phosphate, is presently effective in polypropylene when required for fibre end-uses. Presently, the use of phosphorus- based, halogen-free flame retardants in PP fibres is prevented by the need to have at least 15–20 wt.-% additive presence. Since the latter are char-promoting while all halogen-based systems are essentially non-char-forming in polypropylene, the way forward for a halogen-free, char-forming flame retardant conferring acceptable levels of retardancy at additive levels, 10 wt.-% will require either completely new FR chemistry or the development of a suitably synergistic combination based on the understanding reviewed above. So, the use of nanoparticles might offer this opportunity.

In our laboratories, the same approach as in the case of PA-6nano has been used with a poly(vinylsilsesquioxane) (hereafter called FQ-POSS and supplied by hybrid plastics) in PP (loading of FQ-POSS = 10 wt.-%). Multifilament yarns were made via a melt spinning process after extruding virgin PP and PP/FQ-POSS, and then knitted into fabrics.[72] FQ-POSS is thermally stable and no degradation was detected in the processing conditions. No significant difference was observed between the two fabrics with and without POSS except for the touch of PP/FQ-POSS which is softer. Reaction to fire of PP and PP/FQ-POSS knitted fabrics was evaluated by cone calorimetry by oxygen consumption. The peaks of the maximum of RHR of the two fabrics are similar showing that the fire behaviour of PP is not improved using FQ-POSS but the time to ignition is significantly improved because of the higher thermal stability of the nanocomposite. The total heat evolved (THE) is also not modified (about 200 kJ) suggesting that FQ-POSS does not act as a flame retardant but only as a heat stabilizer (via a decrease of the ignitability).

Carbon nanotubes (CNT) can also be used as nanofillers for making nanocomposite textiles. Kashiwagi *et al.*[73] reported the first study on the fire behavior of polymer, carbon nanotube nanocomposites. They showed significant flame retardant effectiveness of polypropylene (PP)/multi-walled carbon nanotubes (MWNT) (1 and 2% by mass) nanocomposites. Based on this, we prepared melt blended, melt spun multifilaments of PP and PP/MWNT.[74] They were then knitted using a double woven rib structure in order to make thick fabrics (Fig. 2.10). No significant difference was observed between the two fabrics except the black colour due to the presence of MWNT made by visual observation and their handle.

Knitted fabrics were evaluated by oxygen consumption, cone calorimetry (Fig. 2.11), and the RHR peak is decreased by 50% for a fraction of nanotubes of only 1 wt.-% but the time to ignition of the nanocomposite is shorter (135 s vs. 60s). The first explanation is the fibrillation of the yarn observed after the melt spinning as in the case of PA-6nano, which can create a 'wick effect' whereby fibrils act like small flame sources at the surface of the fabric and reduce the time to ignition. The second explanation was highlighted by Kashiwagi *et al.*,[75] in that below a certain concentration of carbon nanotubes (around 2%), the radiative ignition delay time is shortened due to an increase in the radiation in-

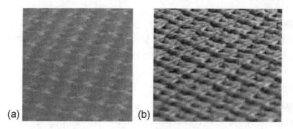

2.10 Knitted fabrics of PP (a) and PP/MWNT (b).

depth absorption coefficient following the addition of carbon nanotubes. The IR transmission spectrum of PP/MWNT does not show significant transmission bands compared to pure PP. Thus, all the external thermal radiant flux of the cone calorimeter is absorbed very near the sample surface and the surface temperature rises quickly, becoming high enough to initiate thermal degradation of PP and to generate enough degradation products to promote the ignition. In our case, we have probably the concurrent action of those two effects accelerating the ignition of the fabric.

Recently, Horrocks and his group used nanoclays (organomodified montmorillonite clay) to make PP/clay fibres.[76] They showed that the dispersion of clay was achieved at the nanoscale before and after melt spinning. They reported that filament samples had both sufficiently acceptable textile properties to enable their knitting into fabric samples and increased modulus. The burning behaviour of fabric samples was observed by LOI and by cone calorimetry but the clay presence did not confer flame retardancy. However they noticed that it did change the burning character and encouraged some char formation (Fig. 2.12).

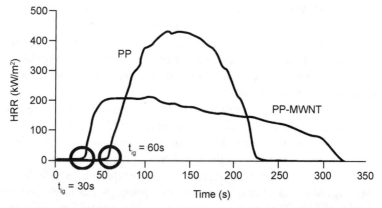

2.11 HRR values versus time of virgin PP fabric and PP/MWNT fabric (external heat flux = 35 kW/m² – samples put between two steel frames – RHR reproducibility within ±10% – average of three replicated experiments).

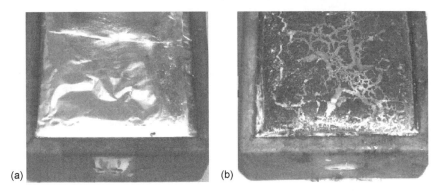

2.12 Cone residues of PP/clay samples showing the formation of char in the presence of clay (b) compared to the virgin PP (a) (from ref. 76).

The examples commented on above show that the nanocomposite approach alone does not provide enough flame retardancy to PP fabric. That is why Horrocks and his group combined nanoclays with conventional flame retardants as they did in the case of polyamides.[77] Flame retardants used include ammonium polyphosphate and a hindered amine stabilizer known to have flame retarding characteristics in polypropylene.[70] They reported that the flammability of polypropylene is reduced by the addition of small amounts of clay in conjunction with a conventional phosphorus-containing and a hindered amine flame retardant. The authors also suspect a P-N synergism to exist and the LOI value for the best formulation is 22 vol.-% compared to 19 vol.-% for neat PP, with only 6 wt.-% total loading.

2.4.5 High performance fibres

High performance fibres are driven by special technical functions that require specific physical properties unique to these fibres. They usually have very high levels of at least one of the following properties: tensile strength, operating temperature, heat resistance, flame retardancy and chemical resistance.[78,79] Applications include uses in the aerospace, biomedical, civil engineering, construction, protective apparel, geotextiles and electronic areas. The resistance to heat and flame is then one of the main properties of interest for determining the working conditions of these fibres.[80]

We have recently reviewed the reaction to fire and the heat resistance of the high performance fibres.[81] Only a brief summary and conclusions of our work will be given here. The principal classes of high performance fibres considered were derived from rigid-rod polymers (lyotropic liquid crystalline polymers and heterocyclic rigid-rod polymers), modified carbon fibres, synthetic vitreous fibres, phenolic fibres, poly(phenylene sulphide) fibres and others. We focused our investigations on poly(p-phenylene-2,6-benzobisoxazole) (PBO or Zylon from Toyobo), poly-p-phenylenediamine-terephtalamide (PPTA or Kevlar, DuPont), co-poly(p-phenylene-3,4-oxidiphenylene- terephthalamide) (TECH or Technora, Teijin), poly(2,6-diimidazo[4,5-b:4′,5′-e]pyridinylene-1,4(2,5-dihydroxy)phenylene) (PIPD or M5, Magellan), phenolic fibres (Kynol, Kynol), melamine fibres (Basofil, BASF), oxidized polyacrylonitrile (PAN) and polyamide-imide fibres (Kermel, Rhodia). Three categories of fibres can be distinguished, rated according to their performance. The first group is PBO and PIPD. These fibres have a very low RHR and in the conditions of post-flashover (external heat flux > 50 kW/m^2) would not to be expected to fire spread; also they have high LOI (>50 vol.-%), and they do not evolve smoke. Note that Northolt confirmed the exceptional reaction to fire of M5 fibres.[82] Kynol and recycled oxidized PAN fibres are in the second group because they have a moderated RHR (<150 kW/m^2). The recycled oxidized PAN fibres and Kynol exhibit high LOI values (>30 vol.-%), and they evolve little smoke. The third group is the p-aramid fibres which while having comparatively high RHR values (~300 kW/m^2), contribute to fire growth and evolve smoke; also have high LOI (>27 vol.-%) values. According to the above ranking, the heterocyclic rigid-rod polymers (PBO and PIPD) exhibit the best performance. The two structures of the polymers are highly conjugated and are heteroaromatic. Moreover, they do not have flexible mid-chain groups which may lead to the reduction of their thermal stability. These factors provide high levels of stability to the polymer and promote high flame resistance. In contrast to PBO and PIPD, p-aramid fibres have phenylene groups linked by amide bridges. These bridges (–CONH–) lead to a reduction in thermal stability of the fibres which consequently yield flammable molecules upon heating. Finally, Kynol and oxidized PAN fibres are cross-linked networks with methylene (Kynol) or ether (oxidized PAN) bridges. These flexible groups lead to the reduction in thermal stability as for the p-aramids, but the stabilizing character of the cross-linked network enhances char yields. It follows that the better fire performance of Kynol and oxidized PAN fibres in comparison with p-aramids can be assigned to the formation of higher yields of char. This enhanced char can act as an insulative shield when burning and can protect the substrate.

Recent work on the degradation of PBO supports our previous discussion and suggests a mechanism of degradation.[83,84] It was found that at temperatures below 660 °C, the polymer retains its original conformation and becomes stabilized by enhancement of its crystallinity. The decomposition takes place in a single step and the main changes occur within a very narrow temperature interval (650–660 °C). Formation of polyaramides as intermediates in the decomposition process was detected. These amide bonds subsequently degrade by homolytic breaking, yielding nitriles. The final carbonaceous residue is rich in nitrogen and retains a certain degree of anisotropy, a fact that was explained by the conservation of crystallinity at an intermediate decomposition stage.

The same group investigated a very interesting concept of impregnating m-aramid fibres (poly(m-phenylene isophthalamide or Nomex from Dupont) with phosphoric acid, which is well known to be a flame retardant for char-forming fibres.[85] They found that hydrolysis of the amide bond at low temperatures is promoted by the presence of H_3PO_4. Almost all the products evolved in the degradation of the impregnated Nomex are also found in its unimpregnated counterpart. However, their relative abundance is different and while amine evolution is preponderant in the case of the unimpregnated polymer, it only occurs to a minor extent with H_3PO_4-impregnated Nomex, in which nitriles prevail as the principal products. They suggested reaction mechanisms involving a nonhomolytic process with the formation of nitriles in the presence of phosphoric acid. At temperatures around 850 °C, elemental P_4 is released as a result of the reduction of phosphorus compounds previously linked to the carbonaceous residue. From this, we can expect a flame retardant action both in the condensed phase (via char enhancement) and in the gas phase (via release of phosphorus into the flame).

Nomex was also used as fire blocker and in multiple layers on foam.[86,87] It was found that the fabric layers can reduce the peak heat release rate of the foam considerably. The delayed ignition and reduced PHRR with the increase in fabric layers in the foam/fabric combination is due to the reduced rate of heating. It is pointed out that Nomex forms an insulating barrier that can resist to external flux as high as 50 kW/m without breaking.

2.4.6 Blends of fibres

A very common type of fabric is represented by blends of cellulosic and thermoplastic fibres (e.g., polyester (PET)/cotton). Very often such combinations of fibres will perform like the more flammable component in the system. Thus, when a thermoplastic fibre is adjacent to a char-former, this will often result in more intense burning than the additive effect of the separate combustible fibres. It will be seen in the following that some synergistic effects can be observed in particular combinations when not using thermoplastics.

PET/cotton has attracted greater attention than the other blends because of its apparent flammability-enhancing interaction in which both components participate. However, because of the observed interaction, only halogen-containing coatings and back-coatings find commercial application to blends, especially when the synthetic fibre is present at over 50% of the blend.[20] Thus some treatments on cotton-rich blends can achieve acceptable flame retardancy (e.g., methylolated phosphonamide finish or Pyrovatex CP, Ciba). Our approach by plasma treatment (see Section 2.3.1) confirms this trend[47] with a blend containing 56% of cotton exhibiting a rise in LOI value from 19 vol.-% for the virgin blend to 27 vol.-% for the treated blend and an M-rating from M4 to M2.

Cotton/nylon blends have been commonly used as the material for protective clothing, but these blend fabrics are not flame retardant-finished because of the unavailability of effective flame retardant finishing systems. As mentioned in Section 2.3.1, the combination of a hydroxyl-functional flame retardant organophosphorus oligomer together with the mixture of DMDHEU and TMM can be used as the binders for flame retardant finishing of cotton. Yang et al.[88] investigated this treatment for 50/50 cotton/nylon fabric. It was shown that the FR treatment is bound to the nylon fabrics mainly through the formation of an oligomer/TMM cross-linked polymeric network, thus becoming durable to multiple launderings. The cotton/nylon blend treated with this FR shows high levels of flame retardant performance and excellent laundering durability and fabrics passed a vertical flammability test even after 50 home laundering cycles.

While high performance fibres such as aramids are heat and flame resistant, the handle of these fibres is generally poor.[89,90] It is likely that combining natural fibres like wool with high performance fibres in blended knitted structures will enhance handle of the fabric and generate possible synergistic flame retardancy effects. In a previous work,[91] we have shown that blends of wool with PPTA (blends made by mixing yarns) improved the FR performance and the thermal stability of the whole fabric. The suggested mechanism was that the molten char of wool coats PPTA fibres. This hinders the diffusion of oxygen to the PPTA fibre, which is very sensitive to oxygen and thermal oxidative degradation. We have noticed that in nitrogen flow (i.e. pure pyrolysis), the TG curve is shifted to higher temperatures by as much as 100 °C compared to that in flowing air (i.e., main step of degradation starts at 600 °C against 500 °C) and the residue is much higher (40 wt.-% against 0.5 wt.-% at 1500 °C). So, the char coating protects PPTA fibres and they are degraded at higher temperatures. The flame retardancy of the blend is therefore enhanced. The combination of wool char with PPTA (or degraded PPTA) is then decomposed via exothermal reactions. We have also investigated other high performance fibres including PBO and Tech combined with wool. In those cases, no synergistic effect was detected.

2.5 Flame retardant inorganic man-made fibres

Man-made inorganic fibres have been traditionally associated with a number of glass and speciality glass fibres and also, refractory ceramic fibres. All these fibres are truly flame retardant since they do not burn even at very high temperatures,[93] hence there are few studies, in terms of reaction to fire, in the literature. Carbon, on the other hand, also classed by some as an inorganic fibre, is indeed flammable under certain conditions like most organic polymer fibres.

Carbon fibres can be prepared from a large number of polymeric precursors. The three main ones used commercially in order of decreasing current use are PAN, coal-based pitch and rayon.[92] The major application of carbon fibres is in polymeric matrix composites; however, they are also used in metal matrix and carbon matrix composites. Carbon fibres are extremely resistant to high temperatures: their melting temperature is 4000 °C. They can also be considered as flame resistant since they will burn only at very high temperatures or in high oxygen-containing atmospheres. They hence represent a choice material for applications at extremely high temperatures, for example in the filtration of molten iron.

Glass fibres exhibit a wide range of glass compositions to suit many textile fibre applications.[93,94] They are made from various compositions and they have softening points in the range 650–970 °C. At temperatures above 850 °C, these fibres partially devitrify and form polycrystalline material that melts at 1225–1360 °C, which is high enough to contain fires for several hours. A relatively new glass-like fibre has appeared on the market recently; the basalt fibre, made from a volcanic rock found on the surface of the Earth's crust (containing 40–60% SiO_2).[95] Continuous fibres of 10–12 μm in diameter can be obtained in the form of roving containing different numbers of elementary fibres, which exhibit somewhat better strength than glass fibres, but their price is presently 10–20% higher. These fibres exhibit a series of excellent properties including high modulus and excellent heat resistance (melting at 1450 °C), heat and sound insulating properties and vibration dampening properties. They also exhibit superior fire resistance, for example knitted basalt fibres resist an open flame applied to its surface at 1200 °C for several hours, while typical E-glass fabric is pierced by the same flame in few seconds.

Ceramic fibres are mostly used as refractory fibres in applications over 1000 °C and are characterized by a polycrystalline structure rather than an amorphous one.[93] These fibres have exceptionally high temperature characteristics and different compositions result in end-use temperatures varying from about 1050 °C or higher for the kaolin-based products to 1425 °C and above for the zirconium-containing materials. This group includes mainly alumina-, silica-, boron- and SiC-based fibres. Oxide fibres (Table 2.3) such as silica and alumina, which have oxidation stability and insulating properties, are used for heat insulators, such as the thermal protection system on the upper part of the Space Shuttle.[96] They also exhibit effectively zero flammability.

The high-temperature stability of SiC-based fibres is well-known, and therefore these materials have been investigated for application to high-temperature, structural materials.[96] Si-C-O fibres prepared from polycarbosilane (Nippon Carbon Co.) and Si-Ti-C-O fibres prepared from polytitanocarbosilane (UBE Industry) are continuous fine fibres that have high tensile strengths and possess high heat resistance.[97] The application of these fibres is in refractory materials and they are also expected to be useful as reinforcement fibres for plastics, metals and ceramics. SiC fibres are also prepared from polycarbosilane by electron beam irradiation and curing in argon (compared to Si-C-O which is prepared by thermo-oxidation and curing in argon); exhibit higher heat resistance temperature than Si-C-O fibres.

Boron fibres are also classed as ceramic fibres. They are produced via chemical vapour deposition (CVD) using the hydrogen reduction of boron

Table 2.3 Physical properties of various oxide fibres (from ref. 96)

Fibre	Manufacturer	Composition (wt.%)	Young's modulus E (GPa)	Diameter (μm)	Density (g cm^{-3})
Almax	Mitui Mining	α-Al_2O_3	320~340	10	~3.6
Altex	Sumitomo	$15SiO_2$-$85Al_2O_3$	200~230	9~17	~3.2
Fiber FP	Du Pont	α-Al_2O_3	380~400	~20	3.9
Nextel 312	3M	$24SiO_2$-$14B_2O_3$-$62Al_2O_3$	150	10~12	2.7~2.9
Nextel 440	3M	$28SiO_2$-$2B_2O_3$-$70Al_2O_3$	220	10~12	3.05
Nextel 480	3M	$28SiO_2$-$7B_2O_3$-$70Al_2O_3$	220	10~12	3.05
Nextel 550	3M	$27SiO_2$-$73Al_2O_3$	193	10~12	3.03
Nextel 610	3M	Al_2O_3	373	10~12	3.75
Nextel 720	3M	$15SiO_2$-$85Al_2O_3$	260	10~12	3.4
Nextel Z-11	3M	$32ZrO_2$-$68Al_2O_3$	76	10~12	3.7
PRD-166	Du Pont	$80Al_2O_3$-$20ZrO_2$	360~390	14	4.2
Saftil	ICI	$4SiO_2$-$96Al_2O_3$	100	~20	2.3
Safikon	Safikon	Al_2O_3	386~435	3	3.97
Sumica	Sumitomo Chemical	$15SiO_2$-$85Al_2O_3$	250	75~225	3.2

trichloride on a tungsten filament in a glass tube reactor. They possess desirable mechanical and physical properties including high tensile strength, high stiffness and low density. The fibres start to degrade at 400 °C in air and that is why they are often covered by a refractory silicon or boron carbide to prevent oxidative degradation.

2.6 Future trends

In this chapter, we have reviewed the recent developments in the flame retardancy of fibres and textiles. Numerous recent works have been published to enhance FR properties of cotton and to find novel solutions that combine laundering durability, handle and performance associated with an easy process and low cost. The mechanism involved is almost always char enhancement including intumescence. Some of these look promising, but efforts remain to render them competitive compared to the well-established commercial products (e.g., Pyrovatex, Ciba and Proban, Rhodia). The other natural fibres have not attracted much attention except for some combination of them as nonwoven blended fabrics. Older treatments (e.g., Zirpro for wool) continue to be used. New developments have appeared in the synthetic fibre area including new biopolymers such as PLA and high performance fibres such as M5 and Zylon. These latter are inherently flame retardant but biopolymers are not. One new development has appeared in the literature for PLA using the potential effect of nanoparticles as flame retardant but the performance is still marginal. Otherwise the traditional treatments have been used for the other biopolymers (e.g., lyocell) with reasonable success. Blends of fibres and, in particular, the combination of natural and high performance fibres associated with the technique of hybrid core yarn production seems to be a very promising way because it associates low flammability of the high performance fibre with good handle and aesthetics of the natural fibre at an acceptable cost with enhanced mechanical properties. This process is ready to use and should be developed at the industrial scale in the next future.

Increasing efforts have been undertaken in attempts to study and realize the potential of nanocomposite fibres. This is a relatively new technology in the field of flame retardancy and it has been applied to a number of textile fibres (PA-6, PP and PET) only since 2000. The mechanism of protection is the formation of a mineral layer combined with char promotion but the protective layer is not efficient enough to provide the highest standard of protection. This technology gives the best results when combined with conventional flame retardants which then lead to synergistic effects. This last approach, as in the case of bulk polymers, should be developed by scientists in the near future. We can also expect developing fibres with multifunctional properties as, for example, the association of carbon nanotubes with conventional flame retardants to obtain flame retardant fibres with antistatic properties. These routes of investigations are still very broad and we believe that there will be new developments in this direction.

The last approach suggested in this paper is to use plasma treatment (polymerization-induced by plasma to create a thin film at the surface of the fabric). FR performance is significantly enhanced and is associated with laundering durability, but further developments are still needed to achieve very low flammability. Synergistic effects have been observed on bulk polymer nanocomposites and the same effect might be expected in the case of textiles. Thus there may be opportunities for investigations based on these observations.

2.7 References

1. M.M. Hirschler and T. Piansay, *Fire Mater.*, 2007, **31**(6), 373–386.
2. K. Rohr, *Products First Ignited in US Home Fires*, National Fire Protection Association, Quincy, MA, April 2005.
3. Fire Statistics, United Kingdom, 2004, OPDM, London.
4. S. Bourbigot and X. Flambard, *Fire Mater.*, 2002, **26**, 155.
5. D. Chen, Y.Z. Wang, X.P. Hu, D.Y. Wang, M.H. Qu and B. Yang, *Polym. Deg. Stab.*, 2005, **88**, 349.
6. J. Troitzsch, *Flammability Handbook*; Ed. J. Troitzsch, Hanser Verlag, Munich, 2003.
7. R.B. LeBlanc, The history of flammability and flame resistance of textiles. *AATCC Review*, 2001, **1**(2), 27–31.
8. EN 1103. Textiles Burning Behaviour Fabrics for Apparel Detailed Procedure to Determine the Burning Behaviour of Fabrics for Apparel. European Committee for Standardization: Brussels, 1995.

9. R.M. Rossi, G. Bruggmann and Rolf Stämpfli, *Fire Mater.*, 2005, **29**, 395–406.
10. C. Lee, I.Y. Kim and A. Wood, *Fire Mater.*, 2002, **26**, 269–278.
11. J.M. Cavanagh, D.A. Torvi, K.S. Gabriel, G.A. Ruff, *Fabric Architecture*, 2006, **18**(5), 62–64.
12. S. Davis S, K.M. Villa, Development of a Multiple Layer Test Procedure for Inclusion in NFPA 701: Initial Experiments, NISTIR 89-4138. National Institute of Standards Technology: Gaithersburg, MD, August 1989.
13. National Highway Traffic Safety Administration, DOT. 49 CFR Ch. V (10-1-97 Edition). 571.302 Standard No. 302; *Flammability of Interior Materials* 628–630.
14. M. Spearpoint, S.M. Olenick, J.L. Torero and T. Steinhaus, *Fire Mater.*, 2005, **29**, 265–282.
15. M. Roslon, S. Olenick, D. Walther, J.L. Torero, A.C. Fernandez-Pello, H. Ross, Micro-gravity ignition delay of solid fuels. *AIAA-Journal*, 2001, **39**, 2336–2342.
16. B.K. Kandola, A.R. Horrocks, K. Padmore, J. Dalton and T. Owen, *Fire Mater.*, 2006, **30**, 241–255.
17. A.R. Horrocks, *Rev Prog Color*, 1986, **16**, 62-101.
18. G.P. Nair, *Colourage*, 2003, **50**(4), 45–48.
19. A.R. Horrocks, B.K. Kandola, P.J. Davies, S. Zhang, S.A. Padbury, *Polym. Deg. Stab.*, 2005, **88**, 3–12.
20. A.R. Horrocks, *Fire Retardant Materials*, Edited by A.R. Horrocks and D. Price, Woodhead Publishing Ltd, Cambridge, 2001, pp. 128–181.
21. A.R. Horrocks, *Polym. Deg. Stab.*, 1996, **54**, 143–154.
22. R. Dombrowski, *AATCC Review*, 2003, **3**, 13–16.
23. S. Giraud, S. Bourbigot, M. Rochery, I. Vroman, L. Tighzert, R. Delobel and F. Poutch, *Polym. Deg. Stab.*, 2005, **88**(1), 106–113.
24. P.J. Wakelyn, W. Rearick, J. Turner, *American Dyestuff Reporter*, 1998, **87**, 13–21.
25. E.D. Weil. In Kirk, *Other Encyclopedia of Chemical Technology*, vol. 10, 4th edn, Grayson M (ed.). Wiley: New York, 1995, 976–998.
26. W. Wu, C.Q. Yang, *J. Fire Sci.*, 2004, **22**, 125–142.
27. C.Q. Yang and X. Qiuz, *Fire Mater.*, 2007, **31**(1), 67–81.
28. W. Wu and C.Q. Yang, *Polym. Deg. Stab.*, 2007, **92**(3), 363-369.
29. E. Lecoeur, I. Vroman, S. Bourbigot, T.M. Lam, R. Delobel, *Polym. Deg. Stab.*, 2001, **74**, 487–492.
30. E. Lecoeur, I. Vroman, S. Bourbigot, R. Delobel, *Polym. Deg. Stab.*, 2006, **91**, 1909-1914.
31. NF P 92-503. Essais de réaction au feu des matériaux, essai au brûleur électrique applicable aux matériaux souples d'une épaisseur inférieure à 5 mm. Afnor; 1975.
32. S.M. Mostashari, M.A. Zanjanchi and O. Baghi, *Combustion, Explosion, and Shock Waves*, 2005, **41**(4), 426–429.
33. Y. Halpern, M. Mather, R.H. Niswander, *Ind. Org. Chem. Prod. Res. Dev.*, 1984, **23**, 233–238.
34. P.J. Davies, A.R. Horrocks, A. Alderson, *Polym. Deg. Stab.*, 2005, **88**, 114–122
35. A.R. Horrocks, P.J. Davies, A. Alderson, B.K. Kandola, *Advances in the Flame Retardancy of Polymeric Materials: Current perspectives presented at FRPM'05*; Ed. B. Schartel, Books on Demand GmbH (BoD), Norderstedt, in press.
36. S. Giraud, S. Bourbigot, M. Rochery, I. Vroman, L. Tighzert, R. Delobel, *Polym. Deg. Stab.*, 2002, **77**, 285–297.
37. S. Giraud, S. Bourbigot, M. Rochery, I. Vroman, L. Tighzert, R. Delobel, F. Poutch, *Polym. Deg. Stab.*, 2005, **88**, 106–113.
38. D. Saihi, I. Vroman, S. Giraud, S. Bourbigot, *React. Funct. Polym.*, 2005, **64**, 127–138.

39. D. Saihi, I. Vroman, S. Giraud, S. Bourbigot, *React. Funct. Polym.*, 2006, **66**, 1118–1125.
40. S. Bourbigot, M. Le Bras, S. Duquesne and M. Rochery, *Macromol. Sci. Eng.*, 2004, **289**(6), 499-511.
41. J.W. Gilman, T. Kashiwagi and J.D. Lichtenhan, *SAMPE J.*, 1997, **33**, 40.
42. J.W. Gilman, *App. Clay Sci.*, 1999, **15**(1–2), 31.
43. S. Bourbigot, S. Duquesne and C. Jama, *Macromol. Symp.*, 2006, **233**(1), 180–190.
44. E. Devaux, M. Rochery and S. Bourbigot, *Fire Mater.*, 2002, **26**, 149-154.
45. L.A. White, *J. Appl. Polym. Sci.*, 2004, **92**, 2125–2131.
46. M.J. Tsafack and J. Levalois-Gruetzmacher, *Surf. Coat. Tech.*, 2006, **201**(6), 2599.
47. A. Vannier, S. Duquesne, S. Bourbigot, R. Delobel, C. Magniez and M. Vouters, Proceeding of International Conference on Textile Coating & Laminating, Nov. 2006, 8–29, Barcelona (Spain).
48. P.J. Davies, A.R. Horrocks and M. Miraftab, *Polym. Int.*, 2000, **49**, 1125–1132.
49. L. Benisek, *Text. Res. J.*, 1981, **51**, 369.
50. A.R. Horrocks and P.J. Davies, *Fire Mater.*, 2000, **24**, 151–157.
51. A.R. Horrocks and S. Zhang, *Textile Res. J.*, 2004, **74**(5), 433–441.
52. L. Benisek, J. Andrews and I. Bentec, *Melliand International*, 2005, **11**(4), 337–339.
53. M. Holmes, *Plastics, Additives and Compounding*, 2004, **6**(6), 42–43.
54. Li H-X, Wang J-G, *Jilin Engineering College Acta*, 1991, **12**(2), 68–71.
55. H. Zhou, *Guangxi Textile Science Technology*, 1995, **24**(2), 35–39.
56. Q. Yao, Y. Cao, *Silk*, 1989, **11**, 24–26.
57. W.B. Achwal, *Colourage*, 1987, June, 16–30
58. G. Kako, A. Katayama, *J. Sericulture Sci. Japan*, 1995, **64**(2), 124–131.
59. J.P. Guan and G.Q. Chen, *Fire Mater.*, 2006, **30**, 415–424.
60. R. Kozlowski, B. Mieleniak, M. Muzyczek and A. Kubacki, *Fire Mater.*, 2002, **26**, 243–246.
61. X. Flambard, S. Bourbigot, R. Kozlowski, M. Muzyczek, B. Mieleniak, M. Ferreira, B. Vermeulen, F. Poutch, *Polym. Deg. Stab.*, 2005, **88**, 98–105.
62. M.E. Hall, A.R. Horrocks, H. Seddon, *Polym. Deg. Stab.*, 1999, **64**, 505–510.
63. E.T.H. Vink, K.R. Rabago, D.A. Glassner, B. Springs, R.P. O'Connor, J. Kolstad, P.R. Gruber, *Macromol. Biosci.*, 2004, **4**, 551–564.
64. S. Solarski, F. Mahjoubi, M. Ferreira, E. Devaux, P. Bachelet, S. Bourbigot, R. Delobel, M. Murariu, A. Da Silva Ferreira, M. Alexandre, P. Degée and P. Dubois, *J. Mater. Sci.*, 2007, **42**(13), 5105–5117.
65. D.Q. Chen, Y.Z. Wang, X.P. Hu, D.Y. Wang, M.H. Qu, B. Yang, *Polym. Deg. Stab.*, 2005, **88**, 349–356.
66. W. Liu, D.Q. Chen, Y.Z. Wang, D.Y. Wang, M.H. Qu, *Polym. Deg. Stab.*, 2007, **92**, 1046-1052.
67. X. Almeras, A. Vannier, P. Vandendaele, S. Duquesne, S. Bourbigot, R. Delobel, M. Ortiz, G. Gupta and E. Pivotto, Proceeding of the 46th Doornbirn man-made fibers congress, Austrian man made fibers Institute, 2007, Vienna.
68. S. Bourbigot, E. Devaux and X. Flambard, *Polym. Deg. Stab.*, 2001, **75**(2), 397–402.
69. A.R. Horrocks, B.K. Kandola, P.J. Davies, S. Zhang, S.A. Padbury, *Polym. Deg. Stab.*, 2005, **88**, 3–12.
70. S. Zhang and A.R. Horrocks, *Prog. Polym. Sci.*, 2003, **28**, 1517–1538.
71. A.R. Horrocks, Flame-retardant finishes and finishing. In: D.H. Heywood, editor. *Textile finishing*. Bradford: Society of Dyers and Colourists, 2003, pp. 214–50.
72. S. Bourbigot, M. Le Bras, X. Flambard, M. Rochery, E. Devaux, J. Lichtenhan, *Fire Retardancy of Polymers: New Applications of Mineral Fillers*, M. Le Bras, S.

Bourbigot, S. Duquesne, C. Jama and C.A. Wilkie (eds.), Royal Society of Chemistry, 2005, pp. 189-201.
73. T. Kashiwagi, E. Grulke, J. Hilding, R.H. Harris Jr., W.H. Awad, J. Douglas, *Macromol. Rapid Commun.*, 2002, **23**, 761.
74. S. Bellayer, S. Bourbigot, X. Flambard, M. Rochery, J.W. Gilman, E. Devaux, *Proceedings of the 4th AUTEX Conference*, ENSAIT, Roubaix 2004, O-3W1.
75. T. Kashiwagi, E. Grulke, J. Hilding, K. Groth, R. Harris, K. Butler, J. Shields, S. Kharchenko, J. Douglas, *Polymer*, 2004, **45**, 4227–4239.
76. A.R. Horrocks, B.K. Kandola, G. Smart, S. Zhang, R. Hull, *J. Apl. Polym. Sci.*, 2007, **106**(3), 1707–1717.
77. S. Zhang, A.R. Horrocks, R. Hull, B.K. Kandola, *Polym. Deg. Stab.*, 2006, **91**, 719–725.
78. T. Hongu, G.O. Philips in *New Fibres*, Woodhead Publishing Limited, Cambridge, 1997.
79. D.J. Sikkema, *J. App. Polym. Sci.*, 2002, **83**, 484–488.
80. E.M. Pearce, E.D. Weil, V.Y. Barinov, in *Fire and Polymers'*, ACS Symposium Series 797, ed. G.L. Nelson and C.A. Wilkie, American Chemical Society, Washington, DC, 2001, pp. 37–47.
81. S. Bourbigot, X. Flambard, *Fire Mater.*, 2002, **26**, 155–168.
82. M.G. Northolt, D.J. Sikkema, H.C. Zegers and E.A. Klop, *Fire Mater.*, 2002, **26**, 169–172.
83. K. Tamargo-Martinez, S. Villar-Rodil, J.I. Paredes, A. Martnez-Alonso, J.M.D. Tascon, *Chem. Mater.*, 2003, **15**(21), 4052–4059.
84. K. Tamargo-Martinez, S. Villar-Rodil, J.I. Paredes, A. Martnez-Alonso, J.M.D. Tascon, *Polym. Deg. Stab.*, 2004, **86**, 263–268.
85. F. Suarez-Garcia, S. Villar-Rodil, C. G. Blanco, A. Martinez-Alonso, J.M.D. Tascon, *Chem. Mater.*, 2004, **16**(13), 2639–2647.
86. T.M. Kotresha, R. Indushekara, M.S. Subbulakshmia, S.N. Vijayalakshmia, A.S. Krishna Prasada, K. Gaurav, *Polym. Testing*, 2005, **24**, 607–612.
87. T.M. Kotresha, R. Indushekara, M.S. Subbulakshmia, S.N. Vijayalakshmia, A.S. Krishna Prasada, K. Gaurav, *Polym. Testing*, 2006, **25**, 744–757.
88. H. Yang, C.Q. Yang, *Polym. Deg. Stab.*, 2005, **88**, 363–370.
89. T. Hongu, G.O. Philips, in *New Fibres*, Woodhead Publishing Limited, Cambridge, 1997.
90. H.H. Yang, ed., *Kevlar aramid fiber*, John Wiley & Sons, Chichester, 1993.
91. S. Bourbigot, X. Flambard, M. Ferreira, F. Poutch, *J. Fire Sci.*, 2002, **20**(1), 3–22.
92. J.B. Donnet and R.C. Bansal, in *Carbon Fibres*, 2nd edn, Marcel Dekker Inc, New York, Chapter 1.
93. In *Advanced Inorganic Fibres: Processes, Structures, Properties, Applications*, ed. Wallenberger F.T., Kluwer Academic Publisher, London, 2000.
94. In *Man-Made Vitreous Fibres*, ed. Eastes W., TIMA Inc, New York, 1993.
95. T. Czigány, *Express Polym. Lett.*, 2007, **1**(2), 59.
96. T. Ishakawa, *Adv Polym Sci.*, 2005, **178**, 109–144.
97. K. Okamura, T. Shimoo, K. Suzuka and K. Suzuki, *J. Ceram. Soc. Jpn*, 2007, **114**(6), 445–454.

3
Developments in phosphorus flame retardants

S V L E V C H I K, Supresta LLC, USA and
E D W E I L, Polytechnic University of Brooklyn, USA

Abstract: Environmental considerations have been a main factor in driving an increasing R&D effort on phosphorus flame retardants, although sometimes efficiency, lower density and better light stability are significant factors.

Patents dealing with flame retardancy of polycarbonate and its blends are especially numerous. Oligomeric aromatic aryl phosphates, notably resorcinol bis(diphenyl phosphate) and bisphenol A bis(diphenyl phosphate) are included in the claims of many patents, and have been found broadly useful because of thermal stability, low volatility, and efficiency.

There is a substantial number of patents dealing with dialkylphosphinic acid salts, and, more recently, with hypophosphite salts which are useful in polycarbonates and poly(butylene terephthalate). Especially when synergized by melamine salts, they appear to be effective in nylons.

Electronic plastics, especially printed wiring boards, appear to be the largest market for flame retardant polymeric materials. Largely driven by waste disposal regulations in Europe and 'green' marketing strategies by OEMs, interest has increased in non-halogen flame retardant systems, particularly in Europe and Asia. Many patents and publications have appeared on epoxy systems in which 9,10-dihydro-9-oxa-10-phosphaphenanthrene 10-oxide is reacted into the polymer. Another product which reacts into the polymer, and has some property and processing advantages is oligomeric resorcinol methylphosphonate.

The low scorch flame retardant pentabromodiphenyl ether, widely used in flexible polyurethane foams, was banned and withdrawn from the market; a search became urgent for an environmentally acceptable replacement. Both halogenated and non-halogenated products have been developed for this application; at present a halogen-phosphorus combination is finding some acceptance, but halogen-free phosphorus compounds are in serious development.

Key words: phosphorus flame retardants, polycarbonate, flame retardant polymers, halogen-free phosphorus compounds.

3.1 Introduction

Flame retardants based on halogen, particularly bromine, have been important in flame retardancy and continue to be important. However, environmental concerns and scrap disposal issues have stimulated strong interest in halogen-free alternatives. There is a growing patent and non-patent literature on non-halogen flame retardants and particularly on phosphorus-based flame retardants.

Phosphorus-based flame retardants have been known and commercially produced for a long time. However, as of today, the application of phosphorus-based flame retardants has been limited to specific polymers or specific classes of polymers. This can be explained by the mechanism of action of phosphorus flame retardants, which is mostly related to char formation. In order to promote charring of the polymer, the flame retardant should be able to react with the polymer during decomposition. Although phosphorus possesses also a gas phase mode of action, the benefits of this mechanism are not used to the full extent in commercial flame retardant systems. Some recent developments in phosphorus flame retardants are focused on designed formulations to make use of synergism between gas phase and condensed phase action. New phosphorus-based thermally and hydrolytically stable molecules are also in active development. Special interest is directed to co-monomer type flame retardants which effectively incorporate into the polymer network. Our review covers recent phosphorus flame retardant developments for several classes of plastics and foams.

3.2 Main types of phosphorus flame retardants

Phosphorus compounds in flame retardancy originated with ammonium phosphate on canvas in the 18th century, and the phosphorus flame retardants evolved with the advent of plastics into a wide range of inorganic and organic phosphorus additives and reactives.

Ammonium and melamine polyphosphates, and related phosphorus acid amine salts with improved stability and water resistance are useful tools for char-forming and intumescent systems such as for polyolefins. Elemental red phosphorus, when properly stabilized and coated, is a highly efficient flame retardant for molded nylons, epoxies and certain other plastics.

Phosphoric acid esters comprise a broad group of flame retardant additives. Triethyl phosphate is used in thermosets, while triaryl phosphates and alkyl diphenyl phosphates are used mainly in vinyls. Oligomeric aryl phosphates have become important in ABS-polycarbonate or HIPS-polyphenylene oxide blends. Several chloroalkyl phosphate liquids continue to be the dominant additives in polyurethane foams. A halogen-free oligomeric phosphate additive is gaining usage in that application. A reactive oligomeric phosphate diol is also commercially available.

Aluminum or other metal salts of organic phosphonic and phosphinic acids as well as certain salts of hypophosphorous acid are useful additives in polyamides and polyesters. Phosphinic acid esters can be reacted into polyester fibers as efficient built-in flame retardants. Certain phosphonic acid esters are effective for post-treating of polyester fabrics and can also be used as additives in certain thermoplastic resins. Some phosphonic esters with reactive hydroxyl groups are useful for coreacting with aminoplastics for flame retardancy of cellulosics. An oligomeric phosphonate ester is showing promise as a reactive component of epoxy printed wiring boards.

A durable cotton finish which is based on reaction of tetrakis(hydroxymethyl)phosphonium salts with urea and ammonia dominates the flame retardant cotton textile market. Strengths and weaknesses of phosphorus flame retardants relative to other flame retardant systems (mostly halogen) can be summarized as following:

Advantages:
- Lower density compared with halogen types
- No need for antimony oxide synergist
- Less likely to be persistent and bioaccumulative in the environment
- Effective in promoting a char barrier in charrable polymers
- Better photostability than most halogen types
- Less tendency to intensify smoke obscuration
- Less acid gas evolution on burning
- No halodioxin formation even on poor incineration of the plastics
- More likelihood of biodegradability compared with halogens
- Avoid negative public image, 'greener'.

Disadvantages:
- Difficulty in making high %P compounds, so may be less efficient
- No good general synergist
- Tendency to be hydrophilic and possibly cause moisture uptake
- May hydrolyze to give acids which damage the plastic or spoil electrical properties
- May be less thermally stable and thus limit processing temperature
- Thermal and hydrolytic stability may impede recycling
- Some toxicity questions related to reactivity
- Cost may be greater at equivalent effectiveness.

3.3 Polycarbonate and its blends

In recent years, patents on flame retardant systems for polycarbonates and blends thereof have become the largest segment of all flame retardancy patents. Aromatic phenyl phosphate oligomers, especially resorcinol bis(diphenyl phosphate) and bisphenol A bis(diphenyl phosphate) have found extensive use because of thermal stability, efficiency and low volatility.[1]

Aromatic phosphates are not much used in plain PC because of reduction in clarity, tendency to stress-crack and somewhat reduced hydrolytic stability, but are very widely used in PC/ABS blends. Oligomeric aromatic phenyl phosphates (mainly diphosphates) are finding broader application than monophosphates because of better thermal stability and lower volatility. Resorcinol bis(diphenyl phosphate) available from Supresta LLC under the trade name Fyrolflex® RDP is a mixture of oligomers with two to five phosphorus atoms, but with the distribution heavily towards the diphosphate.[1] In commercial PC/ABS blends where ABS content normally does not exceed 25%, RDP gives a V-0 rating at 8–12 wt.% loading.[2,3]

Poly(tetrafluoroethylene) (PTFE) is a necessary ingredient in the formulation, usually at <0.5 wt.%, to retard dripping.[4] RDP has some hydrolytic instability, which may reduce the long term aging performance of PC/ABS and may cause a problem in recycling. This shortcoming of RDP can be alleviated by adding acid scavengers such as epoxies, oxazolines, or ortho esters.[5] Inorganic co-additives such as highly dispersed silica[6] or talc[7] can also improve physical properties, notably high temperature dimensional stability. Co-addition of organoclay[8] can result in improved flame retardancy by decreasing afterflaming time.

Bisphenol A bis(diphenyl phosphate), marketed by Supresta as Fyrolflex® BDP and also available from other manufacturers is thermally and hydrolytically more stable then RDP. However, it is more viscous than RDP and thus more difficult to handle. It has slightly lower phosphorus content and as result is usually less active than RDP. In some formulations BDP also shows higher HDT.[9] PC/ABS with ABS at ≤ 25 wt.% usually needs ≥ 12 wt.% BDP plus a small co-addition of PTFE in order to assure a V-0 rating.[10,11] Although BDP is hydrolytically very stable, further improvement in long term stability and consequently improved aging performance of PC/ABS can be accomplished by adding epoxy[12] or oxetane[13] acid scavengers. Co-addition of a filler such as wollastonite fibers,[14] carbon fibers[15] or a highly charring polymer such as a phenoxy resin,[16] improves impact resistance and also cuts the afterflame time.

The gas phase activity of aromatic phosphates has been disputed in the literature. For example, it is well established[17] that RDP is more efficient in PC/ABS than the more volatile triphenyl phosphate (TPP). Moreover, a combination of RDP or BDP with TPP[18] or even a combination of BDP and RDP[19] are more effective flame retardants than each component alone since the total flame retardant loading can be reduced to 10 wt.% while still preserving the V-0 rating. Probably, this result is because of the combination of condensed phase and gas phase actions of the less volatile and the more volatile phosphates. RDP or BDP which are mostly condensed phase active additives tend to cause PC to produce more char, decreasing the fuel supply to the flame and decreasing flame temperature. TPP, which has gas phase activity, becomes more effective in the gas phase with decreasing flame temperature.

Resorcinol bis(di-2,6-xylyl phosphate), (RXP) is another commercial diphosphate, exclusively made in Japan. The steric hindrance provided by the 2,6-xylyl groups causes this product to have higher hydrolytic stability than BDP. Its fire retardant efficiency is similar to that of BDP, and it provides a V-0 rating in PC/ABS at 12–16 wt.% loading.[20] In contrast to RDP and BDP, RXP is a solid, a feature which is preferred by some compounders. Its higher cost limits large-scale use.

Many phosphoramides are high melting solids, and there is some expectation that they would favor high heat distortion temperature (HDT) of PC/ABS. Recently, GE patented a series of bisphosphoramidates with a piperazine bridging unit.[21] These products can provide a V-0 rating at 12 wt.% in PC/ABS and as expected they also provide high HDT. A series of new bisphosphates with one or more phenyl groups replaced with morpholine rings, effective in PC/ABS (4:1) at 12 wt.%, was recently patented by Cheil (Korea).[22]

Cyclic phenoxyphosphazenes are thermally and hydrolytically stable phosphorus-nitrogen products. A blend consisting of tri- and tetra-phosphazenes with some large rings was effective in PC/ABS at 12–15 wt.%.[23] Some cyclophosphazenes may be finding commercial use in Asia. Bisphosphates (RDP or BDP) or monophosphates (TPP) are said to be synergistic with cyclic phosphazenes.[24]

3.4 Polyesters and nylons

Halogen-free flame retardants are of interest in polyesters and nylons, and some new ones have recently been introduced to the market, but certain non-halogen phosphorus-based additives, thermosol finishes, or coreactant systems have been use for many years in PET textiles.[25] The requirements for polyesters and nylons are stringent because of high processing temperatures, sensitivity to hydrolytic degradation catalyzed by possible acids, the need for long-term dimensional stability, and avoidance of exudation ('blooming'). These requirements have eliminated many phosphorus-based flame retardants aside from certain highly thermally stable materials.

In the late 1970s and early 1980s various metal salts of dialkylphosphinates were prepared at and tested in PET by Pennwalt[26] and Hoechst.[27] Later, Ticona and Clariant tested zinc, aluminum or calcium dialkylphosphinate salts in glass-filled nylons and PBT. The Al or Ca salts of ethylmethylphosphinic acid were found to give V-0 at 15 wt.% in plain PBT, at 20 wt.% in glass-filled PBT, and 30 wt.% in glass-filled nylons.[28–30] Later, Clariant evidently found a lower cost route to the diethylphosphinate salts[31,32] and they commercialized the aluminum salt as Exolit® OP 930 (DEPAL), which also is useful in thermoset resins.[33,34]

Although salts of dialkylphosphinic acid by themselves are only moderately efficient in nylons they were found to be synergistic with nitrogen-containing products such as melamine cyanurate,[35–37] melamine phosphate or melamine polyphosphate.[37] Based on this synergism, Clariant commercialized Exolit® OP 1311 for nylon 6 and Exolit® OP 1312 (also containing a stabilizer) for nylon 6.6.[38,39] These products provide UL-94 V-0 ratings in glass-filled nylons at 15–20 wt.% loading. Similarly Italmatch discovered that aluminum hypophosphite $Al(HPO_2)_3$ (Phoslite® IP-A) alone or in combination with melamine cyanurate can effectively flame retard glass filled PBT.[40] For example, 7.5 wt. Phoslite® IP-A and 7.5 wt. melamine cyanurate provide a V-0 rating. In spite of the high industrial interest, academic laboratories studies on the mode of flame retardant action of these phosphinic salts have been few.[41]

The adduct of 9,10-dihydro-9-oxa-10-phosphaphenanthrene 10-oxide (DOPO) and an itaconic acid ester (probably the bis[2-hydroxyethyl] ester) is the reactive flame retardant commercially used in polyester fibers as a coreactant.[42–45] This fabric is commercial in Japan from Toyobo as HEIM[46] and the bis(hydroxyethyl) ester reagent is available from Schill and Seilacher as their Ukanol® FR 50/1. This product is efficient in PET fibers at low loading (0.3–0.65% P).

Recently the fire retardant performance and hydrolytic stability of PET fibers containing this adduct were compared to fibers (Trevira® CS) containing phosphinate phosphorus in the main chain in the form of $-OC(=O)CH_2CH_2P(=O)-(CH_3)O-$ units.[46] Both fibers had almost the same properties and flame retardancy, but the type with the phosphinate unit in the main chain was hydrolyzed about two times faster than the side-chain phosphinate type, and this led to a more significant loss of tensile strength. In general, the flame retardancy of phosphorus-containing PET is mostly achieved by enhanced melt flow and melt drip, presumably catalyzed by polyphosphoric acid produced in the process of oxidative degradation during combustion.

Cyclic phosphonates with high phosphorus content and low volatility known as Antiblaze® 19 (N or NT; now Amgard® cu) (mixture of oligomers, mainly a diphosphonate of lower viscosity) and Antiblaze® 1045 (Amgard®) (mixture of oligomers of higher viscosity, mainly a triphosphonate) from Rhodia have been extensively used, mainly Antiblaze® 19 (N or NT), in PET fibers.[47] The phosphonate is applied as a water solution to the surface of the fiber and then heated to the temperature at which fibers become soft and the surface absorbs the additive (the thermosol process similar to that used for disperse dyeing). Applications of these phosphonates (particularly Antiblaze® 1045) in injection molded unfilled PBT are also reported in the literature. A combination of 10 wt.% Antiblaze® 1045 with 15 wt. % melamine cyanurate[48] or 12 wt.% Antiblaze® 1045 and 8 wt.% melam (a melamine thermal condensate) gave V-0 ratings in 30% glass-filled PBT. The flame retardant synergism between Antiblaze® 1045 and melamine was recently studied by Balabanovich et al.[49] The optimum ratio of Antiblaze® 1045 with melamine was found to be 2:3. Thermogravimetric analysis showed interaction between the two additives and also between the additives and the PBT. Infrared showed that the additives shift the thermal decomposition of PBT toward formation of amides instead of acids. Dehydration of aromatic amides cools the flame and forms volatile terephthalodinitrile, which forms crystalline particles in the flame zone (which may favor recombination of free radicals in the flame), providing likely gas phase contributions to flame retardancy. It is also likely that Antiblaze® 1045 interacts with melamine to form a glassy phosphorus-nitrogen solid residue, also providing a flame retardancy mode of action in the condensed phase.

Despite efforts to flame-retard thermoplastic polyesters or nylons with organic phosphates, it has been difficult to find a commercial solution. Perhaps this is because of the tendency of these polymers to generate combustible gases and to have little tendency to char, where the main flame retardant effects of organic phosphates would be exerted. Also, many additives affect physical properties and stability.

Recently Levchik et al.[50–52] reported the fire behavior of PBT flame-retarded with RDP or BDP and with a phenolic resin (novolac) charring agent. At 10 wt.% BDP unfilled PBT reached a V-2 rating, however at 15 wt.% BDP or higher, some of the flame retardant exuded to the surface. A combination of 10 wt.% novolac and 15 wt.% RDP or BDP reached a V-0 rating, and surprisingly, the novolac improved the migration performance of the aryl phosphates, since all specimens containing novolac gave no exudation in 30 days at 70 °C. To increase the flame retardancy in glass-filled PBT, TPP was added to the formulation to provide some gas phase mode of action.[53] It was found that glass-filled PBT with 10 wt.% BDP, 5 wt.% TPP and 10 wt.% novolac reached a V-0 rating.

An ultimate condensation product of urea or melamine and phosphoric acid is a very high melting solid, phosphorus oxynitride $(PON)_m$. There is even a report of a direct (and potentially very low cost) synthesis of phosphorus oxynitride from ammonium phosphate.[54] Phosphorus oxynitride has been suggested for use in polyesters in combination with melamine phosphate, melamine cyanurate, ammonium polyphosphate or calcium diethylphosphinate.[55] For example 15 wt.% of $(PON)_m$, with 20 wt.% calcium diethylphosphinate gave a V-0 rating in glass-filled PBT. Interestingly, an ultimate condensation product of P_2S_5 and dicyandiamide, which is apparently a sulfur-containing analog of phosphorus oxynitride is even more efficient than $(PON)_m$ and gives a V-0 rating in glass-reinforced PBT at 20 wt.% without the help of a synergist.[56]

A combination of $(PON)_m$ with Fe_2O_3 in nylon 6 and PBT gave a UL94 V-0 rating at a relatively low loading.[57,58] A mechanistic study showed that $(PON)_m$ is an efficient char promoter; however, the char can be penetrated by combustible pyrolysis products and is also not a good thermal insulator. Partial substitution (5 wt.%) of $(PON)_m$ with Fe_2O_3 improved the barrier character of the char and led to a V-0 rating. Another mode of the flame retardant action of $(PON)_n$ is postulated[59] to be its ability to form a low melting non-combustible glass on the polymer surface.

There is commercial usage of red phosphorus especially in Europe and Asia as a very efficient flame retardant for polyesters and nylons.[60] This usage is despite concern regarding evolution of traces of phosphine (from interaction with moisture) and the red color, which is difficult to mask even with large amount of colorants. For safe handling, red phosphorus is available, such as from Italmatch, as masterbatches with a variety of polymers.[61] Red phosphorus is particularly useful in glass-filled nylon 6.6 where a high processing temperature (typically about 280 °C) excludes the use of less stable phosphorus compounds.[62] In glass-filled nylons or polyesters, V-0 ratings can be reached with as little as 6–12 wt.% phosphorus. The use of red phosphorus in flame retardancy was reviewed by one of the present authors.[63]

Many industrial studies have been done to find synergists for use with red phosphorus. It has been found helpful to combine red phosphorus with phenolic resins.[62,64] It is believed that under burning conditions the red phosphorus-phenolic combination forms a cross-linked network which eliminates flaming drips by reducing melt flow. A typical nylon 6.6 formulation which contains 25 wt.% glass fibers, 7 wt.% red phosphorus and 5 wt.% phenolic resin gives a V-0 rating. A formulation of 6% red phosphorus, 5% layered clay and 4% polyolefin compatibilizer gives V-0 in 15% glass fiber reinforced nylon 6.6.[65]

Levchik *et al.*[66] studied thermal decomposition in nitrogen of nylon 6 flame-retarded with red phosphorus. The cumulative formation of phosphoric esters in the condensed phase was indicated by infrared and ^{31}P solid state NMR. Similar to Kuper *et al.*[67] these authors could not exclude possible interaction of red phosphorus with trace O_2 and absorbed moisture as well as H_2O formed during the thermal decomposition of the nylon.[68] An EPR signal was noted in the decomposing mixture of nylon 6 and red phosphorus attributable to products from the depolymerizing red phosphorus interacting with the polymer and Levchik *et al.*[66] hypothesized possible free radical interaction between nylon 6 and red phosphorus. Furthermore, the stable free radical concentration in the carbonizing materials[69] was significantly higher in the solid residues produced in the presence of red phosphorus. The observation that red phosphorus is much more active in oxygen- or nitrogen-containing polymers also is an argument for possible direct high-temperature interaction between phosphorus and polymer.

3.5 Epoxy resins

There are several ways to incorporate phosphorus species in an epoxy resin network and some are being commercialized.[70] The P-H bond can add to the epoxy group, and this reaction can be used to attach hydrogenphosphonates or phosphinates to an epoxy resin. The commercial example of this type of reactant is 9,10-dihydro-9-oxa-10-phosphaphenanthrene 10-oxide (DOPO, see above). It is sold in Japan, China and Europe and it can be used as a part of a curing system for halogen-free flame retardant epoxies.[71] However, it has a shortcoming because of its monofunctionality and can be used only in combinations with other curing agents.

By reacting DOPO with benzoquinone, a difunctional phenolic product can be made. This dihydric phenol product is available in Japan from Sanko. It can be incorporated in an epoxy resin by use of a chain-extension process similar to that customarily used with tetrabromobisphenol A. Although this chemistry provides good physical properties and the necessary level of flame retardancy, it seems to be not finding broad usage because of price.

Phosphine oxide structures have been proposed to impart flame retardancy to curing agents, because phosphine oxides are thermally and hydrolytically very stable. Many studies have been reported in the literature concerning curing of epoxy resins with bis(aminophenyl)methylphosphine oxide, as well as thermal and combustion performance of the fire retardant epoxies.[72-78] Because of the strong nucleophilic character of the amino groups, this compound is able to cure epoxy at as low as 150 °C. Recent Great Lakes patents show a broad range of tris(hydroxyphenyl) or bis(hydroxyphenyl)alkylphosphine oxides;[79] however, probably for synthesis cost reasons they seem to emphasize a mixture of tri-, di- and monofunctional 4-hydroxyphenylphosphine oxides.[80]

Bayer patents show di- and tri-functional 4-aminophenyl phosphates[81] prepared by transesterifying triphenyl phosphate with para-aminophenol. These products are said to be useful in combination with non-reactive aromatic phosphates and aluminum hydroxide (ATH). It appears that a rather high loading is needed to achieve a V-0 rating.

Clariant's aluminum diethylphosphinate, Exolit® OP 1230, perhaps originally developed for polyesters and nylons (see above) now seems to find a commercially promising application in epoxy-based printed wiring board laminates. It can be used as a fine dispersion together with ATH.[82] Advantages have been claimed for combinations with melamine polyphosphate[83] or zinc borate.[84]

Supresta LLC has introduced an oligomeric organophosphonate flame retardant with curing agent properties, Fyrol® PMP, which was specially designed for use in epoxy resins for electrical and electronic applications.[85-87] It is poly(1,3-phenylene methylphosphonate) with terminal aromatic hydroxyl groups. Fyrol® PMP is a low-melting solid, mp 45–55 °C. The product is rich in phosphorus (17.5%) and quite thermally stable with weight loss starting only above 300 °C as measured by TGA in nitrogen. Fyrol® PMP has a unique means of curing epoxy resins by insertion of epoxy groups into phosphonate ester linkages.[88] The reactivity of Fyrol® PMP coincides with the normal curing reactions of epoxies, therefore there is no need to change typical curing or chain extension cycles as commonly used industrially. From 20 to 30 wt.% of PMP provides V-0 ratings in epoxy laminates. The loading of Fyrol® PMP can be decreased to 10 wt.% when it is used with a flame retardant filler such as ATH. Depending on epoxy resin used, the glass transition temperature T_g can vary from 130 to 175 °C. The co-addition of a benzoxazine resin can raise the T_g to above 200 °C. Because of the specific curing mechanism, epoxy laminates made with Fyrol PMP show very high thermal stability with T_d (5% wt. loss) in the range of 360–390 °C and pass delamination tests at 288 °C. Hydrolytic stability of the laminates exceeds two hours of steam exposure and dipping in the solder bath at 288 °C as measured by the Pressure Cooker Test. Fyrol® PMP is a feasible substitute for TBBA in printed wiring boards, and it

3.6 Polyurethane foams

An oligomeric ethyl ethylene glycol phosphate additive containing 19% phosphorus was introduced by Supresta LLC (now ICL-IP) as Fyrol PNX.[90,91] PNX is halogen-free and it has been especially of interest in Europe, particularly with respect to low-fogging low-volatiles-emission requirements of the automotive industry. By the MVSS 302 test, this oligomer is approximately 40–50% more efficient than the chloroalkyl phosphates. In regard to volatile organic content (VOC), PNX compares well with other automotive foam flame retardants, but because of its low use level, the overall effect is low VOC for the resultant foam. PNX has been recommended for automotive applications in combination with alkylphenyl phosphates, which improve the flame retardant performance of PNX and also decrease the additive viscosity.[92] PNX outperforms traditional products in small-scale ignition tests for upholstered furniture (such as the California 117A and D tests). PNX is approximately 50–60% more efficient than traditional chlorinated alkyl phosphates. The oligomer was also found to be synergistic with tris(dichloroisopropyl) phosphate.[93]

In manufacturing of flexible polyurethane foams, if the foam reaches an excessively high temperature during the last stages of foaming after the addition of the water, 'scorch' can result. Scorch is, at the least, a discoloration of the interior of the slab or bun, and more seriously the loss of mechanical properties because of polymer degradation. Some of the commonly used flame retardants can aggravate scorch. Pentabromodiphenyl oxide mixed with an aromatic phosphate ester became widely used for some years as a flame retardant which did not provoke scorch. Because of the widespread detection of pentabromodiphenyl oxide in the environment, its subtle biological activity in laboratory animals, and its likely bioaccumulation, its continued use was banned in Europe and some American states, and its production and use was discontinued. New non-scorching flame retardants, both bromine-containing (but more biodegradable) and halogen-free types are now replacing pentabromodiphenyl ether.

Examples of the halogen-free types are monofunctional phosphates made by Daihachi by reacting cyclic neopentyl acid phosphates with propylene oxide or ethylene oxide[94] to produce monohydric alcohols. These additives are claimed to permit flexible foams to pass the MVSS 302 automotive test at 8.5 parts per hundred of polyol, with little or no scorch. However, because of the alcohol functionality, the foams require reformulation, not required by the non-reactive flame retardants.

Triaryl phosphates, such as isopropylphenyl diphenyl phosphate, are finding commercial use in flexible PU foams, sometimes admixed with a bromine-containing additive, other than the now-discontinued pentabromodiphenyl ether.[95] Great Lakes Chemical (now Chemtura) has Reofos® NHP, a low viscosity aryl phosphate, which can meet the MVSS 302 test for hot-molded automotive seating, and is claimed to be non-fogging.[96] Supresta LLC has introduced a new highly stable halogen-free phosphate ester Fyrol® HF-4.[97] This product is designed for low-scorch application by providing high oxidative stability, which is needed for production of PU foams in hot and humid summer weather.

Mechanistic studies of the scorch chemistry sponsored by Supresta at the University of Turin[98,99] showed that scorch is largely the result of oxidation of aromatic amino groups arising from hydrolysis of isocyanate groups which became isolated in the PU network. Aminophenylamido structures, are known to be readily oxidized to highly colored quinoneimines. It was also shown that some aliphatic chloroalkyl phosphates or halogen-free aliphatic phosphates can alkylate an aminoaryl group by nucleophilic displacement of the chloride or phosphate anion or both.[100] It is known that the more electron-rich alkylaminophenyl groups are more readily oxidized than the unalkylated aminophenyl groups.[101] The formation of the chromophoric groups is aggravated, amongst other factors, by flame retardants with alkylating abilities such as the chloroalkyl or alkyl phosphates, and this alkylation reaction and subsequent oxidation also adds to the exotherm. Those flame retardants most reactive towards arylamino groups tend to be those which aggravate scorch. Phosphorus or halogen compounds that cannot alkylate amino groups do not aggravate scorch; examples are aryl phosphates and bromoaromatic compounds.

3.7 Polyolefins

This major class of polymers forms little char upon exposure to fire and thus is particularly challenging to flame retard by phosphorus compounds that work by char enhancement. The successful flame retardants for polyolefins have usually been halogen types, most often synergized by antimony oxide, or endothermic types used at high loadings, like ATH or magnesium hydroxide. There has, however, been much effort to find ways to utilize phosphorus flame retardants in polyolefins. The entire topic of flame retardancy of polyolefins has been the subject of a journal review article by the present authors.[102]

A very limited approach is to introduce enough halogen into the phosphorus compound so that in effect it works as a halogen compound. The sole success along these lines is tris(tribromoneopentyl) phosphate, available from Ameribrom (Israel Chemicals Ltd.) as FR-370. Indeed, FR-370, which can be melted into the polypropylene prior to fiber spinning and which does not require antimony oxide, may be uniquely suitable as a flame retardant additive for polypropylene fiber, according to Zhang and Horrocks.[103] A special grade of FR-370 is available for fiber production. FR-370 is probably the main component of Ameribrom/ICL's SaFRon® 5371, a white powder blend with 63% Br, mp 108–181 °C, which allows polypropylene to reach V-2 rating at about 4.8% without the use of antimony oxide. SaFRon 5371 also has excellent UV stability.[104] It can be used to flame retard outdoor or indoor stadium seats and in production of fibers for polypropylene carpets. A patent shows a combination of FR-370 with a free-radical initiator exemplified by 2,3-dimethyl-2,3-diphenylbutane (Akzo-Nobel's Perkadox® 30) and a hindered amine stabilizer.[105] The role of the free-radical initiator is probably to lower the molecular weight of the polypropylene, thus encouraging melt drip; the bromine compound may enable non-flaming drips, or possibly the free-radical initiator may activate the bromine compound. It is not clear what contribution, if any, that the phosphorus atom makes to the overall flame retardant action of this compound.

A more extensive body of work, with some commercial outcome, relates to the incorporation of char-forming ingredients, along with a phosphorus compound which can catalyze char formation, and often with a blowing agent to produce intumescence. This approach has its roots in the technology of intumescent coatings, where a typical formulation is ammonium polyphosphate, a pentaerythritol char former, and typically melamine as a blowing agent, in a resin binder which can be thermoplastic or a fairly soft thermoset. As applied to polyolefins, there have been many attempts to find optimum char formers and sometimes the phosphorus structure is prereacted with the char forming structure.

The broad subject of intumescent flame retardants, mostly but not entirely phosphorus-based systems, and not limited to polyolefins, is discussed in a book by Le Bras et al.[106] and also a more recent review.[107] Most of the published work on this topic is academic. However, industrial producers of wire and cable insulating compounds have pursued this approach[108] to achieve lower density, improved processability and improved flame retardance. Commercial applications have been limited by higher cost (vs. bromine additives), moisture sensitivity, and processing difficulties. Much effort has been made in this area by Albright & Wilson, Clariant, Celanese, Montedison, Hoechst, Asahi Denka and others.

Great Lakes Chemical (now Chemtura) introduced Reogard® 2000, a halogen-free intumescent flame retardant for polypropylene and other thermoplastics. It is useful in extruded profiles and electrical parts to meet a V-0 rating with good in impact strength and heat distortion. Patents on intumescent flame retardants from this research group[109–110] suggest that it may contain pentaerythritol phosphate and a melamine phosphate.

Relatively water-insoluble ammonium polyphosphate is available in two crystal forms (type II is least soluble) and varieties are available with thermoset resin surface coatings to further provide water resistance, for example from Clariant as Exolit® 462 and from Budenheim in their FR CROS product group.

Over several decades, compounded flame retardants based on finely divided insoluble ammonium polyphosphate together with char-forming nitrogenous resins, have been developed for polyolefins, ethylene-vinyl acetate, elastomers and coatings. Many of these are patented by Hoechst, Celanese, Montedison, and Albright & Wilson, Asahi Denka, and others. In most cases, the char-forming component is not explicitly identified. Patents suggest that an early Exolit IFR uses tris(hydroxyethyl)isocyanurate as the char former.[111,112] However, this additive is rather too water-soluble, and some improvements involve esterifying the hydroxyethyl groups, for example by reaction with pentaerythritol spirobis(chlorophosphate) as shown in a Hoechst patent.[113] Another combination with APP is Himont's (or Enichem's) Spinflam® MF82 where the char former is an oligomer with triazine rings linked by diamine.[114] Other char-forming components in these APP-containing mixtures are believed to be urea-formaldehyde or melamine-formaldehyde condensation products.

More recent APP-char-forming mixtures are Exolit® AP 750 or AP 751 (preferred for talc or glass reinforced PP). Levels of addition in the 25–30% range can reach V-0 in polyolefins. Smoke is much lower compared to halogen-antimony systems and these phosphorus-based additives are also better than most bromine systems in regard to UV stability. One limitation is that these relatively hydrophilic systems are not satisfactory in very thin films such as <0.5 mm or in long contact with warm water. Occasional contact with water, such as rain, at ambient temperatures presents no difficulty. Roof sheathing and stadium seating applications are claimed to be successful. Electrical applications are the largest use. A proprietary improvement, Exolit AP 760, has been introduced and this product appears especially suited for cable ducts and trays. A char forming ethyleneurea-formaldehyde condensation product can be purchased separately and used with a phosphorus additive such as APP.

Melamine phosphates also have been originally developed for intumescent coatings but have found some use in polyolefins. Non-coating applications of the melamine phosphates (including the pyrophosphate) were reviewed by Weil *et al.*[115] A more recent introduction to the melamine phosphate family is melamine polyphosphate, Ciba's Melapur® 200. In an intumescent formulation in a polyolefin, the melamine phosphates such as Melapur 200 have been shown to have an advantage over ammonium polyphosphate by causing less mold deposition and having better water resistance.[116]

Ethylenediamine phosphate, a solid, mp 250 °C, slightly soluble in water, was introduced by Albright Wilson as Amgard® EDAP and now available as Broadview Technology's Intumax® AC-3 or from Unitex, principally as an additive for polyolefins, EVA and PVC. Unlike ammonium phosphates or melamine phosphates, it is self-intumescent and it forms its own char.[117] However, it is also synergized by melamine or melamine pyrophosphate, and is available as a blend with these, respectively from Unitex Corp. and Broadview Technologies. Some further synergists, such as phase transfer catalysts (quaternary ammonium salts) or spirobisamines may further enhance the action of EDAP and its melamine pyrophosphate combinations as claimed in Broadview patent applications.[118]

Various inorganic synergists are also reported for these phosphorus-based intumescent systems in polyolefins, for example talc or zinc borate.[119]

Other phosphorus-based intumescent systems for polyolefins have recently been in development, for example by Asahi Denka[120] which, from patents, would seem to involve piperazine pyrophosphate or polyphosphate and melamine pyrophosphate. A recent product, with increased stability, is their FS-2 (ADK STAB FP-2200).[121] FS-2 is said to be effective in polypropylene at about 20%, and in LDPE, HDPE or EVA at about 30%. It is stable enough to permit extrusion and molding at 220–240 °C. It appears to be better than previous intumescent systems in regard to water resistance, processability, and mechanical properties. It is more stable to heat and shear than the earlier (presumably) related product of Asahi Denka, FP-1 (ADK Stabilizer FP-2000®).

Pre-reacted products Budit 3118 and 3118F, having the phosphate structure attached to a polyol such as pentaerythritol are available from Budenheim.[122] The acid phosphate groups can be neutralized by a nitrogen compound such as a melamine. These products can be used as intumescent additives in coatings and polypropylene sheets and fibers. They can allow smoother surfaces by avoiding the particulate characteristics of APP.

Chemtura's (formerly Great Lakes') Reogard® 1000 and 2000 have been commercialized for use in polypropylene. They may contain pentaerythritol bicyclic phosphate (PEPA) as a 2:3 mixture with melamine phosphate plus a small amount of a quaternary-treated montmorillonite, as disclosed in a patent application by Great Lakes' researchers.[123] At 19% of the PEPA-melamine phosphate mixture with 0.8% (amount critical) of the montmorillonite, a V-0 rating was obtained in polypropylene.

Another pre-reacted system which was marketed by Great Lakes Chemical for use in polypropylene is the bis(melamine) salt of pentaerythritol bis(acid phosphate), at one time marketed as Char-Guard® 327. This still seems to be of interest in China, as shown by a recent study.[124]

Triaryl phosphates are liquid plasticizers used mostly in PVC, and will normally exude from non-elastomeric polyolefins. However, they can find some use in EPDM, for which Chemtura suggests use of the more highly isopropylated triphenyl phosphate, Reofos 95. An EPDM formulation from FMC for cable sheathing suggests use of 15 phr triaryl phosphate, 10 phr paraffin oil, 150 phr ATH, 25 phr MDH, and 25 phr magnesium carbonate. Crosslinking by dicumyl peroxide increases the flame resistance.

Recently, GE has patented polyolefin blends with specified types of polyphenylene oxide and aryl phosphate flame retardants. Having likely relevance to GE's 'ecoimagination' initiative, these may be non-halogen competitors (the Flexible Noryl family of blends) to PVC or fluoropolymers for wire and cable insulation. There are several GE patents and published patent applications dealing with this technology; a representative one is Kosaka *et al.*[125]

Recently, Italmatch introduced new colorless inorganic phosphorus additives, the Phoslite series. One of these, Phoslite® B361 is effective at as low as 1% for reaching a V-2 rating in PP and thermoplastic polyolefins.[126] The Phoslite series is said to contain phosphorus at a low state of oxidation, yet stable to hydrolysis and to air. At least some of the Phoslite series may contain calcium or aluminum hypophosphite.

Red phosphorus in polyolefins is, despite the lack of charring mechanism, somewhat effective. Surprisingly, it was shown that a V-2 rating at 1.6 mm could be obtained at as low as 2.5% finely-divided red phosphorus (5 μm).[127] To obviate the risk of handling of finely-divided red phosphorus (flammable as a powder!), masterbatches are available (Italmatch) in low density polyethylene, polypropylene and EVA. The masterbatch in nylon 6,6 is also useful as an additive in polypropylene; the nylon provides some char to enhance the flame retardant effect of the phosphorus.

3.8 Future trends

The scientific justification for replacing halogens is not clear cut, and continues to be disputed by experts on both sides of the issue.[128,129] Although some halogen compounds such as the long-discontinued tris(dibromopropyl) phosphate were clearly toxicity risks, other halogen compounds have more ambiguous toxicology. Pentabromodiphenyl ether was clearly widespread at low levels in the environment including mothers' milk, and mouse experiments showed some developmental toxicity, so there was little opposition to its recent discontinuance. Decabromodiphenyl ether, on the other hand, is not so ubiquitous and seems reasonably benign toxicologically. A 10-year risk analysis was quite favorable. It is very favorable in cost, and adequately stable for recycle. Recent RoHS interpretations by the EC would allow its use to continue, but there are continuing disputes in the EC about this at the time of this review. Some North American States are also imposing bans or considering to impose bans on Deca. Even less basis for concern would seem to be the use of tetrabromobisphenol-A which is reacted into epoxy circuit boards, but European WEEE regulations for electronic equipment can cause disposal costs. The users of halogens generally want a 'back up' or a 'green' alternative to hedge against unfavorable developments with the halogens or anti-halogen publicity. Consequently, research on halogen alternatives including phosphorus is quite active, several phosphorus alternatives to bromine compounds have gained market acceptance, and patents on phosphorus flame retardants are becoming numerous.[130]

In the broadest perspective, flame retardants including the phosphorus family will – or should – evolve in the directions of greater efficacy, less damage to polymer properties, freedom from toxicity and more 'environmentally friendly' characteristics, and lower cost contribution. Greater efficiency to the first approximation arises from higher phosphorus content and optimized matching of the flame retardant's decomposition or volatilization temperature to that of the polymer substrate. Less damage to polymer properties can result from lowering the required amount of flame retardant, but also by attention to its physical properties; broadly generalizing, impact loss can be minimized by avoiding insoluble solids, while heat deflection temperature loss can be minimized by avoiding soluble flame retardants – an obvious 'trade off'.

A great deal of progress can be made by the compounders in seeking optimum combinations. Statistical design can be used to develop multi-additive compositions well targeted to the needed product properties. The phosphorus family of flame retardants offers a broad palette of useful materials covering a wide range of melting points and vapor pressures. Combinations of more volatile and less volatile additives seem plausible to invoke both gas phase and condensed phase modes of action.

One area which seems environmentally attractive is the use of reactive flame retardants that become part of the polymer, thus avoiding emissions to water and air. Another environmentally favorable direction is for greater thermal and hydrolytic stability so as to allow recycling of the flame retardant plastic. This is particularly important for phosphate esters, which tend to be hydrolytically less stable than halogen containing flame retardants.

Toxicology must be considered from the beginning in design of safe phosphorus flame retardants – they must be designed to have no serious phosphorylating or alkylating ability toward biologically important molecules. One general strategy in avoiding biological activity and bioaccumulation is to make large molecules that cannot pass through cell membranes – besides reacted-in flame retardants, oligomers can provide this advantage. We can expect to see more oligomeric and polymeric flame retardants, as well as intermediates for condensation and vinyl copolymerization.

Some dormant areas of phosphorus flame retardants seem to be coming alive again. Phosphorus-containing acrylic monomers are being revisited in the UK. Vinylphosphonates need only improved processes and lower cost to become attractive as comonomers.

Inorganic phosphorus will continue to evolve as compounds such as ammonium polyphosphate are improved process-wise and by stronger coating, and salts of phosphorus acids with nitrogen bases will be tailored for intumescent formulations. The recent development of inorganic hypophosphites points to further exploration of inorganic phosphorus compounds with other elements. R&D in China shows evidence of producing useful new inexpensive flame retardants from the very low cost wet-process phosphoric acid. The use of low-cost fertilizer-type ammonium phosphates to produce new P-N compounds for flame retardancy may be a promising direction for more R&D.

To replace vapor-phase-active halogen-antimony systems, vapor-phase-active phosphorus systems must be developed. Basic experiments performed on flame inhibition indicate that phosphorus compounds at the same molar concentration are more effective flame suppressants than bromine or chlorine.[131,132]

3.9 Sources of further information and advice

Test methods (compendium of world-wide methods)

- J. Troitzsch, *Plastics Flammability Handbook*, 3rd Edition, Hanser Publishers, Munich/Hanser-Gardner, New York, 2004.

General review on flame retardancy

- E.D. Weil, 'Flame Retardancy,' *Encyclopedia of Polymer Science and Engineering*, J. Kroschwitz, Ed., John Wiley & Sons, New York, Vol. 10, pp. 21–53 (2004).

Books covering topics in flame retardancy

- V. Babrauskas, *Ignition Handbook*, Fire Science Publishers, Issaquah, WA, 2003.
- *Fire Retardant Materials*, A. R. Horrocks and D. Price, Eds., Woodhead Publishing Ltd., Cambridge, UK, 2001.
- *Fire Retardancy of Polymeric Materials*, A. Grand and C. Wilkie, Eds., M. Dekker, Inc., New York, 2000.
- R.M. Fristrom, *Flame Structure and Processes*, Oxford University Press, Oxford, 1995.
- R.M. Aseeva and G.E. Zaikov, *Combustion of Polymer Materials*, Hanser, Munich, 1986.
- C.F. Cullis and M.M. Hirschler, *The Combustion of Organic Polymers*, Clarendon Press, Oxford, 1981.
- J.W. Lyons, *The Chemistry and Uses of Fire Retardants*, Wiley, New York, 1970.

Reviews by the present authors on individual polymer types

- S.V. Levchik and D. Weil, Combustion and Fire Retardancy of Aliphatic Nylons, *Polym. Int.*, **49**(2000), 1033–1073.
- E.D. Weil and S. Levchik, A Review of Current Flame Retardant Systems for Epoxy Resins, *J. Fire Sci.*, **22**(2004), 25–40.
- E.D. Weil and S. Levchik, Commercial Flame Retardancy of Polyurethanes, *J. Fire Sci.*, **22**(2004), 183–210.
- E.D. Weil and S. Levchik, Current Practice and Recent Commercial

Developments in Flame Retardancy of Polyamides, *J. Fire Sci.*, **22**(2004), 251–264.
- E.D. Weil and S.V. Levchik, Commercial Flame Retardancy of Unsaturated Polyester and Vinyl Resins: Review, *J. Fire Sci.*, **22**(2004), 293–303.
- E.D. Weil and S.V. Levchik, Commercial Flame Retardancy of Thermoplastic Polyesters – A Review, *J. Fire Sci.*, **22**(2004), 339–350.
- S.V. Levchik and E.D. Weil, Thermal Decomposition, Combustion and Fire-retardancy of Epoxy Resins – a Review of the Recent Literature, *Polym. Int.*, **53**(2004), 1585–1610.
- S.V. Levchik and E.D. Weil, Thermal Decomposition, Combustion and Fire-Retardancy of Polyurethanes – a Review of the Recent Literature, *Polym. Int.*, **53**(2004), 1901–1929.
- S.V. Levchik and E.D. Weil, Flame Retardancy of Thermoplastic Polyesters – a Review of Recent Literature, *Polym. Int.*, **54**(2005), 11–35.
- S.V. Levchik and E.D. Weil, Overview of Recent Developments in Flame Retardancy of Polycarbonates, *Polym. Int.*, **54**(2005), 981–998.
- S.V. Levchik and E.D. Weil, Overview of the Recent Literature on Flame Retardancy and Smoke Suppression of PVC, *Polym. Adv. Technol.*, **16**(2005), 707–716.
- S.V. Levchik and E.D. Weil, Flame Retardants in Commercial Use or in Advanced Development in Polycarbonates and Polycarbonate Blends, *J. Fire Sci.*, **24**(2006), 137–151.
- E.D. Weil, S.V. Levchik and P. Moy, Flame and Smoke Retardants in Vinyl Chloride Polymers – Commercial Usage and Current Developments, *J. Fire Sci.*, **24**(2006), 211–236.
- S.V. Levchik, Introduction to Flame Retardancy and Polymer Flammability, in. *Flame Retardant Polymer Nanocomposites*, Eds. A.B. Morgan and C.A. Wilkie, Wiley Interscience, Hoboken, NJ, 2007, pp. 1–29.

Internet information sources

- http://www.fire.nist.gov – a portal to NIST's fire related publications and projects.
- http://fire.nist.gov/bfrpubs/fireall/key/key628.html – a bibliographic database on flame retardants assembled by NIST.
- http://www.specialchem.com – reviews and links on various topics including flame retardants. Product information from Rio Tinto Minerals, Ciba, ICL, Atofina, etc.
- http://www.polymeradditives.com – product information on additives including flame retardants, from Albemarle, Supresta (now ICL-IP), Cytec, GE and others. On-line source for samples.
- http://www.cefic-efra.com – information about flame retardants from the European Flame Retardants Association.

Developments in phosphorus flame retardants   59

- http://www.ebfrip.org – European Brominated Flame Retardant Industry Panel, a trade organization which defends the continued safe use of these products.
- http://www.iaoia.org – International Antimony Oxide Industry Association, which provides current information on risk assessments and regulatory activity on antimony oxides.
- http://www.polymeradditives.com – an e-commerce site, set up originally by Albemarle, Cytec and GE. Other suppliers subsequently joined, such as Atofina.
- http://www.polyone.com/ind/links – Many links to useful industry organizations relevant to flame retardancy.

Directory of Phosphorus Flame Retardant Manufacturers and Distributors (as of early 2007)

- Albemarle Corp., Baton Rouge, LA, USA, www.albemarle.com (organic phosphates).
- Amfine Inc. (sales outlet for Asahi Denka), Upper Saddle River, NJ, USA, www.amfine.com (aryl diphosphates, intumescent phosphate salt).
- Asahi Denka, www.adk.co.jp marketed through Amfine in the US.
- Astaris LLC, St. Louis, MO, USA, www.astaris.com (ammonium polyphosphates).
- Broadview Technologies, Inc., Newark, NJ, USA, www.broadview-tech.com (melamine pyrophosphate, ethylenediamine phosphate, intumescent mixtures).
- Budenheim, Chemische Fabrik, represented in the U.S. by Flame Chk – see below.
- Chemtura (formerly Great Lakes Chemical and Crompton), Middlebury, CT, USA, www.chemtura.com (organic phosphates).
- Ciba Specialty Chemicals, Basel, Switzerland, www.cibasc.com (melamine phosphates, melamine cyanurate).
- Clariant (formerly Hoechst Celanese), Muttenz, Switzerland, www.clariant.com (red phosphorus, ammonium polyphosphates, phosphinate salts).
- Cytec Industries, West Patterson, NJ, USA, www.cytec.com – (melamine phosphates, melamine).
- Daihachi Chemical Industry, Osaka, Japan, www.daihachi-chem.co.jp (organic phosphates).
- Eastman Chemical Products, Kingsport, TN, USA, www.eastman.com (triethyl phosphate).
- Ferro Corp. (Keil Div.), Hammond, IN, USA, www.ferro.com (alkyl diphenyl phosphates).
- Flame Chk, Inc. (representing Chemische Fabrik Budenheim, Germany and Budenheim Iberica, Spain) Medford, NJ, USA, www.flamechk.com – (ammonium polyphosphates, melamine phosphates, intumescent phosphates).

- Italmatch Chemicals, Genova, Italy, www.italmatch.it (red phosphorus, hypophosphites, melamine salts).
- Lanxess Corp. (formerly part of Bayer), Leverkusen, Germany, www.lanxess.com (organic phosphates).
- Rhodia (acquired Albright & Wilson FR products), Paris, France, www.rhodia.com (phosphonates, phosphonium compounds: textile flame retardants).
- Schill & Seilakher, www.4structol.com (reastive phosphinate for epoxy and fibers).
- Supresta LLC, Ardsley, NY (now ICL-IP), www.supresta.com (aryl phosphates, chloroalkyl phosphates, urethane foam additives and reactives).

3.10 Conclusions

From a commercial standpoint, phosphorus flame retardants are in active development driven by their good performance and also by environmental and/or regulatory problems with some halogenated flame retardants. Aryl phosphate oligomers (mainly diphosphates) are gaining wide acceptance in polycarbonate-styrenic blends or polyphenylene oxide-styrenic blends. An oligomeric aliphatic phosphate additive is showing progress in polyurethane foams. A reactive oligomeric resorcinol methylphosphonate and a cyclic phosphinate are both under active development in epoxy printed wiring boards as replacements for tetrabromobisphenol A.

Mode of action research has shown that while phosphorus flame retardants can have important action as char enhancers in the condensed phase, there are increasing indications that vapor phase action can also be significant. The existence of these two different modes of action is a plausible basis for discovery of synergistic combinations.

Some further needs are for increasingly stable phosphorus flame retardants to favor recycling. Some potentially inexpensive but currently unavailable phosphorus-rich inorganic compounds such as phosphorus oxynitride, phosphorus thionitride and phospham have been shown to be active flame retardants but need process development for them to become available.

3.11 References

1. Aaronson AM, 'Oligomeric phosphate flame reatardants', Proc. Conf. *Recent Adv. Flame Retardancy Polym. Mater.*, Stamford, CT, 1996.
2. Catsman P, Govaerts L C and Lucas R, (to GE), *PCT Patent Application WO 00/018844*, 2000.
3. Shibuyu K, Hatchiyu H and Nanba N, (to Asahi Kasei), *US Patent 6417319*, 2002.
4. Nanba N and Nasu H, (to Asahi Kasei), *US Patent 6177542*, 2000.
5. Wroczynski R J, (to GE), *European Patent 0909790*, 2004.

6. Zobel M, Eckel T, Wittmann D and Keller B, (to Bayer), *US Patent 6414107*, 2002.
7. Seidel A, Eckel T, Zobel M, Derr T and Wittmann D, (to Bayer), *US Patent 6737465*, 2004.
8. Morton M L, Khouri F F and Campbell J R, (to GE), *PCT Patent Application WO 99/43747*, 1999.
9. Levchik S V, Bright D A, Moy P and Dashevsky S, 'New developments in non-halogen aromatic phosphates', *J Vinyl Add Technol*, 2000, **6** 123–28.
10. Seidel A, Eckel T, Wittmann D and Kurzidim D, (to Bayer), *US Patent 7067567*, 2006.
11. Eckel T, Seidel A, Gonzalez-Blanco J and Wittmann D (to Bayer), *PCT Patent Application WO 04/015001*, 2004.
12. Burkhardt EW, Bright DA, Levchik S, Dashevsky S and Buczek M, (to Akzo Nobel), *US Patent 6717005*, 2004.
13. Bright D, Weil E and Pirrelli R, (to Ripplewood Phosphorus), *PCT Patent Application WO 04/000922*, 2004.
14. Seidel A, Eckel T, Peucker U and Wittmann D (to Bayer), *European Patent 1355987*, 2006.
15. Tabushi K and Mori B (to Nippon A&L), *US Patent 7063809*, 2006.
16. Zobel M, Seidel A, Eckel T, Derr T and Wittmann D, (to Bayer), *European Patent 1373408*, 2007.
17. Green J, 'Phosphate ester flame retardants for engineering thermoplastics', Proc *ANTEC Conf*, Vol 3, Boston, MA, 1995, pp. 3544–8.
18. Eckel T, Zobel M, Keller B and Wittmann D, (to Bayer), *US Patent 6590015*, 2003.
19. Eckel T, Seidel A, Wittmann D and Peuker U, (to Bayer), *US Patent 6713544*, 2004.
20. Katayama M, Ito M and Otsuka Y, (to Daicel), *US Patent 6316579*, 2001.
21. Campbell JR and Talley JJ, (to GE), *US Patent 5973041*, 1999.
22. Lin J-C, Seo K-H and Yang S-J, (to Cheil), *US Patent 6576161*, 2003.
23. Maruyama K and Motoshige R, (to Mitsubishi), *European Patent 0728811*, 2003.
24. Lim JC, Lee JH and Kwon IH, (to Cheil), *US Patent 6630524*, 2003.
25. DeStio P, 'Flame Retardant Plyesters', Proc Conf *Recent Adv Flame Retardancy Polym Mater*, Stamford, CT, 1991.
26. Sandler SR, (to Pennwalt), *US Patent 4180495*, 1979.
27. Herwig W, Kleiner H-J and Sabel H-D, (to Hoechst) *European Patent Application 0006568*, 1979.
28. Kleiner H-J, Budzinsky W and Kirsch C, (to Ticona), *US Patent 5773556*, 1998.
29. Kleiner H-J and Budzinsky W, (to Ticona), *US Patent 5780534*, 1998.
30. Kleiner H-J, Budzinsky W and Kirsch G, (to Ticona), *US Patent 6013707*, 2000.
31. Weferling N and Schmitz H-P, (to Clariant), *US Patent 6242642*, 2001.
32. Weferling N, Schmitz H-P and Kolbe G, (to Clariant), *US Patent 6248921*, 2001.
33. Hoerold S, (to Clariant), *US Patent 6420459*, 2002.
34. Campbell JR, Duffy B, Rude JR, Susarla P, Vallance MA, Yaeger GW and Zarnoch KP, (to GE), *US Patent 7101923*, 2006.
35. Jenewein E, Kleiner H.-J, Wanzke W and Budzinsky W (to Clariant), *US Patent 6365071*, 2002.
36. Klatt M, Leutner B, Nam M and Fisch H, (to BASF), *US Patent 6503969*, 2003.
37. Schlosser E, Nass B and Wanzke W, (to Clariant), *US Patent 6255371*, 2001.
38. Nass B, Schacker O, Schlosser E and Wanzke W, 'Compounding of phosphorus

containing flame retardants', Proc *Spring FRCA Conf.*, San Francisco, 2001, pp. 167–78.
39. Dietz M, Hörold S, Nass B, Schacker O, Schmitt E and Wanzke W, 'New environmentally friendly phosphorus based flame retardants for printed circuit boards as well as polyamides and polyesters in E&E applications', Proc Conf *Electronics Goes Green 2004*, Berlin, 2004, pp. 771–6.
40. Costanzi S and Leonardi M, (to Italmatch), *PCT Patent Application WO 05/ 121232*, 2005.
41. Toldy A, Szabo A, Anna P, Szep A, Bertalan Gy, Marosi Gy, Krause W and Hörold S, 'Flame retardant mechanism and application of synergistic combinations of phosphinates', Proc Conf *Recent Adv Flame Retardancy Polym Mater*, Stamford, CT, 2004.
42. Endo S, Kashihara T, Osako A, Shizuki T and Ikegami T, (to Toyobo), *US Patent 4127590*, 1978.
43. Endo S, Kashihara T, Osako A, Shizuki T and Ikegami T, (to Toyobo), *US Patent 4157436*, 1979.
44. Rieckert H, Dietrich J and Keller H, (to Schill and Seilacher), *German Patent Application 19711523*, 1997.
45. Seo Y-I, Kang C-S, Choi T-G and Song J-M, (to Kolon), *US Patent 6610796*, 2003.
46. Sato M, Endo S, Araki Y, Matsuoka G, Gyobu S and Takeuchi H, 'The flame retardant polyester fiber: improvement of hydrolysis resistance', *J Appl Polym Sci*, 2000, **78**, 1134–8.
47. Weil ED, 'Flame retardants based on phosphorus', Proc Conf *Recent Adv Flame Retardancy Polym Mater*, Stamford, CT, 1991.
48. Van der Spek PA, Bos MLM., Roovers WAS, van Gurp M and Menting HNAM, (no assignee), *US Patent 6767941*, 2004.
49. Balabanovich AI, Levchik GF, Levchik SV and Engelmann J, 'Fire retardant synergism between cyclic diphosphonate ester and melamine in poly(butylene terephthalate)', *J Fire Sci*, 2000, **20** 71–83.
50. Levchik SV, Bright DA and Alessio GR, (to Akzo Nobel), *US Patent*, 2003.
51. Levchik SV, Bright DA, Alessio GR and Dashevsky S, 'Synergistic action between aryl phosphates and novolac resin in PBT', *Polym Degrad Stab*, 2002, **77** 267–72.
52. Levchik SV, 'Halogen-free approach in flame retardancy of thermoplastic polyesters', Proc Con. *Recent Adv Flame Retardancy Polym Mater*, Stamford, CT, 2002.
53. Levchik SV, Bright DA, Dashevsky S and Moy P, 'Application and Mode of Fire-retardant Action of Aromatic Phosphates', *Specialty Polymer Additives. Principles and Application*, Eds Al-Malaika S, Golovoy A and Wilkie CA, Blackwell Science, Oxford, 2001, pp. 259–69.
54. Leger JM, Chateau C, Haines J and Marchand R, 'Cristobalite-, quartz- and moganite-type phases in phosphorus oxynitride PON under high pressure', Proc Conf *Sci Technol High Pressure, AIRAFT-17*, Hyderabat, India, 2000, pp. 531–4.
55. Brand A, Seiz C and Engelmann J, (to BASF), *PCT Patent Application WO 02/ 096976*, 2002.
56. Freudenthaler E, Brand A, Sterzei H-J, Engelmann J and Klatt M, (to BASF), *European Patent 1292638*, 2004.
57. Levchik GF, Grogoriev YuV, Balabanovich AI, Levchik SV and Klatt M, 'Phosphorus-nitrogen containing fire retardants for poly(butylene terephthalate)', *Polym Int*, 2000, **49**, 1095–100.

58. Balabanovich AI, Levchik SV, Levchik GF, Schnabel W and Wilkie CA, 'Thermal decomposition and combustion of γ-irradiated nylon 6 containing phosphorus oxynitride and phospham', *Polym Degrad Stab*, 1999, **64** 191–5.
59. Weil ED, 'Meeting FR goals using polymer additive systems,' Proc Conf *Improved Fire- and Smoke-Resistant Materials for Commercial Aircraft Interiors*, Washington, DC, 1995, pp 129–50.
60. Weil ED, 'Recent developments in phosphorus-based flame retardants', Proc 3rd Beijing Int Sym, *Flame Retardants and Flame Retardant Mater*, Beijing, 1999, pp 177–83.
61. N. Gatti, 'New red phosphorus masterbatches find new application areas in thermoplastics', *Plas. Add Compound*, 2002, Apr, 34–7.
62. Huggard MT, 'Fire retardant polyester fibers and other goods containing a new fire retardant', Proc Conf *Recent Adv Flame Retardancy Polym Mater*, Stamford, CT, 1994.
63. Weil ED, 'New approach to flame retardancy', Proc Conf *Recent Adv Flame Retardancy Polym Mater*, Stamford, CT, 2000.
64. Huggard MT, 'Phosphorus flame retardants – activation "synergists"', Proc Conf *Recent Adv Flame Retardancy Polym Mater*, Stamford, CT, 1992.
65. Klatt M, Grutke S, Heitz T, Rauschenberger V, Plesnivy T, Wolf P, Wuensch J and Fischer M, (to BASF), *PCT Patent Application WO 98/36022*, 1998.
66. Levchik SV, Levchik GF, Balabanovich AI, Camino G and Costa L, 'Mechanistic study of combustion performance and thermal decomposition behavior nylon 6 with added halogen-free flame retardants', *Polym Degrad Stab*, 1996, **54** 217–22.
67. Kuper G, Hormes J and Sommer K, 'In situ x-ray absorption spectroscopy at the k-edge of red phosphorus in polyamide 6,6 during a thermo-oxidative degradation', *Macromol Chem Phys*, 1994, **195** 1741–53.
68. Levchik SV, Weil ED and Lewin M, 'Thermal decomposition of aliphatic nylons', *Polym Int*, 1999, **48** 532–57.
69. Lewis IC and Singer LS, 'Electron spin resonance and the mechanism of carbonization' *Chemistry and Physics of Carbon*, Vol. 17, Eds Walker PL Jr and Thrower PA, Marcel Dekker, New York, 1985 pp. 1–88.
70. Weil ED and Levchik SV, 'A review of current flame retardant systems for epoxy resins', *J Fire Sci*, 2004, **22** 25–40
71. Lengsfeld H, Altstaedt V, Sprenger S and Utz R, 'Flame-retardant curing. Halogen-free modification of epoxy resin', *Kunststoffe*, 2001, **91** 37–39.
72. Chin W-K, Shau M-D and Tsai W-C, 'Synthesis, structure and thermal properties of epoxy-imide resin cured by phosphorylated diamine', *J Polym Sci Polym Chem*, 1995, **33** 373–9.
73. La Rosa AD, Recca A, Carter JT and McGrail PT, 'An oxygen index evaluation of flammability on modified epoxy/polyester systems', *Polymer* 1999, **40** 4093–8.
74. Shau M-D and Wang T-S, Synthesis, structure, reactivity and thermal properties of new cyclic phosphine oxide epoxy resins cured by diamines', *J Polym Sci Polym Chem Ed*, 1996, **34**, 387–96.
75. Levchik SV, Camino G, Luda MP, Costa L, Muller G, Costes B and Henry Y, 'Epoxy resin cured with aminophenylmethylphosphine oxide 1: Combustion performance', *Polym Adv Technol*, 1996, **7**, 823–30.
76. Levchik SV, Camino G, Costa L and Luda MP, 'Mechanistic study of thermal behavior and combustion performance of carbon fiber – epoxy resin composites fire-retarded with phosphorus-based curing system', *Polym Degrad Stab*, 1996, **54**, 317–22.

77. Levchik SV, Camino G, Luda MP, Costa L, Muller G and Costes B, 'Epoxy resin cured with aminophenylmethylphosphine oxide II. Mechanism of thermal decomposition', *Polym Degrad Stab*, 1998, **60** 169–83.
78. Tchatchoua C, Ji Q, Srinivasan SA, Ghassemi H, Yoon TH, Martinez-Nunez M, Kashiwagi T and McGrath JE, 'Flame resistant epoxy networks based on aryl phosphine oxide containing diamines', *Polym Prepr*, 1997, **38**(1) 113–14.
79. Hanson MV and Timberlake LD, (to PABU Services), *US Patent 6733698*, 2004.
80. Timberlake LD, Hanson MV, Bradley E and Edwards EB, (to PABU Services), *US Patent 6887950*, 2005.
81. Janke N, Mauerer O and Pieroth M, (to Bayer), *US Patent 6774163*, 2004.
82. Hoerold S, (to Clariant), *US Patent 6420459*, 2002.
83. Knop S, Sicken M and Hoerold S, (to Clariant), *EP Patent Application 1403309*, 2004.
84. Knop S, Sicken M and Hoerold S (to Clariant), *EP Patent Application 1403310*, 2004.
85. Levchik SV, Dashevsky S, Weil ED and Yao Q, (to Akzo Nobel) *PCT Patent Application WO 03/029258*, 2003.
86. Levchik SV and Buczek M, (to Akzo Nobel), *European Patent 1570000*, 2006.
87. Levchik SV and Piotrowski AM, (to Akzo Nobel), *PCT Patent Application WO 04/113411*, 2004.
88. Wu T, Piotrowski AM, Yao Q and Levchik SV, 'Curing of epoxy resin with poly(*m*-phenylene methylphosphonate)', *J Appl Polym Sci*, 2006, **101** 4011–22.
89. Levchik SV and Wang CS, 'P-based flame retardants in halogen-free laminates', *OnBoard Technology*, 2007, (Apr), 18–20.
90. Bradford LL, Pinzoni E and Wuestenenk J, 'Clearing the fog about the effects on fogging of common liquid fire retardants in flexible foam' Proc *Polyurethanes Expo '96;* Las Vegas, NE, 1996, pp 358–61.
91. Blundell C, Bright DA and Halchak T, 'New flame retardants for flexible polyurethane foams', Proc *UTECH 2003*, The Hague, NL, 2003.
92. Bradford LL, Pinzoni E, Williams B and Halchak T, (to Akzo Nobel), *European Patent 1218433*, 2003.
93. Bradford Ll, Pinzoni E, Williams B and Halchak T, (to Akzo Nobel), *US Patent 6262135*, 2001.
94. Tokuyasu N and Matsumura T, (to Daihachi), *US Patent 6127464*, 2000.
95. Fallon SB, Rose RS and Phillips MD, (to Chemtura), *US Patent 7008973*, 2006.
96. Rose RS, Buszard DL, Philips MD and Liu FJ, (to Pabu Services), *US Patent 6667355*, 2003.
97. Stowell J, Piotrowski A, Pinzoni E, Williams B and Wuestenenk J, 'Halogen-free flame retardant solutions for flexible polyurethane foam', Proc *PFA Conf*, Baltimore, MD, 2007.
98. Luda MP, Bracco P, Costa L and Levchik SV, 'Discolouration in fire retardant flexible polyurethane foams. Part I. Characterisation', *Polym Degrad Stab*, 2004, **83**, 215–20.
99. Levchik SV, Luda MP, Bracco P, Nada P and Costa L, 'Discoloration in fire-retardant flexible polyurethane foams', *J Cellular Plast*, 2005, **41** 235–50.
100. Bissell ER, 'A novel synthesis of 1,4-diarylpiperazine', *J Heterocyclic Chem*, 1977, **14**, 535–6.
101. Babich HA, Stern A amd Munday A, 'In vitro cytotoxicity of methylated phenyldiaminates', *Toxicol Lett*, 1992, **63**, 171–3.
102. Weil ED and Levchik SV, *J Fire Sci*, 2008, **26**, 5–43.

103. Zhang S and Horrocks AR, 'A review of flame retardant polypropylene fibers', *Progr Polym Sci*, 2003, **28**, 1517–38.
104. Geran T, Finberg I, Resnick G, Hini S, Plewinsky D and Bar Yaakov Y, 'Development of fire retarded plastics with reduced or no presence of antimony trioxide', Proc Conf *Recent Adv Flame Retardancy Polym Mater*, Stamford, CT, 2003.
105. Bar-Yakov Y and Hini S, (to Bromine Compounds), *US Patent 6737456*, 2004.
106. Le Bras M, Camino G, Bourbigot S and Delobel R, eds, *Fire Retardancy of Polymers: The Use of Intumescence*, Royal Society of Chemistry Special Publication 224, Springer Verlag, Cambridge, UK, 1998.
107. Bourbigot S, Le Bras M, Duquesne S and Rochery M, 'Recent advances for intumescent polymers', *Macromol Mater Eng*, 2004, **289**, 490–511.
108. Cogen JM. Jow J, Lin TS and Whaley PD, 'New approaches to halogen free polyolefin flame retardant wire and cable compounds', Proc *52nd IWCS/Focus International Wire & Cable Symposium*, Philadelphia, PA, 2004.
109. Chyall LJ, Hodgen HA, Vyverberg FJ and Chapman RW, (to PABU Services), *US Patent 6905693*, 2005.
110. Chyall LJ, Hodgen HA, Vyverberg FJ and Chapman RW, (to PABU Services), *US Patent 6632442*, 2003.
111. Nalepa R and Scharf D, (to Clariant), *US Patent 5204392*, 1993.
112. Nalepa R and Scharf D, (to Clariant), *US Patent 5204393*, 1993.
113. Staendeke H, (to Hoechst), *US Patent 5484830*, 1996.
114. Pernice R, Checchin M, Moro A and Pippa R, (to Enichem), *US Patent 5514743*, 1996.
115. Weil ED and McSwigan B, 'Melamine phosphate flame retardants', *Plastics Compounding*, 1994, (May–June) 31–9.
116. Imanishi S, (to Daicel), *US Patent 6921783*, 2005.
117. Goin CL and Huggard MT, 'AMGARD EDAP a new direction in fire retardants', Proc *Recent Adv Flame Retardancy Polym Mater*, Stamfort, CT, 1991
118. Rhodes MS, Israilev L, Tuerack J and Rhodes PS, (no assignee), *US Patent 6733697*, 2004
119. Amigouet P and Shen K,'Talc/zinc borate: potential synergisms in flame retardant systems,' Proc *Flame Retardants 2006*, Interscience Commun, London, 2006, pp. 155–62.
120. Kurumatani H, Yamaki A and Kimura R, (to Asahi Denka), *European Patent Application 1516907*, 2005.
121. Kamimoto T, Murase H, Yamaki A, Nagahama M, Kimura R, Funamizu T and Zingde G, 'A highly efficient intumescent flame retardant for polyolefins,' Proc *International Conference on Polyolefins*, Houston, TX, 2004.
122. Futterer T, Nägerl H-D, Götzmann K, Mans V. and Tortosa E, 'New intumescent flame retardants based on APP and phosphate esters,' Proc Conf *Recent Adv Flame Retardancy Polym Mater*, Stamford, CT, 2002.
123. Chyall LJ, Hodgen HA, Vyverberg FJ and Chapman RW, (to PABU Services), *US Patent 6632442*, 2003.
124. Liu Y and Wang Q,'Catalytic action of phosphotungstic acid in the synthesis of melamine salts of pentaerythritol phosphate and their synergistic effects in flame retarded polypropylene,' *Polym Degrad Stab* 2006, **91** 2513–19.
125. Kosaka K, Li X, Mhetar V, Tenenbaum J and Yao W, (to GE), *European Patent Application 1704183*, 2006.
126. Costanzi S, (to Italmatch), *PCT Patent Application WO 07/010318*, 2007.

127. Gatti N and Costanzi S, 'Is red phosphorus an effective solution for flame proofing polyolefin articles?' Proc Conf *Flame Retardants 2004*, Interscience Commun., London, UK, 2004, pp. 133–8.
128. Weil ED, 'An attempt of a balanced view on the halogen controversy', Proc Conf *Recent Adv Flame Retardancy Polym Mater*, Stamford, CT, 1999.
129. Weil ED, 'An attempt of a balanced view on the halogen controversy – update 2001', Proc Conf *Recent Adv Flame Retardancy Polym Mater*, Stamford, CT, 2001.
130. Weil ED, 'Patent activity in the flame retardant field', Proc Conf *Recent Adv Flame Retardancy Polym Mater*, Stamford, CT, 2005.
131. Babushok V and Tsang W, 'Influence of phosphorus-containing fire suppressants on flame propagation', Proc *3rd Int Conf Fire Research Engineering*, Chicago, IL, 1999, pp. 257–67.
132. Babushok V and Tsang W, 'Inhibitor Rankings for Alkane Combustion', *Combust Flame*, 2000, **124** 488–506.

4
Halogen-free flame retardants

Y-Z WANG, Sichuan University, China

Abstract: The chapter first discusses the challenges from replacing halogen-containing flame retardants by halogen-free flame retardants. It then discusses how to overcome the possible low flame retardant efficiency, serious melt-dripping behaviors, deterioration in mechanical properties, and so on, which are caused by this replacement.

Key words: halogen-free flame retardant, catalysis in flame retardancy, nanotechnology, microencapsulation, synergist.

4.1 Introduction

Halogen-containing flame retardants, including about 50–100 kinds of halogen-containing compounds that cover most of the market requirements, are one of the most commonly adopted flame retardant groups due to their highly effective flame retardancy and low price. However, some halogenated flame retardants have been claimed to be a source of toxic halogenated dibenzodioxins and dibenzofurans, which have more recently greatly limited their wide usage. Continuing pressures regarding environmental and toxicity issues have generated demands for some halogen-containing flame retardants to be replaced by halogen-free flame retardants. Owing to the numerous available chemicals and polymer systems, it is impossible to present all their different combinations in a chapter. Therefore, this chapter will focus only on the challenge of how to replace halogen flame retardants by halogen-free flame retardants and some successful examples will be presented.

4.2 Challenges posed by replacing halogen-containing flame retardants

With regard to the question of whether it is possible to replace halogen-containing flame retardants, the answer is not in the short term. Obviously, replacing halogenated flame retardants by halogen-free flame retardants is a long-term process and needs much basic research and technological development. The challenges of finding alternatives include overcoming their possible low flame retardant efficiency, high cost of flame-retardant products, serious melt-dripping behaviors and deterioration in mechanical properties.

4.2.1 Lower flame-retardant efficiency

The efficiency of flame retardants varies with different polymers and some bromine-containing compounds especially are very efficient flame retardants for the most flammable polymers such as the polyolefins in which only about 10 wt% of flame retardant can impart high levels in polyethylene or polypropylene, e.g. UL-94 V-0 ratings and LOI values of 30 or more. However, the same flame-retardant level can only be achieved by adding much more than 10 wt% of halogen-free flame retardants, and typically, 30–50 wt% of flame retardants have to be added, especially for inorganic flame retardants. Improved flame-retardant efficiency in polyolefins is provided by use of intumescent flame-retardant (IFR) additives.[1-4] A typical example is from the work by Wang et al.:[5] a flame retardant (ER) based on the esterification of melamine phosphate (MP) and pentaerythritol (PER) can impart good flame retardancy and non-dripping to polyethylene (PE) by combining with ammonium polyphosphate (APP) via a reactive extrusion technology.

Another example is the flame retardation of acrylonitrile-butadiene-styrene copolymer (ABS) in which an acceptable flame-retardant effect can be achieved by adding 60 wt% of alumina trihydrate (ATH) although only 20 wt% or less of halogen-containing flame retardants can obtain the same flame retardant results[6] and it is very evident that different flame-retardant mechanisms occur in this system.

The generally low flame-retardant efficiencies of replacements will give rise to increases of cost and to the deterioration of mechanical properties of flame-retardant polymer materials as greater concentrations of flame retardants have to be added in order to obtain an acceptable flame-retardant effect.

4.2.2 Deterioration in mechanical properties

As mentioned above, the addition of flame retardants will generally reduce the mechanical properties of polymeric materials and such deterioration will increase proportionately with the amount of flame retardant present. For example, when magnesium hydroxide (MH) was used as the flame retardant of ABS, 60 wt% of MH has to be added to achieve acceptable flame redundancy and, in this case, the impact and tensile strengths of ABS decrease by more than 50%.[7] Furthermore, the large amounts of MH required aggravate particle agglomeration and further worsen polymer mechanical properties. Szép et al.[8] studied the effects of MH and montmorillonite (MMT) on the flame retardancy and mechanical properties of ethylene-vinyl acetate copolymer (EVA). The results showed that the tensile strength decrease from 23 MPa to 5.2 MPa and the elongation-at-break from 1110% to 143% after introducing MH to EVA. Therefore, how to improve the mechanical properties of flame-retardant systems is a major issue that has to be addressed from the practical point of view.

4.2.3 Increased product cost

With regard to the cost of flame-retardant products, obviously, it is related to both the price and the efficiency of the flame retardant. For some low-cost inorganic flame retardants, high loadings are necessary in order to obtain good flame retardancy and conversely, some highly effective halogen-free flame retardants have high costs. Generally, a more cost-effective flame retardant often results in a cheaper final product with better mechanical properties. However, many halogen-containing flame retardants are cheaper than some potential halogen-free replacements, so it becomes more difficult to replace halogen-containing flame retardants by halogen-free flame retardants when the cost is a major factor.

4.2.4 More serious melt-dripping behaviors

Melt dripping is a common phenomenon for most of melting-processed polymeric materials such as polyethylene (PE), polypropylene (PP), polyethylene terephthalate (PET), ABS, and so on. In fact, serious melt dripping is very dangerous in fires because flaming drips act as secondary ignition sources. Generally, for these polymers, resolution of the melt-dripping problem is more difficult than improving their flame retardancy. In such polymers, a conflict often exists between reducing both melt dripping and polymer flame retardancy together since promotion of melt dripping increases energy removal from the flame zone and confers a physical flame retarding effect. Wang *et al.*[9,10] reported that the halogen-free flame retardant, poly(sulphonyl phenylene phosphonate) (PSPPP) confers good flame retardancy on PET in bulk and fibre form and the flame-retardant mechanism is based solely on melting dripping theory. On the contrary, for anti-dripping polymeric materials, the heat produced cannot be transferred and remains in the burning polymer, which makes the flame retardation of the polymer difficult and this is a major problem when char-forming retardants are present.

All points mentioned above are important for polymeric materials. Hopefully, one halogen-free flame retardant can satisfy all the requirements, although it is very difficult.

4.3 Attempts at successful halogen replacement in special applications

4.3.1 Techniques used to improve flame-retardant efficiency

To improve the flame-retardant efficiency of halogen-free flame retardants, some techniques such as nanotechnology and catalysis technique can be employed.

Nanotechnology

The study on polymer nanocomposites, especially polymer/layered silicate (PLS) nanocomposites has been attractive due to the recent demonstrations of their flame-retardant properties (see Chapters 5 and 6). Various nanoparticles, such as layered silicates, carbon nanotubes, polyhedral oligomeric silsesquioxanes (POSS), and graphite oxide, were used alone or together to prepare flame-retardant polymer nanocomposites. These studies of the flame-retardant properties of polymer nanocomposites mainly demonstrate a significant decrease in the peak heat release rate (PHRR), a change in the char structure and a decrease in the rate of mass loss during combustion in a cone calorimeter.[11–20] Not only was reduced flammability obtained at very low nanoparticle contents (2–5 wt%) without increasing carbon monoxide or smoke yields, but also the physical properties of the polymers were improved simultaneously.

Also, the incorporation of nanodispersed particles with other conventional flame retardants can reduce their contents required in the formulated polymer products, while maintaining or improving flammability performance and enhancing the physical properties. For example, a polymer nanocomposite containing POSS can obviously reduce PHRR, and enhance LOI.[7]

Therefore, the approach of preparing nanocomposites has become another tool for scientists and engineers who are engaged in the research and development of flame-retardant materials.[21] Detailed descriptions about nanotechnology for flame retardation can be seen in Chapters 5 and 6.

Catalysis

Recently, considerable interest has been shown in the application of catalysis in flame retardancy, both in halogen-containing and halogen-free flame-retardant systems.[22–25]

The catalytic effect of divalent and multivalent metal compounds on the flame retardancy of intumescent systems has been well studied by Lewin and Endo.[26] A series of metal compounds, including oxides, acetates, acetyl acetonates, borates and sulphates of Mg, Al, Ca, V, Cr, Mn, Co, Ni, Cu, Zn, Zr, Mo, and Ba, was added into the intumescent systems based on 16.6 wt% ammonium polyphosphate (MP) and 8.4 wt% pentaerythritol (PER) in 75 wt% polypropylene (PP). The addition of 0.1–0.5 wt% of the cations of Mn and Zn was found to increase the limiting oxygen index (LOI) of the intumescent systems by 7–9 units, and to improve the vertical burning test (UL-94) grade from V-2 to V-0.[26,27] Another organic metal salt, nickel formate ($Ni(HCOO)_2$), was used as a catalyst to improve the flame retardancy of intumescent systems comprising of PP, APP and PER. LOI values of the composites first increased with the concentration of the catalyst until a maximum was reached and then decreased, over the range 0.1–5 wt% metal salt content. The optimal content of nickel formate was 2 wt%, at which the LOI value of the composite reached 30.[28] A similar catalytic effect was also found when supported nickel catalyst (Ni-Cat, supported on silica-alumina, Ni content ~66 wt%) was used as a catalyst to improve the flame retardancy of intumescent systems based on APP, PER and PP. When the total amount of flame retardants in the composites was controlled at 20 wt%, at an optimum content of Ni-Cat (1 wt%), the LOI value of the composite increased from 26.3 to 33.4 and raised UL-94 rating from V-2 to V-0.[29]

Metal chelates and macromolecular chelates were also synthesized and used to improve the flame retardancy of some polymeric materials, in combination with halogen-free flame retardants. The effect of metal chelates, including copper (II), cobalt (II) and nickel (II) complexes with oligo(salicylaldehyde) (metal-OSA) (Fig. 4.1a) and acetylacetonate (metal-AcAc), on the flame retardancy of a new intumescent flame-retardant system based on APP, melamine phosphate and a novel carbonization agent (CA) (Fig. 4.1b) in low-density polyethylene (LDPE) was investigated by Wang et al.[30] Metal chelates exhibited a strongly catalytic role in promoting the formation of carbonaceous charred layers when it used to prepare flame-retarded LDPE together with the carbonization agents, an LOI value of 29.8 and UL-94 V-0 rating could be achieved when 29 wt% CA, APP and MP were added into LDPE in combination with 1 wt% cobalt oligo(salicylaldehyde), CoOSA. Compared with the flame retarding composite without metal chelates, one containing 30 wt% CA, APP and MP, the LOI increased by 3.5 units and the UL-94 rating raised to V-0 from V-2.

The effect of metal chelates on the flame retardancy of the IFR-LDPE system may be explained by the following possible mechanism. It is possible that metal ions accelerate phosphorylation of both the charring agent and –OH groups formed on polyolefin molecules following metal ion oxidation in parallel with cross-linking. LDPE contains approximately two short chain branches per

(a) Metal-OSA (b) CA

4.1 Structure of metal-OSA and CA.

hundred backbone carbon atoms, arising from intramolecular chain transfer during polymerization which gives rise mainly to n-butyl side-chains. The tertiary carbon atoms in LDPE could then undergo peroxidation through metal chelate catalysis. These peroxidized side chains now cleave from the main polymer chains and the latter then cross-links the residual carbon skeletons and this increases the extent of carbonization. It was also suggested that the cross-linking increased the stability of APP and made more phosphorus available for phosphorylation and char formation. The viscosity of the PE melt was also increased, thus reducing the rate of flow to the flaming surface, and the protecting effect of the char was improved.[30]

Macromolecular chelates were synthesized and used to prepare flame retarding PP by Shehata et al.[31] Modified kaolin was prepared by reaction of kaolin with a prepared resin-iron chelate (Fig. 4.2a), and it was found that the composite loaded with a mixture of kaolin and modified kaolin significantly reduced the smoke generation of PP.[31] A macromolecular cobalt chelate (Fig. 4.2b) based on an N-(4-methyl phenyl) acrylamide monomer cross-linked by 4,4′-bis-(acrylamido)diphenylsulphone was added to PP in combination with magnesium hydroxide, where 10–20 wt% magnesium hydroxide was substituted by the prepared cobalt chelate. The total and the peak heat release rates as well

(a) Kaolin/resin-iron chelate   (b) Cobalt chelate

4.2 Structure of macromolecular chelates.

(a) Azocyclohexane

(b) 4,4'-bis(cyclohexylazocyclohexyl)methane

4.3 Structure of azocyclohexane and 4,4'-bis(cyclohexylazocyclohexyl) methane.

as the amount of emitted CO of PP samples were considerably lowered during burning.[32] Reactive-type metal chelates were also studied. Hexadentate Schiff base metal complexes were synthesized and copolymerized with diisocyanates by Chantarasiri *et al.* to prepare metal-containing polyureas, having LOI values as high as 36.4.[33]

Another interesting result was found by Nicolas *et al.*,[34] who synthesized azoalkanes such as azocyclohexane (Fig. 4.3a) and 4,4'-bis(cyclohexylazocyclohexyl)methane (Fig. 4.3b), which act as radical precursors. They showed high flame-retardant efficiency and it was found that the prepared azoalkanes alone were able to effectively offer flame retardancy and self-extinguishing properties to PP films at a very low concentration of 0.25 to 0.5 wt%, due to the interruption of the combustion process by the alkyl radicals generated by the diazene radical precursors.[34]

Although the investigation of catalytic effects in flame retardance systems is still in its initial phase, various organic and inorganic compounds, including metallic oxides, metallic salts, metal chelates, macromolecular chelates, radical precursors as well as nanoparticles,[35] have been chosen as catalysts in combination with conventional flame retardants or used alone. Several interesting catalytic effects have been found and these discoveries have opened up new opportunities to develop a number of halogen-free flame retardance systems of both academic and industrial interest. The combination of nanotechnology and catalysis techniques is especially promising.

4.4 Flame retardancy and anti-dripping properties

Generally, the flame retardancy and anti-dripping of polymers in fires have challenged materials engineers and scientists because some flame-retardant mechanisms depend solely on melting dripping theory. As mentioned above, melt-dripping behavior is very dangerous in a fire because it easily leads to secondary ignition as well as presenting a hazard to immediate personnel.

Usually, a way to resolve this problem is via the formation of an effective carbon layer and the simultaneous increase in melting viscosity once a polymer is ignited or is subjected to high temperature. The rapid increase in melting viscosity can prohibit polymer dripping and the formation of high-quality, surface carbon layers can protect polymers from being further attacked by fire.

Typically, an intumescent system including ammonium polyphosphate as an acid source and blowing agent, pentaerythritol as a carbonific agent and some synergistic agents has been used successfully to improve flame retardancy of polymers, especially for polyolefin, such as PP, PE, EVA, etc. Atikler et al.[36] reported that colemanite as a synergistic agent associated with APP and PER may be employed to enhance flame retardancy of polypropylene (FR-PP). Based upon the experimental results, the LOI value and the amount of residue for the optimum formulation (APP 65%, PER 28% and colemanite 7%) reached 39.3 and 21.4, respectively. Hu and Li et al.[37–39] synthesized triazine polymers (charring agent CA and CFA), respectively, and triazine polymers were used as a charring agent and/or as a foaming agent in intumescent flame retardants. The structures of these triazine polymers are shown in Fig. 4.4. The intumescent flame retardants with an optimum formulation showed good flame retardancy and non-dripping properties for PP and PE.

There are also some studies on both anti-dripping and flame retardancy for polyester. Wang et al.[40] reported that a novel flame-retardant copolyester (PET-co-DDP)/organomontmorillonite (O-MMT) nanocomposite (PDMN) can be synthesized by condensation of terephthalic acid (TPA), ethyleneglycol (EG), 9,10-dihydro-10[2,3-di(hydroxycarbonyl)propyl]10-phosphaphenanthrene-10-oxide (DDP) and O-MMT. It was found that the clay catalyzed carbonaceous char formation, and the reinforcement of the char caused by the clay was responsible for the lowered flammability of these nanocomposites. Meanwhile, MMT in the polyester can provide a framework and the anti-dripping behavior can be improved. Qu et al.[41] prepared phosphorus-containing copolyester (PET-co-DDP)/barium sulphate ($BaSO_4$) nanocomposite by in situ polymerization. The results showed that LOI values of the resulting nanocomposites decreased with increasing the content of $BaSO_4$ nanoparticles, but their anti-dripping behaviors were improved obviously through the UL-94 test. The flammability tests based on the cone calorimetry showed that the introduction of nano-$BaSO_4$ to the copolyester decreased remarkably the heat release rate and effective heat of combustion.

Wang et al.[42] also synthesized a novel intumscent flame retardant, poly (2, 2-dimethylpropylene spirocyclic pentaerythritol bisphosphonate) (DPSPB, Fig. 4.5a) which was then blended with PDMN to prepare PET with both excellent flame retardancy and anti-dripping properties. To evaluate flame-retardant

(a) Charring agent CA

(b) CFA

4.4 Structure of triazine polymers.

4.5 Structure of spirocyclic and caged bicyclic flame retardants.

properties of polymer materials, LOI values and the results of vertical burning tests are given in Table 4.1. As can be seen, the anti-dripping abilities of the nanocomposites are improved obviously after adding DPSPB; i.e., the nanocomposite with over 5 wt% of DPSPB shows little dripping during test and a V-0 rating and LOI = 29 level can be reached for the nanocomposite when DPSPB content is 10 wt%.

Figure 4.6 clearly demonstrates the anti-dripping effect of DPSPB. In the residues of PDMN, there are some charred components. However, the residues of PDMN/DPSPB are almost all the chars instead of formerly molten materials, and it can be seen that the anti-dripping properties have obviously been improved.

A similar intumescent flame retardant, poly(2-hydroxy propylene spirocyclic pentaerythritol bisphosphonate) (PPPBP, shown in Fig. 4.5b), was used to impart flame retardance and dripping resistance to poly(ethylene terephthalate) fabrics by Chen et al.[43] The results indicated that besides the enhancement in flame retardance, the dripping resistance of the treated PET fabrics is observed. FTIR spectra and SEM graphs of the residue of treated PET fabrics show that the structure and composition of the chars yielded during the decomposition of PPPBP have a direct relationship with the enhancement in flame retardance and dripping resistance of the treated PET fabrics.

Gao et al.[44] prepared a phosphorus-nitrogen containing intumescent flame retardant (P-N IFR, shown in Fig. 4.5c) via the reaction of a caged bicyclic

Table 4.1 The LOI values and UL-94 testing results

Sample	DDP	MMT	DPSPB	LOI	UL-94	Drip
	wt%	wt%	wt%			
PET	0	0	0	21.2	–	Heavy
PDMN	0.5	1	0	32.0	V-2	Slow
PDMN/DPSPB	0.5	1	5	28.8	V-2	Slight
PDMN/DPSPB	0.5	1	10	29.0	V-0	Very slight

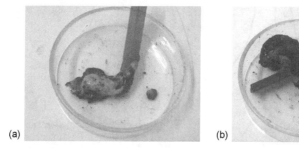

(a)   (b)

4.6 Residues of copolyesters after combustion: common nanocomposite residue of PDMN (left) and the novel nanocomposite residue of PDMN/DPSPB (right).

phosphorus (PEPA) compound and 4,4′-diamino diphenyl methane (DDM) in two steps. The product was added to poly(butylene terephthalate) (PBT) to obtain a halogen-free, flame-retarded polyester. It was shown that the phosphorus-nitrogen containing compound could improve both the flame retardancy and thermal stability more effectively than other P-N flame retardants. Furthermore, it was seen to be a good char-forming agent when incorporated with the co-addition of polyurethane (PU). Good anti-dripping properties and flame retardancy can be obtained in the case of 10% PU and 20% P-N addition to the PBT polymer. Balabanovich et al.[45] reported that a self-extinguishing formulation (V-0 rating) and non-dripping properties were achieved by the addition of 15 wt% APP and 10 wt% 2-methyl-1,2-oxaphospholan-5-one 2-oxide (OP). Probably, the fire-retardant effect was attributed to the condensed-phase mechanism of intumescence. OP-APP was shown to cause ammonolysis and the formation of aromatic nitriles in the high-boiling products and phosphorus ester groups and polyphosphoric acid-phosphorus pentaoxide in the solid residue.

In conclusion, although there are some successful cases for resolving the flame retardancy and anti-dripping question of polymers, there are still needs for a long-term study.

4.5 Flame retardancy and mechanical properties

Additive-type flame retardants, physically incorporated into polymers, provide the most convenient and economical way of introducing flame retardancy into commercial polymers. However, a variety of problems, such as poor compatibility, diffusing out on to the surface of the resulted products and reduction in the mechanical or other physical properties of the polymers, restrict the application of most of the commercially available additive-type flame retardants.[46] To overcome the aforementioned problems, several techniques have been employed and a large number of halogen-free flame retardants have been synthesized.

4.5.1 Encapsulation and microencapsulation

Microencapsulation is a process of enveloping microscopic amounts of matter in a thin film of polymer or metal oxide which forms a thin solid shell. This core/shell structure allows the isolation of the encapsulated substance from the immediate surroundings and thus protects it from any degrading factors such as moisture.[47] Besides the traditional method of treating a filler surface with low-molecular weight coupling agents and surfactants, or by adding macromolecular compatilizer,[48] microencapsulation has appeared as an alternative way to reduce particle surface energy, promote dispersion of the fillers and provide a strong adhesion between the encapsulated substance and the polymer matrix, leading to enhanced mechanical properties and processability of the resulting composites.

Magnesium hydroxide ($Mg(OH)_2$) is an effective flame retardant with a high decomposition temperature, it has smoke suppressibility, and is widely used in thermoplastics. However, the disadvantage of $Mg(OH)_2$ is the high content (more than 60 wt%) required to achieve the desired flame retardant effect, leading to a reduction in the mechanical or other physical properties of the polymers, including reduced elongation-at-break and impact strength. It also leads to increased melt viscosity.[49] Liauw et al.[50] investigated the effect of silane-based filler surface treatment on the mechanical properties of polypropylene/magnesium hydroxide composites, where $Mg(OH)_2$ flame retardant filler was treated with combinations of vinyltriethoxysilane (VS) and dicumyl peroxide (DCP) and melt blended with an impact modified polypropylene matrix. It was found that due to the encapsulation of the filler particles by the cross-linked elastomeric phase of the matrix, the VS/DCP combination afforded the best balance of strength and toughness at 65 wt% $Mg(OH)_2$ and led to a significant improvement in the mechanical properties of highly filled impact-modified PP based composites. Chang et al.[51] prepared polystyrene (PS)-encapsulated $Mg(OH)_2$ by in situ polymerization of styrene on the surface of $Mg(OH)_2$ in a high-speed mixer, and the effects of PS-encapsulated $Mg(OH)_2$ on the properties of HIPS composites were studied. It was demonstrated that by comparing with the composites containing untreated $Mg(OH)_2$, the rheological and flame-retardant properties of those containing PS-encapsulated $Mg(OH)_2$ were found to be significantly improved due to a better dispersion of the encapsulated $Mg(OH)_2$ and a strong adhesion between the encapsulated magnesium hydroxide and PS matrix.

Among the additive-type halogen-free flame retardants, red phosphorus (RP) is an attractive type of powerful flame retardant because of its low cost, high phosphorus content, little smoke, and low toxicity. It is extensively used in thermoplastics, such as polyamide, poly(ethylene terephthalate), and polyolefins. Nevertheless, RP has some disadvantages as a flame retardant, the main ones being color, the poor thermostability and the lack of compatibility with synthetic resins, which adversely affect flame retardant effectiveness and hinder its extensive commercial applications. To overcome these disadvantages, the encapsulation of virgin RP particles with thermosetting resin or metal oxide has been attempted.[52–55] Wu et al.[53] prepared microcapsulated red phosphorus (MRP) with a melamine-formaldehyde resin coating layer by a two-step coating process. The MRP can efficiently improve its themostability, increase the ignition point, and decrease the water absorption. The prepared MRP was used to flame retard linear low-density polyethylene (LLDPE) and polypropylene. In combination with other traditional halogen-free flame retardants, it has been found that MRP is a powerful additive type flame retardant in halogen-free flame-retardant polyolefins. When the total amount of additives was kept constant at 25 wt%, the component NP28 (phosphorus-nitrogen compound) was partially replaced by MRP, it can be seen that the LOI values increased gradually from 30.5 to 36.0 with increasing amount of MRP added into the PP composites. A novel technology was developed by Liu and Wang[54] to prepare microencapsulated red phosphorus with a coating of melamine cyanurate (MCA), which served as both a nitrogen-containing flame retardant and as a solid lubrication agent. MRP was prepared through MCA self-thickening effects during the MCA self-assembly process, and this technology can overcome several drawbacks of current microencapsulation processes. The microencapsulated RP showed nitrogen-phosphorus synergism and further improved flame retardant efficiency of RP due to the introduction of nitrogen by the encapsulation of RP with a nitrogen-containing flame retardant. Flame retardant polyamide 6 (PA6) was also prepared with MRP, and the resulting composites possessed desired flame retardancy with effective char-formation and it also showed satisfactory mechanical properties as the result of the good compatibility between MRP and PA6. This MCA-encapsulated RP was also used to improve the flame retardancy of polyamide 6.6 or PA66 and glass fibre (GF)-reinforced PA66, which also showed better flame retardancy and mechanical performance due to N-P synergistic effects and better compatibility between the MRP and the PA66 matrix.[55]

Microencapsulation was also used to solve the water solubility and migration of phosphate in the polyurethane-phosphate intumescent flame retardant system. Di-ammonium hydrogen phosphate (DAHP) was microencapsulated with polyurethane-urea membrane, a gelatin-polyurea shell and poly(vinyl alcohol)-polyurethane shell respectively. A so-called coacervation and an interfacial polymerization technique were employed. The DAHP microcapsules show characteristics of an intumescent system.[56,57] DAHP microcapsules with polyether-polyurethane and polyester-polyurethane shells respectively present in the polyurea coatings applied to a cotton fabric gave an efficient flame retardant effect.[58] The encapsulation of the small molecular charring agent, PER, by polyurethane was also performed by Liu et al.,[59] which effectively avoids the reaction of PER with melamine phosphate during the compounding with PP at high temperature and also prevents the leaching out of PER. Flame-retarded PP with encapsulated PER shows a much better charring performance and flame retardancy than the flame-retarded PP with virgin PER. A UL-94 V-0 rating at 1.6 mm thickness can be obtained at a 25 wt% flame retardant loading.

Generally, the polymers for the encapsulation of flame retardants are charforming or form an intrinsic intumescent formulation in combination with encapsulated flame retardants. Polymers such as polycarbonate, resol resin, melamine-formaldehyde resin, novolac epoxy, and polyurethane, which promote the formation of compact charred layers in the condensed phase during combustion are typically used. The compact charred layers acts as a barrier and slows down heat and mass transfer between the gas and condensed phases, preventing the underlying polymeric substrate from further burning. In this way encapsulation can both enhance the compatibility and the flame retardant effectiveness of the halogen-free flame retardants. This makes encapsulation one of the most favored approaches to resolving the conflict between flame-retardant efficiency and the need to maximize mechanical properties.

4.5.2 Polymeric additive-type halogen-free flame retardants and intrinsically flame-retardant polymers

To overcome the disadvantages of additive-type, small molecular weight, flame retardants, polymeric additive-type, halogen-free flame retardants have been synthesized and employed as blends to improve the flame retardancy of various polymeric materials, for their better compatibility, miscibility and the potential advantage of retaining the mechanical properties of the polymers.

Poly(sulphonyl phenylene phosphonate) (PSPPP) (Fig. 4.7a) is a polymeric retardant with both phosphorus and sulphur bonded in the main chain. It has been commercially used in polyester fibres, such as the original HEIM fibers of Toyobo Co., Ltd.[60,61]

The polycondensation of phenylphosphonic dichloride (PPD) with 4,4'-sulphonyldiphenol (SDP) under low temperature and high temperature in different solvents was studied by Kim[62] and Masai,[63] respectively. An improvement in the limited oxygen index by 6 units was observed when 7 wt% PSPPP was used as an additive type flame retardant for thermoplastic poly(butylene terephthalate). Using the polycondensation process of Wang,[60] high molecular weight (more than 10,000) PSPPP was prepared by melt polycondensation without solvents. Flame retarded poly(ethylene terephthalate) fibres with an LOI value of 28 and good mechanical properties were obtained at 4 wt% loading of PSPPP.

The combination of poly(sulphonyl phenylene phosphonate) with other additives was also studied by Wang et al.[64] and Balabanovich and Engelmann.[65] These works showed that a synergistic effect was found when PSPPP was combined with potassium diphenyl sulphonate (SSK), polyphenylene oxide (PPO), triphenylphosphate (TPP), or 2-methyl-1,2-oxaphospholan-5-one 2-

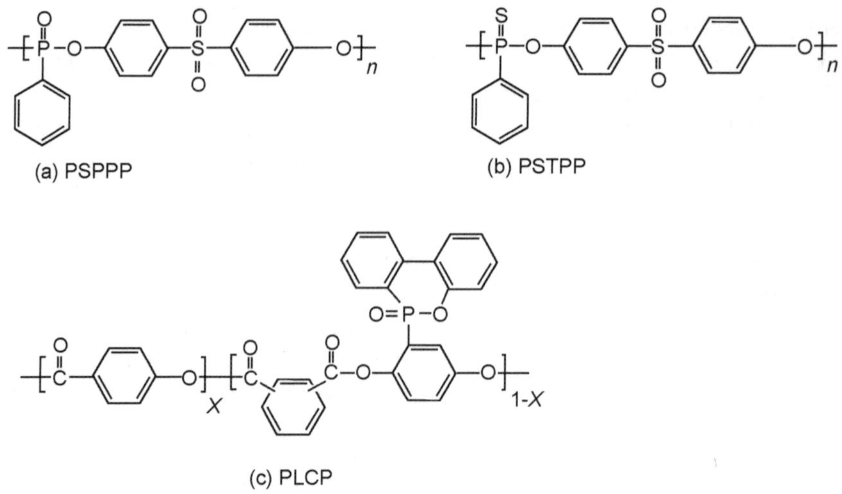

(a) PSPPP

(b) PSTPP

(c) PLCP

4.7 Structure of phosphorus-containing polymeric additive type flame retardants.

oxide.[64,65] A combination of 0.5 wt% SSK and 4.5 wt% PSPPP provided a UL-94, V-0 rating, and increased the LOI from 31.9 to 36.8, compared to the composite with 5 wt% PSPPP. Polysulphonyldiphenylene thiophenylphosphonate (PSTPP) with the similar structure as PSPPP was synthesized from phenylphosphonothioic dichloride (PPTD) and 4,4′-sulphonyldiphenol by polycondensation reaction (Fig. 4.7b). It can be used as the flame retardant for poly(ethylene terephthalate) and PET samples containing PSTPP can reach an LOI value of 29 and a UL-94 V-0 rating when the phosphorus content of the sample is 2.5 wt%.[66,67] Phosphorus-containing poly(ether sulphone) samples were also synthesized and blended with epoxy resin and at a loading of 20 wt%, the LOI value increased from 21.9 to 27.4.[68]

Wholly aromatic thermotropic polyesters are currently receiving considerable attention for their excellent mechanical properties, thermal stability, chemical resistance and low melting viscosity. However, because of their high aromatic contents, this class of polymers generally have high melting temperatures (T_m), which prevent melt processing before thermal decomposition. A series of phosphorus-containing, wholly aromatic thermotropic copolyesters (PLCP) (Fig. 4.7c) with relatively low thermal transition temperatures were prepared from acetylated 2-(6-oxide-6H-dibenz<c,e><1,2>oxa phosphorin-6-yl)-1,4-dihydroxy phenylene, p-acetoxybenzoic acid, terephthalic acid, and isophthalic acid by melt polycondensation. The T_g values of the polymers were rather high, ranging from 183 to 192 °C, because of the rigidity of the main chain and the presence of the bulky substituent on the hydroquinone unit; the T_m values obtained from DSC curves for samples P-20 and P-25 were 290 and 287 °C, respectively (where the number in the sample name indicates the molar fraction of the phosphorus-containing monomer in the reactants). The initial flow temperatures of other samples observed with hot-stage polarizing microscopy were 271–290 °C. All the copolyesters, except P-40, were thermotropic and nematic, showed banded textures at elevated temperatures. The char yields of PLCPs at 640 °C were 41–52%, confirmed that the incorporation of phosphorus into the side groups led to good flame retardancy.[69,70] The synthesized thermotropic liquid crystal copolyester containing phosphorus provides a limiting oxygen index value of 70 and it can be used to prepare *in-situ* reinforced PET composites that have both good flame retardancy and better mechanical properties than pure PET; the melt dripping behavior of PET was also improved.[71,72] This provides a new approach to simultaneous improvement of the flame retardancy, tensile strength and melt dripping reduction of polymers.

The nitrogen-containing polymeric flame retardant, benzoguanamine-modified phenol biphenylene resin (BG-modified PB resin) (Fig. 4.8a) was used as an additive to improve the flame retardancy of phenol-biphenylene-type epoxy resin (PB-type epoxy resin). Adding the BG-modified PB resin in the PB-

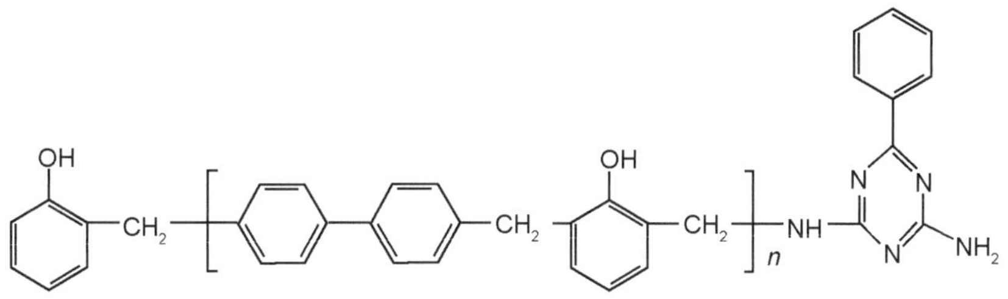

(a) BG-modified PB resin

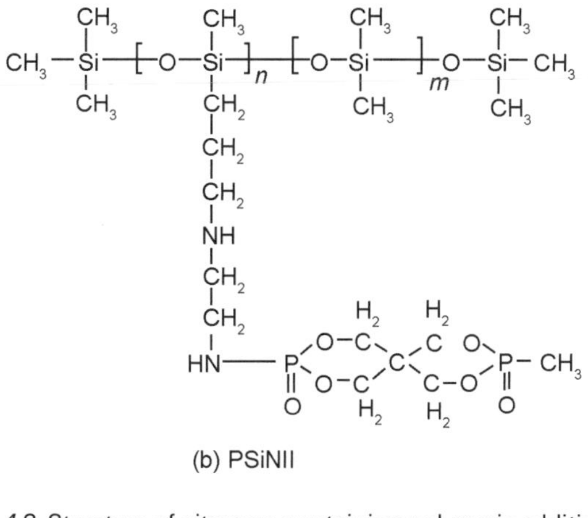

(b) PSiNII

4.8 Structure of nitrogen-containing polymeric additives.

type epoxy resin compound increased the flame retardancy, heat resistance, and glass transition temperature, while the mechanical properties were almost unaffected.[73]

Liu et al.[74] prepared poly(dimethylsiloxane) (PDMS) star polymers having a nanosized silica particle core by reacting silica nanoparticles with monoglycidylether-terminated poly(dimethylsiloxane). This star polymer has an extremely high content of silica and may be used as a flame retardant for polymers.

The polymeric intumescent flame retardant, PSiNII (Fig. 4.8b), containing silicon, phosphorus and nitrogen, has been synthesized and incorporated into PP. A 20 wt% content of PSiNII increased the LOI from 17.4 to 29.5. When the loading amount of PSiNII was kept at 15–25 wt%, the mechanical properties were still acceptably high, which suggested that this polymeric flame retardant could be used to offer flame retardant PP grades with good mechanical properties.[75,76]

Recently, the design of new, intrinsically flame-retarding polymers that do not contain halogen in the polymer structure has become even more attractive due to their permanent flame retardancy and good mechanical properties.

Halogen-free, flame-retardant polyphosphonates were prepared by interfacial polymerization of phenylphosphonic dichloride and 4,4'-bishydroxydeoxybenzoin (BHDB), using dichloromethane as the organic phase and benzyltriphenyl phosphonium chloride as the phase transfer catalyst. The synthesized BHDB-polyphosphonates (Fig. 4.9a) combine the highly desirable property of flame resistance with good solution processability.[77]

Inorganic-organic polymers allow for a unique combination of traditionally features, such as high flexibility and the ability of preserving physical properties over a wide range of temperatures. Highly flame-retardant, siloxane-based inorganic-organic polymers, silicone-based aromatic polyesters and polyamides (Fig. 4.9b) and silicone-based aliphatic polyamides (Fig. 4.9c) were synthesized by a novel biocatalytic approach. All of the synthesized polymers showed good flame-retardant properties.[78] Phosphorus-containing polyaryloxydiphenylsilanes with high flame retardancy, containing both silicon and phosphorus, also have been prepared by Liu et al.[79]

4.5.3 Reactive-type halogen-free flame retardants and the modification of polymers through copolymerization

The main advantage of the reactive approach, in which the existing polymers are modified by reaction, is the maintenance of the physical and chemical properties similar to those of the original polymers.[80]

Incorporating a chemically reactive phosphorus-containing monomer into the polymer chain is one of the most efficient methods of improving the flame retardancy of polyesters. Copolymerization of polyester monomers with 2-

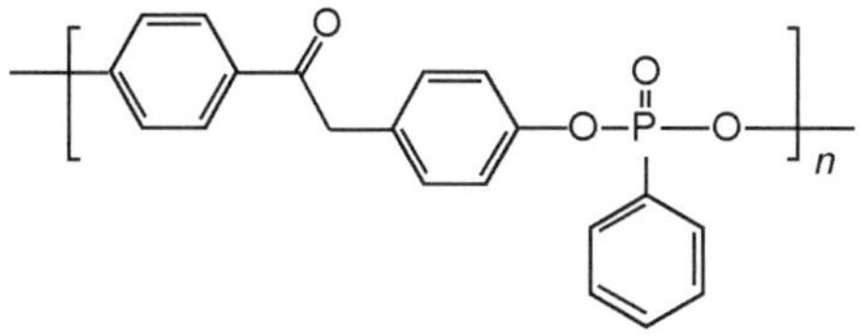

(a) BHDB-polyphosphonate

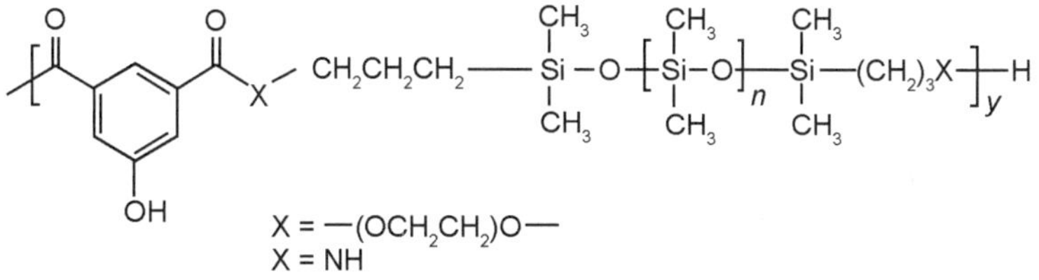

(b) Silicone-based aromatic polyesters and polyamides

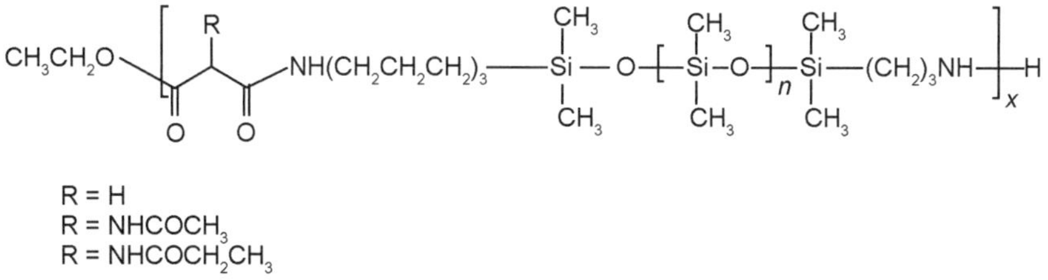

X = —(OCH$_2$CH$_2$)O—
X = NH

(c) Silicone-based aliphatic polyamides

R = H
R = NHCOCH$_3$
R = NHCOCH$_2$CH$_3$

4.9 Structure of intrinsically flame retarding polymers.

carboxyethyl(phenylphosphinic) acid (CEPP), 2-carboxyethyl(methylphosphinic) acid (CEMP) or their cyclic anhydrides (Fig. 4.10a and b) shows excellent flame retardancy at 3–8 wt% content with an overall PET chain structure. The PET fibres made from the phosphorus-containing polyesters have been commercialized as Trevira CS (Trevira GmbH).[60,81] A patent by Birum and Jansen described the synthesis of CEPP in 1978,[82] and several subsequent patents have claimed improvements of the synthesis of CEPP.[83,84] The synthesis of copolyesters of ethylene terephthalate and CEPP was investigated by Asrar et al.[85] who found that CEPP could be copolymerized with TPA and EG in the conventional melt polymerization process used for PET.

Another commercially used, reactive phosphorus-containing monomer in polyester is 9,10-dihydro-10 [2,3-di (hydroxycarbonyl) propyl] 10-phosphaphenanthrene-10-oxide(DDP) (Fig. 4.10c). The fibres have been commercialized by Toyobo Co., Ltd., known as HEIM.[86] Chang et al.[87] studied the synthesis of copolyesters containing the phosphorus linking pendent groups by charging DDP, terephthalic acid, and ethylene glycol in one reactor. The LOI

```
        O
        ‖
    HO—P—CH₂CH₂COOH
        |
        ⌬

    (a) CEPP
```

```
        O
        ‖
    HO—P—CH₂CH₂COOH
        |
        CH₃
    (b) CEMP
```

```
                O
                ‖
         ⌬  P—CH₂——CH—COOH
         ⌬ /          |
            O         CH₂—COOH

          (c) DDP
```

4.10 Structure of CEPP, CEMP and DDP.

value of the phosphorus-containing copolyester is 33.3 at a loading of 0.7 wt% P, while the effect on the mechanical properties of the copolyester is tolerable. The rheological behavior of these phosphorus-containing copolyesters is similar to that of PET. The thermal decomposition behavior in air of DDP-containing copolyesters with the phosphorus linkage as pendent groups, was compared with those of PET and copolyesters with phosphorus linkages in the main chain, which were synthesized from TPA, EG and CEPP. It is shown that the presence of the bulky pendent phosphorus side group in the copolyester tends to decrease the activation energy for decomposition in air, and the decompostion activation energy in air is lower when the phosphorus linkage is on the side chain than in the main chain.[88]

Dopotriol, a trifunctional phosphorus-containing curing agent based on 9,10-dihydro-9-oxa-10-phosphaphenanthrene-10-oxide (DOPO) (Fig. 4.11a), was used to cure diglycidyl ether of bisphenol A (DGEBA). The glass-transition temperatures of the cured epoxy resins increase with the phosphorus content and a UL-94 V-0 rating was achieved with a phosphorus content of 1.87 wt%.[89] Another phosphorus-containing monomer comprising a glycidyl group substituted on to the diglycidyl ether of isobutyl bis(hydroxypropyl) phosphine oxide (IHPOGly) (Fig. 4.11b) was reacted with DGEBA. The resultant phosphorous-containing epoxy resins give a UL-94 V-0 rating and an LOI of 28 when the molar ratio of IHPOGly/DGEBA achieves 1.25/8.75.[90]

Epoxy/amine hybrid resins, which may be attractive for use in high-performance electronic products, were prepared through the *in-situ* curing of bisphenol A epoxy and a nitrogen-containing monomer, hexakis(methoxymethyl) melamine (HMMM) (Fig. 4.11c), with 2 wt% (3-glycidoxypropyl)-trimethoxysilane (GPTMS) as a facial coupling agent. The prepared resins showed good miscibility and high transparency, the LOI values of the epoxy resins increased with the content of HMMM, and gave an LOI value of 32.5 at a loading of 40 wt%.[91]

4.11 Structure of dopotriol, IHPOGly and HMMM.

Recently, silicon-containing monomers have attracted much more attention for use as reactive-type, halogen-free flame retardants, especially the use of polyhedral oligomeric silsesquioxanes (POSS), due to their notable mechanical reinforcing effect.[92,93] Silicon-containing flame retardant epoxy resins have been prepared from silicon-containing epoxides, silicon-containing glycidyl monomers, triglycidyloxyphenyl silane (TGPS), diglycidyloxydiphenyl silane (DGDPS), 1,4-bis(glycidyloxydimethyl silyl)-benzene (BGDMSB) and diglycidyloxymethylphenyl silane (DGMPS), or silicon-containing prepolymers (EpSi) (Fig. 4.12). The prepared epoxy resins had high LOI values and epoxy resins modified by BGDMSB showed a high LOI value of 33.5 at a loading of 12.8 wt% silicon. LOI values of DGMPS modified epoxy resins increased from 24 for a standard commercial resin to 36 for silicon-containing resin containing 8 wt% silicon.[94,95]

Borates are effective flame retardants which yield impenetrable glass coatings when they thermally degrade. The chemical modification of polymer chains with boron-containing, flame-retardant monomers leads to a significant improvement in flame retardancy. For example, bis(benzo-1,3,2-dioxaborolanyl) oxide and bis(4,4,5,5-tetramethyl-1,3,2-dioxaborolanyl)oxide (Fig. 4.13) were used to prepare boron-containing novolac resins by chemically modifying a commercial novolac resin. The catechol derivative resins have higher LOI values and char yields as a consequence of the higher boron content. When novolac resins were modified with bis(benzo-1,3,2-dioxaborolanyl)oxide, the LOI value increased from 24.6 to 38.2 at a 3.8 wt% boron loading.[96,97] Boron-containing epoxy-novolac resins were synthesized by modifying epoxy-novolac resins with bis(benzo-1,3,2-dioxaborolanyl)oxide or bis(4,4,5,5-tetramethyl-1,3,2-dioxa-borolanyl)oxide. The LOI values for bis(benzo-1,3,2-dioxa-

4.12 Structure of silicon-containing glycidyl monomers and EpSi.

(a) bis(benzo-1,3,2-dioxaborolanyl)oxide

(b) bis(4,4,5,5-tetramethyl-1,3,2-dioxaborolanyl)oxide

4.13 Structure of boron-containing flame retardant monomers.

borolanyl)oxide modified epoxy-novolac resins increase as a consequence of the presence of boron.[98]

In conclusion, a number of strategies exist for developing halogen-free flame retardants having high efficiencies. The principal ones discussed above include nanotechnology, metal chelate catalysis and formulation design, each of which can improve the flame-retardant efficiency of the flame retardants and offer a possibility of decreasing the content of flame retardants in the polymer matrix. They may also be beneficial to the mechanical and other properties of the materials in some specific applications.

4.6 Future trends

With the beginning of 21st century, the sustainability of polymeric materials becomes more important than before due to the issues of environmental protection and health. It is very clear that more and more halogen-based flame retardants will be replaced by halogen-free flame retardants in the future.

However, as mentioned above, the replacement will be faced with many challenges, and we have to maintain a balance of acceptable flame retardant, mechanical and physical properties while achieving environmentally friendly polymeric materials. A feasible and practical strategy is to follow two routes. Firstly, we must consider that the existing halogen-based flame retardants are impossible to replace completely and so we need to be selective in their use, reduce their amounts during use and enhance their sustainability. Secondly and in parallel, development of novel, halogen-free, flame-retardant systems, which may be phosphorus- and/or nitrogen-containing organic compounds, inorganic compounds, especially nano-materials and other systems will continue.

Except for inorganic flame retardants, phosphorus- and nitrogen-containing flame retardants are still the halogen-free flame retardants with the largest amount of uses. There are many well-developed phosphorus-containing flame retardants and a wealth of knowledge about why these compounds are effective In addition, the synthetic chemistry of phosphorus has been well studied so that the synthetic routes for many compounds are available. The nitrogen-containing flame retardants, such as melamine-based derivatives, are also receiving more attention due to their good flame-retardant efficiency and low cost. A synergistic flame retardation of different flame retardant elements and systems is always desirable, and an especially acceptable flame retardancy can be obtained when nanoparticles are used together with some conventional flame retardants.

Inorganic flame retardants and nanotechnology will be widely applied in future research in halogen-free flame retardants. In some cases, the addition of nanoparticles can enhance both the flame retardancy and the mechanical properties of polymers. However, for some systems, it will be difficult for polymer nanocomposites to be applied in commercial applications because only the heat release rates of materials reduce and formulations still fail to achieve the UL-94 V-0 ratings. Therefore, practical levels of flame retardancy cannot be obtained by incorporating small amounts of nanomaterials such as MMT, CNT, LDH alone to polymer materials. A promising approach is to combine the nanotechnology with conventional flame-retardant technology, that is, to use nanomaterials together with conventional flame retardants. In this way, a satisfactory flame-retardant polymer system with good flame retardancy, mechanical properties, a reasonable price and with acceptable environmental sustainability, will be able to be developed.

It is probable that there are as-yet flame-retardant additives to be discovered, some of which will be new halogen-free compounds. In this respect, we can be assured of further exciting research results that address the challenge of replacing halogen-containing flame retardants.

4.7 Acknowledgements

This work was supported by the National Science Foundation of China (20674053, 20274027, 50173016, 59603007) and the National Science Fund for Distinguished Young Scholars (50525309). The contribution of my students, Mr De-Yi Wang and Xin-Guo Ge to this chapter was greatly appreciated.

4.8 Sources of further information and advice

The sources of further information include the following references:

- http://plas.specialchem.com.cn/tc/flame-retardants-regulations/
- Lu S Y and Hamerton I, 'Recent developments in the chemistry of halogen-free flame retardant polymers', *Prog. Polym. Sci.*, 2002, **27**, 1661–1712.
- Levchik S V and Weil E D, 'A review of recent progress in phosphorus-based flame retardants', *J. Fire Sci.*, 2006, **24**, 345–364.

and books:

- Horrocks A R and Price D (eds), *Fire retardant materials*, Woodhead Publishing Ltd, Cambridge, 2001.
- Grand A F and Wilkie C A (eds), *Fire retardancy of polymeric materials*, Marcel Dekker Inc, New York, 2000.
- Wang Y Z, *Flame-retardation design of PET fibers*, Sichuan Scientific Publishers, Chengdu, China, 1994.

4.9 References

1. Bras M L, Bugajny M, Lefebvre J M and Bourbigot S, 'Use of polyurethanes as char-forming agents in polypropylene intumescent formulations', *Polym. Int.*, 2000, **49**, 1115–1124.
2. Zhu W M, Weil E D and Mukhopadhyay S, 'Intumescent flame-retardant system of phosphates and 5,5,5′,5′,5″,5″-hexamethyltris (1,3,2-dioxaphosphorinane-methan)amine 2,2′,2″- trioxide for polyolefins', *J. Appl. Polym. Sci.*, 1996, **62**, 2267–2280.
3. Ravadits I, Toth A, Marosi G, Márton A and Szép A, 'Organosilicon surface layer on polyolefins to achieve improved flame retardancy through an oxygen barrier effect', *Polym. Degrad. Stab.*, 2001, **74**, 419–422.
4. Li R S, Halogen-containing flame retardant additives effect on ABS. *Science Today*, 2007, **18**, 106–107.
5. Wang D Y, Liu Y, Wang Y Z, Artiles C P, Hull T R and Price D, 'Fire retardancy of a reactively extruded intumescent flame-retardant polyethylene system enhanced by metal chelates', *Polym. Degrad. Stab.*, 2007, **92**, 1592–1598.
6. Dai P B, 'Flame retardant ABS', PhD Dissertation, Sichuan University, 2007.
7. Dai P B, Wang D Y, Ge X G, Wang Y Z, 'Effect of modified intumescent flame retardant via surfactant/polyacrylate latex on property of intumescent flame retardant ABS composites', *J. Macromol. Sci., Phys.*, (in press).

8. Szép A, Szabó A, Tóth N, Anna P and Marosi G, 'Role of montmorillonite in flame retardancy of ethylene-vinyl acetate copolymer', *Polym. Degrad. Stab.*, 2006, **91**, 593–599.
9. Wang Y Z, Zheng C Y and Yang K K, 'Synthesis and characterization of polysulphonyl diphenylene phenyl phosphonate', *Polym. Mater. Sci. Eng.*, 1999, **1**, 53–56.
10. Wang Y Z, Zheng C Y and Wu D C, 'Flame-retardant action of polysulphonyl-diphenylene phenylphosphonate on PET', *Acta Polym. Sin.*, 1996, **4**, 439–445.
11. Gilman J W, Jackson C L, Morgan A B, Harris R, Manias E, Giannelis E P, Wuthenow M, Hilton D and Phillips S H, 'Flammability properties of polymer-layered-silicate nanocomposires. Polypropylene and polystyrene nanocomposites', *Chem. Mater.*, 2000, **12**, 1866–1873.
12. Zhu J, Morgan A B, Lamelas F J and Wilkie C A, 'Fire properties of polystyrene-clay nanocomposites', *Chem. Mater.*, 2001, **13**, 3774–3780.
13. Zhu J, Uhl F M, Morgan A B and Wilkie C A, 'Studies on the mechanism by which the formation of nanocomposites enhances thermal stability', *Chem. Mater.*, 2001, **13**, 4649–4654.
14. Zanetti M, Camino G, Canavese D, Morgan A B, Lamelas F J and Wilkie C A, 'Fire retardant halogen-antimony-clay synergism in polypropylene layered silicate nanocomposites', *Chem. Mater.*, 2002, **14**, 189–193.
15. Zanetti M, Kashiwagi T, Falqui L and Camino G, 'Cone calorimeter combustion and gasification studies of polymer layered silicate nanocomposites', *Chem. Mater.*, 2002, **14**, 881–887.
16. Zhang J G and Wilkie C A, 'Preparation and flammability properties of polyethylene-clay nanocomposites', *Polym. Degrad. Stab.*, 2003, **80**, 163–169.
17. Kashiwagi T, Harris R H, Zhang X, Briber R M, Cipriano B H, Raghavan S R, Awad W H and Shields J R, 'Flame retardant mechanism of polyamide 6-clay nanocomposites', *Polymer*, 2004, **45**, 881–891.
18. Song L, Hu Y, Lin Z H, Xuan S Y, Wang S F, Chen Z Y and Fan W C, 'Preparation and properties of halogen-free flame-retarded polyamide 6/organoclay nanocomposite', *Polym. Degrad Stab.*, 2004, **86**, 535–540.
19. Beyer G, 'Short communication: carbon nanotubes as flame retardants for polymers', *Fire Mater.*, 2002, **26**, 291–293.
20. Kashiwagi T, Du F M, Douglas J F, Winey K I, Harrisjr R H and Shields J R, 'Nanoparticle networks reduce the flammability of polymer nanocomposites', *Nat. Mater.*, 2005, **4**, 928–933.
21. Gilman J W and Kashiwagi T, in *Polymer-clay nanocomposites* (eds Pinnavaia T J and Beall G W), John Wiley & Sons, New York, 2000, pp. 193–206.
22. Lewin M, 'Synergism and catalysis in flame retardancy of polymers', *Polym. Advan. Technol.*, 2001, **12**, 215–222.
23. Nodera A and Kanai T, 'Thermal decomposition behavior and flame retardancy of polycarbonate containing organic metal salts: effect of salt composition', *J. Appl. Polym. Sci.*, 2004, **94**, 2131–2139.
24. Tian C M, Qu H Q, Wu W H, Guo H Z and Xu J Z, 'Metal chelates as synergistic flame retardants for flexible PVC', *J. Vinyl Addit. Technol.*, 2005, **11**, 70–75.
25. Jang J, Kim J and Bae J Y, 'Synergistic effect of ferric chloride and silicon mixtures on the thermal stabilization enhancement of ABS', *Polym. Degrad. Stab.*, 2005, **90**, 508–514.
26. Lewin M and Endo M, 'Catalysis of intumescent flame retardancy of polypropylene by metallic compounds', *Polym. Advan. Technol.*, 2003, **14**, 3–11.

27. Lewin M, 'Unsolved problems and unanswered questions in flame retardance of polymers', *Polym. Degrad. Stab.*, 2005, **88**, 13–19.
28. Chen X C, Ding Y P and Tang T, 'Synergistic effect of nickel formate on the thermal and flame-retardant properties of polypropylene', *Polym. Int.*, 2005, **54**, 904–908.
29. Song R J, Zhang B Y, Huang B T and Tang T, 'Synergistic effect of supported nickel catalyst with intumescent flame-retardants on flame retardancy and thermal stability of polypropylene', *J. Appl. Polym. Sci.*, 2006, **102**, 5988–5993.
30. Xie F, Wang Y Z, Yang B and Liu Y, 'A novel intumescent flame-retardant polyethylene system', *Macromol. Mater. Eng.*, 2006, **291**, 247–253.
31. Shehata A B, Hassan M A and Darwish N A, 'Kaolin modified with new resin-iron chelate as flame retardant system for polypropylene', *J. Appl. Polym. Sci.*, 2004, **92**, 3119–3125.
32. Shehata A B, 'A new cobalt chelate as flame retardant for polypropylene filled with magnesium hydroxide', *Polym. Degrad. Stab.*, 2004, **85**, 577–582.
33. Chantarasiri N, Chulamanee C, Mananunsap T and Muangsin N, 'Thermally stable metal-containing polyureas from hexadentate schiff base metal complexes and diisocyanates', *Polym. Degrad. Stab.*, 2004, **86**, 505–513.
34. Nicolas R C, Wilén C E, Roth M, Pfaendner R and King III R E, 'Azoalkanes: a novel class of flame retardants', *Macromol. Rapid Comm.*, 2006, **27**, 976–981.
35. Chen K H and Hsien C H, 'Flame retardant composition', US Pat. 20050197440, 2005.
36. Atikler U, Demir H, Tokatlı F, Tıhmınlıoğlu F, Balköse D and Ülkü S, 'Optimisation of the effect of colemanite as a new synergistic agent in an intumescent system', *Polym. Degrad. Stab.*, 2006, **91**, 1563–1570.
37. Hu X P, Li W Y and Wang Y Z, 'Synthesis and characterization of a novel nitrogen-containing flame retardant', *J. Appl. Polym. Sci.*, 2004, **94**, 1556–1561.
38. Hu X P, Li Y L and Wang Y Z, 'Synergistic effect of the charring agent on the thermal and flame retardant properties of polyethylene', *Macromol. Mater. Eng.*, 2004, **289**, 208–212.
39. Li B and Xu M J, 'Effect of a novel charringefoaming agent on flame retardancy and thermal degradation of intumescent flame retardant polypropylene', *Polym. Degrad. Stab.*, 2006, **91**, 1380–1386.
40. Wang D Y, Wang Y Z, Wang J S, Chen D Q, Zhou Q, Yang B and Li W Y, 'Thermal oxidative degradation behaviours of flame-retardant copolyesters containing phosphorous linked pendent group/montmorillonite nanocomposites', *Polym. Degrad. Stab.*, 2005, **87**, 171–176.
41. Qu M H, Wang Y Z, Liu Y, Ge X G, Wang D Y and Wang C, 'Flammability and thermal degradation behaviors of phosphorus-containing copolyester/BaSO$_4$ nanocomposites', *J. Appl. Polym. Sci.*, 2006, **102**, 564–570.
42. Wang D Y, Ge X G, Wang Y Z, Wang C, Qu M H and Zhou Q, 'A novel phosphorus-containing poly(ethylene terephthalate) nanocomposite with both flame retardancy and anti-dripping effects', *Macromol. Mater. Eng.*, 2006, **291**, 638–645.
43. Chen D Q, Wang Y Z, Hu X P, Wang D Y, Qu M H and Yang B, 'Flame-retardant and anti-dripping effects of a novel char-forming flame retardant for the treatment of poly(ethylene terephthalate) fabrics', *Polym. Degrad. Stab.,* 2005, **88**, 349–356.
44. Gao F, Tong L F and Fang Z P, 'Effect of a novel phosphorus-nitrogen containing intumescent flame retardant on the fire retardancy and the thermal behaviour of poly(butylene terephthalate)', *Polym. Degrad. Stab.*, 2006, **91**, 1295–1299.
45. Balabanovich A I, Balabanovich A M and Engelmann J, 'Intumescence in

poly(butylene terephthalate): the effect of 2-methyl-1,2-oxaphospholan-5-one 2-oxide and ammonium polyphosphate', *Polym. Int.*, 2003, **52**, 1309–1314.

46. Lu S Y and Hamerton I, 'Recent developments in the chemistry of halogen-free flame retardant polymers', *Prog. Polym. Sci.*, 2002, **27**, 1661–1712.

47. Giraud S, Bourbigot S, Rochery M, Vroman I, Tighzert L, Delobel R and Poutch F, 'Flame retarded polyurea with microencapsulated ammonium phosphate for textile coating', *Polym. Degrad. Stab.*, 2005, **88**, 106–113.

48. Lu M F, Zhang S J and Yu D S, 'Study on poly(propylene)/ammonium polyphosphate composites modified by ethylene-1-octene copolymer grafted with glycidyl methacrylate', *J. Appl. Polym. Sci.*, 2004, **93**, 412–419.

49. Chang S Q, Xie T X and Yang G S, 'Morphology and mechanical properties of high-impact polystyrene/elastomer/magnesium hydroxide composites', *J. Appl. Polym. Sci.*, 2006, **102**, 5184–5190.

50. Liauw C M, Lees G C, Hurst S J, Rothon R N and Ali S, 'Effect of silane-based filler surface treatment formulation on the interfacial properties of impact modified polypropylene/magnesium hydroxide composites', *Compos. Part A – Appl. S.*, 1998, **29**, 1313–1318.

51. Chang S Q, Xie T X and Yang G S, 'Effects of polystyrene-encapsulated magnesium hydroxide on rheological and flame-retarding properties of HIPS composites', *Polym. Degrad. Stab.*, 2006, **91**, 3266–3273.

52. Kim J, Yoo S, Bae J Y, Yun H C, Hwang J and Kong B S, 'Thermal stabilities and mechanical properties of epoxy molding compounds (EMC) containing encapsulated red phosphorous', *Polym. Degrad. Stab.*, 2003, **81**, 207–213.

53. Wu Q, Lu J P and Qu B J, 'Preparation and characterization of microcapsulated red phosphorus and its flame-retardant mechanism in halogen-free flame retardant polyolefins', *Polym. Int.*, 2003, **52**, 1326–1331.

54. Liu Y and Wang Q, 'Preparation of microencapsulated red phosphorus through melamine cyanurate self-assembly and its performance in flame retardant polyamide 6', *Polym. Eng. Sci.*, 2006, **46**, 1548–1553.

55. Liu Y and Wang Q, 'Melamine cyanurate-microencapsulated red phosphorus flame retardant unreinforced and glass fiber reinforced polyamide 66', *Polym. Degrad. Stab.*, 2006, **91**, 3103–3109.

56. Saihi D, Vroman I, Giraud S and Bourbigot S, 'Microencapsulation of ammonium phosphate with a polyurethane shell part I: Coacervation technique', *React. Funct. Polym.*, 2005, **64**, 127–138.

57. Saihi D, Vroman I, Giraud S and Bourbigot S, 'Microencapsulation of ammonium phosphate with a polyurethane shell, part II. Interfacial polymerization technique', *React. Funct. Polym.*, 2006, **66**, 1118–1125.

58. Giraud S, Bourbigot S, Rochery M, Vroman I, Tighzert L, Delobel R and Poutch F, 'Flame retarded polyurea with microencapsulated ammonium phosphate for textile coating', *Polym Degrad. Stab.*, 2005, **88**, 106–113.

59. Liu M F, Liu Y and Wang Q, 'Flame-retarded poly(propylene) with melamine phosphate and pentaerythritol/polyurethane composite charring agent', *Macromol. Mater. Eng.*, 2007, **292**, 206–213.

60. Wang Y Z, *Flame-retardation design of PET fibers*, Sichuan Scientific Publishers, Chengdu, China, 1994.

61. Levchik S V and Weil E D, 'Flame retardancy of thermoplastic polyesters – a review of the recent literature', *Polym. Int.*, 2005, **54**, 11–35.

62. Kim K S, 'Phosphorus-containing polymers. I. Low temperature polycondensation of phenylphosphonic dichloride with bisphenols', *J. Appl. Polym. Sci.*, 1983, **28**,

1119–1123.
63. Masai Y, Kato Y and Fukui N (to Tokyo Spinning Co., Ltd.), 'Fireproof, thermoplastic polyester-polyaryl phosphonate composition', US Pat. 3719727, 1973.
64. Wang Y Z, Yi B, Wu B, Yang B and Liu Y, 'Thermal behaviors of flame-retardant polycarbonates containing diphenyl sulphonate and poly(sulphonyl phenylene phosphonate)', *J. Appl. Polym. Sci.*, 2003, **89**, 882–889.
65. Balabanovich A I and Engelmann J, 'Fire retardant and charring effect of poly(sulphonyldiphenylene phenylphosphonate) in poly(butylene terephthalate)', *Polym. Degrad. Stab.*, 2003, **79**, 85–92.
66. Ban D M, Wang Y Z, Yang B and Zhao G M, 'A novel non-dripping oligomeric flame retardant for polyethylene terephthalate', *Eur. Polym. J.*, 2004, **40**, 1909–1913.
67. Wang Y Z, Ban D M, Wu B and Zhao H, 'A new halogen-free flame-retardant and its application in preparing no-dripping flame-retardant polyesters', ZL 02113512.6, 2002.
68. Braun U, Knoll U, Schartel B, Hoffmann T, Pospiech D, Artner J, Ciesielski M, Döring M, Graterol R P, Sandler J K W and Altstädt V, 'Novel phosphorus-containing poly(ether sulphone)s and their blends with an epoxy resin: thermal decomposition and fire retardancy', *Macromol. Chem. Phys.*, 2006, **207**, 1501–1514.
69. Wang Y Z, Chen X T and Tang X D, 'In-situ reinforced and flame-retarded polyester composites with thermotropic liquid crystal copolyesters containing phosphorus', ZL02113315.8, 2002.
70. Wang Y Z, Chen X T and Tang X D, 'Synthesis, characterization, and thermal properties of phosphorus-containing, wholly aromatic thermotropic copolyesters', *J. Appl. Polym. Sci.*, 2002, **86**, 1278–1284.
71. Wang Y Z, Chen X T, Tang X D and Du X H, 'A new approach to simultaneous improvement of the flame retardancy, tensile strength and melt dripping of polyethylene terephthalate', *J. Mater. Chem.*, 2003, **13**, 1248–1249.
72. Du X H, Wang Y Z, Chen X T and Tang X D, 'Properties of phosphorus-containing thermotropic liquid crystal copolyester/poly(ethylene terephthalate) blends', *Polym. Degrad. Stab.*, 2005, **88**, 52–56.
73. Iji M, Kiuchi Y and Soyama M, 'Flame retardancy and heat resistance of phenol-biphenylene-type epoxy resin compound modified with benzoguanamine', *Polym. Advan. Technol.*, 2003, **14**, 638–644.
74. Liu Y L and Li S H, 'Poly(dimethylsiloxane) star polymers having nanosized silica cores', *Macromol. Rapid Comm.*, 2004, **25**, 1392–1395.
75. Li Q, Jiang P K and Wei P, 'Thermal degradation behavior of poly(propylene) with a novel silicon-containing intumescent flame retardant', *Macromol. Mater. Eng.*, 2005, **290**, 912–919.
76. Li Q, Jiang P K and Wei P, 'Studies on the properties of polypropylene with a new silicon-containing intumescent flame retardant', *J. Polym. Sci. Pol. Phys.*, 2005, **43**, 2548–2556.
77. Ranganathan T, Zilberman J, Farris R J, Coughlin E B and Emrick T, 'Synthesis and characterization of halogen-free antiflammable polyphosphonates containing 4,4'-bishydroxydeoxybenzoin', *Macromolecules*, 2006, **39**, 5974–5975.
78. Kumar R, Tyagi R, Parmar V S, Samuelson L A, Kumar J, Schoemann A, Westmoreland P R and Watterson A C, 'Biocatalytic synthesis of highly flame retardant inorganic-organic hybrid polymers', *Adv. Mater.*, 2004, **16**, 1515–1520.

79. Liu Y L, Chiu Y C and Chen T Y, 'Phosphorus-containing polyaryloxydiphenylsilanes with high flame retardance arising from a phosphorus-silicon synergistic effect', *Polym. Int.*, 2003, **52**, 1256–1261.
80. Price D, Bullett K J, Cunliffe L K, Hull T R, Milnes G J, Ebdon J R, Hunt B J and Joseph P, 'Cone calorimetry studies of polymer systems flame retarded by chemically bonded phosphorus', *Polym. Degrad. Stab.*, 2005, **88**, 74–79.
81. Wu B, Wang Y Z, Wang X L, Yang K K, Jin Y D and Zhao H, 'Kinetics of thermal oxidative degradation of phosphorus-containing flame retardant copolyesters', *Polym. Degrad. Stab.*, 2002, **76**, 401–409.
82. Birum G H and Jansen R F (to Monsanto Company), 'Production of 2-carboxyethyl(phenyl)phosphinic acid', US Pat. 4081463, 1978.
83. Hazen J R (to Hoechst Celanese Corp.), 'Process for production of 3-(hydroxyphenyl-phosphinyl)propanoic acid', US Pat. 4769182, 1988.
84. Kim J H, Ihm D W and Lee S S (to Saehan Industries Inc.), 'Method for preparation 3-(hydroxyphenylphosphinyl)propanoic acid', US Pat. 6090976, 2000.
85. Asrar J, Berger P A and Hurlbut J, 'Synthesis and characterization of a fire-retardant polyester: copolymers of ethylene terephthalate and 2-carboxyethyl(phenylphosphinic) acid', *J. Polym. Sci. Pol. Chem.*, 1999, **37**, 3119–3128.
86. Sato M, Endo S, Araki Y, Matsuoka G, Gyobu S and Takeuchi H, 'The flame-retardant polyester fiber: improvement of hydrolysis resistance', *J. Appl. Polym. Sci.*, 2000, **78**(5), 1134–1138.
87. Chang S J and Chang F C, 'Synthesis and characterization of copolyesters containing the phosphorus linking pendent groups', *J. Appl. Polym. Sci.*, 1999, **72**, 109–122.
88. Zhao H, Wang Y Z, Wang D Y, Wu B, Chen D Q, Wang X L and Yang K K, 'Kinetics of thermal degradation of flame retardant copolyesters containing phosphorus linked pendent groups', *Polym. Degrad. Stab.*, 2003, **80**, 135–140.
89. Cai S X and Lin C H, 'Flame-retardant epoxy resins with high glass-transition temperatures from a novel trifunctional curing agent: dopotriol', *J. Polym. Sci. Pol. Chem.*, 2005, **43**, 2862–2873.
90. Ribera G, Mercado L A, Galià M and Cádiz V, 'Flame retardant epoxy resins based on diglycidyl ether of isobutyl bis(hydroxypropyl)phosphine oxide', *J. Appl. Polym. Sci.*, 2006, **99**, 1367–1373.
91. Wu C S and Liu Y L, 'Preparation and properties of epoxy/amine hybrid resins from in situ polymerization', *J. Polym. Sci. Pol. Chem.*, 2004, **42**, 1868–1875.
92. Ebdon J R, Hunt B J and Joseph P, 'Thermal degradation and flammability characteristics of some polystyrenes and poly(methyl methacrylate)s chemically modified with silicon-containing groups', *Polym. Degrad. Stab.*, 2004, **83**, 181–185.
93. Liu Y L and Chang G P, 'Novel approach to preparing epoxy/polyhedral oligometric silsesquioxane hybrid materials possessing high mass fractions of polyhedral oligometric silsesquioxane and good homogeneity', *J. Polym. Sci. Pol. Chem.*, 2006, **44**, 1869–1876.
94. Mercado L A, Galia M and Reina J A, 'Silicon-containing flame retardant epoxy resins: Synthesis, characterization and properties', *Polym. Degrad. Stab.*, 2006, **91**, 2588–2594.
95. Mercado L A, Reina J A and Galià M, 'Flame retardant epoxy resins based on diglycidyl-oxymethylphenylsilane', *J. Polym. Sci. Pol. Chem.*, 2006, **44**, 5580–5587.
96. Martín C, Ronda J C and Cádiz V, 'Boron-containing novolac resins as flame

retardant materials', *Polym. Degrad. Stab.*, 2006, **91**, 747–754.
97. Martín C, Ronda J C and Cádiz V, 'Development of novel flame-retardant thermosets based on boron-modified phenol–formaldehyde resins', *J. Polym. Sci. Pol. Chem.*, 2006, **44**, 3503–3512.
98. Martín C, Lligadas G, Ronda J C, Galià M and Cádiz V, 'Synthesis of novel boron-containing epoxy–novolac resins and properties of cured products', *J. Polym. Sci. Pol. Chem.*, 2006, **44**, 6332–6344.

5
Nanocomposites I: Current developments in nanocomposites as novel flame retardants

C A WILKIE, Marquette University, USA and
A B MORGAN, University of Dayton Research Institute, USA

Abstract: This chapter discusses the variety of nano-dimensional materials that can be used as fire retardants with and without other traditional flame retardant additives. Emphasis is placed on the 1-dimensional (layered) materials, such as Montmorillonite, since this is where the majority of the work has been done. The evaluation of the type of nanoparticle dispersion as well as fire retardancy assessment is covered for both nano-dimensional materials used alone and used in combination with other materials.

Key words: polymer-clay nanocomposites, mechanisms of fire retardancy, evaluation of fire retardancy, combinations of nano-dimensional materials with putative fire retardants.

5.1 Introduction

Composites are broadly defined as two materials combined to yield a stronger material or part. For polymeric materials, fibre reinforced polymers are a widely used composite in electronic, aerospace, building, and transportation applications. However, like all polymeric materials, these fibre reinforced composites are flammable, even if the fibre replaces some of the total fuel load in the composite part. Indeed, the flammability of fibre reinforced polymers has generated additional research and regulatory scrutiny as these materials replace more and more metal or inorganic materials.[1,2]

Traditional composites are composed of three components: bulk polymer, reinforcing part (fibre type, dimensions and alignment; weave type and rigidity), and interfacial polymer. Bulk polymer dominates most of the properties while the reinforcing part enhances properties, or gives some additional property to the entire part/structure that would not be present with either component in use by itself. The interfacial polymer (the polymer bridging the gap between reinforcement and bulk polymer) is only a very minor component but it plays an important role. Specifically, the interfacial polymer allows the two dissimilar components to hold together and yield an effective part/structure. A poor interface will yield early structural failure when the part is under mechanical load, or will simply fail to deliver the expected enhancement of properties. To this end there is a great deal of research, both scientific and empirical, on the development of good interfaces for traditional polymers.

A newer type of composite material is becoming more widespread, the polymer nanocomposite. Unlike the traditional composite described above, a polymer nanocomposite is composed of only two components: nanoparticle and interfacial polymer. As with a normal composite, two dissimilar materials are combined to yield a material with superior properties, but the polymer nanocomposite has a very different structure due to the nanoscale size of the 'reinforcing' particle. The term reinforcing is put in quotes as in some cases the nanoparticle may not actually reinforce the polymer matrix, rather it may enhance properties by changing the properties of the interfacial polymer, or by changing the bulk polymer property in ways completely unexpected.

A polymer nanocomposite, when properly formed, has nanoparticles evenly dispersed throughout the polymer such that the polymer chains present an interface only with the nanoparticle; no bulk polymer exists. The type of nanoparticle geometry and amount greatly affect nanoparticle dispersion and nanocomposite formation, and there are numerous references on the theory and resulting structures of these nanoparticles,[3,4] some of which will be discussed later in this chapter. It must be made clear that, unlike a traditional composite where bulk polymer properties dominate, in a polymer nanocomposite the interfacial polymer properties still have a major effect on properties, but the nanoparticle plays an equally important role. As will be described later in this chapter, the type of nanoparticle greatly determines whether or not flammability reductions will be obtained, as well as other property enhancements in the mechanical, thermal, and electrical regimes.

The purpose of this chapter is to highlight the enhanced fire properties which may be ascribed to nanocomposite formation. The observation that is most commonly made is that the peak heat release rate (PHRR), which is a measure of the size of the fire, is significantly reduced for many polymers upon nanocomposite formation. The peak heat release rate is one of the parameters that may be measured using the cone calorimeter, which is the usual laboratory evaluation technique. This reduction in PHRR is considered to be related to the change in mass loss rate so that the heat evolved is changed because the mass of material which burns at any moment is decreased. It is also observed that the total heat released is constant, which means that all of the polymer will eventually burn but at a reduced rate, thus burning is not prevented but only delayed.

Another significant observation is that nanocomposites seem to ignite more easily than does the respective virgin polymer. While there is no consensus on the meaning of this as of yet, a few ideas have been posited:

1. the clay usually contains an organic modifier which will undergo degradation and perhaps this is the cause of the early degradation; and
2. perhaps the clay particles have a higher ability to absorb heat than does the polymer and this will create hot spots throughout the polymer (since the clay

is extremely well-dispersed) so that ignition occurs at many places at the same time.

In order to evaluate the first possibility, it would be necessary to have the same polymer with a well-dispersed clay containing a degradable and volatile surfactant and another in which the surfactant cannot degrade and volatilize and this may not be possible, since a surfactant-free clay will not disperse well in organophilic polymers. Another approach is to change the amount of surfactant; since the normal commercial organically modified clays contain about 30 wt% surfactant and the nanocomposite typically contains 3 wt% or less organically modified clay, there is less than 1 wt% surfactant present in the typical nanocomposite. In one study the clay amount has been varied between 0.1 wt% and 5 wt%, which means that the organic content varied between 0.03 wt% and 1.5 wt% and no systematic increase in the time to ignition has been seen.[5] The purpose of this work was not to study the time to ignition but the data has been reported; a more systematic study in which all of the parameters are controlled must be undertaken to verify this observation. The second suggested process – or some other – seems more likely, especially since an earlier time to ignition has been observed in the presence of other additives as well. The pertinent thermodynamic properties of the additives and the polymers must be available in order to determine if this is really a possibility or not. This may prove difficult in the case of the organically modified clays, since these may be dependent upon the aggregation, or lack thereof, of the clay platelets. Since this early time to ignition is always cited as one of the ways in which nanocomposites do not live up to their expectations, it is important for the future of the field that this be well understood.

5.2 Brief history of polymer nanocomposites in flame retardancy applications

Nanocomposite science, if we can use that term, had its beginning with the work of Blumstein more than forty years ago.[6–15] At that time, it was not realized that they had nano-sized dimensions, rather Blumstein synthesized polymers in the presence of clays and observed some unusual properties. The realization that these materials were in some way special began with work from scientists at Toyota who synthesized polyamide 6 in the presence of 5 mass% clay and discovered an increase of 40% in tensile strength, 68% in tensile modulus, 60% in flexural strength, and 126% in flexural modulus, while the heat distortion temperature increases from 65 to 152 °C and the impact strength is lowered by only 10%.[16] In addition to these enhanced mechanical properties, one also finds that nanocomposites are optically clear and that they exhibit enhanced barrier properties so that diffusion of, for example, oxygen through the material is greatly slowed.

5.3 Nanoparticles for polymer nanocomposite use

Nanoparticles used in polymer nanocomposite synthesis can be widely variable, as most of the periodic table is available for use in construction of various nanoparticles. However, just about all nanoparticles can be classified by geometry, rather than by chemistry. Using this classification scheme, each type of nanoparticle can be described by the number of nanoscale dimensions present: 1D, 2D and 3D. These broad geometrical classes are layered materials (1D), tubes/rods (2D), and spherical/colloidal solids (3D). More details on specific chemistries that have been used for flame retardant polymer nanocomposites in each of these classes will be described below. Before proceeding with this discussion, however, it should be pointed out that this list is not comprehensive, and while most polymer nanocomposites are likely to exhibit some reductions in flammability, the majority of the published nanocomposite materials have never been tested for flammability properties to prove this hypothesis.

5.3.1 1D Nanocomposites (layered materials)

Layered materials are the most commonly used type of nanoparticle in polymer nanocomposite synthesis. Primarily these layered materials have been layered silicates, or clays. Most clays are 2:1 smectite layered silicates, meaning that there are 2 SiO_4 tetrahedral layers sandwiching 1 MO_6 octahedral layer, where M is most commonly aluminium or magnesium. The most commonly used layered silicate in polymer nanocomposite use is montmorillonite (MMT), which has the general formula of $M_x(Al_{4-x}Mg_x)Si_8O_{20}(OH)_4$. Isomorphous substitution of lower charge ions for higher charge means that the clay layers have a negative charge which is balanced by cations in the gallery space between the layers. Other types of clays that have been used to produce polymer nanocomposites are laponite $[Na_{0.7}[(Si_8Mg_{5.5}Li_{0.3})O_{20}(OH)_4]^{0.7}]$,[17] hectorite $[M_x(Mg_{6-x}Li_x)Si_8O_{20}(OH)_4]$,[17] saponite $[M_xMg_6(Si_{8-x}Al_x)O_{20}(OH)_4]$ and magadiite $[NaSi_7O_{13}(OH)_3 4(H_2O)]$,[17] although the last example is a layered silica rather than a layered silicate.

Another type of layered material that can be used is the layered double hydroxide (LDH), which has a very different structure than the layered silicates; the typical LDH has the Brucite [naturally occurring $Mg(OH)_2$] structure in which some aluminium ion substitutes for magnesium ion to give an excess of positive charge on the clay layers which must be balanced by anions in the gallery space. The idealized formula of an LDH is $[M^{2+}_{1-x}M^{3+}_x(OH)_2]^{x+}A_x^- zH_2O$. Since the range of metals which can be used is quite large, including magnesium, aluminium, iron, cobalt, copper, nickel, zinc, lithium, and many others, one has the opportunity to change the metal (M) in order to obtain the desired result. Also, the range of anions (A^-) which can be incorporated is very large, which means that one can also study the effect of the anion on a particular property. Related to the layered double hydroxides are the hydroxy double salts (HDS), with a general formula, $[(M^{2+}_{1-x},Me^{2+}_{1+x})(OH)_{3(1-y)}]^{+}A^{n-}_{(1+3y)/n}\bullet zH_2O]$; these are similar to the LDH systems except that the metal hydroxide layer contains two divalent materials, M^{2+} and Me^{2+}. Double layered hydroxides have the added advantage of being able to release significant amounts of water upon reaching burning temperatures, making it possibly more effective than just clay alone in flame retardant polymer nanocomposite applications.[18–20]

In addition to the above-mentioned layered materials, there also exists a wealth of other layered inorganic materials which have not yet been investigated on a serious level for polymer nanocomposite formation. They will, no doubt, prove difficult to disperse but they may prove more effective than the currently used materials if the correct combination of structural features can be obtained. These include such materials as the various layered sulfates, and more commonly layered phosphates such as zirconium phosphate.[21–24]

A final layered material to consider in nanocomposite synthesis is the carbon-based material, graphite, which is an example of a one-dimensional layered material that has been used to prepare nanocomposites. There are a wide variety of so-called graphite intercalation compounds, in which various materials, including alkali metals, halogens and sulfuric acid have been inserted.[25,26] Typically, upon insertion of the intercalant, the spacing between the graphite layers increases. Work has been reported using potassium-graphite, graphite-sulfuric acid (expandable graphite) and graphite oxide.[27–33]

The two inorganic layered materials (smectite clays and layered double hydroxide (LDH)/hydroxide layered salts (HDS)), are both considered to be clays but they have different chemical structures and thermal properties. Despite those differences, both require an organic 'treatment' for them to become miscible with a polymer matrix to yield a polymer nanocomposite. This treatment is either a cationic organic molecule (in the case of the smectites) or an anionic organic molecule (in the case of layered double hydroxides and hydroxy layered salts). It is this organically functionalized layered material that is actually used in nanocomposite synthesis, not the inorganic particle, as the inorganic particle will not produce good polymer nanocomposite structures. There are some exceptions to this rule, namely when the polymer matrix in use is very polar, but these are specialized cases and so when considering layered materials in nanocomposite synthesis or for flame retardancy applications, one must consider organically functionalized nanoparticles, or organoclays.

The typical organic treatment that is applied to a smectite is an ammonium salt, but this has been generalized as 'onium' salts, which may include phosphonium and related species. In order to render the gallery space sufficiently organophilic, it is usually stated that there must be at least one long chain with a minimum of twelve carbons.[34–36] In fact, the chain is usually longer, 16 or 18

5.1 A few of the commercially available surfactants used for smectites.

carbons, and, depending upon the mode of preparation of the nanocomposite, two may be required. The commercially available organically-modified clays typically contain either one or two 16 or 18 carbon chains and the other substituents may be methyl, in which case the organo clay is quite non-polar, or they may contain hydroxyethyl groups, in which case there is an increased polarity of the organoclay. The polarity of the surfactant is an important factor in the mutual miscibility of the polymer and the organoclay.[37] A few of the commercially available surfactants used to provide good polymer miscibility are shown in Fig. 5.1.

The thermal stability of phosphonium-substituted organoclays is a little higher than for ammonium-substituted clays; imidazolium-substituted clays have an even higher thermal stability.[38–40] The thermal instability of the organoclays is due to the Hofmann elimination reaction which occurs at about 200 °C for the ammonium substituted clays and brings about the loss of an alkene and an amine, leaving a surfactant with a proton as the counter ion, as shown in Fig. 5.2.

There have only been a very few instances where other cations have been used, most notably a substituted tropylium ion.[41] This system has excellent thermal stability but it is difficult to disperse in the polymer.

For the LDH and HDS systems, where the clay layers bear a positive charge, one must do anion exchange to make these materials organophilic enough to enable nanocomposite formation. The anion exchange capacity of these materials is quite high, typically more than 200 meq per 100 g, for comparison, with montmorillonite the cation exchange capacity is about 100 meq per 100 g, and it is quite difficult to fully exchange the anion. The typical anion that is used is either a carboxylate or a sulfonate.

5.2 The Hofmann elimination reaction as applied to a clay-bound surfactant.

5.3.2 2D Nanocomposites (tubes/rods)

While layered silicates are still the most commonly studied class of nanoscale fillers used in polymer nanocomposite fabrication, carbon nanotubes and nanofibres are quickly becoming more commonly used. Carbon nanotubes and nanofibres require some additional explanation here, as while they are similar in shape, they do have different morphologies and properties. Carbon nanotubes come in two general forms: single-wall carbon nanotubes (SWNT) and multi-wall carbon nanotubes (MWNT). Both of these forms are composed of a graphitic sheet, rolled up into a tube. In the case of SWNT, this is a single sheet, and in the case of MWNT, this is multiple sheets, hence the name 'multi-wall' as can be seen from high resolution TEM images. Carbon nanofibres have some variability as well, but the most commonly produced is a vapour-grown carbon nanofibre (VGCNF). VGCNF has been described as a series of stacked graphitic 'cups', or a long strip of graphitic carbon rolled up into a helix to give a tube-like shape. It is also sometimes described as a conical carbon nanofibre (CCNF); Fig. 5.3 gives some ideas on the scale and geometry of these materials.

The carbon-based tube structure of MWNT/SWNT and VGCNF materials leads to some additional property enhancements in polymer nanocomposites that

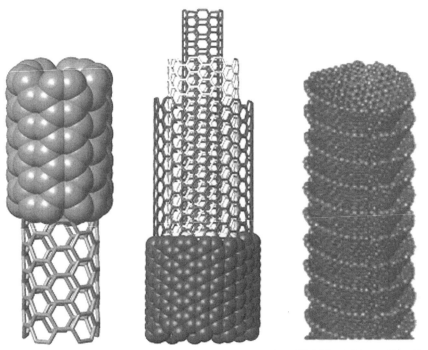

5.3 Schematics of single-wall carbon nanotubes (SWNT – left), multi-wall carbon nanotubes (MWNT – middle) and vapour grown carbon nanofibres (VGCNF – right).

are not typically seen in layered silicate (clay) nanocomposites. One of the most common improvements is electrical conductivity. Since the graphitic carbon structure provides electrical conductivity in the nanotube/nanofibre structure, setting up a network structure throughout the polymer matrix in effect creates a conductive pathway for electrons through the polymer matrix, yielding electrical conductivity to a polymer which may normally be non-conductive. This network structure that is set up in polymer matrices with the nanotubes/fibres is noteworthy as it also is responsible for the improvements in mechanical,[42] thermal[43] and flammability[44] properties that have been measured or observed for these types of polymer nanocomposites.

In addition to the above mentioned nano-materials, there are a wealth of other layered materials which also adopt tube/rod shapes which have either not been investigated with polymers or have only been very lightly investigated. These include such materials as WS_2, MoS_2, TiO_2, BN, and others.[45]

5.3.3 3D Nanocomposites (spheres/colloids)

An example of a three dimensional nano-material is polyhedral oligomeric silsesquioxane (POSS), with the general formula $(RSiO1.5)_n$; a structure of a POSS is shown in Fig. 5.4. From one to all eight of the R groups can be reactive but the most common is the presence of only a single reactive group, denoted as X in the figure. The diameter is on the order of 15 X and they are easily incorporated into the polymer matrix, especially if the X group contains a polymerizable substituent.

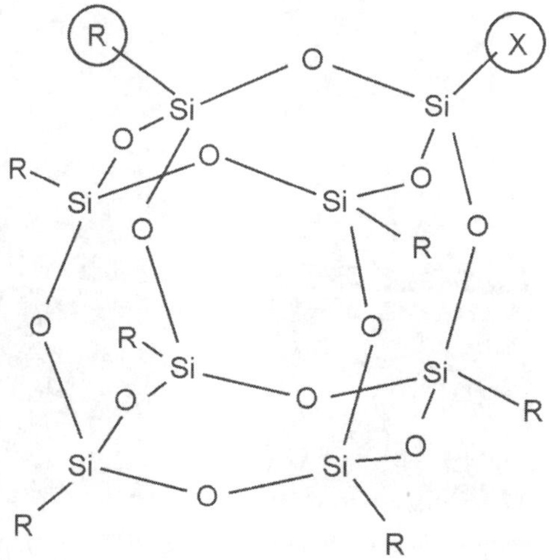

5.4 A generalized structure of a POSS material.

Current developments in nanocomposites as novel flame retardants   103

5.4 Polymer nanocomposite formation

The polymer nanocomposites may either be formed by a polymerization step, in which one permits the monomer (and/or solvent) to penetrate between the clay layers, followed by a polymerization, or by a blending operation in which one begins with the polymer and, either in the melt or in solution, the polymer is permitted to insert between the clay layers. As a generalization, one can say that the requirements on the organoclay are more severe for the blending operation, since it will be more difficult to insert the larger polymer than the smaller monomer. In the case of styrene, only one long chain is required on the surfactant for polymerization but two are required for blending.

It is usually considered that there are three morphologies possible for nanocomposites: immiscible (also known as a microcomposite), intercalated and exfoliated (also known as delaminated). In an immiscible system, the clay is not well-dispersed and rather one finds large bundles of clay in various locations and only polymer in others; this is the situation that obtains when a conventional filler is used. An intercalated system shows good overall dispersion of the clay in the polymer, there are no, or at most only a few, locations where the clay is aggregated and the registry between the clay layers is maintained. In an exfoliated system, once again the degree of dispersion is excellent with no congregations of clay particles at one spot, and the registry that exists between the clay layers in the native clay has been lost and the layers are randomly dispersed. These situations are depicted in Fig. 5.5.

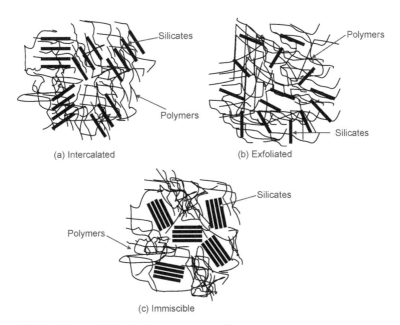

5.5 Possible morphologies for nanocomposites.

The morphology seems to be more important for enhanced mechanical and barrier properties than it does for fire retardancy. Most work has not shown a difference in cone calorimetry measurements between intercalated and exfoliated systems. Since it is very rare to have any nanocomposite that is purely one morphology, it is somewhat hazardous to state which morphology is best. The assignment of a morphology to a nanocomposite, as will be covered shortly, is not easy and so many are mis-identified in the literature.

5.4.1 Determination of morphology

The most commonly used method to determine the morphology of a nanocomposite is through the use of the combination of X-ray diffraction (XRD) and transmission electron microscopy (TEM). In a pristine organoclay, the regular structure means that the material will diffract and hence one can obtain an XRD trace which, by the use of Bragg's Law, can give the d-spacing, which is the distance between the clay layers and also the size of the gallery space. In a very simplistic system, as depicted in Fig. 5.6, the d-spacing will be unchanged for a microcomposite, increased for an intercalated nanocomposite and no peak will be seen for an exfoliated nanocomposite. As noted above, this is a great oversimplification since disordering in a microcomposite may occur, which also means that the peak vanishes; the dangers of using only XRD to evaluate morphology have been described.[46]

The normal complement to XRD is TEM, which permits the imaging of the material so that the morphology can be identified. Since only a very small portion of the material is imaged, the results may be misleading unless the material is statistically sampled or image analysis is performed to better define the morphology. Since this seems to be only rarely done, for the obvious reason that TEM is quite demanding in terms of instrumentation and analysis labour, many

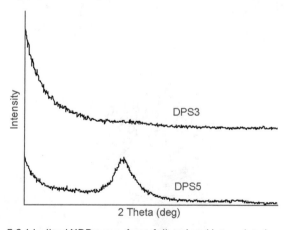

5.6 Idealized XRD traces for exfoliated and intercalated nanocomposites.

Current developments in nanocomposites as novel flame retardants 105

times nanocomposites are not correctly identified. It is doubtful if a nanocomposite which is purely one morphology has ever been obtained; most often two, or even all three, types will be present in a single material. The TEM images of a largely intercalated and a largely exfoliated nanocomposite are shown in Fig. 5.7.

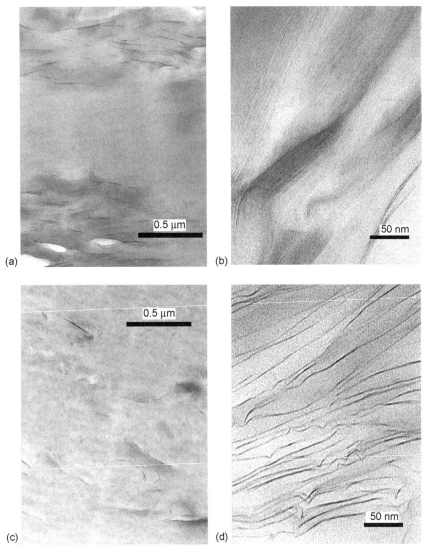

5.7 TEM Images of polymer nanocomposites: (a) and (b) an intercalated polystyrene nanocomposite; (c) and (d) a delaminated polystyrene nanocomposite. On the left is the low magnification image which shows if the clay is well-dispersed while on the right is the higher magnification image which shows the actual state of the clay dispersion. Reprinted with permission from *Chem Mater*, 13, 3774–3780 (2001). Copyright 2001, American Chemical Society.

A variety of other techniques have been identified for determining morphology, including infrared spectroscopy,[47] nuclear magnetic resonance spectroscopy,[48] atomic force microscopy[49] and rheological measurements.[50–53] There is a large need for a bulk measurement of morphology and all of these techniques sample a much larger amount of sample than does TEM and thus they may provide a better description of the morphology but they must be calibrated against the XRD + TEM standard.

5.5 Polymer nanocomposite flammability – measured effects without conventional flame retardants

Material flammability can be difficult to define precisely, as the degree of flammability is determined by which test method was used to measure that flammability and each test measures a different aspect of flammability. For example, limiting oxygen index (LOI)[54] measures flammability by determining the percentage oxygen necessary to sustain a candle-like flame whereas the cone calorimeter[55,56] measures flammability by oxygen consumption calorimetry, providing heat release rates as a function of time. Therefore one must understand how the particular test is measuring the flammability of a material before one can state what the real flammability of a material is. From the above example, it should be obvious that one cannot compare the flammability of two different materials tested by two different techniques; flammability of any material must be compared using the same test data. However, this becomes problematic in that most fire tests are based upon addressing the flammability of a material in a particular fire risk scenario, and so the scientific usefulness of regulatory tests in describing material flammability becomes suspect. Certainly one should have regulatory test data, (such as the UL 94 protocol or ASTM E-84/NFPA 262 Steiner Tunnel Test) to compare a new material to an old one for eventual real-world fire safety applications, but when it comes to scientific understanding, there are only a few tests from which to choose in understanding material flammability. The most commonly used technique is cone calorimetry.

The cone calorimeter (ASTM E-1354/ISO-5660) is an oxygen consumption calorimetry test used to mimic real-world fire scenarios in a bench scale setting by converting the oxygen consumption rates into heat release rates.[57,58] It does this very well, but does have some difficulty in realistically measuring the effects of dripping polymers on fire growth, as well as how sample orientation affects heat release.[59,60] Further, cone calorimeter data is directly affected by sample thickness and heat flux, and so data collected at one sample thickness and one heat flux cannot be compared to data collected at a different heat flux and sample thickness.[61–63] Despite these shortcomings, the cone calorimeter remains the standard test in determining the flammability performance of a material in a scientific manner, and this is why most polymer nanocomposite flammability studies have utilized the cone calorimeter as the key tool for measuring the flammability of these new flame retardant materials.

The reduction in the peak heat release rate (PHRR) is dependent upon the heat flux which is used but, it is also very polymer dependent and it may also depend upon which nano-material is used. For example, with poly(methyl methacrylate) (PMMA) the reduction is in the range of 20–30% while polystyrene (PS), high impact polystyrene (HIPS), acrylonitrile-butadiene-styrene terpolymer (ABS) and ethylene-vinyl acetate copolymer (EVA) are in the range of 40–70% and the simple polyolefins, polyethylene (PE) and polypropylene (PP) are 20–50%.[64] Similar reductions were obtained with graphite as the nano-material.[27–30] When an HDS is used as the nano-material, the dispersion is not as good as seen with the smectite or graphite, and there is no change in the PHRR with PMMA.[65] Likewise, an LDH does not give as good a dispersion as is seen with the smectites or graphite and the reductions in PHRR are not as good as those seen with these other nano-materials but they are still far beyond that which should be observed when the nano-material is not well-dispersed.[66] The heat release rate curves for polystyrene and poly(methyl methacrylate) nanocomposites are shown in Figs 5.8 and 5.9, showing the variability of the effect for the two polymers; the maximum reduction for PS is 46% while it is 30% for PMMA.[67]

Despite dramatic reductions in peak HRR, this alone is not enough to provide flame retardancy in regulatory tests.[68,69] To date, no polymer nanocomposite has shown any success in passing a regulatory test when the material is used by

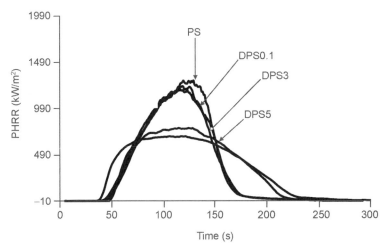

5.8 Heat release rate curves for a polystyrene nanocomposite with varying amounts of clay – the numbers are the mass% of clay present. (DPS contains a surfactant with two styryl moieties.) Reprinted with permission of John Wiley & Sons, Inc., from J. Polym. Sci.: Part A: Polym. Chem., 2003, S. Su and C.A. Wilkie.[67]

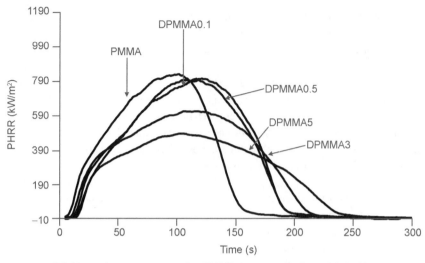

5.9 Heat release rate curves for PMMA nanocomposites with varying amounts of clay – the number are the mass% of clay present. (DPMMA contains a surfactant with two methyl methacrylate moieties.) Reprinted with permission of John Wiley & Sons, Inc., from *J. Polym. Sci.: Part A: Polym. Chem.*, 2003, S. Su and C.A. Wilkie.[67]

itself. This result does not state that polymer nanocomposites are not flame retardant materials. However, the reductions in peak HRR certainly do suggest that these are flame retardant materials, but when considering the way in which they provide flame retardancy, which will be elaborated upon in the next section, it becomes obvious that reducing mass loss rate is not enough to provide effective fire protection in all fire risk scenarios. In regulatory tests like the UL-94 vertical burn test,[70] once the polymer nanocomposite is exposed to flame, it burns very slowly, does not drip, but always burns up the entire bar, leaving behind a char in the shape of the original UL-94 burn bar. In the LOI test, polymer nanocomposites provide no improvement in LOI, which reinforces evidence for the condensed phase mechanism discussed in the next section.

5.6 Mechanism of flammability reduction

Polymer nanocomposites, whether based upon layered materials, nanotubes, or colloidal particles all reduce polymer flammability in a similar manner. While there are certainly exceptions to this brought on by very specific polymer-nanoparticle chemical reactions at elevated temperature, in general all nanoparticles reduce mass loss rate by slowing the rate at which polymeric fuel is pyrolyzed, which in turn lowers the heat release rate as the polymer burns. The underlying phenomena which is responsible for this lowered mass loss rate is the formation of protective barriers as the polymer nanocomposite decomposes, which is responsible for the lowered mass loss rate. How the protective barrier is formed and how it lowers mass loss rate is nanoparticle specific, however, and some general examples will be described for layered silicates and for carbon nanotube/nanofibre materials.

For layered silicates, a protective clay-rich barrier is formed shortly after exposure to heat/flame.[71] Recent work has suggested that the barrier may actually form prior to exposure to the flame and may arise because of the surface free energy of the clay in the polymer; it was shown by annealing the nanocomposite at relatively low temperatures, certainly below that at which burning would commence, that clay appears at the surface.[72,73] This protective barrier then slows fuel pyrolysis as polymer decomposing underneath this layer must percolate past the barrier, and this same barrier may also provide thermal protection to the underlying material.[74] This barrier only slows the rate of pyrolysis, it does not stop it and eventually essentially all of the polymer mass will be combusted under continual heat exposure. This is why total heat release is unchanged for polymer nanocomposite materials and why once a polymer nanocomposite is ignited, it will not easily self-extinguish without the help of other flame retardant additives.[68,69] The protective layer forms due to loss of the polymeric material, which results in de-intercalation of polymer between clay layers and eventually collapse of the clay layers. This effect is illustrated in data from a recent publication[74] which shows how the clay collapses as a function of heat exposure time in a gasification apparatus, which is much like a cone calorimeter except the samples are pyrolyzed in the presence of nitrogen, thus allowing clear study of condensed phase flame retardancy effects. In Fig. 5.10, the mass loss rate of the nanocomposite was sampled as a function of time by stopping the pyrolysis process and recovering the degraded samples. These samples were then sampled further by depth into the sample and x-ray diffraction data was collected on each of the layers of each sample (Fig. 5.11).

Some XPS studies looking at the concentration of silicon on the surface of a decomposing polymer clay nanocomposite[75–77] further support the conclusion that a clay-rich protective barrier is formed which provides the observed flammability reduction for this class of nanocomposite. An alternative explanation is that the clay acts as a barrier wherever it is within the polymer. The degrading radicals are prevented from escaping for a short time and this time is long enough to permit radical recombination reactions to occur, which leads to stretching the time of burning as new polymer is produced.[64] It should be noted that this mechanism also occurs with LDH and HDS systems, but with additional chemical effects from the endothermic decomposition of the hydroxides and subsequent condensed phase (charring/crosslinking) chemical reactions that are not prevalent in the layered silicate/clay systems.[18]

For carbon nanotube and nanofibre composites, the flammability reduction mechanism is brought about by the formation of a network structure of the entangled nanotubes and fibres, which like the clays, slows the mass loss rate.[78–80]

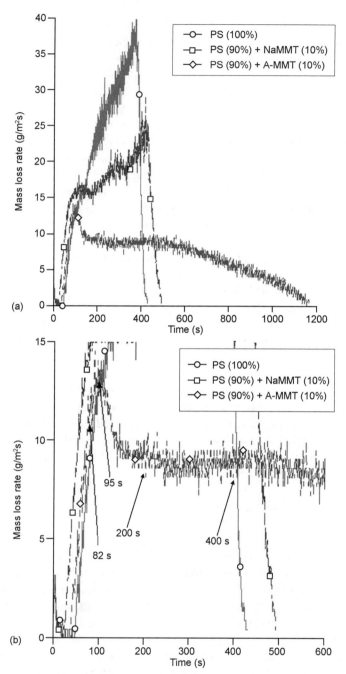

5.10 Gasification of PS nanocomposite and time that chars were sampled and analyzed. Reprinted with permission of John Wiley & Sons, Inc., from *Polym. Adv. Tech.*, 2006, JW Gilman, RH Harris, Jr., JR Shields, T Kashiwagi and AB Morgan.[74]

Current developments in nanocomposites as novel flame retardants 111

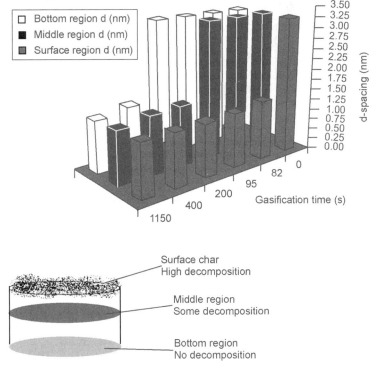

5.11 Clay d-spacing as a function of gasification time and depth in the sample. Reprinted with permission of John Wiley & Sons, Inc., from Polym. Adv. Tech., 2006, JW Gilman, RH Harris, Jr., JR Shields, T Kashiwagi and AB Morgan.[74]

In Fig. 5.12, the heat release rates of several PMMA + nanotube nanocomposites are presented, showing the significant differences of nanotube loading on flammability reduction. Alongside this HRR plot are pictures of the chars from these nanocomposites, showing the increasing amount of char formation, and density of char, as single wall nanotube (SWNT) loading is increased. SEM analysis of the char shows the network structure that led to the reinforced char and lowered heat release rate.

POSS materials have also been used to make nanocomposites and subsequently been studied for fire retardancy. One patent reports significant reduction in the PHRR for several materials which contain POSS.[81] These materials have also been examined in textiles and, while there is no reduction in the PHRR, the time to ignition is increased.[82] Finally, several recent studies have appeared on the thermal stability of POSS containing polymers, which may be a precursor to fire retardancy studies.[83-85] The mechanism of flame retardancy seems to be similar to those described for layered materials and nanotubes in that the POSS slows the rate of mass loss. It has been described in the literature that POSS materials slow the mass loss rate through creation of a network

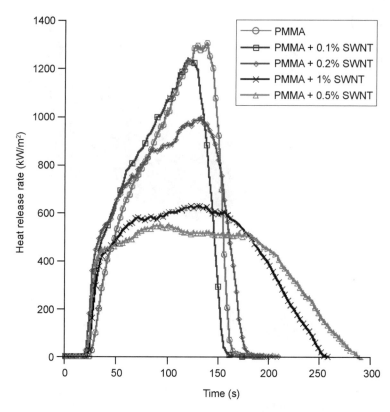

5.12 Heat release rates, char residues, and SEM char analysis of PMMA + SWNT nanocomposite. Adapted from ref. 78.

structure by hydrogen bonding between silica groups on the POSS molecule, but the mechanism of flammability reduction brought by POSS nanocomposites has not been as well studied as those of layered silicates and nanotubes, so there may be other explanations possible for these materials.[86–89]

5.7 Polymer nanocomposites combined with conventional flame retardants

Nanocomposites offer advantages in that they improve mechanical and barrier properties, reduce dripping upon burning and reduce the heat release rate, but there are also some disadvantages to their usage. The most notable seems to be that the presence of clays may make polymers less stable to ultraviolet irradiation.[90–95] Despite this, they are still often pursued since they have the potential for bringing a far better balance of properties (mechanical, thermal and flammability) to commercial and experimental polymer formulations where conventional flame retardants are used.

Current developments in nanocomposites as novel flame retardants 113

Since polymer nanocomposites by themselves only lower the heat release rate but do not self-extinguish, they have been unable to pass existing regulatory tests. There is, however, one example of a self-extinguishing epoxy-LDH nanocomposite.[18] Therefore the best approach is to use the polymer nanocomposite in combination with a conventional flame retardant to achieve regulatory fire safety performance. Indeed, this approach has been a success and has resulted in many successful formulations and materials which were recently covered in a review paper.[96] The conventional flame retardants that have been combined with nanocomposite formation include halogen-based, phosphorus-based and mineral-based ($Al(OH)_3$ and $Mg(OH)_2$) systems. The general approach has been to use the clay nanocomposite to replace part of the conventional flame retardant material on a greater than 1:1 weight ratio such that more conventional flame retardant is removed from the formulation than is replaced by the nanofiller. The resulting flame retardant nanocomposite yields a synergistic level of flame retardancy, making the flame retardant more effective than it normally would have been. Indeed, the use of a nanocomposite has resulted in systems that pass the regulatory rating where they would have failed before using just conventional flame retardant alone.[97,98] This synergistic performance is primarily noted for UL-94V systems, lowering of peak HRR in the cone calorimeter, and some flame spread tests for wire and cable, but not always for other tests, such as LOI. It is also seen broadly for almost all flame retardants (halogenated, phosphorus-based, mineral fillers, intumescents) except for two reported examples. The first was for a flame retardant that promoted polymer depolymerization/dripping as a means to pass a test. This notable example is melamine cyanurate for polyamide-6, where the strong anti-dripping behaviour brought by the clay nanocomposite actually made flammability worse for the polyamide-6 + melamine cyanurate system, which normally would have obtained a V-0 rating.[99] The other example of clay-flame retardant antagonism was in an epoxy + phosphorus flame retardant system where the clay caused poorer performance in the fire performance index and/or rises in peak heat release rate.[100] In this case the cause of the antagonism is not clear, but it may be that the clay caused some of the phosphorus flame retardant to be dispersed poorly or absorbed on the clay surface rather than present in the polymer matrix where it was needed. Despite these two examples, it is a testament to the strength of using a polymer-clay nanocomposite for flame retardant applications in combination with other flame retardants *because* only two examples of antagonism have been found. Indeed, if the clay nanocomposite brings improved fire performance and a better balance of mechanical properties, why not use it as part of a flame retardant system? Chapter 6 explores and expands upon many of the above examples and related issues in regard of the potential for nanocomposites as components within flame retardant formulations for use in a number of specific applications.

5.8 Applications for flame retardant polymer nanocomposites

Not surprisingly, flame retardant materials are used wherever an application exists that has a fire risk scenario. This means that wherever a polymeric material is used near a potential fire source (spark, open flame, or heat source), some level of fire protection will be needed so that the polymeric material does not contribute to additional flame spread or damage. The level of fire protection needed in a flame retardant material is also determined by the same fire risk scenario for which the use of the flame retardant material was originally required, and so the degree of flame retardancy will vary from material to material. Relating back to the section on how polymer nanocomposites behave by themselves in various fire tests, these same fire tests determine the level of fire safety required, and therefore, what the flame retardant formulation will be. It is important to note that there is no universal flame retardant system; what works for one polymer in one fire test will likely not work for another polymer in another test, and in many cases, even for the same polymer in another test. This point is highlighted to ensure that the reader is aware of the practice to design the flame retardant system to pass a test, not to provide universal fire safety. If the latter objective is truly the goal of the researcher, then it appears that the lower base flammability of a polymer nanocomposite is a step in the right direction.

Since polymer nanocomposites provide inherently lower flammability, they have been used successfully to produce commercial flame retardant formulations in two notable examples: wire/cable jacketing and general UL-94 V-rated plastics. In regards to wire/cable jacketing, the very first commercial flame retardant nanocomposite was produced by Kabelwerk Eupen in Europe where the clay was combined with alumina trihydrate (ATH) in ethylene-vinyl acetate copolymer (EVA) to yield a material with far better mechanical strength and flammability performance than the EVA system with ATH alone.[101,102] The superior flammability performance was noted in the UL-1666 vertical riser test where the nanocomposite containing cable jacket had slower flame spread and better char strength was noted for the nanocomposite material at the end of the test. For general UL-94 V-rated plastics, PolyOne has released a series of UL-94 V-rated polyolefins which contain organoclay nanocomposite technology under the Maxxam FR tradename. Other than this common knowledge and the fact that one can buy these materials, very little is known about these commercial materials from PolyOne other than what is reported on the manufacturer's website.

With the two above commercial examples in mind, and the previous section on polymer nanocomposite technology combined with conventional flame retardants, it should be obvious that polymer nanocomposites can be used in just about any flame retardant application where it makes economic sense to do so, or where superior fire safety may be desirable.

Current developments in nanocomposites as novel flame retardants   115

With regard to economics, some nanoparticles used in nanocomposite synthesis can actually be cheaper than the flame retardant they partially replace. This means that since the polymer nanocomposite typically replaces conventional flame retardants at a greater than 1:1 ratio, cost savings may actually be achieved with a flame retardant polymer nanocomposite formulation. Of course, as should be clear from the section on nanocomposite synthesis, polymer nanocomposite technology is not a drop-in solution. Therefore any polymer nanocomposite flame retardant formulation will have additional up-front research and development costs associated with it, perhaps more so than a traditional flame retardant formulation.

When considering superior fire safety performance, since nanocomposites lower several heat release parameters when combined with additional flame retardants, a good strategy to consider addressing strict fire safety standards would be to add a polymer nanocomposite to an existing formulation *without* replacing any of the conventional flame retardant. This will improve the flammability resistance further, and could impart additional fire safety performance in the form of anti-drip effects and enhanced thermal protection for the underlying material. With this in mind, it is very likely that there will be an increase in flame retardant nanocomposites used in mass transportation, aerospace, and military applications where fire performance standards/codes are set to very high levels. One obvious example would be in the use of fire safe composites used in naval applications. For example, in the US the DDG-1000 Zumwalt class destroyer, where most of the hull and superstructure will be composed of polymer composite and fire safety will be of utmost importance. Another example is in commercial aviation, such as composite airframes, like the ones used in the Boeing 787 and Airbus A380. These composites already pass the FAA burn through tests (otherwise they would have never reached commercialization), but the addition of polymer nanocomposite technology to these composites could improve their performance, if they are not already being used. Likewise, the use of polymer nanocomposite technology in aerospace applications may be dictated by other material property requirements, such as lightning strike resistance. By utilizing a conductive nanoparticle technology, the aerospace nanocomposite material will have enhanced flame retardancy *and* resistance to lightning strike when in flight, something that would not have been possible with existing conventional composite technology.

5.9   Future trends and conclusions

It is very likely in the near future that additional commercial polymer nanocomposites combined with conventional flame retardants will be introduced, but the formulations introduced will likely be focused on organoclay systems in polymers that melt below 200 °C. The reasons for this go back to the thermal instability of existing commercial organoclays above 200 °C, and the fact that organoclays are cost effective enough to replace flame retardant without causing major price increases for the final system. Other nanocomposite technologies (POSS, MWNT, SWNT, VGCNF) are still too expensive for commodity, or even specialty, use to justify their price. Only in applications that demand some additional property enhancement, such as electrical conductivity in a composite brought by carbon nanotubes/fibres, will flame retardant systems that utilize these more expensive nanoparticles be commercialized. Also, until improved thermal stability for organoclays is obtained, it is unlikely that flame retardant systems in engineering plastics will be introduced.

With this prediction, however, it is likely that future research in this field will focus on addressing these shortcomings. Since there had already been success in addressing the thermal stability of organoclays with imidazolium chemistry,[38] it is expected that future research will investigate imparting additional functionality to the imidazolium structure. Such functionality may be sites for polymer grafting, or perhaps flame retardant molecules that impart charring chemistry evenly dispersed throughout the interfacial polymer in the nanocomposite. On the commercial side it is hoped that sufficient interest in imidazolium chemistry will lead to demand and eventual economies of scale that would lower the price of imidazolium compounds and concurrently result in commercial imidazolium organoclays. Predicting market forces, however, are beyond the scope of this chapter and the authors' abilities. Similarly, additional breakthroughs in technology will be needed for the costs of carbon nanotubes/fibres and POSS to come down to reasonable levels so that this technology may become more widespread.

In the area of new research for nanocomposite technology, either for flame retardant applications or other areas, combinations of nanoparticles may begin to be utilized. Keeping in mind the mechanism of flame retardancy brought by layered silicates, a combination of an organoclay and another nanoparticle, such as POSS, may provide an enhanced barrier for thermal protection which would enable significant reductions in peak HRR with thinner nanocomposite components. Likewise, organoclays combined with carbon nanotubes/fibres may also impart improved thermal barriers and even lower mass loss rates which may enable even lower loadings of conventional flame retardants to be used. Of course, since flame retardancy is only one of many needed requirements in modern material applications, it may be that other nanoparticles may be combined with a layered silicate to impart additional properties that the layered silicate is incapable of providing. For example, the combination of carbon nanotubes/fibres for electrical conductivity and layered silicates for flammability may result in a superior aerospace polymer system which can resist lightning strike damage as suggested earlier.

To some extent the polymer nanocomposite field is a mature one, but new advances are occurring every day, and there are application gaps that need to be bridged before polymer nanocomposite technology becomes more widespread.

It is our belief, however, that the field will continue along at its current rapid interest rate in academia, and will continue to grow, albeit slowly, in the industrial sector until the issues of clay thermal stability and high cost of other nanofillers are addressed. To conclude, polymer nanocomposite technology today is a nearly universal flame retardant system that can be combined with other conventional flame retardants to provide superior fire safety protection and balance of properties. Certainly additional exceptions to the observed synergistic effects of polymer nanocomposites combined with other flame retardants will be discovered, but for now this technology is a potent tool for the material scientist to use in addressing the fire safety of plastics.

5.10 Acknowledgements

The help of Professor E. Manias from Penn State University in obtaining the pictures of the carbon nanotubes is acknowledged as is the assistance of Dr Takashi Kashiwagi of NIST-BFRL in showing how SWNT materials reduce flammability.

5.11 Sources of further information and advice

Clay-containing Polymeric Nanocomposites by L. A. Utracki, Rapra Technology 2004, ISBN 978-1859574379.

Polymer-Clay Nanocomposites (Wiley Series in Polymer Science) by T. J. Pinnavaia and G. W. Beall (eds), John Wiley & Sons, 2001. ISBN 978-0471637004.

'Polymer/layered silicate nanocomposites: a review from preparation to processing' Ray, S. S.; Okamoto, M. *Prog. Polym. Sci.* **2003**, 28, 1539–1641.

'Polymer-layered silicate nanocomposites: preparation, properties and uses of a new class of materials' Alexandre, M.; DuBois, P. *Mater. Sci. Eng*, **2000**, R28, 1–63.

Flame Retardant Polymer Nanocomposites by A. B. Morgan and C. Wilkie (eds), John Wiley & Sons, Hoboken, NJ, 2007. ISBN 978-0-471-73426-0.

5.12 References

1. 'Effect of fibre type on fire and mechanical behaviour of hybrid composite laminates' Kandola, B. K.; Myler, P.; Herbert, K.; Rashid, M. R. *SAMPE Fall Technical Conference Proceedings* Dallas, TX Nov 6–9, 2007.
2. 'The ASTM E84 tunnel test and fibre reinforced plastics as interior finish – an ethical crossroad' Schroeder, R.; Williamson, R. B. *Fire and Materials 2005 Conference Proceedings* San Francisco, CA Jan 31–Feb 1, 2005.
3. 'Numerical identification of the potential of whisker- and platelet-filled polymers' Gusev, A. A. *Macromolecules* **2001**, 34, 3081–3093.
4. 'Micromechanics of nanocomposites: comparison of tensile and compressive elastic moduli, and prediction of effects of incomplete exfoliation and imperfect alignment on modulus' Brune, D. A.; Bicerano, J. *Polymer* **2002**, 42, 369–387.
5. 'Studies on the mechanism by which the formation of nanocomposites enhances thermal stability' Zhu, J.; Uhl, F. M.; Morgan, A. B.; Wilkie, C. A. *Chem. Mater.* **2001**, 13, 4649–4654.

6. 'Polymerization of adsorbed monolayers. I. Preparation of the clay-polymer complex' Blumstein, A. *J. Polym. Sci. Part A.* **1963**, 3, 2653–2664.
7. 'Polymerization of adsorbed monolayers. II. Thermal degradation of the inserted polymer' Blumstein, A. *J. Polym. Sci. Part A.* **1963**, 3, 2665–2672.
8. 'Polymerization of adsorbed monolayers. III. Preliminary structure studies in dilute solution of the insertion polymers' Blumstein, A.; Billmeyer, Jr., F. W. *J. Polym. Sci.* Part A-2. **1966**, 4, 465–474.
9. 'Association I: Two-dimensionally crosslinked poly(methyl methacrylate)' Blumstein, A.; Blumstein, R. *J. Polym. Sci. Polym. Lett.* **1967**, 5, 691–696.
10. 'Branching in poly(methyl methacrylate) obtained by γ-ray irradiation' Blumstein, A.; *J. Polym. Sci. Polym. Lett.* **1967**, 5, 687–690.
11. 'Tacticity of poly(methyl methacrylate) prepared by radical polymerization within a monolayer of methyl methacrylate adsorbed on montmorillonite' *J. Polym. Sci. Polym. Lett.* **1968**, 6, 69–74.
12. 'Polymerization of adsorbed monolayers. IV. The two-dimensional structure of insertion polymers' Blumstein, A.; Blumstein, R.; Vanderspurt, T. H.; *J. Colloid Surf. Sci.* **1969**, 31, 236–247.
13. 'Polymerization of adsorbed monolayers. V. Tacticity of the insertion poly(methyl methacrylate)' Blumstein, A.; Malhotra, S. I.; Watterson, A. C. *J. Polym. Sci. Part A-2* **1970**, 8, 1599–1615.
14. 'Polymerization of monolayers. VI. Influence of the nature of the exchangeable ion on the tacticity of insertion (poly(methyl methacrylate)' *J. Polym. Sci. Part A-2* **1971**, 9, 1681–1691.
15. 'Polymerization of monolayers. VII. Influence of the exchangeable cation on the polymerization rate of methylmethacrylate monolayers adsorbed on montmorillonite' Malhotra, S.L.; Parikh, K. K.; Blumstein, A. *J. Colloid Surf. Sci.* **1972**, 41, 318–327.
16. 'Mechanical properties of a nylon 6 hybrid' Kojima, Y.; Usuki, A.; Kawasumi, M.; Okada, A.; Fukushima, Y.; Kurauchi, T.; Kamigaito, O. *J. Mater. Res.* **1993**, 8, 1185–1189.
17. 'Synthetic, layered nanoparticles for polymeric nanocomposites (PNCS)' Utracki, L. A.; Sephr, M.; Boccaleri, E. *Polym. Adv. Tech.* 2007, **18**, 1–37.
18. 'Preparation and flame resistance properties of revolutionary self-extinguishing epoxy nanocomposites based on layered double hydroxides' Zammarano, M.; Franceschi, M.; Bellayer, S.; Gilman, J. W.; Meriani, S. *Polymer* **2005**, 46, 9314–9328.
19. 'Synthesis, flame-retardant and smoke-suppressant properties of a borate-intercalated layered double hydroxide' Shi, L.; Li, D.; Wang, J.; Li, S.; Evans, D. G.; Duan, X. *Clays and Clay Minerals* **2005**, 53, 294–300.
20. 'New nanocomposites constituted of polyethylene and organically modified ZnAl-hydrotalcites' Contantino, U.; Gallipoli, A.; Nocchetti, M.; Camino, G.; Bellucci, F.; Frache, A. *Polym. Degrad. Stab.* **2005**, 90, 586–590.
21. 'Thermal and mechanical properties of poly(ethylene terephthalate)/lamellar zirconium phosphate nanocomposites' Brandao, L. S.; Mendes, L. C.; Medeiros, M. E.; Sirelli, L.; Dias, M. L. *J. App. Polym. Sci.* **2006**, 102, 3868–3876.
22. 'Epoxy nanocomposites based on the synthetic alpha-zirconium phosphate layer structure' Sue, H. J.; Gam, K. T.; Bestaoui, N.; Spurr, N.; Clearfield, A. *Chem. Mater.* **2004**, 16, 242–249.
23. 'Effect of nanoplatelet aspect ratio on mechanical properties of epoxy nanocomposites' Boo, W-J.; Sun, L.; Warren, G. L.; Moghbelli, E.; Pham, H.;

Clearfield, A.; Sue, H-J. *Polymer* **2007**, 48, 1075–1082.
24. 'Covalently linked nanocomposites: poly(methyl methacrylate) brushes grafted from zirconium phosphate' Burkett, S. L.; Ko, N.; Stern, N. D.; Caissie, J. A.; Sengupta, D. *Chem. Mater.* **2006**, 18, 5137–5143.
25. 'Graphite intercalation compound,' Rüdorff, W. *Adv. Inorg. Chem. Radiochem.* **1959**, 1, 223–266.
26. 'Interstitial compounds of graphite' Hennig, G. H. *Prog. Inorg. Chem.* **1959**, 1, 125–205.
27. 'Polystyrene/graphite nanocomposites: effect on thermal stability' Uhl, F. M.; Wilkie, C. A. *Polym. Degrad. Stab.* **2002**, 76, 111–122.
28. 'Preparation of nanocomposites from styrene and modified graphite oxides' Uhl, F. M.; Wilkie, C.A. *Polym. Degrad. Stab.* **2004**, 84, 215–226.
29. 'Formation of nanocomposites of styrene and its copolymers using graphite as the nanomaterial' Uhl, F. M.; Yao, Q.; Wilkie, C. A. *Polymers Adv. Tech.* **2005**, 16, 533–540.
30. 'Expandable graphite/polyamide-6 nanocomposites,' Uhl, F. M.; Yao, Q.; Nakajima, H.; Manias, E.; Wilkie, C. A. *Polym. Degrad. Stab.* **2005**, 89, 70–84.
31. 'Preparation and characterization of polystyrene/graphite oxide nanocomposite by emulsion polymerization' Ding, R.; Hu, Y.; Gui, Z.; Zong, R.; Chen, Z.; Fan, W. *Polym. Degrad. Stab.* **2003**, 81, 473–476.
32. 'Flammability and thermal stability studies of styrene–butyl acrylate copolymer/ graphite oxide nanocomposite' Zhang, R.; Hu, Y.; Xu, J.; Fan, W.; Chen, Z. *Polym. Degrad. Stab.* **2004**, 85, 583–588.
33. 'Mechanistic aspects of nanoeffect of poly(acrylic acid)-GO composites: TGA-FTIR studies on thermal degradation and flammability of polymer layered graphite oxide composites' Wang, J.; Han, Z.; in *Fire and Polymer IV: materials and concepts for hazard prevention,* Eds., C.A. Wilkie and G.L. Nelson, ACS Symposium Series #922, 2005, pp. 172–184.
34. 'Molecular dynamics simulation of organic-inorganic nanocomposites: layering behavior and interlayer structure of organoclays' Zeng, Q. H.; Yu, A. B.; Lu, G. Q.; Standish, R. K. *Chem. Mater.* **2003**, 15, 4732–4738.
35. 'Poly(propylene)/organoclay nanocomposite formation: Influence of compatibilizer functionality and organoclay modification' Reichert, P.; Nitz, H.; Klinke, S.; Brandsch, R.; Thomann, R.; Mulhaupt, R. *Macromol. Mater. Eng.* **2000**, 275, 8–17.
36. 'The influence of silicate modification and compatibilizers on mechanical properties and morphology of anhydride-cured epoxy nanocomposites' Zilg, C.; Thomann, R.; Finter, J.; Mulhaupt, R. *Macromol. Mater. Eng.* **2000**, 280/281, 41–46.
37. 'The relationship between the solubility parameter of polymers and the clay-dispersion in polymer/clay nanocomposites and the role of the surfactant' Jang, B. N.; Wang, D.; Wilkie, C. A. *Macromolecules* **2005**, 38, 6533–6543.
38. 'Polystyrene-clay nanocomposites prepared with polymerizable imidazolium surfactants' Bottino, F. A.; Fabbri, E.; Fragala, I. L.; Malandrino, G.; Orestano, A.; Pilati, F.; Pollicino, A. *Macromol. Rapid Commun.* **2003**, 24, 1079–1084.
39. 'Polymer/layered silicate nanocomposites from thermally stable trialkyl-imidazolium-treated montmorillonite' Gilman, J. W.; Awad, W. H.; Davis, R. D.; Shields, J.; Harris, R. H., Jr.; Davis, C.; Morgan, A. B.; Sutto, T. E.; Callahan, J.; Trulove, P. C.; DeLong, H. C. *Chem. Mater.* **2002**, 14, 3776–3785.
40. 'Melt-processable syndiotactic polystyrene/montmorillonite nanocomposites' Wang, Z. M.; Chung, T. C.; Gilman, J. W.; Manias, E, *J. Polym. Sci. Part B: Polym. Phys.* **2003**, 41, 3173–3187.

41. 'A carbocation substituted clay and its styrene nanocomposite' Zhang, J.; Wilkie, C. A. *Polym. Degrad. Stab.* **2004**, 83, 301–307.
42. 'Mechanical reinforcement of polymers using carbon nanotubes' Coleman, J. N.; Khan, U.; Gun'ko, Y. K. *Advanced Materials* **2006**, 18, 689–706.
43. 'Thermal stability of polypropylene/carbon nanofibre composite' Chatterjee, A.; Deopura, B. L. *J. App. Polym. Sci.* **2006**, 100, 3574–3578.
44. 'Nanoparticle networks reduce the flammability of polymer nanocomposites' Kashiwagi, T.; Du, F.; Douglas, J. F.; Winey, K. I.; Harris, R. H.; Shields, J. R. *Nature Materials* **2005**, 4, 928–933.
45. 'Inorganic menagerie' Halford, B. *Chem. Eng. News* **2005** (August 29), 30–33.
46. 'Characterization of polymer-layered silicate (clay) nanocomposites by transmission electron microscopy and x-ray diffraction: a comparative study' Morgan, A. B.; Gilman, J. W. *J. App. Polym. Sci.* **2003**, 87, 1329–1338.
47. 'An infrared method to assess organoclay delamination and orientation in orgqanoclay polymer nanocomposites' Ijdo, W. L.; Kemnetz, S.; Benderly, D. *Polym. Eng. Sci.* **2006**, 46, 1031–1039.
48. 'Investigation of nanodispersion in polystyrene-montmorillonite nanocomposites by solid state NMR' Bourbigot, S.; Gilman, J. W.; VanderHart, D. L.; Awad, W. H.; Davis, R. D.; Morgan, A. B.; Wilkie, C. A. *J. Polym Sci Part B: Polym Phys.* **2003**, 41, 3188–3213.
49. 'Silicon-methoxide-modified clays and their polystyrene nanocomposites' Zhu, J.; Start, P.; Mauritz, K. A.; Wilkie, C. A. *J. Polym. Sci. Part A: Polym. Chem.* **2002**, 40, 1498–1503.
50. 'A rheological method to compare the degree of exfoliation of nanocomposites' Wagener, R.; Reisinger, T. J. G. *Polymer* **2003**, 44, 7513–7518.
51. 'Rheological characterization of polystyrene-clay nanocomposites to compare the degree of exfoliation and dispersion' Zhao, J.; Morgan, A. B.; Harris, J. D. *Polymer* **2005**, 46, 8641–8660.
52. 'Soft glassy dynamics in polypropylene-clay nanocomposites' Treece, M. A.; Oberhauser, J. P. *Macromolecules* **2007**, 40, 571–582.
53. 'Ubiquity of soft glassy dynamics in polypropylene-clay nanocomposites' Treece, M. A.; Oberhauser, J. P. *Polymer* **2007**, 48, 1083–1095.
54. Limiting Oxygen Index (LOI): ASTM D2863-97 'Standard test method for measuring the minimum oxygen concentration to support candle-like combustion of plastics (oxygen index)'.
55. Cone Calorimeter: ASTM E1354 'Standard heat method for heat and visible smoke release rates for materials and products using an oxygen consumption calorimeter'.
56. 'Reaction-to-fire tests – Heat release, smoke production and mass loss rate – Part 1: Heat release (cone calorimeter method)' ISO/FDIS 5660-1.
57. 'Specimen heat fluxes for bench-scale heat release rate testing' Babrauskas, V. *Fire and Materials* **1995**, 19, 243–252.
58. 'Heat release rate: The single most important variable in fire hazard' Babrauskas, V.; Peacock, R. D. *Fire Safety Journal* **1992**, 18, 255–272.
59. 'Full scale flammability measures for electronic equipment' Bundy, M.; Ohlemiller, T. *NIST Technical Note 1461* **August 2004**, US Dept. of Commerce.
60. 'Fire retardant foam flammability' Gilman, J. W.; Ohlemiller, T. J.; Shields, J. R. *Association for the Polyurethanes Industry Proceedings – Polyurethanes 2005 Technical Conference & Trade Fair – Infinite Possibilities,* October 17–19 2005, Houston, TX.
61. 'Comprehensive fire behaviour assessment of polymeric materials based on cone

calorimeter investigations' Schartel, B.; Braun, U. *e-Polymers* **2003**, No. 13. http://www.e-polymers.org/papers/schartel_010403.pdf.
62. 'Some comments on the use of cone calorimeter data' Schartel, B.; Bartholmai, M.; Knoll, U. *Polym. Degrad. Stab.* **2005**, 88, 540–547.
63. 'Cone calorimeter analysis of ul-94 v rated plastics: qualitative correlations and heat release rate understanding' Morgan, A. B.; Bundy, M. *Fire and Materials* **2007**, 31, 257–283.
64. 'The relationship between thermal degradation behavior of polymer and the fire retardancy of polymer/clay nanocomposites' Jang, B. N.; Costache, M.; Wilkie, C. A. *Polymer* **2005**, 46, 10678–10687.
65. 'Thermal stability and degradation kinetics of poly(methyl methacrylate)/layered copper hydroxyl methacrylate composites' Kandare, E.; Deng, H.; Wang, D.; Hossenlopp, J.M. *Polym. Adv. Tech.*, **2006**, 17, 312–319.
66. 'The influence of carbon nanotubes, organically modified montmorillonites and layered double hydroxides on the thermal degradation of polyethylene, ethylene-vinyl acetate copolymer and polystyrene' Costache, M. C.; Heidecker, M. J.; Manias, E.; Camino, G.; Frache, A.; Beyer, G.; Gupta, R. K.; Wilkie, C. A. *Polymer* **2007**, 48, 6532–6545.
67. 'Exfoliated poly(methyl methacrylate) and polystyrene/ nanocomposites occur when the clay cation contains a vinyl monomer' Su, S.; Wilkie, C. A. *J. Polym. Sci: Part A: Polym. Chem.* **2003**, 41, 1124–1135.
68. 'Layered silicate polymer nanocomposites: new approach or illusion for fire retardancy? Investigations of the potentials and the tasks using a model system' Bartholmai, M.; Schartel, B. *Polymers for Advanced Technologies* **2004**, 15, 355–364.
69. 'Some comments on the main fire retardancy mechanisms in polymer nanocomposites' Schartel, B.; Bartholmai, M.; Knoll, U. *Polym. Adv. Technol.* **2006**, 17, 772–777.
70. UL-94/ASTM D3801: Test for flammability of plastic materials for parts in devices and applications.
71. 'Flammability properties of polymer-layered silicate nanocomposites. polypropylene and polystyrene nanocomposites' Gilman, J. W.; Jackson, C. L.; Morgan, A. B.; Harris, R.; Manias, E.; Giannelis, E. P.; Wuthenow, M.; Hilton, D.; Phillips, S. H. *Chem. Mater.* **2000**, 12, 1866–1873.
72. 'Nanocomposites at elevated temperatures: migration and structural changes' Lewin, M.; Pearce, E. M.; Levon, K.; Korniakov, A.; Mey-Marom, A.; Zammarano, M.; Wilkie, C. A.; Jang, B.N. *Polym. Adv. Tech.* **2006**, 17, 226–234.
73. 'Additional evidence for the migration of clay upon heating of clay-pp nanocomposites from x-ray photoelectron spectroscopy (XPS)' Hao, J.; Lewin, M.; Wilkie, C. A.; Wang, J. *Polym. Degrad. Stab.* **2006**, 91, 2482–2485.
74. 'A study of the flammability reduction mechanism of polystyrene-layered silicate nanocomposite: Layered silicate reinforced carbonaceous char' Gilman, J. W.; Harris, R. H.; Shields, J. R.; Kashiwagi, T.; Morgan, A. B *Polym. Adv. Technol.* **2006**, 17, 263–271.
75. 'An XPS study of the thermal degradation and flame retardant mechanism of polystyrene-clay nanocomposites' Wang, J.; Du, J.; Zhu, J.; Wilkie, C. A. *Polym. Degrad. Stab.* **2002**, 77, 249–252.
76. 'An XPS investigation of thermal degradation and charring on poly(vinyl chloride)-clay nanocomposites' Du, J.; Wang, D.; Wilkie, C. A.; Wang, J. *Polym. Degrad. Stab.* **2003**, 79, 319–324.

77. 'Additional XPS studies on the degradation of poly(methyl methacrylate) and polystyrene nanocomposites' Du, J.; Wang, J.; Su, S.; Wilkie, C. A. *Polym. Degrad. Stab.* **2004**, 83, 29–34.
78. 'Flammability properties of polymer nanocomposites with single-walled carbon nanotubes: effects of nanotube dispersion and concentration' Kashiwagi, T.; Du, F.; Winey, K. I.; Groth, K. M.; Shields, J. R.; Bellayer, S. P.; Kim, H.; Douglas, J. F. *Polymer* **2005**, 46, 471–481.
79. 'Thermal and flammability properties of polypropylene/carbon nanotube nanocomposites' Kashiwagi, T.; Grulke, E.; Hilding, J.; Groth, K.; Harris, R.; Butler, K.; Shields, J.; Kharchenko, S.; Douglas, J. *Polymer* **2004**, 45, 4227–4239.
80. 'Nanoparticle networks reduce the flammability of polymer nanocomposites' Kashiwagi, T.; Du, F.; Douglas, J. F.; Winey, K. I.; Harris, R. H.; Shields, J. R. *Nature Materials* **2005**, 4, 928–933.
81. 'Preceramic additives as fire retardants for plastics,' Lichtenham, J. D.; Gilman, J. W. US Patent 6,362,279, March 26, 2002.
82. 'Polyhedral oligomeric silsesquixones: application to flame retardant textiles' Bourbigot, S.; Le Bras, M.; Flambard, X.; Rochery, M.; Devaux, E.; Lichtenham, J. D. in *Fire Retardancy of polymers New applications of mineral Fillers,* Eds., Le Bras, M.; Wilkie, C. A.; Bourbigot, S.; Duquesne, S.; Jama, C. Royal Society of Chemistry, Cambridge, 2005, pp. 189–201.
83. 'Octaisobutyl POSS thermal degradation' Fina, A.; Tabuani, D.; Frache, A.; Boccaleri, E.; Camino, G. in *Fire Retardancy of polymers New applications of mineral Fillers,* Eds., Le Bras, M.; Wilkie, C. A.; Bourbigot, S.; Duquesne, S.; Jama, C. Royal Society of Chemistry, Cambridge, 2005, pp. 202–220.
84. 'Polyhedral oligomeric silsesquioxanes (POSS) thermal degradation' Fina, A.; Tabuani, D.; Carnaito, F.; Frache, A.; Boccaleri, E.; Camino, G. *Thermochim Acta* **2006**, 440, 36–42.
85. 'Polypropylene metal functionalized POSS nanocomposites: A study by thermogravimetric analysis' Fina, A.; Abbenhuis, H. C. L.; Tabuani, D.; Frache, A.; Camino, G. *Polym. Degrad. Stab.* **2006**, 91, 1064–1070.
86. 'A novel photocrosslinkable polyhedral oligomeric silsesquioxane and its nanocomposites with poly(vinyl cinnamate)' Ni, Y,; Zheng, S. *Chem. Mater.* **2004**, 16, 5141–5148.
87. 'Synthesis and thermal properties of hybrid copolymers of syndiotactic polystyrene and polyhedral oligomeric silsesquioxane' Zheng, L.; Kasi, R. M.; Farris, R. J, Coughlin, E. B. *J. Polym. Sci. Part A, Polym. Chem.* **2002**, 40, 885–891.
88. 'Polyimide/POSS nanocomposites: interfacial interaction, thermal properties and mechanical properties' Huang, J.C.; He, C. B.; Xiao, Y.; Mya, K. Y.; Dai, J.; Siow, Y. P. *Polymer* **2003** 44, 4491–4499.
89. 'Silicon-based flame retardants' Kashiwagi T.; Gilman, J. W. in Fire Retardancy of Polymeric Materials. Grand, A. F,; Wilkie, C. A., eds. New York, Marcel Dekker, **2000**, Chapt. 10: 353–389.
90. 'Photooxidation of polymeric-inorganic nanocomposites: chemical, thermal stability and fire retardancy investigations' Tidjani, A.; Wilkie, C. A. *Polym. Degrad. Stab.* **2001**, 74, 33–37.
91. 'Photodegradation of polypropylene nanocomposites' Mailhot, B.; Morlat, S.; Gardette, J.-L.; Boucard, S.; Duchet, J.; Gérard, J.-F. *Polym. Degrad. Stab.* **2003**, 82, 163–167.
92. 'Photo-oxidation of polypropylene/montmorillonite nanocomposites. 1. Influence of nanoclay and compatibilizing agent' Morlat, S.; Mailhot, B.; Gonzalez, D.;

Gardette, J.-L. *Chem. Mater.* **2004**, 16, 377–383.
93. 'Photooxidation of polypropylene/montmorillonite nanocomposites. 2. Interactions with antioxidants' Morlat-Therias, S.; Mailhot, B.; Gonzalez, D.; Gardette, J.-L. *Chem. Mater.* **2005**, 17, 1072–1078.
94. 'Photooxidation of ethylene-propylene-diene/montmorillonite nanocomposites' Morlat-Therias, S.; Mailhot, B.; Gardette, J.-L.; Da Silva, C.; Haidar, B.; Vidal, A. *Polym. Degrad. Stab.* **2005**, 90, 75–85.
95. 'Photooxidation of vulcanized EPDM/montmorillonite nanocomposites' Morlat-Therias, S.; Fanton, E.; Tomer, N. S.; Rana, S.; Singh, R. P.; Gardette, J.-L. *Polym. Degrad. Stab.* **2006**, 91, 3033–3039.
96. 'Flame retarded polymer layered silicate nanocomposites: a review of commercial and open literature systems' Morgan, A. B. *Polym. Adv. Technol.* **2006**, 17, 206–217.
97. 'Self-extinguishing polymer/organoclay nanocomposites' Si, M.; Zaitsev, V.; Goldman, M.; Frenkel, A.; Peiffer, D. G.; Weil, E.; Sokolov, J. C.; Rafailovich, M. H. *Polym. Degrad. Stab.* **2007**, 92, 86–93.
98. 'Synergy between conventional phosphorus fire retardants and organically-modified clays can lead to fire retardancy of styrenics' Chigwada, G.; Wilkie, C. A. *Polym. Degrad. Stab.* **2003**, 80, 551–557.
99. 'Preparation and combustion properties of flame retardant nylon-6/montmorillonite nanocomposite' Hu, Y.; Wang, S.; Ling, Z.; Zhuang, Y.; Chen, Z.; Fan, W. *Macromol. Mater. Eng.* **2003**, 288, 272–276.
100. 'Effect of organo-phosphorus and nano-clay materials on the thermal and fire performance of epoxy resins' Hussain, M.; Varley, R. J.; Mathys, Z.; Cheng, Y. B.; Simon, G. P. *J. App. Polym. Sci.* **2004**, 91, 1233–1253.
101. 'Flame retardant properties of EVA-nanocomposites and improvements by combination of nanofillers with aluminium trihydrate' Beyer, G. *Fire and Materials* **2001**, 25, 193–197.
102. 'Flame retardancy of nanocomposites – from research to technical products' Beyer, G. *J. Fire Sci.* **2005**, 23, 75–87.

6
Nanocomposites II: Potential applications for nanocomposite-based flame retardant systems

A R HORROCKS, University of Bolton, UK

Abstract: The potential commercial exploitation of nanocomposite-based flame retardant formulations depends on the ease with which nanoparticulate species can be introduced into polymers and polymer processing systems This depends on the nature and stability of the functionalising groups present to optimise nanodispersion, the effect of added nanoparticles on the rheology of the polymer solution, dispersion or melt and the character of the dispersion and nanocomposite structure in the final polymer. It is the difficulty in overcoming these challenges that probably explains why only very few commercially viable nanocomposite flame retardant products are as yet in the marketplace. This chapter discusses these issues in detail prior to reviewing the potential for exploitation in the bulk, fibre, textiles, film, coating, composite and foam application areas. In all cases it is evident that applications will most likely involve combinations of both nanoparticulate and conventional flame retardant to yield overall formulations that require lower total amounts of additive or agent in the final polymer in order to deliver a required degree of flame retardancy or fire performance.

Key words: nanoparticle, nanocomposite, nanoclay, nanodispersion, functionalising group, rheology, flame retardant, bulk polymer, fibre, textile, film, coating, composite, foam.

6.1   Introduction

Chapter 5 provides an overview of the present position regarding the influence that adding nanodispersed inert particles to polymers has on the fire performance of the resulting nanocomposite structures. This chapter also recorded the recent history of nanocomposite development and how this was driven by the initially observed increases in tensile strength and modulus which led to improvements in properties of engineered plastics for the automotive industry.[1,2] The subsequent observation that nanocomposites show improved ablation performance together with thermal and fire stability[3,4] has caused the significant expansion in research reviewed in Chapter 5. The motivation for this has been the need to develop new flame retardants which have equivalent or better performance, improved environmental sustainability and reduced costs when compared with the current armoury of commercially available flame retardants for polymeric substrates. However, this improved fire performance is usually concerned with reduction in heat release properties only and ease of ignition and times to self-extinguish are usually adversely affected. In fact, use of simple techniques such as limiting oxygen index show that the addition of nanoclays and other nanoparticles such as fumed silica alone do not significantly increase LOI values[5–7] unless their presence modifies the burning behaviour and melt dripping character of the polymer as observed with various montmorillonite clays dispersed in nylon 6 and 6.6 polymers.[8] Chapter 5 has also shown that the flame retardant potential of nanodispersed particles in polymeric matrices more likely lies in their ability to function additively and even synergistically with other flame retardants or the matrix itself, if inherently flame retardant. In this way, nanodispersed flame retardant formulations may be more effective than conventional ones or as effective but at significantly reduced overall additive concentrations. This has real consequences for fire retardant polymer applications in terms of possible reduced additive costs, improved physical and mechanical properties and environmental sustainability as well as improved overall fire performance. While such advantages will obviously have possible applications across the whole spectrum of polymer product types, they will be especially interesting to those applications where flame retardant minimisation and polymer physical and mechanical property optimisation are of crucial importance. Prime examples will be fibres and textiles, films, foams and composites where not only high specific surfaces are often evident but also product physical and mechanical properties are of major importance to their success in a variety of applications.

6.2 Applications and processing challenges for nanocomposite polymers

Essential issues to be considered when considering the upscaling of an otherwise successful, experiment of bench-scale nanocomposite-based formulation are nanoparticle compatibility with the polymer matrix and other additives present, the ability to maintain a particle dispersion during all processing stages and related temperatures, their influence on rheology and the possible compromise required to adjust component concentration levels to optimise both processibility and flame retardancy. In the case of nanoparticles, the need to maximise the level of nanodispersion (i.e., in terms of simple nanodispersion, intercalation and/or exfoliation) is of paramount importance although the maximisation of other flame retardant species present, whether as a microdispersed solid or liquid, must also be attempted.

Currently there are various significant methods for the preparation of nanocomposites in polymeric matrices including sol-gel formation,[9] *in-situ* polymerisation,[10] intercalative polymerisation,[1] solution blending and melt intercalation.[2] In all of these the above process-dependent factors will be crucial in enabling successful small-scale systems to be scaled up to full commercial levels.

6.2.1 Functionalising group type and stability

Of primary importance for nanocomposite structure optimisation is the compatibility of largely organophobic nanoparticulates with surrounding polymeric matrices and this is largely determined by the nature of the functionalising groups present on the former. In the case of clays, for example, hydrophobic, long chain aliphatic substituents within the quaternised functionalising complex will encourage nanoclay intercalation and exfoliation within a non-polar polymer matrix typified by the polyolefins and polystyrene. Substituents with variously polar side groups such as –OH, –NH– or NH_2 will encourage nanodispersion in polar (e.g., polymethyl methacrylate, ethylene-vinyl-acetate (EVA) copolymers) and hydrogen-bonded polymers (e.g., polyamides 6 and 6.6 and poly(vinyl alcohol).Typical examples of functionalising groups used by Southern Clay Products Inc. and competing clay products from Sud-chemie AG, Germany are shown in Table 6.1 along with polymers with which they are respectively compatible. The respective densities listed for the Cloisite clays show how the presence of functionalising species, especially those with bulky side groups, reduces clay density by increasing gallery spacings. Similar ranges of functionalised clays are commercially available from companies such as Nanocor Inc, USA and Elementis Specialities Inc., USA. Surprisingly, even unfunctionalised clays may nanodisperse in very polar polymers and recent work in our own laboratories suggests that this may be possible with nylons 6 and 6.6.[8,11]

However, and as noted initially by Gilman and coworkers,[12,13] quaternised ammonium salts with aliphatic side chains tend to decompose at temperatures in the 200–250 °C region and so will degrade during the compounding and processing of most conventional melt-processed polymers like PA6, PA6.6, PET, PS, etc. From our own recent studies, Figs 6.1(a) and 6.1(b) show the respective TGA and DTA responses in air for the unfunctionalised Cloisite Na^+ and functionalised Cloisite 10A, 15A, 25A and 30B clays having functionalising group structures defined in Table 6.1.[14] As expected, the unfunctionalised sodium montmorillonite shows very little weight loss up to 600 °C above which some dehydroxylation occurs accompanied by a small mass loss as seen from Fig. 6.1(a). However, there is a high residual mass at 800 °C and the related DTA response is featureless (see Fig. 6.2(b)). All the organically modified clays, however, show two stages of weight loss – an initially double-peaked (in the temperature range of 235–293 °C and 307–348 °C) and then a single-peaked DTG maxima (575–605 °C) – reflected by the respective DTA exotherms in Fig. 6.1(b). The first stages are considered to be due to decomposition and oxidation in air of respective organic components of the clays and the second a consequence of the dehydroxylation of the clay layers.[15] Although the TGA data shown is in air, and the behaviour under nitrogen might be expected to indicate some improved thermal stability, it is clear that functionalising groups of these

Table 6.1 Typical characteristics of commercial clays

Clay	Treatment	Modifier conc. (meq/100 g clay)	d spacing (Å)	Density (g/cm^3)	Compatible polymer examples
Southern Clays Inc. products					
Cloisite Na$^+$	None	93	11.7	2.86	PVOH, Nylons 6 and 6.6
Cloisite 10A	CH_3-N$^+$(CH$_3$)-CH$_2$-⌬, HT	125	19.2	1.90	PET, PBT, PS
Cloisite 15A	CH_3-N$^+$(CH$_3$)-HT, HT	125	31.5	1.66	PLA, EVA, PS
Cloisite 25A	CH_3-N$^+$(CH$_3$)-CH$_2$.CH(C$_2$H$_5$).(CH$_2$)$_3$.CH$_3$, HT	95	18.6	1.87	PLA, PMMA, PS
Cloisite 30B	CH_3-N$^+$(CH$_2$CH$_2$OH)-T, CH$_2$CH$_2$OH	90	18.5	1.98	EVA, epoxy, PC, PBT
Sud-chemie AG products					
Nanofil 2	Organic intercalated clays	–			PET
Nanofil 5					Grafted PE/PP
Nanofil 9					PS, PA6, PA6.6
Nanofil SE3000	Modified nanoclays	–			PE/PP
Nanofil SE3010					PA6, PA6.6, ABS, PS, PC
Elementis Specialities					
Bentone 107	dimethyl, dehydrogenated tallow quaternary ammonium modified bentonite clay				Low polarity: PE, PP, PS
Bentone 108	Organically modified hectorite clay				Low & medium polarity: PE, PP, PS, PC, ABS
Bentone 109	Organic derivative of a smectic clay			1.7	

Note: where, HT is hydrogenated (~65% C_{18}; ~30% C_{16}; ~5% C_{14}), anion: sulphate; T is tallow (~65% C_{18}; ~30% C_{16}; ~5% C_{14}), anion: chloride in 10A, 15A and 30B; sulphate in 25A

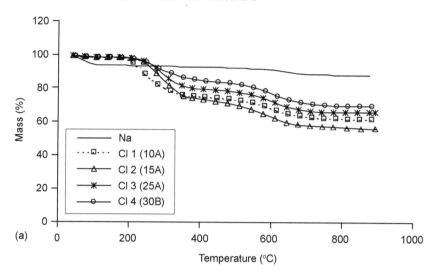

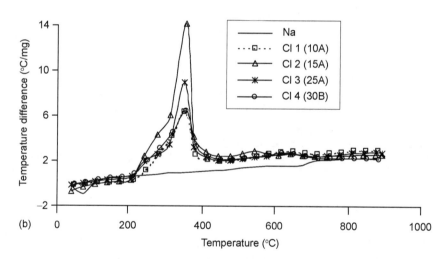

6.1 (a) TGA and (b) DTA responses for all Cloisite clays in air;[14] (reproduced by permission from The Royal Society of Chemistry).

clays would most likely be degraded during normal melt polymer processing as suggested above.

For polymers that may be compounded and processed below 200 °C such as EVA,[16] the more simple quaternary ammonium functionalising groups may be used (e.g., using dimethyl- distearylammonium salts) to generate nanocomposite polymers. Very recent work by Camino and coworkers[17] has studied the effects of thermal degradation on functionalised clays (Cloisite 30B, Southern Clay, Inc., and Nanofil 784, Nanocor, Inc.) under conditions required for PA6 processing by heating them in the range 200–250 °C. These clays contain functionalities based on methyl, tallow, bis-2-hydroxyethyl quaternary ammonium chloride and protonated ω-aminododecanoic acid, respectively, this latter being polyamide-reactive. After heat treatment of clays alone up to 250 °C, the reactivity of functionalising groups determines their stability with the latter functionalising group being more reactive than the former tallow-based moiety. While both preheated clays produce nanocomposite morphologies when polymerised *in-situ*, those containing Cloisite 30B were still exfoliated and more so than for 784-containing composites. In fact, following preheating, both functionalised clays at 350 °C enabled only microcomposite formation. These results suggest, therefore, that while functionalities may be thermally labile above 200 °C, such decomposition is insufficient to prevent nanocomposite formation when temperatures as high as 250 °C are experienced. Melt compounding both clays at 245 °C with PA6 produced exfoliated and intercalated morphologies for 30B and 784 clays respectively.

Gilman and his coworkers[12,13] have subsequently shown that layered silicate nanoparticles functionalised with higher temperature stable groups such as imidazolium derivatives and crown ethers can increase stability to temperatures in the range 262–343 °C under nitrogen compared with a typical alkyl ammonium-based salt such as dimethyl dioctadecyl ammonium bromide which starts to degrade at 225 °C. Figure 6.2 shows the TGA responses in nitrogen of dimethyl hexadecyl-imidazolium (DMHDIM) salts with different anions (Cl^-, Br^-, BF_4^-, PF_6^-) and montmorillonite (MMT) clay ion-exchanged with dimethyl hexadecyl-imidazolium (DMHDIM) salt. The dramatic increase in the thermal stability of dimethyl hexadecyl-imidazolium-intercalated montmorillonite clays is evident as is a dependence of salt stability on the type of anion present. Halide ions are seen to destabilise salts and so it is important to remove all halide residue that may contaminate the intercalated product after the ion exchange. More recently, this same group has synthesised and characterised imidazolium salts that may be used to functionalise clays for dispersion in epoxy resins.[18] The reason for this research was to address concerns that aerospace epoxy-based nanocomposite resins might show poor thermal stability as a consequence of high service life temperature exposure. The resulting montmorillonite clays functionalised with either 1-hexadecyl-2-methylimidazolium or 1-hexadecyl-3-(10-hydroxydecyl)- 2-methylimidazolium cationic groups showed good thermal stabilities up to 300 °C. The former was designed not to react with an epoxy matrix whereas the latter enables the hydroxyl group in the imidazolium moiety to catalyse polymerisation in the gallery space between the silicate layers. In terms of morphology, the reactive, functionalised clay showed slightly improved nanodispersion using XRD, TEM and for the first time, laser scanning confocal microscopy.

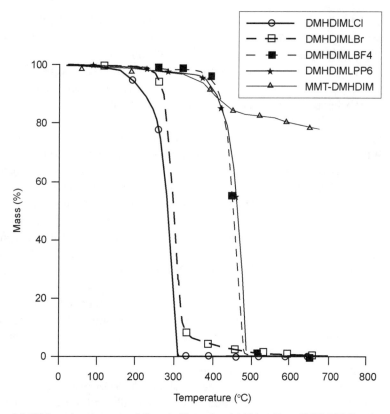

6.2 TGA under nitrogen of dimethyl hexadecyl-imidazolium (DMHDIM) salts with different anions (Cl^-, Br^-, BF_4^-, PF_6^-) and montmorillonite (MMT) clay treated with dimethyl hexadecyl-imidazolium (DMHDIM) salt[13] (Reprinted from *Thermochimica Acta*, 409(1), W H Awad, J W Gilman, M Nyden, R H Harris, Jr., T E Sutto, J Callahan, P C.Trulove, H C DeLong and M Fox, Thermal degradation studies of alkyl-imidazolium salts and their application in nanocomposites, 3–11, 2004 with permission from Elsevier).

6.2.2 Effect on rheology

Generally, the addition of a nanodispersed phase will increase the viscosity of a polymer melt under a given shear stress and temperature, although shear stress sensitivity may also be increased as noted for nylon 6 and shown in Fig. 6.3.[19] A similar behaviour has been reported by Sinha Ray and Okamoto[20] for molten polylactide (PLA)/layered silicate nanocomposites at 175 °C. These changes have implications on processability efficiencies for high throughput processes such as melt extrusion of filaments. Hence, if nanodispersed particles are present, there may have to be an upper limiting concentration determined by the need to compromise between added property and reduced extrusion efficiency. In addition, any increase in melt viscosity will most likely reduce the ease of melt

Applications for nanocomposite-based flame retardant systems 131

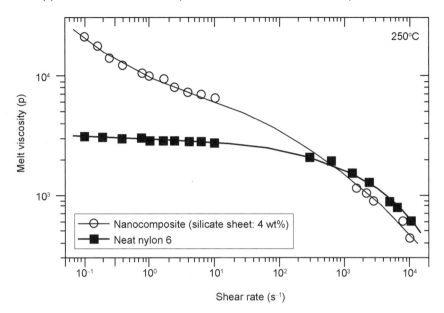

6.3 The shear rate dependence of melt viscosity of nylon 6 in the presence and absence of a nanoclay[19]; (Copyright John Wiley & Sons Limited. Reproduced with permission).

blending, although the associated increased shear stresses at higher extrusion rates may offset this factor and partly restoring process efficiency.[21] In addition to such physical effects, in some polymers like poly(ethyleneterephthalate), sensitised degradation may occur in the presence of species like nanoclays. Matabayas et al.[22] have shown that during melt compounding the inherent viscosity of a high molecular weight PET reduced from 0.98 to 0.48 dL/g as clay increased from 0.36 to 6.7% (expressed as ash%). Davies et al.[23] have also noted that for in-situ-polymerised nylon 6 in the presence of a montmorillonite clay, significant thermal degradation occurs when the compounded melt is subjected to injection moulding processes. Clearly, the effects of adding nanoparticulates to a polymer melt or indeed a solution are quite complex and so will impact on the processability of nanocomposites.

Not unrelated to the direct effect of clays on rheology is the possible influence that compatibilising species present to improve nanodispersion might have. For example, the use of epoxy resin as a compatibiliser in clay-polycarbonate systems, while improving clay dispersion, when molten, pre-cured formulations are analysed using a parallel plate rheometer, the linear and nonlinear dynamic rheological properties are affected.[24] The use of grafted-PP, especially with maleic acid functionality, also can have adverse rheological effects. We have shown that the presence of g-PP at up to 3% w/w level reduces the improvements in tensile properties that the addition of clay at 2.5 w/w% has produced with respect to unmodified PP polymer.[7] Melt flow index reduces at low levels of g-PP addition (1% w/w) probably due to improved clay dispersion but then increases at higher levels (3%) and these latter polymer blends experienced more filament breaks during extrusion.

Rheological effects have also been observed in our own laboratories during the formulation of aqueous copolymeric emulsions for use in textile back-coating formulations.[25] Here the addition of either a nanoclay (5% (w/w) Cloisite 15A (Southern Clay products, Inc., see Table 6.1), with respect to coating solids) or fumed silica (up to a maximum of 17% w/w with respect to coating solids) modifies the paste rheology considerably with the latter especially producing significant viscosity changes and hence difficulty in maintaining uniform and reproducible coating applications. The high concentrations of silica were used because of the need to attempt to introduce assumed high levels of flame retardancy, which in the event proved to be false (see Section 6.3). However, this does illustrate one adverse effect of nanoparticle concentration as has the work of Matayabas and Turner[21] on PET degradation above.

6.2.3 Dispersion

Dispersion and morphology of dispersed clays and other particles at the nano-level are, of course, essential if fire performance properties are to be optimised[26] and so the challenges of optimising nanodispersion and ensuring that it is unchanged during polymer processing may be considerable. Recently reported work from our own laboratories[7] has shown that compounding nanoclay-polypropylene (PP) and nanoclay-graft PP mixtures more than once prior to screw extrusion into filaments, yields improved dispersion and fibre physical properties. Use of a masterbatch of compatibilised clay-graft PP mixtures and its dilution during melt extrusion also improved dispersion sufficiently to be reflected in resulting fibre properties and a reduced peak heat release rate of knitted fabric samples.

Whether or not a clay should be fully exfoliated is not clear if maximum levels of reduced fire performance are to be achieved. Beyer *et al.*, for example, have shown that by subjecting EVA nanocomposites to higher shear and hence exfoliation levels produces no further reductions in cone calorimetric-determined peak heat release rate (PHRR) values than less than those of sheared samples having higher intercalated nanoclay contents.[27]

As mentioned above, companies like Southern Clays Inc., Sud-chemie AG and Nanocor Inc, produce a range of nanoclays designed to have high levels of dispersion in designated polymers (see Table 6.1) and typically the products like Nanocor's Nanomer I.30P (designed for film grade) and Nanomer I.44PA (for engineering grade applications) are compounded nanoclay dispersions in polyolefin matrices. These are based on amine chemistry and quaternised ammonium

Applications for nanocomposite-based flame retardant systems 133

ion modified montmorillonites respectively, designed for maximum compatibility and dispersion in a polyolefin matrix. They are available as free-flowing powders with mean dry particle sizes of 15–25 microns or as masterbatches containing 40% and 50% (w/w) clay contents, and they are capable of dispersion to the nanoscale in conventional twin screw compounders. The reader is referred to respective company websites in order to review all the increasingly diverse grades available for most commercial polymers.

6.3 Potential application areas

6.3.1 Bulk polymeric components

While the patent literature is populated with claims of the effectiveness of nanofilled polymeric formulations offering improved fire performance, there are few commercial products available at the present time. The currently best known example (as mentioned in Chapter 5) uses nanocomposite technology specifically to improve fire performance of electrical cable sheathings – Kabelwerk Eupen AG now market cables that incorporate nanoclays within ethylene vinyl acetate (EVA)-based sheathings which have the advantage of requiring reduced levels of conventional flame retardant such as alumina trihydrate.[18,28,29] For normal flame retardancy requirements, up to 65% (w/w) alumina trihydrate (ATH) is required which means that the overall cable sheathing physical properties may be significantly reduced. By incorporation of up to 5% of a functionalised nanoclay (with functionalising dimethyl- distearylammonium salts), the ATH content may be reduced to about 45% with a consequent improvement in sheath properties and saving on flame retardant costs. Furthermore, the char formed by the nanofilled ATH-containing formulation is far more rigid and less prone to crack development than the conventional analogue. It is this char that enables cable sheathings to pass tests such as UL-1666 in which indoor cables are subjected to a 145 kW burner and the maximum temperature of fire gases at 3.66 m (12 ft) and flame heights are recorded. The presence of nanoclay reduces the flame height from above 4 m to less than 2 m and maximal flame temperature at 3.66 m from 1054 to 345 °C.[29] This later work by Beyer[29,30] has shown that introduction of multiwalled carbon nanotubes on a weight-for-weight basis (at 5% w/w) into EVA reduces the cone calorimetrically-determined peak heat release values to slightly greater levels than when a functionalised nanoclay is present. However, inclusion of nanoclays into other polymers with potential for use in cables, such as thermoplastic polyurethanes (TPU), PVC and blends of the two, shows mixed results. For instance, while inclusion of 5 phr organoclay with TPU and TPU containing a phosphate ester flame retardant showed reductions in respective PHRR values, in each case the time-to-ignition values decreased and burn-out times increased.[31,32] However, whereas the time-to-ignition of TPU under a 35 kW/m^2 heat flux increased from about 70 s to 85 s in the presence of the phosphate flame retardant, it reduced again to about 70 s following subsequent addition of organoclay. PVC/EVA and PVC/TPU polymer blends, on the other hand showed heat release curves little affected by the introduction of organoclays apart from again reductions in times-to-ignition although their presence in each diminishes the darkening normally seen in PVC compounds.[32] Clearly, whether or not a useful level of flame retardancy is achieved depends on the polymer type, the conventional flame retardant present and the type of nanospecies present.

A second example is the group of nanocomposites for improved flame retardancy based on Nanocor® products developed by Gitto/Global Corporation, Lunenburg, MA, USA.[33] One of these is for heavy-duty electrical enclosure applications. These are typically made of injection-moulded polypropylene and vary in size up to one cubic metre. By using a nanocomposite, the level of the flame retardant additive present may be significantly reduced yielding an overall weight saving of 18%. While their original UL-94 V-0 ratings are maintained, both flexural and tensile moduli are claimed to increase by about 25% without loss of impact resistance. These same Nanocor® nanoclay dispersions have also been shown to reduce the amounts of flame retardants such as magnesium hydroxide (MOH) and decabromodiphenyl oxide-antimony III oxide formulations required to achieve UL94 V(0) levels of flame retardancy in EVA and PP respectively.

These commercial examples provide evidence that inclusion of nanoparticle species can indeed reduce overall flame retardant additive levels required to achieve a desired fire performance level. To corroborate this, work in our own laboratories[11,34–36] with polyamide 6 and 6.6 films and selected phosphorus-containing flame retardants suggests this to be the case when selected functionalised nanoclays are also present and this work is reviewed later.

The situation with other polymers like polystyrene, for example, is less clear. Wilkie and his coworkers[37] have recently shown that while introducing nanoclays and brominated species into polystyrene (PS) produces significant reductions in PHRR values and reduced times to ignition, making the compounded nanocomposites more prone to ignite, their behaviour to a practical burning test such as UL94 is less straightforward. For instance, while the introduction of 3% Cloisite 30B (Southern Clay Products, Inc., see Table 6.1) enables a previously failing compounded copolymeric PS containing 10% dibromostyrene (DBS) to achieve a V-2 pass rating, raising the DBS level to 20% alone yields a V-2 pass which is transformed to a fail following introduction of Cloisite 30B or a quaternised dimethyl-n-hexadecyl-4-vinyl-benzyl ammonium (VB16) functionalised clay. These results are listed in Table 6.2. However for copolymers containing 40% DBS, which alone maintains a V-2 rating, addition of Cloisite 30B raises this to the highest pass rating of V-0. Earlier work by this same group[38] investigated the effects of adding a variety of organophosphate additives and the clay, Cloisite 10A (see Table 6.1) to polystyrene and studied thermal degradative, burning and cone calorimetric

Applications for nanocomposite-based flame retardant systems

Table 6.2 UL94 test results for polystyrene nanocomposites containing either dibromostryene comonomer or phosphate additives[37,38]

Flame retardant comonomer or additive	Clay	PHRR at 35 kWm^{-2}	UL94 result
Pure Polystyrene	Nil	1419	NC
Pure Polystyrene	3% Cloisite 10A	310	–
10% Dibromostyrene	3% Cloisite 30B	–	V-2
20% Dibromostyrene	Nil	–	V-2
20% Dibromostyrene	3% Cloisite 30B	–	NC
20% Dibromostyrene	3% VB16	–	NC
40% Dibromostyrene	Nil	–	V-2
40% Dibromostyrene	3% Cloisite 30B	–	V-0
15% Tricresyl phosphate	Nil	1122	–
30% Tricresyl phosphate	3% Cloisite 10A	378	V-2
30% Tricresyl phosphate	5% Cloisite 10A	342	V-1
30% Tricresylphosphate	10% Cloisite 10A	324	V-1/V-0?
30% Resorcinol diphosphate	Nil	499	–
30% Resorcinal diphosphate	5% Cloisite 10A	110	V-2
30% Resorcinal diphosphate	10% Cloisite 10A	307	V-0/V-1?
30% Trixylyl phosphate	Nil	864	–
30% Trixylyl phosphate	5% Cloisite 10A	313	V-2

properties. Table 6.2 also lists some of their results which show that only when the phosphate flame retardants are present at higher concentrations, respective peak heat release or PHRR values are significantly reduced although the reduction is not as great as that caused by the presence of 3% nanoclay alone with respect to virgin PS. Furthermore, the presence of both a clay and a phosphate causes not only a reduction in PHRR values but also a very significant reduction in total heat released, both compared to the virgin polymer and to a styrene nanocomposite. In the practical UL94 tests, selected phosphate-containing samples showing the best flame retardant effect in terms of their respective ability not to sustain a flame after ignition are also listed in Table 6.2. Here it can be clearly seen that not only does the presence of nanoclay together with a retardant raise the fire performance in terms of UL94 V-rating but also increasing concentration from 3 to 10% promotes a similar increase.

From these studies, while it is clear that nanoclay-flame retardant interactions are not simple and may be dependent upon respective concentrations present, inclusion of nanodispered phases offers the opportunity to reduce overall additive loadings or to improve fire performance at currently acceptable loadings. However, while this has commercial advantages, the effects on rheology, processing efficiencies and resulting properties must not be forgotten. These effects will be especially significant and possibly beneficial when flame retardants, such as metal hydroxides, which are normally present in very high concentrations are reduced and replaced by nanoclays, for example. Hornsby and Rothon[39] have discussed this issue and compounded polymer melt viscosities and shear sensitivities, for example, are determined by hydroxide type, particle size, surface chemistry and concentration. However, Lomakin et al.[40] have recently described the effects of introducing nanoparticulate aluminosilicates (as Cloisite 15A, see Table 6.1) into magnesium hydroxide (or $Mg(OH)_2$)-filled polypropylene. They show that while a 60% w/w MgOH-filled polymer has similar thermal stability to one containing 50% $Mg(OH)_2$ and 10% w/w Cloisite 15A, no effects of improved processibility, were reported. In contrast, Song et al.[41] report that the mechanical properties of polyamide 6.6 are improved if a nanoclay is added in the presence of $Mg(OH)_2$ and red phosphorus as flame retardants. The two flame retardants and nanoclay act synergistically thereby enabling lower concentrations to be used. A similar flame retardant synergism was reported by Fu and Qu[42] who noted that addition of fumed silica to magnesium hydroxide-filled ethylene-vinyl acetate copolymer not only enabled low levels of $Mg(OH)_2$ to be used but an increase in elongation-at-break occurred.

The potential for commercial application of the combination of intumescents and nanoparticles was reviewed initially by Gilman et al.[43] and more recently by Bourbigot, Duquesne and coworkers.[44,45] The possible synergies between micro-dispersed intumescent flame retardants and nano-dispersed species, including zeolites, in addition to oligomeric silsesquixonanes (POSS) and montmorillonate clays, offer obvious opportunities to reduce overall filler contents with beneficial effects on polymer processing and properties although no such examples worthy of application have been reported to date. One of the first reported instances of this was by Bourbigot et al.[46] who showed that inclusion of a polyamide 6-montmorillonite nanocomposite within an EVA matrix with APP as conventional flame retardant enabled a UL94 test rating of V-0 still to be achieved with only two-thirds the concentration of the latter with respect to an equivalent intumescent formulation without any nanoclay present. Subsequently, Vyver-Berg and Chapman[47] reported that combination of functionalised clays at 1–3% w/w in the presence of intumescent flame retardants such as melamine phosphate, ammonium polyphosphate, pentaerythritol phosphate and zinc borate in appropriate combinations enables lower concentrations than normal to be used to achieve UL94 V-0 ratings in polypropylene at total concentrations just below 20% w/w.

6.3.2 Films, fibres and textiles

Minimisation of flame retardant additive concentrations is especially important in synthetic fibres where levels in excess of 10% w/w usually reduce the ease of polymer extrusion and subsequent processing as well as adversely affecting their normally desirable textile properties. The major difference between fibres and bulk polymers, including films and composites, is the small thickness of individual fibres, typically being 15–30 μm in diameter, yielding yarns of 50–100 μm diameter and fabrics having thicknesses varying from as low as 100 μm to several mm. While reported fire performance based on cone calorimetric data of bulk polymers[12,43,48] typically shows that the presence of nanoclays reduces peak heat release rates, they most often reduce times to ignition and extend total burning periods while little affecting the overall heat release of the polymeric substrate. Thus, while slowing down the burning process but encouraging more rapid ignition, they also encourage increased char formation. In fact, in some cases where polymers are not char-formers, some char development has been reported observed[43,48] and this is of especial importance to extremely thermoplastic and negligible char-forming, fibre-forming polymers such as poly(ethyleneterephthalate) and polypropylene.

Bourbigot et al.[49,50] reported the first fire performance studies of nanocomposite polyamide 6 filaments and these were converted into fabric having an area density of 1020 g/m² and thickness 2.5 mm. These fabrics were exposed to 35 kW/m² heat flux in a cone calorimeter and ignition times of 70 s and 20 s and peak heat release rate (PHRR) values of 375 and 250 kW/m² respectively for the normal and nanocomposite polyamide 6 fabrics were recorded. While the latter represents a significant 33% reduction in PHRR, ignition resistance was significantly reduced and total heat release was little, if any, affected. However, thermogravimetric analysis suggested that presence of nanoclay had little effect up to 400 °C but above 450 °C there appeared higher char formation. It was clear that the fibres were not flame retardant in the more accepted sense in that ignition resistance was not increased by inclusion of nanoclays alone. A further problem with fibres and fabrics with respect to bulk polymers is their high specific surface areas and their thermally thin character. This is significant since Kashiwagi et al.[51] has suggested that the effectiveness of nanoclays in reducing PHRR values and related fire performance may be a function of sample or composite thickness. Thinner samples appear to show lower PHRR reductions because of competition between the formation of a surface carbonaceous-silica shield and the volatilisation to fuel of surrounding polymer. In thicker composites, the competition favours ceramic barrier formation while for thin composites, volatilisation dominates.[52] This can be considered as the difference between so-called thick and thin thermal behaviour.[53] In 'thin' textile fabrics it is possible that the 'shield-forming' mechanism observed for bulk polymer nanocomposites may be too slow for effective improvement in fire performance. It is likely, however, that the thickness effect observed by Kashiwagi and coworkers will be influenced by the heat flux since both competing mechanisms are thermally driven but to different extents.

Assuming Kashiwagi's results to be reasonably valid and that a simple negative linear relationship exists between composite thickness and PHRR, in a recent paper,[35] we have suggested that the 33% reduction in PHRR observed by Bourbigot et al.[49,50] is a consequence of the lower heat flux of 35 kW/m². It is proposed that the competing volatilisation and ceramic shield-forming reactions will be influenced by the heat flux but to different extents and that lower heat flux will favour the diffusion of clay particles to the surface and formation a surface clay layer. Thus it may be surmised that nanoclay presence alone in fibres, films and textiles will only be significant at lower heat fluxes. More recent work by Bourbigot et al.[54] has extended his polyamide research to include nanoclays into melt-spun poly(lactic acid), PLA filaments where again at loadings of up to 4% w/w, reductions in PHRR values as much as 38% and increased char yields are recorded at a heat flux of 35 kW/m^2; times-to-ignition are still reduced, however.

One possible means of increasing nanoclay efficiency is the possibility of using char-promoting functionalising groups but since these are present at low concentrations within the particle substrate and the functionalised nanoparticles themselves are introduced only at 2–5% w/w loadings, their effectiveness including possible vapour phase activity might be questioned when present at very low (<<1%) levels in the polymer. However, since the thermal stability of the functionalising species during processing significantly affects the resulting nanoclay behaviour as previously discussed,[12,13] and products of group decomposition have been identified,[17] possible char-promoting or vapour phase effects at such low levels should not be ruled out.

However, as shown above in bulk polymers, combination of nanoparticles with conventional flame retardants may promote overall additive and even synergistic activity. Work in our own laboratories has shown that this is in fact possible, in polyamide 6 and 6.6 films that may be used as models for respective fibres.[8,11,14,34,35] Normally, minimal flame retardant additive contents of about 15–20% w/w are required to render these polyamides flame retardant,[55] levels which are too high for inclusion in conventional synthetic fibres. This is because flame retardant property versus flame retardant concentration effects are rarely linear and more often 'S' shaped because of the need to generate a threshold char level having an extended coherence throughout the polymer.[55–57] This defines for each polymer a critical minimal concentration of a given flame retardant that must be present in a given polymer in order to achieve acceptable flame retardancy. Reduction of this minimal level by use of nanoparticles offers a valuable means of achieving effective flame retardance in fibre-forming polymers.

We have reported both additive and/or synergistic effects of adding selected flame retardants based on ammonium polyphosphate (as Antiblaze MCM, Rhodia), melamine phosphate (as Antiblaze NH, Rhodia), pentaerythritol (PER), pentaerythritol phosphate (as Chemtura formerly Great Lakes NH 1197), cyclic organic phosphate (as Antiblaze CU, Rhodia), intumescent mixtures of APP, PER and melamine (as Antiblaze MPC, Albemarle, formerly Rhodia), ammonia cross-linked polymer of a tetrakis(hydroxyl phosphonium salt-urea condensate, as Proban CC polymer (Rhodia) and related formulations into nylon 6 (N6), and

Applications for nanocomposite-based flame retardant systems 139

6.6 (N66) polymer films (~80 μm thick) in the presence of nanoclays (polyamides supplied by RTP Plastics).[11,14,34,35] Analysis of the flame retardant performance of each FR-nanoclay-nylon film compared with respective flame retardant nylon films, for nylon 6.6 films shows that only for APP, Proban CC and the intumescent Amgard MPC does the presence of nanoclay significantly increase LOI values and that using Lewin's 'Synergistic Effectivity' measure,[58] these demonstrate possible synergy as reported elsewhere.[11,35] The effect of additions of both commercial and Cloisite 30B nanoclay are clearly seen below in the previously published LOI versus FR concentration plots presented in Fig. 6.4.[11,34] Why only a few of the selected flame retardants show positive effects in the presence of the nanoclay is not understood suffice it to say that ammonium polyphosphate is not only the most synergistic but also has a decomposition temperature in the range 250–300 °C and this overlaps the melting point of nylon 6.6 (~265 °C). It is considered that this will encourage flame retardant mechanisms to start alongside polymer fusion. Surprisingly, however, the APP/PER combination does not show synergy and, indeed, shows a slightly less than additive flame retardant effect when both flame retardant and nanoclay are present. However, the intumescent Amgard MPC does show evidence of synergy. The pentaerythritol phosphate-derived species (Chemtura NH 1197) have higher decomposition temperatures than APP, as does melamine phosphate (as Antiblaze NH).[59] The curves in Fig. 6.4 are variants of the typical 'S-shaped' curve[55-57] and from them the reductions in flame retardant concentration to achieve a specific level of flame retardancy may be assessed. These concentrations are shown in Table 6.3.

Thus to achieve LOI values up to 24 vol%, the addition of nanoclay, at a 2% assumed level, enables flame retardant levels to be reduced significantly. At LOI = 25 vol%, similar reductions in concentration are observed for APP and Proban CC polymer-containing films but a converse situation was observed for the Amgard MPC system. Unfortunately, a similar analysis could not be undertaken for the nylon 6 film set because the presence of nanoclay significantly changed the burning manner of cast nylon 6 films in a manner that rendered simple LOI

Table 6.3 Flame retardant concentrations required to achieve defined LOI values in nylon 6.6 films[34]

	LOI = 23		LOI = 24		LOI=25	
	N6.6	N6.6 nano	N6.6	N6.6 nano	N6.6	N6.6 nano
APP	23.8	15.0	28.5	20.1	33.3*	25
MPC	16.3	14.5	20.5	18.0	30.0	>30
Proban CC	20.5	10.5	28.5	17.5	36.3*	25

Note: * Values extrapolated from Fig. 6.4

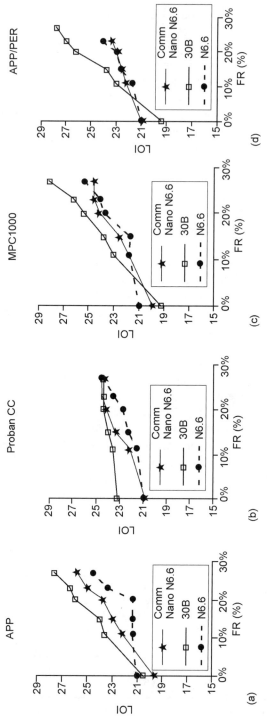

6.4 LOI values for polyamide 6.6 films in the absence and presence of various flame retardant/commercial and Cloisite 30B nanoclay combinations; (a) Ammonium polyphosphate; (b) Proban CC polymer; (c) Intumescent MPC1000 and (d) ammonium polyphosphate/pentaerythritol; 0% FR values for FR and nanoclay-containing films are extrapolated;[34] (reproduced with permission from the Textile Institute).

Applications for nanocomposite-based flame retardant systems 141

value comparison difficult although evidence of synergism was generated as indicated above.[8,11]

Possible effects of clay functionalising groups have also been investigated using Cloisite clays selected from those in Table 6.1. These were introduced with selected flame retardants into cast nylon 6 and 6.6 films. Clay selection was based on observing the greatest increase in LOI that a given clay alone could promote in a cast nylon 6.6 film in the first instance. Cloisite 30B clay fitted this criterion and yielded a cast film LOI value of 28 vol% compared with the clear film value of 21.0. This clay was selected along with the non-functionalised Cloisite Na^+ which was found to disperse quite easily in the polar polyamide films yielding an LOI value of 25.2 vol%. The variations in LOI for 2% clay-containing films were noted to be mainly a consequence of the effect that each clay had on the burning character of the film in an LOI test using vertical geometry. A similar effect was noted for nylon 6 films in which previously reported flame retardant-nanoclay interactions for nylon 6 formulations appeared to be less efficient.[8,11] This lower activity could be a consequence of the lower melting point of nylon 6 of about 215 °C, a temperature too low for most flame retardants to start functioning efficiently.

The LOI results for the polyamide 6.6 films in the absence and presence of the nanoclays, are shown in Fig. 6.4 for Cloisite 30B compared with clay-free and previously reported commercial clay-containing polyamide 6.6 films.[8,34,35] These curves exhibit the expected 'S-shaped' curve behaviour with LOI values increasing significantly only when respective flame retardant concentrations exceed 20% w/w. Introduction of the nanoclays appears to linearise respective LOI vs FR, percentage trends and generally all selected nanoclays have a positive effect on the burning behaviour of the polymer when flame retardant is present.

Clearly, the presence of each clay significantly reduces the concentration required of each flame retardant required to achieve a specific LOI value as shown in Table 6.3 for the commercial nanocomposite nylon 6.6 films and in Fig. 6.5 where for each flame retardant, the percentage reduction is plotted for each clay type in polyamide 6.6 films. Furthermore, the two clays used in this study show significant advantages over the previously studied commercial nanocomposite polyamide 6.6 and that for LOI values of about 23 vol%, levels of retardant approaching only 10% w/w may be applicable and to achieve LOI = 24 vol%, levels of about 15% may suffice.

There have been other attempts to produce nanoclays in the presence of flame retardants in other fibre-forming polymers, such as polypropylene[60] and polyester.[61] In the case of polypropylene, the addition of nanoclay to a flame retardant formulation based on a hindered amine stabiliser and a char-promoting ammonium polyphosphate at concentrations of the order of only about 5% (w/w) does enhance char formation although insufficiently to increase the LOI above 22 vol%.[60] A similar char-enhancing effect of added functionalised

142 Advances in fire retardant materials

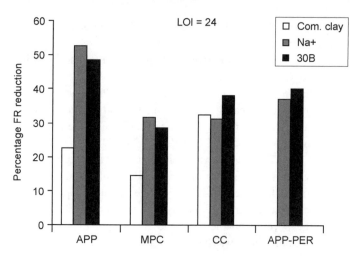

6.5 The percentage reduction of each flame retardant required to produce an LOI value of 24 when present with an unspecified commercial clay and the specified clays Cloisite Na^+ and 30B;[34] (reproduced with permission from the Textile Institute).

montmorillonite clay was observed by Wang *et al.*[61] in a copolymer of poly(ethyleneterephthalate) and a phosphorus-containing monomer in that higher residues were above 450 °C were recorded. Our previously reported work has been extended to observe the effect of nanoclays both alone and in the presence of flame retardants in polypropylene fibres and fabrics.[7,62] Table 6.4 presents tensile and flammability data for polypropylene fibres and fabrics containing Cloisite 20A clay (see Table 6.1) and a maleate-grafted polypropylene (Polybond 3200 (Pb), Crompton Corporation) at various concentrations.[7] All polymer samples had been twice compounded to maximise dispersion prior to fibre extrusion and sample 5 differed from sample 4 in its having been produced as a more concentrated masterbatch before being let down during the extrusion stage. While it is clear that the presence of nanoclay alone (sample 2) promotes an expected improvement in fibre tenacity and modulus, there is also a decrease in peak heat release rate determined by cone calorimetry at a heat flux of 35 kW/m². Addition of the compatibilising maleate-grafted PP reduces the tensile properties as expected and, apart from sample 3, suggests that it causes further reduction in PHRR values. This is associated with the improved dispersion as shown by TEM. While there was insufficient sample to enable LOI values to be obtained, values for cast films indicated that all samples had values within the range 19.6–20.0 confirming the absence of any flame retarding property.

Subsequent work[62] investigated the effect of introducing selected phosphorus-containing flame retardants ammonium polyphosphate APP (Amgard MCM, Rhodia Specialities, UK), melamine phosphate, NH (Antiblaze NH, Rhodia Specialities Ltd., UK) and pentaerythritol phosphate (NH1197, Chemtura), the

Table 6.4 Polypropylene fibre compositions and tensile properties and fabric PHRR values[7]

Sample	Nanoclay 20A (%, w/w)	Graft Pb (%, w/w)	Linear density (tex)	Modulus (N/tex)	Tenacity (cN/tex)	Fabric area density (g/m^2)	PHRR (kW/m^2) at 35 kW/m^2 heat flux
1	0	0	4.9 ± 0.3	4.1 ± 1.3	32.6 ± 2.5	430	525 ± 40
2	2.5	0	4.0 ± 0.8	5.6 ± 1.6	36.4 ± 8.0	400	477 ± 105
3	2.5	1	4.3 ± 0.9	3.7 ± 1.1	22.0 ± 6.0	390	531*
4	2.5	3	4.5 ± 0.1	3.4 ± 0.8	16.9 ± 7.0	430	420 ± 90
5**	2.5	3	3.9 ± 0.4	5.2 ± 0.8	26.3 ± 8.2	390	394*

Note: * only one sample tested; ** sample 5 is produced as a concentrated masterbatch before being let down during extrusion to yield the stated additive concentrations.

hindered amine stabiliser NOR 116 (Ciba)[57] and the bromine-containing tris (tribromopentyl) phosphate (FR 372, DSBG, Israel) and tris(tribromophenyl)cyanurate, FR 245 (DSBG, Israel) species. These were compounded with selected clays (Cloisite 20A and 30B, Bentone 107, Elementis: a bentonite clay modified with dimethyl, dehydrogenated tallow quaternary ammonium ion and a montomorillonite modified with vinyltriphenyl phosphonium bromide) and compatibilisers (Polybond) Pb and polypropylene grafted with diethyl-p-vinylbenzyl phosphonate (DEP)). Extrusion into filaments proved to be challenging because of problems with optimising clay and flame retardant dispersion and this was especially the case when APP was present because of its very poor dispersion and relatively large particle size (25–30 μm). As a consequence, extrusion of these formulations often resulted in broken filaments and reduced tenacities and moduli. Because of the limited fibre and hence, derived fabric quantities available, either compounder extrudate or tape samples were used for LOI examination. A selection of the fibre/film compositions and their thermal and LOI properties are shown in Table 6.5. It is evident that while the melting temperature of the formulations are unaffected by their contents, melt flow indices generally are seen to increase, indicating a general reduction in melt viscosity possibly driven by thermal degradation associated with the extrusion process. Given that clays are introduced at nominal

Table 6.5 Selected combinations of clay, compatibiliser and flame retardant in polypropylene (PP) filaments and tapes[62]

Sample**	DSC melting peak max., °C	Char yield at 800 °C from TGA, %	MFI, g/600 s	LOI vol%
PP	164	0.	25.9	19.2
PP-20A	165	2.0	35.4	–
PP-Pb-20A	165	1.7	26.4	20.2
PP- Pb - 20A - APP	166	3.0	31.2	20.6
PP-E	166	1.6	38.4	20.5
PP-Pb-E	167	3.3	39.4	20.3
PP-Pb-E-APP	168	3.2	24.6	20.6
PP-NOR-Pb-E	167	1.0	33.7	17.5*
PP-NOR-Pb-E-APP	168	3.2	24.0	17.8*
PP-NOR-Pb-E-NH	168	2.9	27.0	17.9*
PP-NOR-Pb-E-1197	168	2.2	34.8	17.8*
PP-NOR- Pb-E-FR245	167	2.5	36.0	17.2*
PP-NOR-Pb-E-FR372	166	1.0	38.4	18.7*

Notes : * LOI samples are coarse filaments from the compounder
**20A, E = Clays, 3% w/w
Pb = Polybond (maleic anhydride grafted PP), 1% w/w
NOR 116 present at 1% w/w
Flame retardants APP, NH, NH1197, FR245 or FR372 present at 5% (w/w).

3% w/w, then char yields at 800 °C in air, at which most organic components will have oxidised are close to the nominal inorganic residue expected and vary within a range 1.0–3.3% when clay is present. LOI values are also unaffected by either the presence of clays and/or flame retardant but then the low concentrations of flame retardants present (5% w/w except for NOR 116 at 1% w/w) would not be expected to raise the LOI values significantly when present alone.[57] However, the burning behaviours of knitted fabrics having the formulations in Table 6.5 may be recorded as times to burn for successive 60 mm distances, see Figs 6.6(a) and (b), after which ignited samples have been timed for flame fronts to reach 60, 120 and 180 mm when subjected to the standard vertical strip test BS 5438:1989:Part 3. It is seen that while 100% polypropylene fabrics burnt their entire length quite rapidly, those containing 3% w/w of each clay alone, showed slightly longer times to reach the 60 mm mark and hence had slower burning rates during the first 60mm of sample. However, only four fabrics (PP, PP-E, PP-Pb-E and PP-Pb-20A) burnt beyond the 60 mm mark and reached the 120 mm mark; of these only pure PP and PP-Pb-E samples burnt the whole length. Thus it might be concluded that either the Polybond compatibiliser or both clays when present have minimal flame retarding activity; in fact the burning rates of PP compared with PP-Pb-E at 180 mm show that the latter has the higher burning rate. The results in Fig. 6.6(a) have been converted to rates for each 60 mm length increment and shown in Fig. 6.6(c).

Nevertheless, samples with flame retardants self-extinguished beyond the 60 mm mark and show longer burning times, see Fig. 6.6(b), and generally lower burning rates. For the Elementis (E) clay-containing formulations, fabric burning rates are in the decreasing order:

PP > PP-E > PP-Pb-E-APP > PP-Pb-E > PP-NOR-Pb-1197 > PP-NOR-Pb-E-FR372

which demonstrates the obvious effect of small amounts of added flame retardant. Conversely, the order for Cloisite 20A-containing formulations is:

PP > PP-Pb-20A-APP > PP-Pb-20A

which suggests that the presence of APP has a deleterious effect. However, samples also burned differently, depending upon the flame retardant used. For instance, samples containing APP (PP-Pb-20A-APP and PP-Pb-E-APP) burned up to the 60 mm mark quickly and within 4 s, although the flames flickered quite significantly, probably due to the poor dispersion of APP. Times to reach the 60 mm point are longer and hence rates of flame spread lower for samples containing NH 1197 and FR372 and they self-extinguished beyond the 60 mm mark after 34 and 45 s, respectively.

It must be remembered that normally APP concentrations above 20% would be required to render PP flame retarded[57] and so the observation that in the presence of a nanoclay that only 5% can cause marked effects in PP fabrics is encouraging.

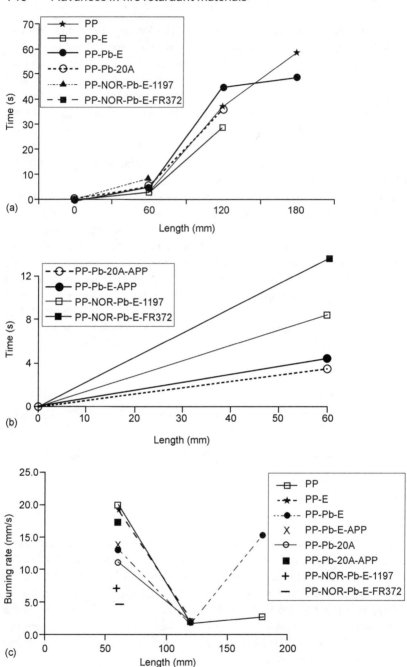

6.6 (a) Times to burn (BS5438:1989:part 3), b) times to burn for the first 60 mm only, and (c) burning rates versus fabric distance burned (see p. 145) for polypropylene fabric samples comprising clays, compatibiliser and flame retardants listed in Table 6.5.

Other recent work in our laboratories[63,64] has shown that poly(acrylonitrile) copolymer having properties suitable for fibres when polymerised in the presence of a functionalised nanoclay, may absorb ammonium polyphosphate during filament extrusion and yield fibres having LOI > 40 vol%. In these fibres, a clear synergy between nanoclay and flame retardant is observed and filament properties are little changed from those acceptable for normal textile applications. Table 6.6 summarises some of the results of this work which show that the introduction of clays at the 1% level has no effect on tenacity (although the initial Young's modulus increased) and the tensile values are comparable to commercial values (typically in the range 2.3–3.5 cN/dtex). More importantly, they act synergistically with the APP present to yield improved flame retardancy quantified as values of ΔLOI_{nano}. Thus it is seen that the addition of Cloisite Na$^+$ in particular is effective in raising the LOI by as much as 5 units above that expected from the APP alone at a similar level. Unfortunately, APP is not durable to water soaking or washing and so introduction of a cross-linkable or insoluble flame retardant would be required to achieve required levels of launderability. Notwithstanding this, the evidence is clear that clays in the presence of a suitable flame retardant benefit the overall fire performance of polyacrylic filaments in a manner similar to that observed in polyamide films.[8,34,35]

Table 6.6 Tensile and limiting oxygen index properties for experimental polyacrylic filaments containing 1% w/w of either Cloisite Na+ or 30B clays subjected to ammonium polyphosphate treatment in bath phosphorus, PL, concentrations 1–6%[63,64]

Dope-blended samples	Tenacity, cN/dtex	P_L (nominal), % w/w	P_F, % w/w	ΔP_F % w/w	LOI, vol%	$\Delta LOI/P_F$	ΔLOI_{nano}, vol%
Control (Courtelle)	2.6	0	0.0	–	19.0	0.0	–
		1	1.2	–	21.0	1.2	–
		3	3.5	–	26.0	2.0	–
		6	6.5	–	36.0	2.6	–
Cloisite Na$^+$, 1%	2.5	0	0.0	0	20.4	0.0	1.4
		1	1.5	0.5	21.8	0.9	0.8
		3	4.4	0.9	31.0	2.4	5.0
		6	6.8	0.3	41.0	3.0	5.0
Cloisite 30B, 1%	2.7	0	0.0	0	19.0	0.0	0.0
		1	1.8	0.6	21.8	1.6	0.8
		3	4.3	0.8	30.0	2.6	4.0
		6	6.5	0	36.6	2.7	0.6

Notes: P_L = % w/w phosphorus in liquour; P_F = % w/w P on fibre; ΔP_F = (P_F nano P_F control); ΔLOI_{nano} = (LOI $_{nano}$ LOI $_{FR\ control}$).

6.3.3 Coatings

Examples of the potential for use of nanoparticulate fillers to enhance the fire performance of polymer coatings have largely been restricted to coatings for textile substrates including back-coatings. Bourbigot et al.[49,50,65,66] have shown that addition of nanoclays and poly(silsosesquioxanes) can reduce the peak heat release rates in polyurethane-coated knitted polyester fabrics as shown in Fig. 6.7. However, the presence of these nanoparticles alone reduced the time to ignition and prolonged the time of burning – exactly the opposite of what is required for flame retarded coated textiles.

More recently, Horrocks et al.[25,59,67] have shown that if a back-coating is to be effective it must have a transferable flame retardant activity from the coating on the reverse face of the textile when ignited from the front face in tests such as BS 5852:Part 1 1979 and 1990 and EN 8191 Parts 1 and 2. The use of purely char-promoting flame retardants within the coating does not allow this to occur unless the retardant species becomes mobile and can diffuse through the fabric to the front face. Furthermore, the addition of a nanoclay to a back-coating polymeric film has been shown to have no beneficial effect when alone.[25] Also, when fumed (nanoparticulate) silica is added with ammonium polyphosphate to the back-coating formulation not only was there an adverse effect noted with respect to formulation rheology but also the flame retardant character as determined by LOI was reduced with increasing silica content. Using a dispersion of the silica equivalent to a maximum 17% w/w within a solid coating, a

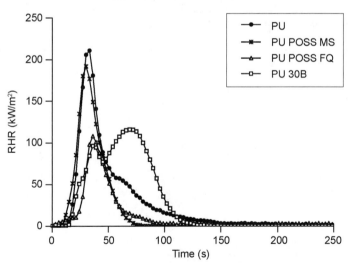

6.7 Rate of heat release curves of PU-nanocomposite coatings on PET knitted fabric samples at 35 kWm^{-2} treated with an octamethyl POSS (POSS MS), a poly(vinylsilsosesquioxane) (POSS FQ) or a nanoclay (Cloisite 30B) at 10% loadings with respect to PU[65]; (Copyright John Wiley & Sons Limited. Reproduced with permission).

Applications for nanocomposite-based flame retardant systems 149

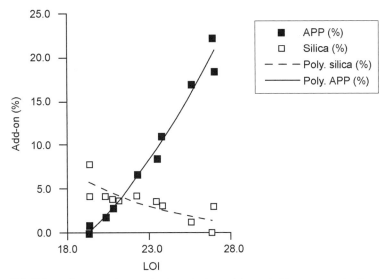

6.8 Effect of ammonium polyphosphate and fumed silica concentrations present in coatings on cotton fabric on the fabric LOI values.[25]

series of Vycar PVC (Noveon Inc., USA) copolymeric dispersions were prepared comprising a constant total retardant concentration (250 parts by mass with respect to 100 parts dry resin mass) with silica:APP molar ratios varying from 1:0 to 0:1 respectively in 0.1 increments. Coated fabric samples, at a nominal 30% dry add-on were prepared but with increasing silica content and associated degradation in rheology, add-ons decreased to below 10% for Si:APP > 0.6:1. Subjecting samples to LOI testing enabled the results in Fig. 6.8 to demonstrate the negative effect that silica has on the overall flame retardancy of such a coated material.

Clearly the potential applications of nanocomposites within the coating area, especially with respect to coated textiles, must be questioned based on the present data available.

6.3.4 Composites

Within the area of rigid, reinforced composites, the major requirements for acceptable fire performance are resistance to ignition, minimal flame spread and heat release rate and low smoke generation. The currently used bromine-based flame retardants, whether as additives or resin comonomeric modifications, present major challenges as a consequence of their increased smoke-generating propensity coupled with questions regarding the environmental concerns related to the use of halogenated flame retardants in general. The search for non-brominated alternatives has led to the potential offered by intumescent-based systems on the one hand and nanocomposite formation on the other. This whole area has been recently reviewed[68,69] and Sorathia[70] (see also Chapter 20) has reviewed the requirements of end-users such as the US Navy and cites recently published attempts by Wilkie et al.[71] to use nanocomposites in combination with nanoclays and phosphorus-based flame retardants such as tricresyl phosphate and resorcinol diphosphate within vinyl ester resin matrices. Interestingly, this latter work indicated that while the presence of clay did not increase times-to-ignition during cone calorimetry, its introduction at 6% w/w level reduced peak heat release rate relative to the pure resin and subsequent decreases were proportional to the amounts of phosphate added.

Work in our own laboratories has investigated the thermal degradation effects[14] of introducing functionalised nanoclays along with phosphorus-containing flame retardants in vinyl ester resins. Subsequent work has reported the effect of nanoclays and flame retardants on cone calorimetric properties.[72] In this work, a typical polyester resin – (Crystic 471 PALV (Scott Bader)) was investigated by DTA-TGA in the presence of a range of clays (Cloisite Na^+, 10A, 15A, 25A and 30B (Southern Clay Products, USA) see Table 6.1) and phosphorus-containing (ammonium polyphosphate (Antiblaze MCM, Rhodia Specialities), melamine phosphate (Antiblaze NH, Rhodia Specialities), dipentaerythritol/melamine phosphate intumescent mixture (Antiblaze NW, Rhodia Specialities)) and alumina trihydrate (ATH). Initial results[14] reported that in the derived unsaturated resin nanocomposites, nanoclays reduce thermal stability and char formation tendency of the resin up to 600 °C. While introduction of different condensed phase active flame retardants increased char formation of the resin above 400 °C, when nanoclays were added char formation was not greatly affected and in fact for ammonium polyphosphate-containing resins, char reduced. Thus, except for APP, the introduction of Cloisite 25A clay appears to have minimal effect on the thermal degradation of a vinyl ester resin containing the above flame retardants suggesting that fire performance may be little influenced by its addition. Furthermore, in our subsequent publication[72] in which only X-ray diffraction was used as a means of understanding whether or not each resin-clay composition had a nanocomposite structure or not, a major conclusion was that inclusion of flame retardants neither influences the level of clay dispersion present nor facilitates nanocomposite formation. The fire performances were derived using cone calorimetry at $50kWm^{-2}$ incident flux and expressed in terms of peak heat release (PHRR), total heat release (THR), fire growth index (FIGRA) and smoke evolution. The differences in these parameters with respect to pure resin behaviour as a consequence of adding clays at 5% levels alone to vinyl ester are shown by revisiting our data[72] and plotting as in Fig. 6.9(a). To a first approximation, each Cloisite clay has a similar PHRR- and FIGRA-suppressing effect, although smoke generation is increased generally. The presence of each flame retardant similarly has a beneficial effect in terms of reducing heat release rate parameters although again smoke is generally increased. The effects of adding clays at 5% w/w and

Applications for nanocomposite-based flame retardant systems 151

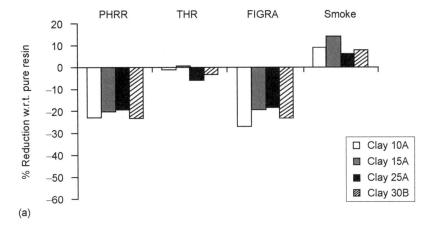

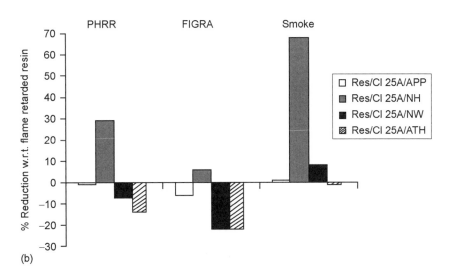

6.9 Cone calorimetric behaviour (a) containing different functionalised Cloisite clays at 5% (w/w) only with respect to that of pure vinyl ester resin of formulations and (b) containing Cloisite 25A clay with each flame retardant with respect to resins containing each respective FR only.[72]

different flame retardants at 20% w/w on cone calorimetric parameters with respect to respective flame retarded resin formulations are shown in Fig. 6.9(b) for Cloisite 25A with each flame retardant. Figure 6.9(b) also shows that whether or not a given clay improves fire performance of an already flame retarded resin depends on the latter type. Thus for Antiblaze NW (melamine phosphate and dipentaerythritol) and ATH in particular, the addition of clay further reduces PHRR and FIGRA while having little effect on smoke formation, whereas melamine phosphate alone (Antiblaze NH) shows a reduced fire performance with increases in PHRR, FIGRA and particularly smoke evident in the presence of Closite 25A. Results also show that all clays in Fig. 6.9(a) behave similarly in the presence of APP in promoting reductions in PHRR, FIGRA and, apart from Cloisite 25A, smoke generation.

In conclusion, while there appears to be no general improvement in fire performance when nanoclays are added to conventionally flame retarded resins, there is evidence that in certain formulations such as APP and ATH, such benefits are observed and this opens opportunities for favourable exploitation of introducing nanoclays and other nanoparticles into flame retardant resin formulations for use in reinforced composites having improved fire properties.

6.3.5 Foams

The challenge of developing novel flame retardant foams containing nanoparticulates will lie not only in selecting and optimising a fire retarding synergistic formulation itself but also in accommodating the expected increased rigidity that probably will accompany formation of a nanocomposite structure. This could render any potential suitable exploitation relevant only to rigid foam applications, although flexible foams for use in less critical areas like packing as opposed to fillings for upholstered products might be feasible. Furthermore, since many foams are based on polyurethane chemistry produced by *in-situ* polymerisations including concurrent inflation, the generation of a truly nano-dispersed phase will be a significant challenge in itself. There is evidence, however, that nanodispersions can be achieved if the functionalising substituents are carefully chosen and contain a significantly high number of hydroxyl groups.[73] Recent research has also shown that in the case of polylactide foams, where high levels of nanodispersion can be achieved by melt processing and physical inflation with an inert gas, layered silicates influence cell dimensions and density by nucleating cell formation.[74] This gives rise to foams having a finer and more uniform cell structure. A similar phenomenon has been observed in clay-filled polystyrene foams[75] where improved fire retardance has also been reported.

Even more recent research by Liggat and coworkers[76] has not only enabled the nanoclay dispersion challenge to be overcome during the formation of flexible polyurethane foams, but also these have achieved flame retardancy levels commensurate with passes to the Crib 5 ignition source defined in BS5852:1990. This, therefore, enables the nanocomposite foams to be acceptable to current UK regulations for upholstered furnishing foams and filling.[77] With a recent patent application,[78] this development shows great promise for imminent commercial exploitation.

6.4 Future trends

It is evident that future successful commercial application of nanocomposites in developing improved fire performance will take the lead from the work of Beyer[16,27–30] and Kabelwerk Eupen AG where the combination of appropriate nanoparticulates and conventional flame retardants (e.g. ATH) enables a reduction in concentration of the latter to be achieved in EVA-based cable sheathings. Furthermore, because of possible synergies between the nano- and microdispersed phases, the total concentration of both components may be reduced below the previously higher concentration of the single flame retardant (ATH) necessary to create the desired levels of fire performance. This would also give rise to improved mechanical properties. There is considerable evidence from the literature that under experimental conditions, similar nanoparticle-conventional flame retardant synergies are observed in other polymer matrices.[79] Although this is by no means a general observation, opportunities do exist for significant reductions in additive concentrations being made on the one hand to achieve given fire retardance or performance levels, while on the other gaining mechanical performance and cost advantages. These may only be realised, of course, if inclusion of nanophases can be achieved with minimal or controllable effects on polymer rheology during processing, a factor which is considered by some authors to be of the most significant challenges.[80] Overcoming these effects will especially be valuable in the film and fibre sectors where minimal flame retardant concentrations are mandatory if the expected tensile properties are to be maintained.

6.5 References

1. Kojima Y, Usuki A, Kawasumi M, Okada A, Kurauchi T, Kamigaito O. Synthesis of nylon 6-clay hybrid by montmorillonite intercalated with ϵ-caprolactam. *J. Polym. Sci., Polym Chem.* 1993; 34(4): 983–986.
2. Usuki A, Kojima Y, Kawasumi M, Okada A, Fukushima Y, Kurauchi T, Kamigaito O. Synthesis of nylon 6-clay hybrid. *J. Mater. Res.* 1993; 8: 1179.
3. Pinnavia T J, Beall G W (eds) *Polymer-clay Nanocomposites.* In Wiley Series in Polymer Science, Chichester and New York: Wiley; 2000.
4. Vaia R A. Structure characterisation of polymer-layered silicate nanocomposites. In: Pinnavia T J, Beall G W (eds) *Polymer-clay Nanocomposites.* In Wiley Series in Polymer Science, Chichester and New York: Wiley; 2000, pp. 229–266.
5. Yngard R, Yang F, Nelson G L. Flame retardant or not: fire performance of polystyrene/silica nanocomposites prepared via extrusion. 14th conference on Advances in Flame Retardant Polymers; 2003; Stamford. Norwalk (CT): Business Communications; 2003.
6. Yang F, Nelson G L. PETg/PMMA/silica nanocomposites prepared via extrusion. Proceedings of the 5th conference on Advances in Flame Retardant Polymers; 2004; Stamford. Norwalk (CT): Business Communications; 2004.
7. Horrocks A R, Kandola B K, Smart G, Zhang S, Hull T R. Polypropylene fibres containing dispersed clays having improved fire performance. Part I: Effect of

nanoclays on processing parameters and fibre properties. *J. Appl. Polym. Sci.* 2007; 106: 1707–1717.
8. Padbury S A. Possible interactions between char-promoting flame retardants and nanoclays in polyamide films, PhD Thesis 2004, University of Bolton, UK.
9. Carrado K A, Xu L, Seifert S, Csencsits R, Bloomquist C A A. Polymer-clay nanocomposites derived from polymer-silicate gels. In: Pinnavia T J, Beall G W (eds) *Polymer-clay Nanocomposites*. In Wiley Series in Polymer Science, Chichester and New York: Wiley; 2000, pp. 47–63.
10. Lann T, Pinnavaia T J. Clay-reinforced epoxy nanocomposites. *Chem. Materials* 1994; 6: 2216.
11. Padbury S A, Horrocks A R, Kandola B K. The effect of phosphorus containing flame retardants and nanoclay on the burning behaviour of polyamides 6 and 6.6. Proceedings of the 14th conference 'Advances in flame retardant polymers'; 2003; Stamford. Norwalk (CT): Business Communications; 2003.
12. Gilman J W, Awad W H, Davis R, Morgan A B, Trulove P C, DeLong H C, Sutto T E, Mathias L, Davies C , Chiraldi D. Improved thermal stability of crown ether and imidazolium treatments for flame retardant polymer-layered silicate nanocomposites. In: *Flame Retardants* 2002, London: Interscience Publications Ltd.; 2002, pp. 139–146.
13. Awad W H, Gilman J W, Nyden M, Harris, Jr. R H, Sutto T E, Callahan J, Trulove P C, DeLong H C, Fox M. Thermal degradation studies of alkyl-imidazolium salts and their application in nanocomposites. *Thermochimica Acta* 2004; 409: 3–11.
14. Kandola B K, Nazaré S , Horrocks A R. Thermal degradation behaviour of flame retardant unsaturated polyester resins incorporating functionalised nanoclays. In: Le Bras M, Wilkie C A, Bourbigot S, Duquesne S, Jama C editors. *Fire Retardancy of Polymers: New Applications of Mineral Fillers*. London: Royal Society of Chemistry; 2005, pp. 147–160.
15. Pramoda K P, Liu T, Liu Z, He C, Sue H-J. Thermal degradation behaviour of polyamide 6/clay nanocomposites. *Polym. Degrad. Stab.* 2003; 81: 47–56.
16. Beyer G. Flame retardant properties of EVA-nanocomposites and improvements by combination of nanofillers will aluminium trihydrate. *Fire Mater.* 2001; 25: 193–197.
17. Monticelli O, Musina Z, Frache A, Bellucci F, Camino G and Russo S. Influence of compatibilizer degradation on formation and properties of PA6/organoclay nanocomposites. *Polym. Degrad. Stab.* 2007; 92: 370–378.
18. Langat J, Bellayer S, Hudrlik P, Hudrlik A, Maupin P H, Gilman Jr J W, Raghavan D. Synthesis of imidazolium salts and their application in epoxy-montmorillonite nanocomposites. *Polymer* 2006; 47: 6698–6709.
19. Yasue K, Katahira S, Yoshikawa M , Fujimoto K. In situ polymerisation route to nylon 6-clay nanocomposites. In: Pinnavia T J, Beall G W (eds) *Polymer-clay Nanocomposites*. In Wiley Series in Polymer Science, Chichester and New York: Wiley; 2000, pp. 111–126.
20. Sinha Ray S, Okamoto M. New polylactide/layered silicate nanocomposites: 6. Melt rheology and foam processing. *Macromol. Mater. Eng.* 2003; 288: 936944.
21. Matayabas J C, Turner S R. Nanocomposite technology for enhancing the gas barrier of polyethylene terephthalate. In: Pinnavia T J, Beall G W (eds) *Polymer-clay Nanocomposites*. In Wiley Series in Polymer Science, Chichester and New York: Wiley; 2000, pp. 207–226.
22. Matayabas J C, Turner S R, Sublett B J, Connell G W, Barbee R B. PCT Int. Appl. WO 98/29499 (to Eastmann Chemical Co., 7/9/98).

23. Davies R D, Gilman J W, VanderHart D L. Processing degradation of polyamide 6 montmorillonite nanocomposites. Proceedings of the 13th conference on Advances in Flame Retardant Polymers; 2002; Stamford. Norwalk (CT): Business Communications; 2003.
24. Wu D, Wu L, Zhang M, Wu L. Effect of epoxy resin on rheology of polycarbonate/ clay nanocomposites. *Europ. Polym. J.* 2007; 43: 1635–1644.
25. Horrocks A R, Davies P J, Alderson A, Kandola B K. The challenge of replacing halogen flame retardants in textile applications: Phosphorus mobility in back-coating formulations. Proceedings of the 10th European conference on Flame Retardant Polymeric Materials (FRPM05), 6–9 September 2005. In: Schartel B (ed.) *Advances in the Flame Retardancy of Polymeric Materials: Current Perspectives at FRPM05*, Herstellung und Verlag, Berlin; 2008, pp. 141–158.
26. Schartel B. Considerations regarding specific impacts of the principal flame retardant mechanisms in nanocomposites. In: Morgan A B, Wilkie C A (eds) *Flame Retardant Polymer Nanocomposites*. Wiley Interscience; New Jersey, USA; 2007, pp. 107–129.
27. Beyer G, Alexandre M, Henrist C, Cloots R, Rulmont A, Jérome R, Dubois P. Preparation and properties of layered silicate nanocomposites based on ethylene vinyl acetate copolymers. *Macomol. Rapid Commun.* 2001; 22: 643–646.
28. Beyer G. Flame retardancy of nanocomposites from research to technical products. *J Fire Sci.* 2005; 23: 75–87.
29. Beyer G. Flame retardant properties of organoclays and carbon nanotubes and their combinations with alumina trihydrate. In: Morgan A B, Wilkie C A (eds) *Flame Retardant Polymer Nanocomposites*. Wiley Interscience; New Jersey, USA; 2007, pp. 163–190.
30. Beyer G. Carbon nanotubes as flame retardants in polymers. *Fire Mater.* 2002; 26: 291–294.
31. Beyer G. Progress with nanocomposites and new nanostructures. In: *Flame Retardants 2006*, London: Interscience Publications Ltd.; 2006, pp. 123–133.
32. Beyer G. Flame retardancy of thermoplastic polyurethane and polyvinyl chloride by organoclays. *J Fire Sci.* 2007; 25: 65–78.
33. Lan T, Qian G, Liang Y, Cho J W. FR application of plastic nanocomposites. Technical paper, Nanocor® Inc, 1500 West Shure Drive, Arlington Heights, IL, USA, http://www.nanocor.com/tech_papers/FRAppsPlastic.asp.
34. Horrocks A R, Kandola B K, Padbury S A. The effect of functional nanoclays in enhancing the fire performance of fibre-forming polymers. *J Text. Inst.* 2005; 94: 46–66.
35. Horrocks A R, Kandola B K, Padbury S A. Effectiveness of nanoclays as flame retardants for fibres. In: *Flame Retardants 2004*, London: Interscience Publications Ltd.; 2004, pp. 97–108.
36. Horrocks A R, Kandola B K, Padbury S A. Interaction between nanoclays and flame retardant additives in polyamide 6 and polyamide 6.6 films. In: Le Bras M, Wilkie C A, Bourbigot S, Duquesne S, Jama C (eds) *Fire Retardancy of Polymers: New Applications of Mineral Fillers*, London: Royal Society of Chemistry; 2005, pp. 223–238.
37. Wang D, Echols K, Wilkie C A. Cone calorimetric and thermogravimetric analysis evaluation of halogen-containing polymer nanocomposites. *Fire Mater.* 2005; 29: 283–294.
38. Chigwada G, Wilkie C A. Synergy between conventional phosphorus fire retardants and organically-modified clays can lead to fire retardancy of styrenics. *Polym.*

Degrad. Stab. 2003; 81: 551–557.
39. Hornsby P R, Rothon R N. Fire retardant fillers for polymers. In: Le Bras M, Wilkie C A, Bourbigot S, Duquesne S, Jama C (eds) *Fire Retardancy of Polymers: New Applications of Mineral Fillers*, London: Royal Society of Chemistry; 2005, pp 19–41.
40. Lomakin S, Zaikov G E, Koverzanova E V. Thermal degradation and combustibility of polypropylene filled with magnesium hydroxide micro-filler and polypropylene nano-filled aluminosilicate composites. In: Le Bras M, Wilkie C A, Bourbigot S, Duquesne S, Jama C (eds) *Fire Retardancy of Polymers: New Applications of Mineral Fillers*, London: Royal Society of Chemistry; 2005, pp. 100–113.
41. Song L, Hu Y, Lin Z, Xuan S, Wang S, Chen Z, Fan W. Preparation and properties of halogen-free flame-retarded polyamide 6/organoclay nanocomposite. *Polym. Degrad. Stab.* 2004; 86: 535–540.
42. Fu M, Qu B. Synergistic flame retardant mechanism of fumed silica in ethylene-vinyl acetate/magnesium hydroxide blends. *Polym. Degrad. Stab.* 2004; 85: 633–639.
43. Gilman J W, Kashiwagi T. Polymer-layered silicate nanocomposites with conventional flame retardants. In: Pinnavia T J, Beall G W (eds) *Polymer-clay Nanocomposites*. In Wiley Series in Polymer Science, Chichester and New York: Wiley; 2000, pp. 193–206.
44. Duquesne S, Bourbigot S, Le Bras M, Jama C, Delobel R. Use of clay-nanocomposite matrixes. In: Le Bras M, Wilkie C A, Bourbigot S, Duquesne S, Jama C (eds) *Fire Retardancy of Polymers: New Applications of Mineral Fillers*, London: Royal Society of Chemistry; 2005, pp. 239–247.
45. Bourbigot S, Duquesne S. Intumescence and nanocomposites: a novel route for flame retarding polymeric materials. In: Morgan A B, Wilkie C A (eds) *Flame Retardant Polymer Nanocomposites*. Wiley Interscience; New Jersey, USA; 2007, pp. 131–162.
46. Bourbigot S, Le Bras M, Dabrowski F, Gilman J W, Kashiwagi T. PA-6 clay nanocomposite hybrid as char-forming agent in intumescent formulations. *Fire Mater.* 2000; 24: 201–208.
47. Vyver-Berg F J, Chapman R W, Great Lakes Chemical Corporation, World Patent WO 01/10944 15 February 2001.
48. Gilman J W. Flammability and thermal stability studies of polymer layered-silicate (clay) nanocomposites. *Applied Clay Science* 1997; 15: 31–49.
49. Bourbigot S, Devaux E, Rochery M, Flambard X. Nanocomposite textiles: new routes for flame retardancy. Proceedings from the 47th International SAMPE Symposium, 2000, 12–16 May 2002; Volume 47, pp. 1108–1118.
50. Bourbigot S, Devaux E, Flambard X. Flammability of polyamide-6/clay hybrid nanocomposite textiles. *Polym. Degrad. Stab.* 2002; 75: 397–402.
51. Kashiwagi T, Shields J R, Harris Jr R H, Awad W A. Flame retardant mechanism of a polymer clay nanocomposite. Proceedings of the 14th conference 'Advances in flame retardant polymers'; 2003; Stamford. Norwalk (CT): Business Communications; 2003.
52. Kashiwagi T, Harris Jr R H, Zhang X, Briber R H, Cipriano B H, Raghavan S R, Awad W H, Shields J R. Flame retardant mechanism of polyamide 6-clay nanocomposites. *Polymer* 2004; 45: 881–891.
53. Drysdale D. *An Introduction to Fire Dynamics*, 2nd Edition. Chichester; John Wiley and Sons: 1999, pp. 212–222.
54. Solarski S, Mahjoubi F, Ferreira M, Devaux E, Bachelet P, Bourbigot S, Delobel R, Murariu M, Da Silva Ferreira A, Alexandre M, Degée P, Dubois P, (Plasticized)

Polylactide/clay nanocomposite textile: thermal, mechanical, shrinkage and fire properties. *J. Mater. Sci.* 2007; 42(13): 5105–5117.
55. Levchik S V, Weil E D. Combustion and fire retardancy of aliphatic nylons. *Polym. Int.* 2000; 49: 1033–1073.
56. Horrocks A R, Price D, Tankard C, unpublished results, 1995 see also Tankard C. Flame retardant systems for polypropylene, M.Phil Thesis 1995, University of Manchester, Manchester, UK.
57. Zhang S, Horrocks A R. A review of flame retardant polypropylene fibres. *Prog. Polym. Sci.* 2003; 28: 1517–1538.
58. Lewin M, Weil E D. Mechanisms and modes of action in flame retardant polymers. Horrocks A R and Price D (eds) *Fire Retardant Materials.* Cambridge: Woodhead Publishing Ltd; 2001, p. 39.
59. Horrocks A R, Wang M Y, Hall M E, Sunmomu F, Pearson J S. Flame retardant textile back-coatings. Part 2: Effectiveness of phosphorus-containing retardants in textile back-coating formulations. *Polym. Int.* 2000; 49: 1079–1091.
60. Zhang S, Horrocks A R, Hull T R, Kandola B K. Flammability, degradation and structural characterization of fibre-forming polypropylene containing nanoclay-flame retardant combinations. *Polym. Degrad. Stab.* 2006; 91: 719–725.
61. Wang D-Y, Wang Y-Z, Wang J-S, Chen D-Q, Zhou Q, Yang B, Li W-Y. Thermal oxidative degradation behaviours of flame-retardant copolyesters containing phosphorous linked pendent group/montmorillonite nanocomposites. *Polym. Deg. Stab.* 2005; 87: 171–176.
62. Smart G, Kandola B K, Horrocks A R, Nazaré S, Marney D. Polypropylene fibres containing dispersed clays having improved fire performance. Part II: Characterisation of fibres and fabrics from nanocomposite PP blends. *Polym. Adv. Technol.*, 2008; 19: 658–670.
63. Hicks J, Flame retardant investigations in acrylic fibre-forming copolymers, PhD Thesis 2005, University of Bolton, Bolton, UK.
64. Horrocks A R, Hicks J, Davies P, Alderson A, Taylor J, presented at the 11th European Conference on Fire Retardant Polymers, Bolton, UK, July 2007, in *Fire Retardancy of Polymers: New Strategies and Mechanisms*, Hull T R and Kandola B K (eds), The Royal Society of Chemistry, Cambridge, UK, 2009: 307–330.
65. Bourbigot S, Devaux E, Rochery M. Polyurethane/clay and polyurethane/POSS nanocomposites as flame retarded coating for polyester and cotton fabrics. *Fire Mater.* 2002; 26: 149–154.
66. Bourbigot S, Le Bras M, Flambard X, Rochery M, Devaux E, Lichtenhan J D. Polyhedral oligomeric silsesquioxanes: applications to flame retardant textiles. In: Le Bras M, Wilkie C A, Bourbigot S, Duquesne S, Jama C (eds) *Fire Retardancy of Polymers: New Applications of Mineral Fillers*, London: Royal Society of Chemistry; 2005, pp. 189–201.
67. Horrocks A R, Davies P J, Kandola B K, Alderson A. The potential for volatile phosphorus-containing flame retardants in textile back-coatings. *J. Fire Sci.* 2007; 25(6): 523–540.
68. Kandola B K, Horrocks A R. Composites. Horrocks A R and Price D (eds) *Fire Retardant Materials.* Cambridge: Woodhead Publishing Ltd; 2001, Chapter 6, pp. 182–203.
69. Horrocks A R, Kandola B K. Flammability and fire resistance of composites. Long A C (ed.) *Design and manufacture of textile composites*. Cambridge: Woodhead Publishing Ltd. 2005, Chapter 9, pp. 330–363.
70. Sorathia U. Improving the fire performance characteristics of composite materials

for naval applications. In proceedings of the conference Fire and Materials, 2005, London: Interscience Communications, 2005, pp. 415–424.
71. Wilkie C A. Fire retardancy of vinyl ester nanocomposites: synergy with phosphorus-based fire retardants. *Polym. Degrad Stab.* 2005; 89: 85–100.
72. Nazaré S, Kandola B K, Horrocks A R. Flame-retardant unsaturated polyester resin incorporating nanoclays. In: special edition *Polymers for Advanced Technologies*, Schartel B and Wilkie C A (eds), 2006; 17: 294–303.
73. Tien Y I, Wei K H. High-tensile-property layered silicates/polyurethane nanocomposites by using reactive silicates as pseudo chain extenders. *Macromolecules* 2001; 34: 9045.
74. Sinha Ray S, Okamoto M. New polylactide/layered silicate nanocomposites: 6. Melt rheology and foam processing. *Macromol. Mater Eng.* 2003; 288: 936944.
75. Han X, Zeng C, Lee L J, Koelling K W, Tomasko D L. Extrusion of polystyrene nanocomposite foams with supercritical CO_2. *Polym Eng Sci.* 2003; 43: 1261.
76. Liggat J J, Daly J H, McCulloch L, MacRithcie E, Petherick R A, Rhoney I. Synthesis, thermal and crib 5 characterisation of nanocomposite flexible foams, presented at the 11th European Conference on Fire Retardant Polymers, Bolton, UK, July 2007.
77. Consumer Protection Act (1987), the Furniture and Furnishings (Fire) (Safety) Regulations, 1988, SI1324 (1988), London, HMSO, 1988.
78. Liggat J J, Pethrick R A and Roney I. Fire retarded flexible nanocomposite polyurethane foams. International Patent Application PCT/GB2005/002600, 12 January 2006 and WO/2006/003421.
79. Morgan A B, Wilkie C A (eds) *Flame Retardant Polymer Nanocomposites*. New Jersey: Wiley Interscience; 2007.
80. Morgan A B, Wilkie C A. Practical issues and future trends in polymer nanocomposite flammability research. In: Morgan A B, Wilkie C A (eds) *Flame Retardant Polymer Nanocomposites*. New Jersey: Wiley Interscience; 2007, chapter 12, pp. 355–400.

7
Flame retardant/resistant textile coatings and laminates

A R HORROCKS, University of Bolton, UK

Abstract: The area of textile coatings and lamination presented here builds on the review of developments within the general area of flame retardant developments in textiles in Chapter 2. In the special case of coated textiles and laminates, the fibre and textile components are often present to the same level in terms of mass ratio as the coating or lamination resin present. Consequently, the burning behaviour of both fibre and resin and their mutual flame retardancy require to be considered.

Coated textiles usually have other main functions such as water repellency and waterproofing and weather resisting properties as well as requiring to be flexible and so contain potentially flammable plasticisers in addition to flame retarding components. Currently used polymers are discussed alongside the flame retardants specifically selected for use in coating and laminating applications. Such flame retardants are often chosen because they have minimal effects on the physical properties of the final products and, in some case, may act as plasticisers if in liquid form. These include phosphorus-containing, halogen-containing and inorganic retardants including magnesium and aluminium hydroxides, zinc stannates and borates. Intumescent systems are particularly useful since their normal water solubility is offset by the encapsulating matrix polymer and their expanding chars yield greater levels of fire protection than the initial thickness of the coated textile or laminate might suggest.

The challenges of replacing the very effective halogen-containing formulations, particularly in back-coatings for furnishing applications are discussed alongside recent research into the possible usefulness of nanoparticulates.

Finally, while a number of coating and lamination technologies are well-established, discussions of novel or smart ways of introducing flame retardancy and improved fire performance are presented. Here the challenge of generating application levels of nanocoatings necessary to promote acceptable flame resisting properties is discussed. This is followed by a review of recent research in which plasma-coatings have been used to improve flame retardancy of the underlying substrates. It is in these areas that perhaps major future innovatory developments will take place.

Key words: textile, fibres, coating, back-coating, lamination, natural and synthetic rubber, poly(vinyl chloride), polyvinyls, formaldehyde resins, polyurethane, acrylics, silicones, polyfluorocarbons, melt polymers, flame retardant, phosphorus, halogen, halogen replacement, inorganic, intumescent, nanoparticles, nanocoatings, plasma.

7.1 Introduction

In Chapter 2, the recent developments for flame retardant textiles have been reviewed with respect to those which relate to specific textile and/or fibre types. Within this review, the author has briefly included any recent advances in the flame retarding of textile coatings. However, because coated textiles include a wide range of materials in which flame retardancy is only one property (e.g., tarpaulins, awnings and outdoor textiles, which require waterproof and weather resistant properties) and also overlaps the area of laminated textile materials (e.g., airbags and seating composites for automotive and other transport applications, decorative textile laminates, etc.), this chapter considers this very wide group of materials and the flame retardant solutions in existence and under development. Some overlap with the contents of Chapter 2 will be inevitable.

In terms of terminology, coating relates to the application of a continuous or discontinuous layer on to the surface or within the structure of the substrate, thereby generating a heterogeneous fibre/polymer composite. The substrate is usually a textile fabric but may also be a yarn or even a single filament in some cases. Lamination, on the other hand, involves the same coating process but to which a second fabric or even solid surface is applied and the coating material acts as the adhesive. While the available coating technologies are quite numerous and varied,[1] from a physical property of the applied polymeric formulation viewpoint they include the following:

- solvent-based systems
- chemically cured systems and
- hot melt processes.

With the need for reduced volatile organic species (VOCs) produced in the workplace and rising costs of solvents, chemically cured and hot melt formulations have become more popular in recent years. Furthermore, the former are often applied as polymer dispersions in aqueous media. In all cases, the selection of the flame retardant, if present, must be made in the knowledge that it must be compatible with the appropriate solvent or aqueous dispersion or be melt stable in the particular polymer present.

The established and lamination coating technologies for textiles have been recently reviewed by Woodruff,[1] although those specifically relevant to flame retardant applications have not been highlighted. However, most types of coating and lamination systems processes incorporate flame retardancy only if required by the application performance requirements. For example, automotive seatings and linings are often laminated structures comprising at least two different textile substrates, a face and backing fabric, in combination with polymer coatings on the reverse face to create mouldability and adherence to the underlying automotive component surface. For European and US markets, the whole composite must pass the internationally accepted US Federal Motor Vehicle Safety Standard FMVSS 302 horizontal burning test in which a clamped

356 mm (14 in) × 102 mm (4 in) specimen not exceeding 12.7 mm (0.5 in) thickness is horizontally mounted and one edge subjected to a burner flame for 15 s (see also Chapter 10). The specimen passes the test if the burning rate is less than 102 mm/s (4 in/min). While this is only a moderate test of flame retardancy, other coated textiles may have to pass far more stringent test conditions. Examples here are simulated leather, coated textile contract furnishing fabrics, which in the UK must pass the ignition criteria defined in BS7176 in which the specimen, mounted around a filling such as polyurethane foam in both seat and back geometries is subjected to a wooden crib (BS 5852: 1990: Source 5). After the crib has been ignited, it burns with an energy output equivalent to two burning sheets of newspaper, about 300 J. The sample must self-extinguish after the crib has burnt out and no evidence of continuing afterflame of smoulder should occur within the total test period of 10 minutes.

These two examples do not represent extremes, and even higher levels of fire resistance may be required in aerospace and military applications. One notable example here is the requirement in commercial aircraft that all internal structures, including textile fabrics laminated to wall panels, must achieve the stringent flammability requirements defined by the US FAA specification FAR 25.853 Part IV Appendix F and described in full in Chapter 20. This regulation covers all commercial aircraft carrying passengers that wish to fly in US and European air space. Such textile/wall material composites may achieve a pass only if the peak release rate is less than 65 kW m^{-2} and, over a 2-minute period, the average heat release rate is also less than 65 kWm^{-2} min^{-1} when exposed to a heat flux of 35 kW m^{-2} in the Ohio State or OSU calorimeter. In such a composite or laminate, the choice of adhesive and back-coating are critical as we have shown in our research into the potential use of exotic animal hair-containing fabrics for use as decorative textiles in executive jets.[2]

In generating respective levels of flame resistance, coated fabrics may comprise a balance of normal non-flame retardant components (e.g., normal polyester or polypropylene yarns and fabrics) in combination with less flammable or inherently flame retardant fibres or fabrics and/or flame retardant coatings. These latter may comprise an inherently flame retardant polymer such as poly(vinyl chloride), PVC, or a flammable one such as an acrylic copolymer or a synthetic rubber in which a flame retardant additive chemical is included. Thus there is a considerable armoury of flame retardant fibres, polymers and additives available in order to ensure that a given coated textile or laminate can achieve a desired level of flame resistance.

7.2 Main types of fire retardant/resistant coatings and laminates

Normally flame retardant coatings are expected to confer a defined level of flame retardancy to the overall coated textile of laminate and their effectiveness is often determined by the flammability of the underlying substrate fibres. In many cases these may be conventional unretarded fibres and blends, e.g. pure cotton, cotton/polyester, 100% polyester, polyamide or polypropylene, etc., or be already flame retarded fibres and blends, e.g. flame retardant (FR) cotton, FR cotton-rich/synthetic fibre blends. However, in the latter case, antagonisms may arise between flame retardant species present in the coating and those present in the underlying fibres/textile and so care must be taken when developing such formulations. The simplest situation is probably that in which the coating contains the flame retardant species and the underlying substrate has no such presence. In the case of inherently flame resistant fibres, antagonisms may still exist if the flame retardant property is conferred by an additive or comonomer unless the underlying chemistries are similar to those already in the flame retardant present in the coating. Thus, for instance, the phosphorus-containing species present in the inherently flame retardant polyester Trevira CS (Trevira GmbH) is similar to many of the phosphorus-containing additives that could be present in a coating formulation. However, if a halogen-based flame retardant is present in either fibre or coating, then it is possible that some antagonistic interactions might occur. Anecdotal evidence suggests that some flame retarded PVC coating formulations when applied to cotton flame retarded with some phosphorus-containing agencies are inferior to the same coating applied to unretarded cotton. Unfortunately, flame retardancy is not an exact science and predicting antagonisms and indeed synergisms is often impossible; empirical formulation is usually the preferred means of achieving a flame retardant, coated textile or laminate having a defined performance in terms of both flammability and other required properties.[3]

Obviously if the underlying fibres are extremely inert, such as glass in woven glass fabrics, then the selection of polymer coating and flame retardant type will be little affected. Coated glass fabrics are particularly applicable to outdoor fabrics such as awnings and here typically PVC plastisols, which are inherently flame retardant, are commonly used.

Typical coating and laminating polymers include the following:[1]

- Natural and synthetic rubbers of which the latter include polyisobutylene (or 'butyl'), styrene butadiene (SBR), poly(butadiene-acrylonitrile) (or 'nitrile'), poly(chloroprene) (or neoprene), chlorosulphonated polyethylene, poly-(fluorocarbon) and silicone elastomers. These may have varying degrees of inherent flame retardancy dependent upon whether halogen, silicon and/or sulphur contents are significant.
- Poly(vinyl chloride) or PVC plastisols and emulsions are widely used to confer the most cost effective balance of both water resistance and flame retardancy because of the high polymer chlorine content. However, environmental concerns regarding their high chlorine and plasticiser contents have encouraged their replacement by alternatives, although at a higher cost.

- Poly(vinyl alcohols) or PVAs have varying degrees of water solubility depending on the degree of saponification of the parent poly(vinyl acetate). They find application in end-uses where wash durability is not of great significance and because of this, flame retardant inclusion not only has to render the very flammable base polymer flame retardant but also possess a similar level of durability.
- Formaldehyde-based resins, including phenol-, urea- and melamine-formaldehydes offer a relatively cheap range of durable, coating polymers in which flame retardants may be introduced. The phenolics have the advantage of relatively high inherent flame resistance.
- Acrylic copolymers (or more simply, 'acrylics') offer a combination of high levels of flexibility and softness as well as some degree of moisture permeability and so find preferred applications in many textile areas where aesthetics are important. Typical applications include curtains and linings, roller and pleated blinds, mattress tickings and bedding to confer dust and microbe impermeability and textile back-coatings as carriers for flame retardants. It is in this last area where significant interest in flame retardancy has lain since the UK furnishing regulations of 1988 were implemented.[4]
- Vinyl acetate copolymers (including cross-linkable varieties) with vinyl chloride and/or ethylene. These are flexible coatings ideal for upholstered furnishings in a manner similar to the acrylics; presence of vinyl chloride will add to the overall flame retardant formulation property.
- Polyurethanes (or PURs) may be applied to textiles by solvents and more recently by hot melt coating and as direct powders. Addition of flame retardants is effective but they have to be selected to suit the application method.
- Silicones offer water repellency and an inherent flame resistance because of their tendency to promote the formation of a silicaceous char and eventually silica. However, when present on synthetic textiles they may prevent melt dripping and so prevent energy being removed from the burning textile as flaming drips. Thus the silicone-coated textile may be quite flammable relative to the textile and resin alone. It is often the case, therefore, that additional flame retardancy is required through inclusion of additives.
- Fluorocarbons are typified by poly(tetrafluoroethylene) (PTFE) although others exist such as fluorinated ethylene polymers (or FEPs) and poly(vinyl fluoride) (PVF) and poly(vinylidene difluoride) (PVDF). All have varying levels of inherent flame resistance which when applied to flame retardant substrates will enhance performance although their presence is often insufficient to fully flame retard a flammable substrate.
- A number of fusible polymers exist that may be applied in powder or hot melt form. These are primarily low, medium and high density polyethylenes, polyamides (e.g., PA6 and PA66), polyesters and polyesters and copolyesters and ethylene-vinyl acetate (or EVA) copolymers. Whilst being inherently

flammable, they offer the opportunity of incorporation of flame retardant additives during the resin compounding stages providing they are stable at the relevant processing temperatures.

Table 7.1 lists these generic polymer types and gives approximate measures of their flammability in terms of limiting oxygen index values[5] noting that for acceptable levels of flame retardancy LOI > 26–27 vol% is required. It should also be noted that actual LOI values depend on sample dimensions, polymer processing history, presence of fillers, etc. Notwithstanding this, it is evident that

Table 7.1 Limiting oxygen index values of typical coating and laminating resins

Polymer or resin	Acronym or trivial name	LOI, vol % oxygen
Natural rubber		19–21
Synthetic rubbers:		
polyisobutylene	butyl rubber	20–21
styrene butadiene	SBR	19–21
poly(butadiene-acrylonitrile)	nitrile rubber	20–22
poly(chloroprene)	neoprene	38–41
chlorosulphonated polyethylene		26–30
poly(fluorocarbon)		>60
silicone elastomers		26–39
Poly(vinyl chloride)	PVC	45–47
Poly(vinyl alcohols) and poly(vinyl acetate)	PVA	19–22
Formaldehyde resins:		
phenolic		21–22
urea		~30
melamine		~30
Acrylic copolymers	acrylics	17–18
Polyurethanes	PURs	17–18
Silicones		≥26
Ethylene-vinyl acetate and related copolymers (emulsions); vinyl chloride presence will increase LOI	EVA; EVA-VC	≥19–20
Poly(fluorocarbons):		
Poly(tetra fluoroethylene)	PTFE	98
Fluorinated ethylene polymer	FEP	~48
Poly(vinyl fluoride)	PVF	23
Poly(vinylidene fluoride)	PVDF	44
Fusible/powders:		
low density poly(ethylene)	LDPE	17–18
high density poly(ethylene)	HDPE	17–18
polyamides	PA6, PA66	24–26
polyesters	PES	20–21
ethylene-vinyl copolymers	EVA	19

certain polymer coating and laminating matrices will have varying levels of inherent flame retardancy (e.g., PVC and other chlorine and fluorine-containing polymers), although the more commonly used polymers and copolymers are quite flammable and so presence of flame retardants is necessary to flame retard both the coating matrix polymer and the underlying textile substrate.

7.2.1 Use of additives

Most flame retardant additives are selected from the range offered across the flame retardant additive field and the reader should consult standard texts and references.[3,6,7] Selecting those for coating and lamination applications is largely dependent upon the following factors apart from flame retardant efficiency:

- compatibility with the matrix polymer; e.g., the flame retardant should mix and disperse well and even dissolve in the polymer if possible at both processing and ambient temperatures;
- have minimal effect on coating/lamination processing efficiency; e.g., the additive should be stable during processing and have minimal effect on rheological properties;
- have minimal effect on overall product properties including aesthetics; e.g., presence of large diameter solid particulate flame retardants will reduce surface lustre and cause unacceptable roughness, additives that are insoluble at ambient temperatures will migrate to the coating surface causing blooming and stickiness.

The specific examples, discussed here, are flame retardants with particular benefits to coating and laminating applications.

Phosphorus-containing agents

Ideally, liquid flame retardants are preferred providing they do not unduly plasticise the polymer film, although in many cases some level of plasticisation is required. Table 7.2 lists typical examples of acceptable flame retardants including the long chain alkyl/aryl-substituted phosphate examples where plasticisation is also required. While Table 7.2 concentrates on single chemical entities, many commercial proprietary flame retardants are formulated mixtures or blends of more than one species which are especially easily achievable when components are liquids. Such blends enable balances of flame retardancy to be achieved while offering acceptable processing and end-product performance.

Where coating or resin layer thicknesses are significant, then solid, particulate agents often based on phosphorus- and nitrogen-containing species such as ammonium polyphosphate and melamine chemistries may be used. These may be part of an overall intumescent system and so are considered in the next section.

Table 7.2 Selected phosphorus-containing flame retardants for use in coating and lamination

Chemical formula/name	Commercial examples	Comments
Triaryl phosphates	Reofos 35-95; Chemtura Phosflex 71B; Supresta*	Proprietary formulations with 7.6–8.0% P
Cresyl diphenyl phosphate	Kronitex CDP; Chemtura	9.1% P
Tricresyl phosphate	Kronitex TCP; Chemtura	8.4% P
Trixylyl phosphate	Kronitex TXP; Chemtura Phosflex 179; Supresta*	7.8% P
Triethyl phosphate	Fyrol TPE; DSBG	17% P
Isodecyl diphenyl phosphate	Phosflex 390; Supresta*	Functions as plasticiser in PVC; 7.9% P
Oligomeric phosphate-phosphonate	Fyrol 51; Supresta	Textile back-coatings; 20.5% P
Cyclic organophospates and phosphonates	Antiblaze CU; Rhodia Pekoflam PES; Clariant Aflammit PE ; Thor	Substantive to PES fibres but may be incorporated in most coating resins; 17% P
Nitrogen-containing polyol phosphate	Exolit OP 920; Clariant	Non-halogen FR for lattices with plasticising effects; 16% P, 9% N

Note: *formerly Akzo

On a final point, P-containing agents usually have efficiencies which depend up on the polymer type and chemistry. This is because they act primarily in the condensed phase by promoting the formation of carbonaceous char and this depends on the polymer matrix structure. Polymers rich in hydroxyl or pendant ester groups favour dehydration and carbonisation under the Lewis acid-driven activity of phosphorus-containing species when heated.[8,9] Consequently, in poor char-formers like poly(ethylene) and polyesters, P-containing agents are not very efficient unless part of an intumescent system.

There is evidence, however, that some phosphorus-containing flame retardants may also function in the volatile phase and this will be discussed with reference to back-coatings for textiles in Section 7.3.3.

Halogen-containing flame retardants

Unlike the phosphorus-containing flame retardants, these are not polymer-specific in that they act primarily in the vapour phase by suppressing the flame chemistry.[10] Bromine-containing agents predominate because not only are they more efficient than similar chlorine-containing species, but also the high atomic weight of bromine ensures that it is present in a high mass fraction within most organo-bromine compounds. Typically for many polymers acceptable levels of flame retardancy are achieved if about 5 wt% bromine is present in the final formulation. In flame retardants, such as the very commonly used decabromodiphenyl ether (decaBDE; see Table 7.3), where bromine contents are as high as 83 wt%, flame retardant presence is often less than 10 wt%, which is quite low compared with most flame retardant polymers containing other additive flame retardants. However, the synergist antimony III oxide (ATO) is usually present.[9] For greatest effectiveness, the molar ratio Br:Sb = 3 (reflecting the possible formation of SbBr$_3$ as an intermediate[10,11]) is commonly used. This equates in the case of decaBDE to a mass ratio Sb$_2$O$_3$:decaBDE = 1:2, thereby ensuring that the total flame retardant concentration present in the polymer may be as high as 15 wt% or so. Recently, a number of tin compounds including zinc stannate (ZS) and zinc hydoxystannate (ZHS) have been shown to be synergistic with halogen-containing flame retardants, but unlike ATO, bromine-containing FR/ZS or ZHS combinations have to be selected for maximum efficiency.[12] These are also briefly discussed in the following section on inorganic flame retardants.

Table 7.3 lists the more commonly used halogen-containing flame retardants which have applications in textile coatings and laminates.

Phosphorus and halogen together in the same molecule often produce additivity and even synergy in terms of flame performance with respect to the contributions of each element present.[13] A number of phosphorus and chlorine-containing flame retardants has been developed and commercially used in

Table 7.3 Halogen-containing flame retardants for coatings and laminates

Chemical formula/name	Commercial examples	Comments
Dibromostryene	Great Lakes DBS; Chemtura	59% Br
Decabromodiphenyl ether	Great Lakes DE-83R; Chemtura FR-1210; DSBG Saytex 102E; Albemarle Myflam and Performax; Noveon	Principal FR for textile back-coatings; 83% Br
Hexabromocyclododecane (HBCD)	Great Lakes CD-75; Chemtura Flacavon H14; Schill & Seilacher FR-1206; DSBG Saytex HP-900; Albemarle	Competes with decaBDE in textile back-coatings; 73% Br
Tetrabromophthalic anhydride and diol	Great Lakes PHT4 and PHT4-DIOL; Chemtura Saytex RB-49; Albemarle	68% Br 46% Br 68% Br
Tetrabromobis-phenol A (TBBA)	Great Lakes BA-59; Chemtura FR-720; DSBG	59% Br
Dedecachloropenta-cyclooctadecadiene	Dechlorane; Occidental	Used in elastomeric coatings (synthetic and silicone); 65% Cl

Table 7.4 Selected phosphorus- and chlorine-containing flame retardants

Chemical formula/name	Commercial examples	Comments
Tris (1,3-dichloroisopropyl) phosphate (TCDP)	Fyrol 38 & FR-2; DSBG	7.1% P, 49% Cl
Oligomeric chloroalkyl phosphate ester	Fyrol 99; DSBG	14% P; 26% Cl
Chlorinated phosphate ester	Antiblaze 78; Albemarle	
Tris (2-chloroisopropyl) phosphate (TCPP)	Fyrol PCF; DSBG	9.5% P; 32.5% Cl

applications often requiring a degree of flexibility such as foams, films, coatings and laminates. Table 7.4 presents a selection of these, which like the phosphorus-containing agents listed in Table 7.2, are high boiling liquids. Because of the halogen present, they are more effective per unit of phosphorus present although, of course, they introduce possible environmental concerns into subsequent formulations.

Inorganic flame retardants

These are typified by compounds such as the hydrated aluminium and magnesium oxides. The former is often referred to as alumina trihydrate (ATH) or aluminium hydroxide and the latter, magnesium hydroxide (MDH).[14] Both release water when heated and this increases the overall endothermicity of the flame retardant polymer and generates water vapour which then dilutes the flame, thereby promoting flame extinction. However, both hydroxides require to be used at high mass concentrations, typically above 50 wt% and then may promote stiffness when used in coatings. They can confer high levels of flame retardancy when present in rigid laminates, however. They have different sensitivities to heat with aluminium hydroxide releasing water (up to 34.6 wt% of initial mass) when heated above 200 °C and so may only be used in low melting fusible polymers such as polyethylene and EVA. On the other hand, magnesium hydroxide, which is more expensive, is stable up to 300 °C and so may be used in many higher-temperature processed polymers such as polypropylene, polyamides and fluorinated copolymers. Neither hydroxide may be used in thermoplastic polyesters since they can catalyse decomposition. When used in textile coatings and thin laminates, particle size control is essential. The coarser grades produced by grinding may have average diameters as high as 35 μm or so, the finer grades, in particular of ATH, are preferred. These are produced by precipitation and can have diameters below 5 μm The finest grades of ATH at about 1 μm may be used in films and for coating fabrics while the coarser grades between 3 and 12 μm find application in polyethylene carpet backing formulations.[15] In order to improve dispersion and rheology, surface-coated variants are generally commercially available.

Other well-established inorganic flame retardants like zinc borate (e.g., Firebrake® ZB, Rio Tinto), while being used primarily in bulk polymeric applications, may be used as an antimony III oxide (ATO) synergist replacement in flexible PVC in waterproof coatings (e.g., tentage, awnings) and carpet backings. Its presence also has a smoke-reducing effect as do ATH and MDH when present in coating formulations. Other inorganic salts used as ATO-replacement synergists include zinc stannate (Flamtard S, William Blythe, UK) and zinc hydroxyl stannate (Flamtard H, William Blythe, UK), both of which have the advantage of very low particle size (1–2 μm) as well as suppressing smoke.

7.2.2 Intumescent systems that form a carbonaceous/vitreous protective layer over the polymer matrix

For coated and laminated textiles requiring high levels of flame barrier properties then it is more usual to incorporate an intumescent system[16,17] within the polymer. These are especially beneficial in polymers such as the polyolefins which lack any char-forming ability and where the intumescent char provides a supportive network preventing melt dripping and restricting the overall burning process. Such formulations may be intumescent in their own right and generate carbonaceous chars independently of the surrounding polymer matrix or they may interact with the matrix so that the flame retardant-polymer together give rise to an expanded, intumescent char when exposed to heat and flame. The majority of these are based on ammonium polyphosphate (APP) and melamine chemistries and selected examples are presented in Table 7.5. All are particulate solids, of which one or more components may be water soluble, and so for water durability they may only be used in hydrophobic polymer matrices which may create dispersion problems during processing. Hence, many commercial particulate examples are coated or microencapsulated either to reduce water solubility and/or to improve polymer matrix compatibility. Furthermore, manufacturers are attempting to reduce particle sizes as shown for APP and melamine phosphates in particular.

While APP is not an intumescent in its own right, it is a powerful char-former when in the presence of oxygen-containing polymers and copolymers. To ensure intumescent action, it used in combination with other agents such as pentaerythritol and melamine.[16] The melamine phosphates shown in Table 7.5 do have a greater degree of inherent intumescent activity since the acid-forming component phosphate is chemically combined with the gas-forming melamine. They also have superior water insolubilities often <1 g/100cm^3 before any subsequent coating or microencapsulation. Particle sizes are often less than normal APP samples and may have particle diameter values of $D_{50} \leq 8\,\mu$m.

170 Advances in fire retardant materials

Table 7.5 Selected intumescent and intumescent component flame retardants

Chemical formula/name	Commercial examples	Comments
Ammonium polyphosphate	*Phase I types*:* Antiblaze MC; Albemarle Exolit AP 412; Clariant FR CROS 480-485; Budenheim	Water solubility ~4 g/100 cm^3
	Phase II* types: Exolit AP 422; Clariant FR CROS 484; Budenheim	Water solubility ~4 g/100 cm^3
	Coated Phase II types: Exolit AP 462 & 463	Microencapsulated version of AP 422; water solubility <0.5 g/100 cm^3
	FR CROS 486; Budenheim FR CROS 487; Budenheim	Silane coated: melamine-formaldehyde (MF) coated: water solubility ~0.1 g/100 cm^3
	FR CROS C30/C40/C60/C70/489; Budenheim	Surface reacted MF, varying particle sizes D_{50} = 7–18 μm; water solubility ≤0.1 g/100 cm^3
Melamine phosphates	BUDIT 310; Budenheim	Dimelamine orthophosphate
	Antiblaze ND; Albemarle	Dimelamine orthophosphate
	BUDIT 311; Budenheim	Dimelamine pyrophosphate
	BUDIT 312; Budenheim	Melamine phosphate
	Antiblaze NH; Albemarle	Melamine phosphate
	Melapur MP; Ciba	Melamine phosphate
	Antiblaze NJ; Albemarle	Melamine pyrophospahte
	Melapur 200; Ciba	Melamine polyphosphate
	BUDIT 3141; Budenheim	Melamine polyphosphate
Other melamine salts	BUDIT 313; Budenheim	Melamine borate
	BUDIT 314/315	Melamine cyanurate
	Melapur MC; Ciba	Melamine cyanurate
Other pentaerythritol derivatives	Great Lakes NH 1197; Chemtura	Phosphorylated pentaerythritol
	Great Lakes NH 1511; Chemtura	Phosphorylated pentaerythritol/melamine salt
Intumescent blends	BUDIT 3077 and related products; Budenheim	
	Antiblaze NW; Albemarle	Melamine phosphate and dipentaerythritol

*Phases I and II refer to different levels of molecular weight, cross-linking and hence crystalline characteristics. Phase I APP variants have much lower degrees of polymerisation and cross-linking and greater water solubilities

Of all flame retardant coating innovations of the last few years, it is probably true to say that those incorporating intumescent flame retardant agents have been the most commonly reported.[2,16–18] Indeed the recent demand for open flame-resistant barrier fabrics in US markets driven by Californian regulations for furnishings (TB 133) and mattresses (TB 129 and 630) and federally by the US Consumer Product Safety Commission (CPSC 16 CFR 1633) for mattresses[19] has encouraged the development of intumescent coatings applied to inherently fire resistant fibre-containing fabrics, including glass which are exemplified by the established Springs Industries products[20] and fabrics from Sandel International Inc.

7.3 Halogen replacement in back-coated textile formulations

One of the major issues that has faced the flame retardant industry for the last 15 years or so, has been the desire to remove from use flame retardant chemicals that have been shown to have an unacceptable level of environmental risk. This whole issue is too complex for this present discussion but an outline has been presented elsewhere.[3] An extensive risk analysis of 16 commonly used flame retardants was undertaken by the US National Academy of Sciences in 2000[21] and current debate continues as shown by following web-sites belonging to organisations such as the US Consumer Product Safety Commission (CPSC), the European Flame Retardants Association (EFRA) and the Bromine Science and Environmental Forum (BSEF). Other chapters within this book (notably Chapters 8 and 9) have covered the environmental challenges posed by halogen-containing flame retardants. In summary, all halogen and more specifically, bromine-containing flame retardants have come under scrutiny, and while some like penta- and octabromodiphenyl ether have been banned, others like decabromodiphenyl ether (decaBDE) and tetrabromobisphenol A (see Table 7.3) have been subjected to risk assessments and have been found to be safe.[22,23] However, in spite of the scientific evidence, the pressures to replace bromine-containing flame retardants and of decaBDE in particular has been and continues to be intense. This is partularly so in the case of textile back-coatings and other coated textiles.

Attempts to reduce the use of brominated flame retardants in textile back-coatings have met with varying degrees of success.[11,24,25] For instance, Fig. 7.1 shows the minimum add-ons applied to back-coated cotton in order for the fabric to achieve a pass to the simulated BS 5852:1979:Part 1 match test. The different conventional formulations cited show a gradual replacement of the decaBDE-antimony III oxide content with a number of bromine-free alternatives including ammonium polyphosphate (APP), a cyclic oligomeric phosphonate (cylicP) (Amgard CU; Rhodia), alumina trihydrate and zinc hydroxystannate. Figure 7.1 shows that these fabrics pass this test when total add-ons are 30 wt% or less with

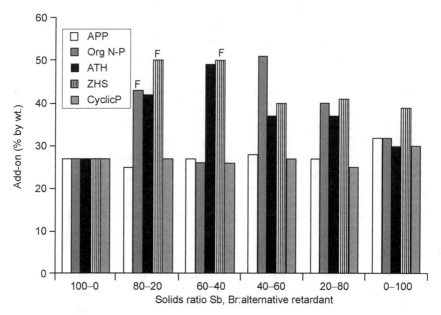

7.1 The minimum percentage add-ons of flame retardants in a back-coating formulation applied to cotton required to achieve a pass to the simulated BS 5852 match test for different formulations (F denotes failure).[11] Key: APP is ammonium polyphosphate, Org N-P is an organonitrogen and phosphorus-containing agent, ATH is alumina trihydrate, ZHS is zinc hydroxyl stannate and CyclicP is a cyclophosphonate agent. (reproduced with permission from Sage Publishing).

respect to substrate.[11] However, while all non-bromine-containing formulations examined pass if present as 100% replacements for the decaBDE-based component, in formulations containing both brominated and non-brominated retardants, only those containing either APP or the cyclic oligomeric agent (CylicP in Fig. 7.1) passed at commercially acceptable add-on levels. In the latter case, however, because it is a liquid, the back-coating formulation was plasticised to the extent of yielding a tacky and unacceptable handle when present at greater than 50 wt% with respect to the original decaBDE-antimony III oxide component concentration. The UK regulations[3] require that flame retarded upholstered fabrics pass the small flame, simulated match test after a water durability test as defined in BS 3651. After application of the 30 min 40 °C water soak, however, both formulations failed to pass the small flame ignition source requirement. The revised version, BS 5852:2006, includes the automatic requirement to submit all samples to be tested to Source 1 to this prior 40 °C water soak.

In a later paper,[26] the performance of selected, less soluble as well as intumescent, phosphorus-based flame retardants was studied. A number were applied in a standard back-coating formulation to 100% cotton and 35% cotton-

65% polyester blend fabrics of typical area densities for furnishing fabrics. Only ammonium polyphosphate-based formulations yielded simulated match passes on both cotton and cotton-polyester substrates and this appeared to be associated with their relatively low temperatures of thermal decomposition behaviour as determined by thermogravimetric analysis. However, the associated poor water durability caused all APP-containing samples to fail after the required 40 °C water soak treatment. A major conclusion from these results was that any phosphorus-based candidate for replacing conventional bromine-containing back-coating formulations would have to decompose and preferably transform to a liquefied state at temperatures well below the ignition temperature of the most flammable fibres present in the supporting fabric. The now-fluid flame retardant could then wet substrate fibre surfaces and diffuse from the back-coating through the fabric and to the front face and prevent ignition by the igniting flame. In the case of cotton fibres which ignite at about 350 °C, this would require the flame retardant component to decompose at about 300 °C or less, a condition shown only by APP. To the author's knowledge, no single phosphorus-based flame retardant fulfils both the low decomposition temperature and high water insolubility criteria at the present time.

Thus in developing a phosphorus flame retardant strategy for the replacement of decaBDE and similar bromine-based formulations, it is evident that the vapour-phase activity of the latter is a key factor in determining their efficiency apart from their excellent insolubility and general intractability. Notwithstanding these prime issues, the outcomes of our previous research[11,26] have led to three strategies that may be proposed to achieve these requirements:

1. the sensitisation of decomposition or flame retarding efficiency of phosphorus-based systems;[27]
2. the reduction in solubility of successful but soluble systems and
3. the introduction of a volatile and possible vapour phase-active, phosphorus-based flame retardant component.[24]

With regard to the first, we have demonstrated that the inclusion of small amounts of certain transition metal salts, notably those of zinc II and manganese II can reduce the onset of decomposition of ammonium polyphosphate from 304 °C to as low as 283 °C in the case of 2 wt% manganese II sulphate addition.[27] When applied in a back-coating formulation with APP, the presence of metal ions increases LOI values slightly (of the order of 1–1.5 vol% for manganese and zinc salts) from 25.1 for APP-only coated cotton to 26.6 vol% in the presence of 2% manganese acetate. However, all coated fabrics still failed the simulated small flame ignition version of BS 5852, which is not perhaps surprising since our earlier research indicated that an LOI value for a coated cotton fabric above 26 and closer to 29 vol% was required for a pass.[11] Furthermore, it was noted that the presence of the transition metal salt reduced the width of the charring area subjected to the flame source when compared with the APP-only sample. It should be pointed out, however, that even if passes had been obtained, the problem of durability to water soaking would still remain.

Recent work by Bourbigot and coworkers, described in Chapter 2, has shown that microencapsulation of otherwise soluble flame retardants like ammonium phosphate with polyurethane shells can improve the durability of coatings containing them.[28] However, the preparation of these microencapsulated agents is not an easy process and different techniques are being developed in order to improve yields.[29,30]

7.4 Role of nanoparticles

The inclusion of nano-particles in coating formulations has been investigated by Bourbigot et al.,[31,32] also discussed in Chapter 2 previously. Briefly, their work showed that both nanoclay and polyhedral oligomeric silsesquioxanes (POSS) when present alone in polyurethane coatings applied to polyester and cotton fabrics, were found to reduce peak heat release values of back-coated fabrics as determined by cone calorimetry. However, neither nano-species increased ignition times nor reduced extinction times. In fact the converse tends to be the case. Thus it was evident that the presence of nanoparticles alone could not impart a flame retarding effect in the conventionally understood sense.

This observation was confirmed in subsequent work in this author's laboratory[24] which showed that the introduction of nanoparticulate clays by themselves had no beneficial effect to a back-coating polymeric film in terms of enhancing its thermal resistance. Furthermore, the introduction of nano-particulate fumed silica to a flame retarded back-coating formulation reduced its effectiveness when a conventional flame retardant like APP was also present. Figure 7.2 shows the effect of adding nanoparticulate fumed silica to a back-coating formulation containing increasing levels of APP. While it must be noted that silica particle dispersion was poor, the presence of higher amounts of silica do not relate to increasing values of limiting oxygen index of a coated fabric fabric.

In other polymers, the incorporation of nanoparticles along with conventional flame retardants indicates that possible synergies may occur, enabling considerably lower amounts of the latter to be used to provide the same level of flame retardant performance. This has been demonstrated by Bourbigot and Duquesne for selected intumescent-montmorillonite clay and -nanosilica polypropylene/polyamide blend formulations such as EVA-APP/PA6-clay and EVA-APP/PA6-nanosilica[33] and also by Beyer for EVA-clay-alumina trihydrate blends for cable applications.[34] We have also demonstrated that the inclusion of nanoclays and APP and similar flame retardants in polyamide 6 and 6.6 films promote synergistic interactions and the opportunity to reduce APP levels by up to 50% while maintaining an acceptable performance as measured by LOI.[35] However, to the author's knowledge there appears to be no published evidence

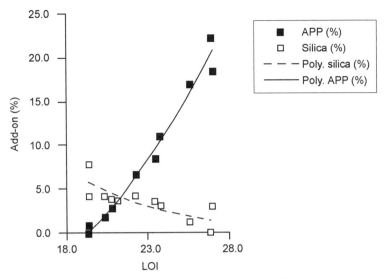

7.2 Effect of ammonium polyphosphate and fumed silica concentrations present in coatings on cotton fabric on the fabric LOI values.[24]

of this potential synergy having been introduced into flame retardant coating formulations for textiles and laminates.

7.5 Introduction of volatile phosphorus-containing species

In Section 7.2.1, the possibility that certain phosphorus-containing flame retardants could act in the vapour phase was briefly mentioned as well as being suggested as a means of developing effective, halogen-free back-coatings in Section 7.3. The literature does not contain much published research that demonstrates clearly that vapour-phase phosphorus activity in fact can occur. For instance, Rohringer et al.[36] have proposed that the relatively superior flame retarding efficiency of tetrakis (hydroxymethyl) phosphonium chloride (THPC)-based flame retardants applied to polyester-cotton blends may be associated with the evolution of volatile phosphine oxides, which then act in the vapour phase and retard the burning polyester component. Day et al.[37] have also provided evidence that the flame retarding efficiency of now-banned tris (2,3-dibromo-propyl) phosphate or 'tris', when applied to polyester, is also a consequence of vapour phase activity of phosphorus species. Hastie and Bonnel[38] used spectroscopic and high pressure sampling mass spectrometry to study possible flame inhibition effects of a number of phosphorus-containing compounds including trimethyl phosphate, phosphoryl choride and triphenylphosphine oxide. When mixed with methane and propane fuels, flame inhibition was noted in diffusion flames burning in air, although in premixed flames (with air), some

P-containing additives could increase flame strength. These same experiments undermined previous considerations that the PO· radical was the predominant species in flames and they proposed that the HPO_2· radical was more significant. It was then suggested that this then interacts with H· and OH· radicals in manner similar to halogen radicals, thus interfering with the main flame propagation reactions as follows:

$$HPO_2\cdot + H\cdot \rightarrow PO\cdot + H_2O$$
$$HPO_2\cdot + H\cdot \rightarrow PO_2\cdot + H_2$$
$$HPO_2\cdot + OH\cdot \rightarrow PO_2\cdot + H_2O$$

Very recent work by Babushok et al.[39] concerning the inhibition of alkane combustion in premixed flames has suggested that in the vapour phase, phosphorus may be more effective than halogen.

In accordance with these findings, our recent work[24,25] initially considered four potentially volatile phosphorus flame retardants selected from their reported boiling or decomposition data. These were the monomeric cyclic phosphate Antiblaze CU (mass loss occurs above 197 °C[26]), tributyl phosphate (TBP) (m.pt. = −80 °C, b.pt. = 289 °C with decomposition), triphenyl phosphate (TPP) (m.pt. 48–52 °C, b.pt. = 244 °C at 10 mm Hg, 5% weight loss at 208 °C) and triphenylphosphine oxide (TPPO) (m.pt. = 156–158 °C), this last being one studied by Hastie and Bonnel.[38] Because interest lay in generating new back-coatings for both polypropylene and cotton fabrics, thermogravimetric studies suggested that TBP would be most suitable because it begins to lose mass, i.e. produces volatiles, at about 150 °C, well below the melting temperature of polypropylene (~165 °C) and the ignition temperature of cotton (~350 °C), although tackiness was anticipated to be a problem. Triphenyl phosphate (TPP) was also selected as the next most volatile agent with volatilisation starting at about 200 °C. Each was combined with an intumescent char-forming agent, Great Lakes NH 1197 (Chemtura) comprising phosphorylated pentaerythritol[26] (see also Table 7.5) in formulations that maintained constant overall flame retardant contents, although in varying volatile:non-volatile phosphorus ratios, namely:

- 250 dry mass units NH1197/100 dry mass units resin
- 200 dry units NH1197, 50 dry units TBP or TPP/100 dry mass units resin
- 150 dry units NH1197, 100 dry units TBP or TPP/100 dry mass units resin.

Formulating novel flame retardant combinations is often fraught with problems which impede or prevent facile sample preparation. Here the liquid tributyl phosphate produced a very tacky coating at dry unit contents beyond 100 parts while TPP, although a solid thereby removing the tackiness problem, tended to agglomerate thus producing a very granular back-coating formulation, which prevented even coating. Nevertheless, formulations were back-coated on to 220 gm^{-2} cotton and 260 gm^{-2} polypropylene fabrics respectively to achieve nominal dry add-ons in the 40–70 wt% range. Table 7.6 shows the LOI and

Table 7.6 Flammability testing results of back-coated cotton and polypropylene fabrics after a 40 °C water soak treatment[25]

Formulation (dry units FR with 100 dry units resin)	Fabric	Dry add-on, %	Simulated BS 5852	LOI, vol%	Comment
NH-1197 (250)	Cotton	41	–	26.1	
NH-1197 (250)	PP	64	–	22.4	
NH-1197 (200) TBP (50)	Cotton	37	Pass	26.7	
NH-1197 (200) TBP (50)	PP	57	Fail	21.5	Visual observation suggests near to pass
NH-1197 (150) TBP (100)	Cotton	52	Pass	26.3	Char length greater than 200:50 analogue
NH-1197 (150) TBP (100)	PP	66	Fail	22.4	
NH-1197 (200) TPP (50)	Cotton	49	–	26.7	
NH-1197 (200) TPP (50)	PP	53	–	21.5	
NH-1197 (150) TPP (100)	Cotton	104	–	26.4	High add-on is a consequence of agglomeration of solids

small-scale simulation test of BS5852 results after a 30 minute 40 °C water-soaking test.

These show that the partial replacement of the char-forming retardant NH 1197 by the volatile TBP and less volatile TPP gave back-coated cotton samples that showed improved performance with the dry mass ratio formulation of 200:50 giving the highest LOI values. It is interesting to note that while the agglomerating effect of TPP at 100 parts presence resulted in a high add-on of 104%, this almost doubled total flame retardant presence with respect to fabric but had minimal effect on the LOI value. This suggests that once the flame retardant presence in the back-coating is sufficient to raise the fabric LOI to just above 26 vol%, this represents an asymptotic maximum value. A similar position may exist for the back-coated polypropylene samples except that this maximum value is just above an LOI value of 22 vol%.

The significance of the simulated match test pass of the NH 1197/TBP-containing coated cotton sample suggested that the presence of the volatile phosphorus-containing component improves flame extinction during front face ignition. This same effect is not obviously seen in the polypropylene fabrics which have relatively low LOI values and excessive thermoplasticity with melting, which is not overcome or supported by the char-promoting elements within the back-coating. In fact the additions of TBP or TPP have little effect on the overall LOI with respect to back-coated PP fabrics containing only NH 1197; however, the 200:50 NH 1197:TBP only just failed the simulated BS 5852 test in spite of an LOI value of only 21.5 vol%.

Further evidence of the volatile phosphorus activity was gained by determining the retention of phosphorus in charred residues from backcoated samples containing the following flame retardants:

- ammonium polyphosphate (Antiblaze MCM, Albemarle)
- melamine phosphate (Antiblaze NH, Albemarle)
- cyclic phosphonate (Amgard CU, Rhodia)
- oligomeric phosphate-phosphonate (Fyrol 51, Supresta)

where the liquid Amgard CU and Fyrol CU species (see Table 7.2) were selected as potentially vapour-phase active flame retardants. In order to produce chars having different thermal histories, back-coated samples of known weight were then placed in a furnace at 300, 400, and 500 and 600 °C for 5 minutes in an air atmosphere. These experiments were not intended to simulate actual combustion conditions but were designed to create chars having residual phosphorus contents that would be dependent on the volatility of the phosphorus-containing moeities present. Table 7.7 summarises the results of char phosphorus content analyses expressed as $\Delta P\%$, the respective phosphorus loss from each char, where ΔP equals the theoretical phosphorus content assuming 100% retention in the char minus the experimental value.[24,25]

It is evident that phosphorus loss is lowest for the fabrics containing the char-promoting Antiblaze MCM (APP) and NH (melamine phosphate) retardants and highest for the Amgard CU and Fyrol 51 liquid components. These two also exhibit the highest coated fabric LOI values suggesting that not only is the phosphorus present volatile, but when released into the flame, it reduces flammability.

Table 7.7 Back-coated fabric LOI values and loss of phosphorus (ΔP) from chars[24]

Flame retardant/ dry parts by weight	Initial add-on, %	LOI, vol%	ΔP, %			
			300 °C	400 °C	500 °C	600 °C
Antiblaze MCM/250	13.9	23.2	0.41	1.35	4.93	3.07
Antiblaze NH/250	11.0	20.8	−0.16	−0.24	0.90	1.87
AmgardCU/250	11.9	26.3	1.91	4.77	10.51	23.95
Fyrol 51/250	16.6	26.1	1.62	2.78	7.64	7.59

7.6 Novel or smart ways of introducing flame retardant coatings to textiles and laminates

Coating levels versus nanotechnological and nanofilm challenges

For a typical retardant to prevent ignition of a typical textile fibre-containing textile, concentrations of between 5 and 20 wt% with respect to the textile are usual. This is because for a phosphorus-containing retardant, for example, phosphorus levels typically between 1 and 3 wt% are required. Most commercial flame retardants contain only 8–20 wt% P (see Table 7.2) which results in chemical levels present on the fabric ranging from as low as 5 up to as high as 25 wt%. With brominated flame retardants in which bromine contents are much higher (see Table 7.3), the levels of flame retardant may be less but then once the additional antimony III oxide synergist is taken into account, total flame retardant concentrations approach similarly high levels. In the particular case of coated textiles, the flame retardant present must not only act on the textile fibres present, but also on the coating resin, which unless it has an inherent flame retardant property (see Table 7.1), will be similarly flammable. Thus levels of flame retardant present in many coatings are often higher than is necessary to be effective on the fabric alone.

Once the level of effective flame retardant has been established to achieve the required flame retardancy, the application requirement may also influence the final concentration and physical form required. In free standing textiles such as curtains, linings and drapes, ignition resistance and self extinction are the sole flame retardant requirements. However, if the flame retardant textile is required to act as a barrier to an underlying surface, such as the filling in upholstered furnishings or an underlying clothing layer in protective clothing, then the flame retardant coating should maintain or even enhance the insulative property of the outer fabric layer, usually by char promotion. For char-forming fibres like the cellulosics and wool, this is quite easily achieved and in back-coated furnishing fabrics comprising these fibres, total dry coating levels of 20–30 wt% of a formulation containing about 2/3 by weight of flame retardant and 1/3 by weight of resin are typical.[3] However, if the fibres are thermoplastic and possibly fusible, e.g. polyester, polyamide and polypropylene, then the coating formulation must be char-promoting in its own right and be able to support the melting/shrinking substrate fibres in order to maintain an effective flame barrier. This is why back-coatings for polyester and polypropylene furnishing fabrics, for example, are applied at levels typically in the 50–100 wt% range.

Thus any novel means of applying flame retardant coatings must be able to achieve such high levels of application. Recent reviews[40–43] highlight the possibilities of conferring films and coatings at nanodimensions on to fibre and textile surfaces in order to achieve high levels novel effects such as hydrophobicity, soil release, self-cleaning, bioactivity, etc. Methods cited include:

- self-assembly of nanolayer films[41]
- surface grafting of polymer nanofilms,[42] and
- synthesis of smart switchable hybrid polymer nanolayers.[42,43]

Even assuming that the conferred nanofilms possessed the required flame retarding functions and efficiencies of conventional coatings, it is likely that none of these will be relevant to the present argument because of the need to achieve high loadings.

The possibility does exist, however, of reducing coating thickness while retaining overall constant levels, if the coating, instead of being applied on textile fabric surfaces is applied only to component fibre surfaces thereby exploiting their very high specific surface areas. In the case of the application of fluorocarbons at about 0.6 wt% to a typical polyester fibre of 10 dtex (~30 μm diameter), the surface layer thickness is calculated to be above 50nm.[39] At microfibre dimensions (~10 μm diameter), the surface layer thickness on the increased fibre surface area reduces to about 10 nm and at sub-microfibre dimensions, even thinner films are theoretically possible. However, flame retardant coatings will be required to be present at ten to twenty times these fluorocarbon concentrations yielding much thicker theoretical film thicknesses as well as problems associated with interfibre adhesion and occlusion of fibre interstices.

Generation of heat reflective finishes at the nanolevel

Coated textiles and laminates are physically quite thin materials when compared with more conventional ones such as bulk polymers. In fire science terms, they are also more often to be defined as thermally thin materials[44] in which the temperature of the surface is assumed to equal the temperature of the interior during heat exposure. In normal flame retardant textiles and coated fabrics and laminates, unless they are quite thick (>3–5 mm), the ability to form a thick, surface insulating char is limited and the underlying fibres soon reach temperatures approaching that of the igniting source (>500 °C) when they degrade and may ignite. Even the most inherently flame resistant fibres such as the poly(meta- and para-aramids), poly(benzimidazole), semicarbons, etc.,[45] are only able to offer a thermal barrier during sustained high heat exposures for limited periods.

However, if we are able to convert a thermally thin textile into one showing thermally thick behaviour, its overall fire protective character will increase and many conventional coatings, especially those comprising intumescent additives, attempt to do this. It is highly unlikely that nanocoatings could promote a similar effect unless they could offer a heat shield property of unusual efficiency.

In the area of heat protective textiles,[46] use is made of the deposition of reflective metal films on to fabric surfaces to reduce the effects of heat radiation from a fire source and it is in this area that nanofilm and nanocoating deposition may have opportunities.

7.7 Plasma-initiated coatings

Plasma technology offers a possible means of achieving novel nanocoatings having the desired thermal shielding effects, although the literature is sparse with regard to reported examples. Shi has demonstrated that low pressure, radio frequency discharge plasma treatment of a number of polymer surfaces including poly(ethylene terephthalate) in the presence of gaseous (CF_4/CH_4) leads to flame retardation.[47] Later studies in which ethylene-vinyl acetate copolymers were plasma-exposed for times up to 15 minutes followed by immersion into acrylamide, gave very high yields of surface grafted poly(acrylamide) and LOI values approaching 24 vol% at 47 wt% grafting levels.[48] The more recent studies of low pressure argon plasma graft polymerisation by Tsafack, Hochart and Levalois-Grützmacher report the successful grafting of phosphorus-containing acrylate monomers (diethyl(acryloyloxyethyl)phosphate (DEAEP), diethyl-2-(methacryloyloxyethyl)phosphate (DEMEP), diethyl-(acryloyloxymethyl)phosphonate (DEAMP) and dimethyl(acryloyloxymethyl)phosphonate (DMAMP)) to polyacrylonitrile (PANA) fabrics (290–300 g/m^2).[49,50] In the presence of a grafting agent, ethyleneglycoldiacrylate (EGDA), graft yields were optimised as high as 28 wt% resulting in limiting oxygen index values up to 26.5 vol%, although after accelerated laundering this reduced to 21 vol%. Fabric samples were first immersed in a solution of monomer in ethanol followed by plasma exposure.

These reported techniques would not be expected to provide nanofilms since the type of grafting achieved may be perhaps best considered as a variation of established polymer surface and textile-grafting procedures[51] and the high yields, 28 wt% in the case of grafted DMAMP,CH_2=CH.CO.O.CH_2.P(CH_3)$_2$, would explain both the level of flame retardancy but poor launderability achieved. Grafted polymer layers tend to have ill-defined physical structures and may not be firmly bonded to the underlying polymer surfaces thereby giving poor durability in terms of cleansing and abrasion resistance.

When extended to cotton (120 and 210 g/m^2), low pressure argon plasma graft polymerisation of these same acrylate monomers,[52] again yielded grafted fabrics having elevated LOI values as high as 26.0 vol% in the case of DMAMP. However, even higher and more acceptable levels of flame retardancy were achieved only if synergistic nitrogen was also present in grafts. This was demonstrated following the grafting of the phosphoramidate monomers, diethyl(acryloyloxyethyl)phosphoramidate (DEAEPN) and acryloyloxy-1,3-bis(diethylphosphoramidate)propane (BisDEAEPN). These yielded LOI values of 28.5 and 29.5 vol% respectively at levels of 38.6 (= 3.36%P) and 29.7 (= 3.29%P) wt%. Launderability was improved when the cross-linking agent, ethyleneglycoldiacrylate (EGDA), was present at high concentration. In the case of BisDEAEPN, after a simulated laundering, graft level reduced to 26.7 wt% and the LOI to 25.0 vol%. The improved durability achieved here is probably associated with the greater reactivity of the plasma-activated cellulose chains compared with those generated on PAN fibre surfaces. In this work, the challenge of achieving the high flame retardant agent levels has been realised and, while no grafted film thicknesses have been reported, they are probably within the micron range and not the nanometer range.

Marosi et al.[53] have used plasma treatment of a polyethylene substrate surface treated by vinyltriethoxysilane and by organoboroxo-siloxane (OBSi), and an OBSi-containing intumescent flame-retarded compound (IFR-OBSi) based on polypropylene, ammonium polyphosphate and pentaerythritol in attempts to improve the oxygen barrier properties of this intumescent coating. Plasma treatment did in fact reduce the oxygen permeability of the coating by one order of magnitude, although the effects that this has on the fire barrier properties are not reported. The possibility that plasma deposition of silicon-based films might improve the flame retardancy of underlying polymer surfaces has been reported by Jama et al.[54] Here normal and nanocomposite polyamide 6 films were activated by a cold nitrogen plasma and then transferred to a reactor containing 1,1,3,3-tetramethyldisiloxane (TMDS) vapour in an oxygen carrier gas for 20 minutes. This remote plasma-assisted polymerisation is similar to that used by Tsafack et al. above except that the monomer is in the vapour phase prior to polymeric deposition. Thermogravimetry showed that increasing the oxygen flow rate considerably increases the thermal stability in air of deposited coatings as the increasingly oxygenated polysiloxane coating transformed to a silica-based structure at about 800 °C. This gives the opportunity for a thermal barrier effect coupled with a moderate increase in flame retardancy of a coated polyamide 6 film and a surprising increase in the flame resistance of the nanocomposite polyamide 6 films, as determined by limiting oxygen index. Figure 7.3 shows that LOI values exceeded 45 vol% for the latter. Char residues mirror respective LOI trends and analysis shows that those from the coated nanocomposite films are largely silica-based while those for coated normal polyamide 6 films are essentially polysiloxane-like. The presence of the nanoclay at 2 wt% appears to have synergised the formation of silica from the plasma-generated coating. The thermal barrier efficiency of the coated nanocomposite films was demonstrated by cone calorimetric analysis under an incident heat flux of 35 kW/m^2. Here the peak heat release rates (PHRR) of plasma-coated nanocomposite films were reduced in intensity by 25% compared to the uncoated films. A subsequent paper[55] demonstrated that on scaling up the experiments using a larger low-pressure plasma source and reactor, thereby enabling larger and more consistently coated samples to be produced. Of particular interest to the present discussion is that the film thicknesses obtained in the earlier and smaller reactor were about 48 μm in thickness whereas those from the larger reactor reduced to only 1.5 μm thickness and the coated nanocomposite polyamide 6 films continued to yield LOI values as high as 48 vol%. Furthermore, if film thickness was increased above 1.5 μm, the LOI reduced to a

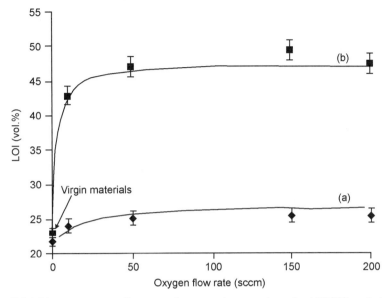

7.3 LOI versus oxygen flow rate of remote plasma polymerised TDMS coated on (a): polyamide 6 and (b): nanocomposite polyamide 6 films[54] (Reproduced with permission from the American Chemical Society).

constant value of about 42 vol%. Once again and as expected, cone calorimetry showed that PHRR values were advantageously reduced for uncoated nanocomposite films (PHRR = 1972 kW/m^2) with respect to normal polyamide films (PHRR = 1102 kW/m^2); in other words, the presence of clay in the films reduced PHRR values by 44%. The additional plasma coating on the nanocomposite PA6 films reduced PHRR by a further 59% to a value of 807 kW/m^2. Analysis of residues after cone calorimetric exposure demonstrated that the coated nanocomposite film had transformed to a silica-like structure and it is this that creates the thermal shielding effect.

In conclusion, it should be noted that plasma technological modification of fibre and textile surface has a history spanning about 40 years and although it has gained commercial significance within industrial sectors such as microelectronics and more recently in improving paint/coating adhesion to plastics for automotive and other applications, its adoption by the textile industry has been slow.[56] One of the main reasons for this is that the majority of successful plasma applications occurred using low pressure plasma and it is only recently that atmospheric pressure plasma technologies have been developed which are considered to be more appropriate to continuous processing of textile fabrics.[57] The desire to use atmospheric pressure plasma increases further the challenge of achieving high levels of surface deposition since plasma polymerisation can be best controlled in low pressure plasmas which have more well-defined plasma zones.[58] Hegemann[59] also states a preference for low-pressure plasma systems, at least at the research level, because the greater mean free paths of ions within such plasma enable greater penetration depths within textile materials, hence the potential for more cohesive nanocoatings. Furthermore, plasma metallisation sputtering techniques, currently used to confer conductive nanolayers on textile surfaces but with the potential for thermally reflective coating deposition, favour the use of low-pressure plasmas. However and notwithstanding these arguments, it is most likely that any commercial plasma process acceptable for the textile industry will have to be based on atmospheric pressure technologies and so future research efforts should be cognisant of this requirement, especially given that the established non-thermal plasma processes previously feasible at low pressures have been successfully transferred to atmospheric pressure conditions as evidenced by the current (2007) range of Dielectric Barrier Discharge, arc-jet, microwave and hybrid sources available.[56,60]

7.8 References

1. Woodruff F A, 'Coating, laminating, flocking and prepregging'. In *Textile Finishing*, Heywood D (ed.), Society of Dyers and Colourists, Bradford (2003), pp. 447–525.
2. Kandola B K, Horrocks A R, Padmore K, Dalton J and Owen T, 'Comparison of cone and OSU calorimetric techniques to assess behaviour of fabrics used for aircraft interiors', *Fire Mater.*, 30 (4), 241–256 (2006).
3. Horrocks A R, 'Flame retardant finishes and finishing'. In *Textile Finishing*, Heywood D (ed.), Society of Dyers and Colourists, Bradford (2003), pp. 214–250.
4. Consumer Protection Act (1987), the Furniture and Furnishings (Fire) (Safety) Regulations, 1988, SI1324 (1988), HMSO, London.
5. Horrocks A R, Price D and Tunc M, 'The burning behaviour of textiles and its assessment by oxygen index methods'. *Text Prog.*, **18**, 1–205 (1989).
6. Horrocks A R and Price D (eds), *Fire Retardant Materials*, Woodhead Publishing, Cambridge (2001).
7. Proceedings of conferences *Recent Advances in Flame Retardant Polymeric Materials*, Lewin M (ed.), volumes 1–18, Business Communications Company, Norwalk, CT, USA (1990–2007).
8. Lyons J W, *The Chemistry and Uses of Flame Retardants.* Wiley Interscience, London (1970).
9. Weil E D, 'Phosphorus-based flame retardants'. In *Flame Retardant Polymeric Materials, Vol. 3.* Lewin M et al. (eds), Plenum, New York, 103–131 (1978).
10. Hastie J W, 'Molecular basis for flame inhibition', *J Res. Nat. Bureau Stds.*, **77A** (6), 733 (1973).
11. Wang M Y, Horrocks A R, Horrocks S, Hall M E, Pearson J S and Clegg S, 'Flame retardant textile back-coatings. Part 1: Antimony-halogen system interactions and the effects of replacement by phosphorus-containing agents'. *J Fire Sciences*, **18** (4), 243–323 (2000).
12. Cusack P, Hornsby P, 'Zinc stannate-coated filler: novel flame retardant materials and smoke suppressants for polymeric materials', *J of Vinyl and Additive Technology*, **5**(1), 21–30 (1999).
13. Weil E D, 'Additivity, synergism and antagonism in flame retardancy'. In *Flame Retardancy of Polymeric Materials*, Kuryla W C and Papa A J (eds), Dekker, New

York, 185–243 (1975).
14. Horn W E, 'Inorganic hydroxides and hydroxycarboinates: their function and use as flame retardants'. In *Fire Retardancy of Polymeric Materials*, Grand A F and Wilkie C A (eds), New York, Marcel Dekker, 285–352 (2000).
15. Weil E D and Levchik S V, 'Flame retardants in commercial use or development for polyolefins', *J Fire Sci.*, **26** (1), 5–44 (2008).
16. Camino G and Lomakin S, 'Intumescent materials'. In Horrocks A R and Price D (eds). *Fire Retardant Materials*, Woodhead Publishing, Cambridge, 318–336 (2001).
17. Le Bras M, Camino G, Bourbigot S and Delobel R (eds), *Fire Retardancy of Polymers: The Use of Intumescence*. Royal Society of Chemistry, London (1998).
18. Horrocks A R, 'Developments in flame retardants for heat and fire resistant textiles – the role of char formation and intumescence', *Polym Deg Stab*, **54**, 143–154 (1996).
19. Consumer Product Safety Commission. Final Rule for the Flammability (Open Flame) of Mattress Sets (4010); http://www.cpsc.gov/LIBRARY/FOIA/foia06/brief/briefing.html Jan. 13, 2006.
20. Tolbert T W, Dugan J S, Jaco P and Hendrix J E, Springs Industries Inc., US Patent, 333174, 4 April 1989; EP19890310591, 14 December 1994.
21. National Academy of Sciences, 'Toxicological risks of selected flame-retardant chemicals', Sub-committee on Flame-retardant Chemicals of the United States National Research Council, Washington, DC; National Academy Press, Washington (2000).
22. European Union Risk Assessment Report for bis(pentabromodiphenyl) ether, European Chemicals Bureau,2003 (closed 26 May 2004): www.bsef.com.
23. European Union Risk Assessment Draft Report for 2,2'6,6'-tetrabromo-4,4'-isopropylidene (or tetrabromobisphenol-a), European Chemicals Bureau, May 2005 (to be finalized end 2007): www.bsef.com.
24. Horrocks A R, Davies P, Alderson A and Kandola B, 'The challenge of replacing halogen flame retardants in textile applications: Phosphorus mobility in back-coating formulations'. In proceedings of 10th European Meeting of Fire Retardant Polymers, FRMP'05, Berlin, 6–9 September 2005; *Advances in the Flame Retardancy of Polymeric Materials: Current perspectives presented at FRPM'05*, Schartel B (Ed.), Norderstedt, Germany, 141–158 (2007).
25. Horrocks A R, Davies P, Alderson A and Kandola, 'The potential for volatile phosphorus-containing flame retardants in textile back-coatings', *J Fire Sciences*, **25** (6), 523–540 (2007).
26. Horrocks A R, Wang M Y, Hall ME, Sunmonu F, Pearson J S, 'Flame retardant textile back-coatings. Part 2: Effectiveness of phosphorus-containing retardants in textile back-coating formulations', *Polymer Int*, **49**, 1079 (2000).
27. Davies P J, Horrocks A R and Alderson A, 'The sensitisation of thermal decomposition of APP by selected metal ions and their potential for improved cotton fabric flame retardancy', *Polym Deg Stab*, **88**, 114–122 (2005).
28. Giraud S, Bourbigot S, Rochery M, Vroman I, Tighzert L, Delobel R and Poutch, F, 'Flame retarded polyurea with microencapsulated ammonium phosphate for textile coating', *Polym Degrad Sta*, **88** (1), 106–113 (2005).
29. Saihi D,Vroman I, Giraud S and Bourbigot, S, 'Microencapsulation of ammonium phosphate with a polyurethane shell part I: Coacervation technique', *React Funct Polym*, **64** (3), 127–138 (2005).
30. Saihi D, Vroman I, Giraud S and Bourbigot S, 'Microencapsulation of ammonium phosphate with a polyurethane shell. Part II. Interfacial polymerization technique',

React Funct Polym, **66** (10), 1118–1125 (2006).
31. Bourbigot S, Devaux E, Rochery M and Flambard X, 'Nanocomposite textiles: New Routes for flame retardancy. In proceedings 47th International SAMPE Symposium, May 12–16, 2002, pp. 1108–1118.
32. Devaux E, Rochery M and Bourbigot S, 'Polyurethane/clay and polyurethane/POSS nanocomposites as flame retardaed coating for polyester and cotton fabrics', *Fire Mater*, **26** (4–5), 149–154 (2002).
33. Bourbigot S and Duquesne S, 'Intumescence and nanocomposites: a novel route for flame retarding polymeric materials'. In *Flame Retardant Polymer Nanocomposites*, Morgan A B and Wilkie C A (eds), Wiley Interscience, John Wiley and Sons, New Jersey, 131–162 (2007).
34. Beyer G, 'Flame retardant properties of nanoclays and carbon nanotubes and their combinations with alumina trihydrate'. In *Flame Retardant Polymer Nanocomposites*, Morgan A B and Wilkie C A (eds), Wiley Interscience, John Wiley and Sons, New Jersey, 163–190 (2007).
35. Horrocks A R, Kandola B K and Padbury S A, 'The effect of functional nanoclays in enhancing the fire performance of fibre-forming polymers', *J Text Inst.*, **94**, 46–66 (2003).
36. Rohringer P, Stensby T and Adler A, 'Mechanistic Study of Flame Inhibition by Phosphonate- and Phosphonium-Based Flame Retardants on Cotton and Polyester Fabrics', *Text. Research J.*, **45**, 586 (1975).
37. Day M, Ho K, Suprunchuk T and Wiles D M, 'Flame retardant polyester fabrics – a scientific examination', *Canadian Textile J.*, **99** (5), 39 (1982).
38. Hastie J W and Bonnell D W, 'Molecular chemistry of inhibited combustion systems', Report NBSIR 80-2169, National Bureau of Standards, Gaithersburg, MD., USA. (1980).
39. Babushok V and Tsang W, 'Inhibitor rankings for alkane combustion', *Combust. Flame*, **123**, 488–506 (2000).
40. Stegmaier T, Dauner M, von Arnim V, Scherrieble A, Dinkelmann A and Planck H, 'Nanotechnologies for coating and structuring of textiles'. In *Nanofibers and Nanotechnology in Textiles*, Brown P J and Stevens K (eds), Woodhead Publishing, Cambridge, 409–427 (2007).
41. Hyde G K and Hinestroza J P, 'Electrostatic self-assembled nanolayer films on cotton fibres'. In *Nanofibers and Nanotechnology in Textiles*, Brown P J and Stevens K (eds), Woodhead Publishing, Cambridge, 428–447 (2007).
42. Luzinov I, 'Nanofabrication of thin polymer films'. In *Nanofibers and Nanotechnology in Textiles*, Brown P J and Stevens K (eds), Woodhead Publishing, Cambridge, 448–469 (2007).
43. Minko S and Motornov M, 'Hybrid polymer nanolayers for surface modification of fibers'. In *Nanofibers and Nanotechnology in Textiles*, Brown P J and Stevens K (eds), Woodhead Publishing, Cambridge, 470–492 (2007).
44. Drysdale D, *Introduction to Fire Dynamics. Second edition.* John Wiley and Sons, Chichester, 193–232 (1999).
45. Horrocks A R, Eichhorn H, Schwaenke H, Saville N and Thomas C, 'Thermally resistant fibres'. In Hearle J W S, *High Performance Fibres*, Cambridge, Woodhead Publishing, 289–324 (2001).
46. Horrocks A R, 'Thermal (heat and fire) protection'. In *Textiles for Protection*, Scott R (ed.), Woodhead Publishing, Cambridge, 398–440 (2005).
47. Shi L S, 'Investigation of surface modification and reaction kinetics of PET in CF_4-CH_4 plasmas', *J Polym Eng*, **19**, 445 (1999).

48. Shi L S, 'An approach to the flame retardation and smoke suppression of ethylene–vinyl acetate copolymer by plasma grafting of acrylamide', *React Funct Polym*, **45**, 85–93 (2000).
49. Tsafack M J, Hochart F and Levalois-Grützmacher J, 'Polymerization and surface modification by low pressure plasma technique', *Uer Phys J Appl Phys.*, **26**, 215–219 (2004).
50. Tsafack M J and Levalois-Grützmacher J, 'Plasma-induced graft-polymerization of flame retardant monomers onto PAN fabrics', Surface and Coatings Technology, 200, 3503–3510 (2006).
51. Bhattacharya A and Misra B N, 'Grafting: a versatile means to modify polymers. Techniques, factors and applications', *Prog Polym Sci*, **29**, 767–814 (2004).
52. Tsafack M J and Levalois-Grützmacher J, 'Flame retardancy of cotton textiles by plasma-induced graft-polymerization (*PIGP*)', *Surface and Coatings Technology*, **201** (6), 2599–2610 (2006).
53. Ravadits I, Tóth A, Marosi G, Márton A and Szép A, 'Organosilicon surface layer on polyolefins to achieve improved flame retardancy through an oxygen barrier effect', *Polym Degrad Stab.*, **74**, 419–422 (2001).
54. Jama C, Queidei A, Goudmand P, Dessaux O, Le Bras M, Delobel R, Bourbigot S and Gilman J W, 'Fire retardancy and thermal stability of materials coated by organosilicon thin films using a cold remote plasma process', *ACS Symposium Series*, 797, 200–213 (2001).
55. Queìdei A, Mutel B, Supiot P, Dessaux O, Jama C, Le Bras M and Delobel R, 'Plasma-assisted process for fire properties improvement of polyamide and clay nanocomposite-reinforced polyamide: A scale-up study'. In *Fire Retardancy of Polymers. New Applications of Mineral Fillers*, Le Bras M, Wilkie C A, Bourbigot S, Duquesne S and Jama C (eds), Royal Society of Chemistry, London, 276–290 (2005).
56. Shishoo R (ed.), *Plasma Technologies for Textiles*, Woodhead Publishing, Cambridge (2007).
57. Herbert T, Atmospheric-pressure cold plasma processing technology'. In *Plasma Technologies for Textiles*, Shishoo R (ed.), Woodhead Publishing, Cambridge, 79–128 (2007).
58. Hegemann D, 'Plasma polymerization and its application to textiles', *Ind Fibre Text Res*, **31**, 99–115 (2006).
59. Hegemann D and Balazs D J, 'nano-scale treatment of textiles using plasma technology'. In *Plasma Technologies for Textiles*, Shishoo R (ed.), Woodhead Publishing, Cambridge, 158–180 (2007).
60. Stegmaier T, Dinklemann, von Arnim V and Rau A, 'Corona and dierlectirc barrier discharge plasma treamtnet of textiles for technicalk applications'. In *Plasma Technologies for Textiles*, Shishoo R (ed.), Woodhead Publishing, Cambridge, 129–157 (2007).

8
Environmentally friendly flame resistant textiles

P J WAKELYN, National Cotton Council of America (retired), USA

Abstract: Virtually all common textiles will ignite and burn. Mandatory and voluntary cigarette and open-flame ignition regulations have been developed to address unreasonable fire risks associated with textile products that require them to be treated with and/or contain flame retardant chemicals to make them flame resistant. There are also national and state laws and potential legislation that regulate the flame retardants that can be used as well as ignition sources (e.g., candles, cigarette lighters, and reduced ignition propensity cigarettes). The focus is on the environmental/ecotoxicological implications of the use of flame retardant chemicals and desirable properties of environmentally friendly flame retardant chemicals. Some examples of potentially environmentally friendly flame retardant applications for textiles are given.

Key words: textiles, flame resistance, toxicity, environmental issues, inherently flame resistant fibers.

8.1 Introduction

Virtually all common textile fabrics will ignite and burn. Textile fabrics burn by two distinctly different processes:

1. *Flaming combustion* (e.g., that caused by an open flame source, such as a match, cigarette lighter or candle), requires that the polymer undergoes decomposition to form small, volatile organic compounds that constitute the fuel for the flame. The combustion of fiber-forming polymers is a very complex, rapidly changing system that is not yet fully understood and may be too complex to understand fully. For many common polymers, this decomposition is primarily pyrolytic with little or no thermo-oxidative character; and
2. *Smouldering* or *glowing combustion* (e.g., that caused by a cigarette or by radiant heat from a remote fire), on the other hand, involves direct oxidation of the polymer and/or chars and other non-volatile decomposition products. Smouldering spreads slowly but can change into flaming combustion.

Since smoulder ignition and open flame ignition are different mechanisms, they usually require different flame retardant treatments to be addressed, and treatments to control open-flame ignition can adversely affect smoulder resistance.[1] Some textile products, e.g. mattresses and upholstered furniture, are a potential fire risk for both smoulder and open-flame ignition.

Mandatory and voluntary standards have been developed by national and state governments, industry groups, and independent standard setting organizations to address unreasonable fire risks to the public from the flammability of textiles. There are various ways to determine compliance or pass/fail (P/F) with the various standards: heat release (peak heat release rate, heat release rate, total heat release), flame spread, char length, mass/weight loss %, ease of ignition (ignitability, time to ignition), etc. To meet flammability regulations, many textile products are treated with and/or contain flame retardant chemicals to make them flame resistant. The designation of flame resistance of a textile is test method dependent. Therefore, the test that the material passes should always be specified when claims of flame resistance are made.

Making a textile flame resistant is complex. What chemical treatment is used for the textile product and how it is used to meet the various flammability standards depends on several factors: performance, cost, and meeting consumer expectations. In addition, the consumer expects textile home furnishings products and apparel to remain unchanged in terms of aesthetics, price, performance and care requirements. The mode (i.e., horizontal, vertical, 45° ⌊) in which the textile product is required to be tested also influences the flame retardant chemicals used, if any chemical treatment is necessary. All of these requirements impact the manufacturer's ability to make textiles both commercially acceptable and flame resistant. The hazard and risk to the public from death, injury and property loss from fire should be balanced with the risk to human health and the environment that is associated with the use of flame retardant chemicals. The benefits for fire prevention should outweigh the risks to health and environment.

This chapter is not intended to be a review of the current developments in flame retardant textiles but will focus upon the environmental/ecotoxicological implications of their use. The references section contains a number of references that will enable the reader to understand the basic science and applications technologies of rendering fibers both heat and flame resistant more fully.[2–9]

8.2 Key environmental/ecotoxicological issues

Exposure can occur at many points in the life-cycle of a flame retardant chemical,[10–12] exemplified by occupational exposure during manufacturing/industrial operations (i.e., in the workplace); exposure to consumers during use; and exposure to the public and the environment from releases that occur from product disposal or from manufacturing facilities. Exposure is typically characterized by pathways and routes, because there are multiple ways people and the environment can be exposed to chemicals. Different routes of exposure, e.g. the primary routes of exposure are dermal absorption, ingestion (oral), and inhalation, can impact whether or not a hazardous substance is a significant risk. An *exposure pathway* is the physical course by which a chemical comes into contact with the individual and/or environment while an *exposure route* is how the chemical penetrates the individual and/or environment.

Both *hazard* – anything that can cause harm, e.g. chemicals that are persistent, bioaccumlative, and toxic by various exposure routes, and *risk* – the chance, high or low, that someone will be harmed by the hazard – are discussed in this chapter. Hazard is determined by the innate nature of the product, whereas risk is a function of both hazard and exposure probability. For example, a compound may be a hazardous substance but an insoluble polymer (unless it breaks down to the monomer or contains unreacted monomer) formed from or with this compound, or if this compound is covalently bonded to the fabric, may not be a risk because there is little or no chance of exposure. Thus, if there is no or very little exposure, then the most hazardous chemicals may pose minimal risk to consumers and the environment. On the other hand, if exposure is not well characterized or understood, it must not be assumed that there is no risk. All of these considerations must be balanced with an understanding of how the public perceives and accepts these factors.

8.3 Desirable properties of environmentally friendly flame retardant chemicals

Flame retardant chemicals should not be persistent, bioaccumulative, or toxic to humans, other animals and ecosystems in general. They should not cause adverse health effects to consumers or adversely affect the general environment (i.e., air, water, soil). There is much concern about the use of flame retardant chemicals in textiles of all kinds.[12-14] Any product that humans can be exposed to either by direct contact, e.g. by children chewing or sucking, dermal or inhalation and through breakdown in the environment and ingestion by animals is of concern.

Persistence is how readily the chemical breaks down in the environment and how long the chemical remains in the environment. For example, some chemicals are detected in the environment for years after use, others can break down to toxic metabolite or harmless substances. Bioaccumulation is whether the chemical accumulates in human and animal tissues, e.g. some flame retardants have been found in fatty tissues and breast milk.

Characterization of toxicity includes hazard identification, dose-response assessment and exposure assessment.[10,11] *Hazard identification* includes determinations as to whether causal relationships exist between the dose of a flame retardant chemical and an adverse health effect. Adverse health end-points include, human (epidemiological studies (both cross-sectional and prospective and retrospective longitudinal), clinical observations, and case reports) and laboratory animal data on neurotoxicity, immunotoxicity, reproductive and developmental toxicity, dermal and pulmonary toxicity, carcinogenicity, and other local and systemic effects. Also *in vitro* data should be reviewed to determine the potential for genotoxicity as well as other toxic effects and to understand the mechanisms of toxic action. Toxicokinetic studies should be reviewed to understand the absorption, distribution, metabolism, and excretion of the flame retardant chemicals. For *dose-response assessment*, the relationships between increases in the dose of a flame retardant chemical and changes in the magnitude of the incidence or severity of toxic effects should be made. *Exposure assessment* should be made but unfortunately, exposures to flame retardant chemicals in treated residential furniture fabrics and for most textile products have not been studied well and there are no quantitative measurements of exposures under relevant exposure conditions. However, it can be assumed that human exposure to flame retardant-treated fabric in homes can occur potentially via skin contact (dermal), ingestion (oral – specifically for infants or children who might suck or chew on fabric), inhalation of particles generated during abrasion of surface fibers, and inhalation of vapors off-gassing from treated fabric.

For many durable flame resistant textiles their potential ecotoxicological character is only an issue during manufacture and post consumer/end-of-life disposal.

Brominated and chlorinated flame retardants

Brominated and chlorinated flame retardants are of particular concern,[10–13] since they have been reported to form dioxins and furans on combustion and many can be persistent, bioaccumulative, and/or toxic to humans. It is a complex argument to compare the breakdown products of materials that are often not called on to perform their primary function, i.e., breakdown in a fire situation, and the effect of the chemicals present during normal service lives on human health and the environment. If brominated or chlorinated compound are used in backcoating to convey flame resistance to fabrics, antimony oxides are normally used with them to obtain effective flame resistance. The presence of antimony may be connected with the formation of dioxins. The antimony oxides may reduce to antimony metal in the hot carbon-rich degraded polymer and the metalic antimony is known to be a good catalyst for cyclizations, etc.

In the USA, the state of California legislature in 2007–08 considered legislation (Assembly Bill 706 [13]) that would ban all brominated and chlorinated flame retardants. Several USA states[13–16] and the EU have banned pentabromo diphenyl ether (pentaBDE) and octabromo diphenyl ether (octaBDE).[14,17] Manufacturers of pentaBDE and octaBDE voluntarily stopped their production in 2004, after these chemicals were found in the environment in fish and human breast milk.[10,12,14] Great Lakes Chemical Corporation (now Chemtura Corporation), was the only USA manufacturer of pentaBDE and octaBDE.[10,14] In the USA in 2007, the State Legislatures of Washington and Maine[15,16] passed bans on all forms of flame retardants known as polybromo diphenyl ethers (PBDEs), including decabromodiphenyl ether (decaBDE). In the USA state of Maine,[15] the use of decaBDE in mattresses and furniture was banned, effective on 1 January 2008. The new law in the USA State of Washington[16] prohibits the use of decaBDE in mattresses from 2008 onwards and will outlaw the manufacture and sale of decaBDE-containing residential upholstered furniture in 2011 provided a safer, technically feasible substitute is found for making the items flame resistant. In 2006, Sweden banned the use of decaBDE, effective 1 January 2007.[14,17,18] In parallel, there has been extensive testing in the EU and elsewhere of decaBDE that indicates that it does not readily breakdown to octa- or other lower isomeric bromodiphenyl ethers (BDEs). Others have reported otherwise.[12] The Swedish government's decision to ban decaBDE contradicts a 10-year EU scientific assessment (environment and human risk assessment reports) which did not identify any significant risk for human health or the environment.[17,19] On that basis, the EU Competent Authorities agreed on 26 May 2004 to finalize the risk assessment with no restrictions in the use of decaBDE. The EU Competent Authorities also agreed during the same meeting that issues relating to the environmental findings of decaBDE in Europe should be addressed by the initiation of a voluntary industry monitoring programme. This was complemented by a further voluntary programme of industrial emissions control in partnership with decaBDE user industries in Europe.[19,20]

Despite the lack of a complete toxicity database on decaBDE and hexabromocyclododecane (HBCD), the 2000 NAS report[11] concluded that they can be used on residential furniture with minimal risk even under the worst-case assumptions. If decaBDE or HBCD are used in backcoatings for upholstered furniture fabrics, antimony trioxide normally is used with them to obtain effective flame resistance. The use of antimony synergists is not limited to furniture but is used in a large number of FR applications. The NAS report[11] recommended on the basis of the hazard indices for non-cancer effects and/or the potential for cancer, that exposure studies needed to be conducted on antimony trioxide and pentoxide to determine whether further toxicity studies need to be conducted. So there is concern about the use of decaBDE and HBCD for backcoating furniture fabrics until toxicity assessments of antimony trioxide and pentoxide indicate that there are not toxicity concerns. HBCD has been classified as a substance of 'very high concern', the uses of which would have to be authorized by the EC, according to the European Chemicals Agency (ECHA).

Selected flame retardant chemicals

The 2000 NAS study[11] reviewed potential non-cancer and cancer effects of 16 selected FR chemicals that could be used to make furniture flame resistant. It concluded that eight chemicals could be used on residential furniture with minimal risk even under the worst-case assumptions and for eight chemicals exposure studies needed to be conducted to determine whether additional toxicity studies were needed. The chemicals that were considered acceptable are discussed in Section 8.5 in examples where they are used and in the above paragraph. The chemicals that needed additional study, included chlorinated paraffins, calcium and zinc molybdates, organic phosphonates (dimethyl hydrogen phosphate), tris (monochloropropyl) phosphates, tris (1,3-dichoropropyl-2) phosphate, and aromatic phosphate plasticizers (tricresyl phosphate). Some of these are potential replacements for pentaDBE for making polyurethane foam flame resistant[10] (see Section 8.5.2).

Cotton research needs

There is a need for improved flame retardant systems both for mild treatments to meet the US general wearing apparel standard[21] and for more severe vertical flame test requirements.[22,23] The aesthetics of the fabric, including minimizing the impact of the finish on hand, odor, yellowing, and dye (shade change, etc.), process requirements, toxicity, and overall cost must all be considered. Some specific research needs include:[9,24]

1. cheaper, durable, non-toxic chemicals that are easier to apply and that can be applied during typical fabric finishing steps;
2. decreased toxicity, e.g., by replacement of brominated and chlorinated flame retardant chemicals that are used with antimony oxides;
3. improved coatings/backcoatings;
4. environmentally friendly finishes and especially formaldehyde free); and
5. flame retardant technical barriers.

Improved systems for cotton/polyester blend textiles also are needed. Cotton is reactive and hydrophilic and polyester is inert and highly hydrophobic. Cotton is a char former on heating but generally maintains some structural integrity. Polyester melts and flows at temperatures above 260 °C. If the two fibers are blended and heated, the molten polyester tends to 'wick' on the cotton char resulting in the phenomenon of 'scaffolding' (i.e., molten polyester is supported by the charred cellulose) causing a more complex flammability problem to control.[25–28] Scaffolding can also occur in layered structures where one layer of thermoplastic is adjacent to a non-thermoplastic, such as flame resistant cotton. This can be a particular problem with flame resistant blackout curtains where the curtain is polyester and the blackout lining is a non-thermoplastic coated fabric. Because of these effects, it is difficult to predict the flammability of cotton/polyester blends on the basis of knowledge of the behavior of the individual component fibers.

8.4 Examples of potentially environmentally friendly flame retardant applications for cotton-based textiles

8.4.1 Apparel and sleepwear

Thermal analysis studies indicate that cotton ignites at ~360–425 °C.[29] Cotton does not melt but forms a char on burning. Thermoplastic fibers melt at lower temperatures, generally below 300 °C and burn at higher temperatures, usually >500 °C. The major factors that influence the flammability of cotton materials and other textiles are airflow, relative humidity of the fabric, the amount of oxygen available, physical factors, e.g. geometry, area density, thickness, etc., chemical factors such as inorganic impurities, heat source type and rate of heating. Washing in hard water with powdered detergents is known to make some fabrics/garments more flammable.[29] The color, i.e., dyes used and depth of shade, can also affect the flammability of a garment. This was demonstrated by a recall by the US CPSC[30] that involved sweaters (84% cotton/16% polyester) in red only, the other colors passed the USA general apparel test (16 CFR 1610[21]). The dye effect may be particularly pronounced with metallized dyes and pigments.

The US general wearing apparel *45° angle test* (1s surface ignition; 16 CFR Part 1610[21]) is a less severe standard than the *vertical flame test* (bottom edge ignition for 3.0 ± 0.2 s; 16 CFR Part 1615/1616[22,23]) used for children's sleepwear. Usually untreated 100% cotton fabrics that are 88.2 g/m^2 (>2.6 oz/yd^2) can meet 16 CFR Part 1610. Surface flash of cotton fabrics in the general apparel test is also a consideration. For raised surface fabrics and fabrics, typically about 88 g/m^2 and below, usually only a mild flame retardant treatment is necessary for the fabric to meet this test or the cotton may be blended with polyester or acrylic.[24,31] For fleece fabrics, denser constructions of 100% cotton with lower naps can be made which may consistently pass the flammability test without treatment.[31] However, denser constructions may not be acceptable in terms of aesthetics or cost.

The US CPSC 16 CFR Part 1615/1616[22,23] vertical flame test for determining children's sleepwear flammability is a rather severe open flame test method. Cotton fabrics require flame retardant treatments to pass the test. Thermoplastic fibers, e.g. polyester, nylon and polypropylene, pass the test because they melt and drip away from the flame. Fiber blends involving core spun yarns can also be used. There are many non-durable treatments, but the test method requires flame retardant treatments that are durable to 50 hot (>60 °C) launderings (water wash/dry cycles). Durable finishes are frequently requested by retailers and apparel companies in order to expand the variety of 100% cotton children's sleepwear, which is in turn highly desired by consumers. Garments that meet the tight-fitting requirements of the standard are exempt from the USA and Canadian sleepwear regulations. These tight-fitting exemptions limit the flexibility of some designers, and such children's sleepwear is not often appealing to large numbers of parents and children. Currently, there are relatively few commercially available flame retardant (FR) chemistries that are durable under the types of conditions required. Some of the reasons include low commercial availability of the chemicals, costs, safety concerns, process control issues, and difficulty in application.[31]

For certain fibers, e.g. wool and silk, and delicate fabrics, treatments with more limited durability would be more appropriate for general wearing apparel because the test and the cleaning requirements are less severe. Resistance to dry-cleaning and hand washing also are typically required here.

The main FR finishes used on cotton are phosphorus-based.[2,9,32] One of the problems with typical phosphorus-based FR treatments on fleece is they often require add-on levels that alter the aesthetic properties of the fleece, resulting in a fabric that is stiff or matted and often has unpleasant odors. Some common types of dyes used on cotton may be affected by pH or oxidation/reduction procedures that are used during the FR treatments.

Durable phosphorous-based flame retardant treatments for cellulosics

1. *Precondensate/NH_3 process* (eg Proban®, Rhodia). The full chemistry of these systems, based on the chemistry of tetrakis (hydroxyl methyl) phosphonium salt (typically chloride as THPC) precondensates followed by ammonia curing, has been reviewed elsewhere.[2,9] When the finishing treatment is complete, an insoluble cross-linked polymer exists in the fibrils of the cotton fibers and is not combined chemically with component cellulosic –OH groups. The highly reducing environment of the ammonia cure stage can cause dye shade changes unless vat dyes are used. However, the flame retardant is durable to in excess of 50 home and commercial launderings and will confer flame retardancy to most 100% cotton and cotton-rich cotton/synthetic fiber blends. Because the product is a condensed product with one of the component chemicals being formaldehyde, gradually it potentially depolymerizes to release trace amounts of formaldehyde, which are normally below levels of concern for apparel for children. The concern about formaldehyde release from furniture is not from FR-chemicals in upholstery fabrics and other textiles used but from composite wood products.[33]

   Despite the lack of a complete toxicity database on THP salts, the 2000 NAS report[11] concluded that it can be used on residential furniture with minimal risk even under the worst-case assumptions.

2. *Reactive phosphorus-based flame retardants.* These compounds react with the –OH groups in the cellulose and can be used for cotton alone and for cotton blends with low synthetic fiber content. They are typically applied by a pad/dry/cure method in the presence of a phosphoric acid catalyst. N-methylol dimethyl phosphonopropionamide (MDPPA) is an example, typified by the Pyrovatex (Ciba®) range of products where R=CH_3 and n=2 in the formula below.

$$RO-\underset{\underset{RO}{|}}{\overset{\overset{O}{\|}}{P}}-[CH_2]_n-\overset{\overset{O}{\|}}{C}-NH-CH_2-OH$$

The acidic curing conditions used can degrade fibers and alter dye shades. Furthermore, Pyrovatex treated textiles gradually can become less flame resistant over time, due to acid hydrolysis, if it is not washed regularly. Because of this Pyrovatex is no longer recommended by the manufacturer (Ciba®) in the USA for use on child's sleepwear. This flame retardant was also assessed as being low risk in the 2000 NAS study.[11]

3. *Phosphate-phosphonate ester FR in conjunction with a wrinkle-resist resin.* Effective applications of this flame retardant, which was originally sold under the name Fyrol 51® (Akzo) and has previously been used for foams and paper, have been reported on fleece.[35] The commercial product has been renamed Fyroltex HP® and is a purer version of the original product. This chemical has a simple application procedure and much less negative effect on the aesthetics of 100% cotton fleece than other durable finishes. This treatment can also be used for cotton and cotton blend carpets that need to pass the burning tablet (methenamine pill) test but has not been used on any retail product in the USA. However, its durability to laundering is inferior to either the previously described THP salt condensates or the reactive phosphorus-based finishes.

4. *Polycarboxylic acids.* Published results[24,32,34,35] indicate that several different carboxylic acids, including maleic, malic, succinic and 1,2,3,4-butanetetracarboxylic acid (BTCA) are effective for improving the flame resistance of carpets and raised surface apparel such as fleece that require mild FR treatment to pass apparel and carpet flammability tests. These carboxylic acids are applied by pad/dry/cure with different catalyst combinations (e.g., sodium hypophosphite or sodium phosphate) and also effective non-formaldehyde containing durable press/easy care agents for cotton fabrics. However, for applications using carboxylic acids there are concerns related to the pH of the treated samples and the effect on dye shade as well as possible strength loss of the fabric.

8.4.2 Military uses

Core spun yarns

The USA military requires significant quantities of flame resistant fabric for tents and uniforms. An important specification, and a difficult one to attain, is FR-finished fabrics with tear strength of at least 2.5 kg. With long staple, combed cotton, it is possible to meet the required tear strength in the greige fabric, but not to retain that strength level in the FR-finished fabric. With core-spinning technology and using only 10% (by weight of the yarn/fabric) gel-spun polyethylene (PE) staple fiber (only in the yarn core), an almost 100% cotton surface fabric that meets the required military specification of tear strength in the filling direction and considerably exceeds it in the warp direction can be produced.[36]

Many cotton fabrics treated with flame retardant and easy-care finishes cannot meet high strength performance standards required by the military and other end uses. Stronger, more durable fabrics can be produced from predominately cotton yarns that are reinforced with high-tenacity manufactured fibers through both intimate blending and filament-core yarns. Military uniforms and tents have been produced from fabrics made with yarns containing about 70% cotton in combination with polyaramids such as Nomex® and Kevlar® (Du Pont), nylon, glass, or polyethylene that were treated with flame retardant (tetrakis (hydroxyl methyl) phosphomium chloride-urea precondensate) and durable-press finishes (dimethylodihydroxyethyleneurea).[37] The military uniforms and tents produced in this research reportedly met flame resistance requirements, preserved the softness, absorbancy, breathability and other desirable properties of the cotton, and had the required strength and durability because of the manufactured fiber content.[37] The environmental footprint of the resulting fabrics and textile products made from these fabrics has not been determined.

8.5 Potentially environmentally friendly flame retardant applications for mattresses, bedclothes and upholstered furniture

8.5.1 Cover and barrier fabrics

In addition to the flame retardant performance, other factors must be considered when choosing a chemical system/means for meeting standards for barriers or for fabrics and other components which may be in intimate contact with the user. There are especially toxicological concerns with some flame retardant chemicals that could be used to meet standards for soft furnishings, as evidenced by the finding of the 2000 NAS study[11] that reviewed potential non-cancer and cancer effects of selected FR chemicals that potentially would be used for upholstered furniture. Despite the lack of a complete toxicity database, the NAS study concluded that eight chemicals (including HBCD, decaBDE, and THP salts) could be used on residential furniture with minimal risk even under the worst-case assumptions and for eight chemicals (including antimony trioxide and pentoxide, dimethyl hydrogen phosphate, tris (1,3-dichloropropyl-2) phosphate and chlorinated parafins; see Section 8.3) exposure studies needed to be conducted to determine whether additional toxicity studies were needed. Manufacturers are required to show a proper concern for the safety of any applied finish in terms of its potential hazard to consumers and public environments. The tendency of a fiber to char (particularly natural fibers) or melt (mainly thermoplastic fibers) can play a vital role in determining the choice of finish and the requirements of specific test methods. Other issues for treatments include effects on physical characteristics, such as hand (soft feel of the fabric), strength, and whiteness/yellowness, and on dyes or pigments. Finally, cost, ease of manufacturing and compatibility with current processes might also be important.

Engineered batting products as fire blocking barriers

Engineered cotton batting properly treated with boric acid ($\simeq 10\%$) and/or blended with inherently flame resistant fibers (e.g., enhanced FR-modacrylic, FR-polyester, Visil, etc.) is both cigarette (principally smoulder) resistant and open flame resistant and can be used like any other padding material.[38–41] Boric acid and borates are safe in consumer products such as futons, mattresses and upholstered furniture, since they have low intrinsic toxicity and the exposure to borates in these products is limited (i.e., toxic doses to humans are unattainable from use of boric acid in mattresses, futons and upholstered furniture).[41] In addition, despite the lack of a complete toxicity database on zinc borate, the NAS report[11] concluded that it can be used on residential furniture with minimal risk even under the worst-case assumptions.

Engineered cotton batting can be used as a drop-in component of fire blocking barriers in mainstream soft furnishings, i.e., mattresses, bedding, and upholstered furniture.[38,39,41] Cotton batting, either boric acid treated or blended with modacrylic should be helpful in meeting the various cigarette resistance and open flame resistance regulations for mattresses, futons, and upholstered furniture. These compete with other nonwoven materials (see further discussions in Section 8.5) not containing cotton as barriers under the fabric/ticking layer to prevent involvement of the filling materials or as filling materials in 'top-of-the-bed' products.

Interior fabric fire blocking barriers

Cotton and other fibers for use in flame barriers generally can have FR treatments and may be blended with inherently FR fibers. This can often reduce the costs of competive products comprising 100% inherently FR fibers as described above for batting. Aramid fibers, and other non-thermoplastic inherently FR fabric or fabrics topically treated with various durable flame retardants can be used as interior barriers to meet the US CPSC open flame mattress standard (16 CFR 1633).[40,42,43]

Core spun yarns made with cotton and inherently fire resistant fibers, as described above, can also be used to make fire barriers for mattresses and furniture.[42,43] A commercially available example of a patented core spun yarn product that can be used as an open-flame barrier fabric for compliance with the US CPSC open flame regulation for mattresses (16 CFR 1633) is Firegard® Brand Products produced by Springs Creative Products Group. This product has a glass fiber core and the cotton is either untreated or treated with a phosphorous compound (see the Firegard® website for more information). It can be used for woven tickings and interliners for contract and residential mattresses, borders and box springs, for knit interliners for luxury and viscoelastic foam quilting and for knitted ticking with a flame barrier built right in.

Cover fabric fire barriers

Here the cover fabric itself provides the barrier and can include fabrics made of leather, wool, polyvinyl chloride (PVC)-coated substrates, polyester microfibers or flame retardant cotton. Topical treatments and backcoatings are commonly used to render these fabrics flame and smoulder resistant although, of course, inherently FR fibers can also be used. The UK standard BS 5852 requires cover fabrics to act as barriers except for cellulosic fiber fabrics.

For most commercial backcoatings, brominated flame retardants and principally decaBDE, are used with antimony trioxide and an acrylic binder and it is possible to obtain the desired barrier effects on most fibers and blends with this system. Requirements for a limited durability or preventing retardant components migrating to the surface of a textile in use limit the choice of chemistry. Even so, backcoating formulations are carefully tailored to suit a given cover fabric's proerties and requirements. Such formulations are closely guarded by both formulator and customer. There is a direct connection between the insolubility, for instance, of decaBDE and its environmental tendency to persist. However, this must be balanced against the tendency of such intractable retardant species bonded within a resin matrix to be released during normal use in the first place. Recent evidence suggests that releases of decaBDE and HBCD from backcoated fabrics are negligible during use.[44] However, since there are studies that indicated that decaBDE can be contaminated with traces of pentaDBE and octaDBE and that decaBDE and HBCD are persistent and bioaccumlative (i.e., found in human tissue), there is still concern about these chemicals as discussed in Section 9.3 as well as potential concern about end-of-life disposal.

Intumescent systems, which often use phosphorus chemistry, have been suggested as possible alternatives but an intumescent based on a saccharide or similar polyol to give char and a phosphorus oxy-anion-containing salt to donate an acid does not meet typical water soak/wash durability requirements and loadings tend to be higher than required with brominated flame retardants.

Finally, and while the use of more environmentally acceptable phosphorus chemistry does not offer the possibility of treating all synthetic fibers, it usually produces excellent charring effects on cellulosic fibers. Horrocks *et al.*[45] have studied such systems during recent years in attempts to develop new non-halogen systems for fabric back-coatings. More recently, they have shown that while intumescents alone cannot fulfil both the flame retardant and durability requirments, addition of a volatile phosphorus-containing species can.[46]

8.5.2 Filling materials

Polyurethane foam

Since pentaBDE is no longer used in the USA to make flexible PU foam flame resistant, replacement species need to be developed.[10] Tris (dichloropropyl) phosphate (TDCP; chlorinated Tris), Firemaster 550 (which may be ecotoxic[10]), proprietary bromine-based and chlorine-based compounds (halogenated aryl esters and tetrabromophthlate diol diester), proprietary organophosphates (e.g., dimethyl hydrogen phosphate and chloroalkyl phosphates) are the main chemicals used in the USA currently.[62] Some of these were also studied in the recent NAS Risk Assessment[11] and it was recommended that they needed further toxicity studies before they could be used on residential furniture with confidence.

Melamine, a principal retardant component for flexible PU foam, is not usually used alone but when blended with other species can be an effective flame retardant system. Melamine blended with pentaBDE was a very good flame retardant system, but when blended with phosphate esters is not as efficient. Melamine is used in the USA in flexible PU foam for so-called California 'TB 117' and 'TB 117+' foams.[47] In Europe and the UK in particular, so-called combustion modified foams pass the UK regulatory requirements for domestic furnishing filling foams (BS 5652:1979:part 2, Source (or crib) 5). Melamine is typically the main FR ingredient, although chlorinated phosphates are frequently blended with it. If halogenated chemicals cannot be used, there are currently not many commercially available options except, for example, melamine. Some reactive organophosphates may be promising alternatives (see Section 8.3 Selected FR Chemicals[10,11]). The question of meeting performance requirements, particularly flexibility and elasticity, when using high loadings of inert additives such as alumina trihydrate, is not easily resolved. Alumina trihydrate was studied in the recent NAS Risk Assessment[11] and the committee determined that it could be used on residential furniture with minimal risk.

However, for some applications, if a flame retardant barrier material is used in mattresses and upholstered furniture, then the PU foam may not need to be flame retarded or combustion modified.

Cotton batting

Garnetted cotton batting containing about 10% boric acid, ground to a very fine particle size and evenly distributed so there is little or no dust-out or leach-out of the boric acid, passes open flame and smoulder tests, e.g. the US CPSC Part 1632 and Part 1633 tests[41]).

To produce a thermally bonded batting, about 10–20% w/w low melt polyester is usually blended with the cotton fiber in the willow prior to garnetting. Also sometimes an inherently flame resistant fiber, e.g. modacrylic fiber (up to about 30% w/w), is blended with the boric acid treated cotton to enhance flame resistance. The resulting batting containing low melt polyester is then heat-treated to form a thermally bonded batting. This can be used as a batting barrier with either high loft or compacted by a needle punch process in mattresses or upholstered furniture to meet open flame and smoulder (cigarette resistant) tests.

Polyester batting

For polyester batting to be used as filling material in comforters that pass a small open flame tests (e.g., TB 604[48]), it usually has to be made of inherently FR-polyester fiber or made by blending conventional polyester fiber with an inherently flame resistant fiber, e.g. modacrylic fiber (about 30–35% w/w). Thus the ecotoxicological problem of topical treatments is rarely an issue.

8.6 Inherently flame retardant fibers

In these fibers, the flame retardant moiety is either present as an additive or chemically bonded to the fiber-forming polymeric molecules present. In both cases, the flame retardant groupings or species are prevented from migrating out of the fiber and so their potential ecotoxicological character is only an issue during their manufacture and end-of-life disposal.

Modacrylics

The modacrylics have been available for over 50 years and are defined as fiber-forming polymers made from resins that are copolymers of acrylonitrile comprising at least 35% but less than 85% by weight of acrylonitrile units ($-CH_2CH[CN]-$) and another comonomer, such as vinyl chloride, vinylidene choride or vinyl bromide. The usual comonomer is vinylidine chloride and to improve flame resistance antimony trioxide can be added to the resin prior to wet or dry spinning or vinyl bromide can also be used as a co-monomer. Today there are only a few commercially available modacrylic fibers, typified by Kanecaron (Kaneka Corporation) and this is vinylidene chloride/antimony oxide-based.[49] However, because of questions being raised about the presence of both halogen and antimony in these fibers, their economic future may be questionable.

Cellulosic/viscose rayon (regenerated cellulose)

These comprise two main types:

1. A *silica-containing* example like VISIL® (Sateri, Finland) is a speciality cellulosic/viscose fiber for flame retardant applications. It is permanently fire

resistant because of the high polysilicic acid complex content (30–33%) built into the fiber in the manufacturing process.[49] A fire-blocking barrier made using VISIL® fibers can be the fire blocking layer in complete seating units.[40] The use of this inherently fire resistant fiber, either alone or in blends, reduces the need for flame retardants in upholstery fabric or foam. It does not melt or flow when in contact with heat or flame and emits essentially no smoke or toxic fumes. VISIL can also be combined with natural fibers such as wool and cotton. VISIL viscose is considered environmentally benign by its manufacturer and it is biologically degradable.

2. The *phosphorus-containing* Lenzing FR® (Lenzing) is a specialty cellulosic/viscose fiber for flame retardant applications. Like Visil®, it is a regenerated cellulosic fiber derived from wood pulp only it contains an organophosphorus additive (Sandoz 5060 or Clariant Sandoflam 5060[49]) introduced during fiber manufacturing process. It is similar to Proban- and Pyrovatex-treated cotton in that it forms a char when in contact with flame and can offer protection from heat and flame in a variety of different applications. It has thermal insulation properties combined with permanent flame resistance and can be used in fire-blocking barriers to meet USA open flame regulations for mattresses. Unlike flame retarded cotton, it offers the opportunity for blending with other inherently FR fibers such as aramids and modacrylics.

Polyester

During the last 30 years or so, a number of inherently flame retardant polyester fibers have been developed but only one, Trevira CS® (Trevira GmbH), has found significant commercial success. Fidion FR produced by Montefiber is another product. In this fiber, the phosphorus-containing comonomer (organo-phosphorus components, e.g. phosphinic acid comonomers and phosphorous additives) confers the required flame retardant properties without the possibility of loss during service.[49] Trevira CS® upholstery materials can be used to meet international fire protection standards. They are considered by their manufacturer to have positive ecological properties. No fumes or effluent are produced in their manufacture and they are economic to clean. Trevira fiber and filament yarns can qualify for the Oeko-Tex 100 Certificate/eco-label, indicating they are free from toxic substances.

Polypropylene

Both halogen and non-halogen additives are incorporated into the melt prior to melt spinning the polypropylene fiber. Sandoflam 5072 (Clariant) and Ciba® Flamestab® NOR™ 116 are used as the additive.[50]

Aromatic polyamide fibers (aramids)

These fibers have been extensively reviewed elsewhere[50] and their key features only are presented here.

Para- and *meta-aramids* are exemplified by the two principal respective commercial Du Pont examples, Kevlar® and Nomex®. Nomex® is a registered trademark for flame resistant meta-aramid material. It is considered an aromatic polyamide, the *meta* variant of the *para*-aramid Kevlar®, which consists of long molecular chains produced from poly (paraphenylene terephthalamide). It is used as a fabric wherever resistance from heat and flame is required coupled with acceptable aesthetics and wear properties. Both the firefighting and vehicle racing industries use Nomex to create clothing and equipment that can stand up to intense heat. All aramids are heat and flame resistant but Kevlar®, having a *para* orientation, can be molecularly aligned and gives higher strength and modulus. *Meta* aramid polymer molecules have lower symmetry and cannot align efficiently during filament formation and so derived fibers have relatively poorer strength although directly comparable with nylons 6 and 6.6.

The major environmental concerns derive from the difficulty of disposal of both manufacted and end-of-life waste but fortunately their high value encourages recycling. During manufacture, potential toxicities of intermediates have also proved to be a challenge. In particular, Kevlar® is synthesized from the monomers 1,4-phenylene-diamine (*para* phenylenediamine) and terephthaloyl chloride in condensation reaction giving hydrochloric acid as byproduct. The result is a liquid-crystalline behavior and mechanical drawing causing the polymer chains to orient in the direction of the fiber. Hexamethylphosphoramide (HMPA) was originally used as the solvent for the polymerization, but toxicology testing showed it produced tumors in the noses of rats. So DuPont abandoned HMPA and used a different solvent, a mixture of N-methyl-pyrolidone and calcium chloride.

Aramid blend fabrics can be used as fire blocking fabrics in aircraft seats and other upholstery seating, where it may be positioned between an outer fabric and an inner foam core to cover and protect the foam.[42] Yarns, fabrics and garments containing modacrylic, cotton and aramid fibers are also used.[37,43]

Polymelamine fiber

Basofil® (BASF) heat and flame resistant fiber is based on patented melamine technology. Basofil® fiber has the same characteristics typical of other common melamine-formaldehyde- based materials: heat stability, low flammability, high wear performance, solvent resistance and ultraviolet resistance. According to its manufacturer, Basofil® fiber meets all environmental regulations with regard to processing and use. It has a Limiting Oxygen Index (LOI) value of 32, low thermal conductivity, excellent heat dimensional stability and it does not shrink, melt or drip when exposed to flame.

Although rather weak, the fiber can be processed on standard textile machinery for woven, knit, and nonwoven fabrics and can be used to make batting or fabric fire barriers for mattresses and upholstered furniture. Again like the aramids, disposal after use creates an environmental challenge.

8.7 Risk management, risk mediation and risk reduction

Organic bromine compounds continue to be used instead of halogen-free alternatives due to a number of factors – low price and the marketplace prefers tried and tested flame retardants. With increasing pressure on halogenated flame retardants, there is much research underway on more environmentally friendly alternatives and their direct replacement by equally effective formulations is proving to be difficult. There are, however, voluntary risk management programmes that offer producers and users of flame retardant chemical best management practices for minimizing emissions of these chemicals to the environment. One example of a voluntary risk management programs is the Voluntary Emissions Control Action Program (VECAP) for commercial brominated flame retardants (BFRs).[20] There are also risk reduction/pollution prevention programs for alternatives to decaBDE in textile applications.

8.7.1 Risk management programs: VECAP™

Although a European Commission risk assessment of decabromodiphenyl ether (decaBDE) completed in 2004 concluded,[14,19,51] 'there is no need for measures to reduce the risks for consumers beyond those that are being applied already,' there remains a perception that it is a problem because it is detected in the environment in low levels due to its persistence. No other indicators for human health and the environment have been found from 10 years of intensive evaluation. Because of the lack of scientifically defined risk from the use of decaBDE and based upon the realization that the normal guidance from a product safety data sheet was not sufficient to direct proper control of the product in use, the Bromine Science and Environment Forum (BSEF[19]) launched in 2004 the

VECAP[TM20] to reduce levels of decaBDE in the environment. Under the program, manufacturers and users of decaBDE for textiles and plastics formed a voluntary action group that agreed to limit releases of the flame retardant into the environment by providing more data on the use of the chemical and establishing and demonstrating control over the fate of the waste. The forum developed 'best practices' guidance to help firms.[51] Through VECAP[TM], the manufacturers and users of brominated flame retardants work together to establish and implement best practices on managing brominated flame retardants to reduce and prevent emissions to the environment. A VECAP[TM] program specific to decaBDE was introduced in the United Kingdom in 2004, extended to other European Union countries in 2005 and then launched in the United States and Canada in 2006. VECAP[TM] is a proactive industry plan to understand, control and reduce emissions of decabromodiphenyl ether. The Code of Practices for Textiles ('Controlling Emission of Persistent Chemicals by Proactive Commitment to Good Practices'[52]) explains the program for textiles. The EU program is currently being extended to other brominated flame retardants and to additional European countries and VECAP North America (USA and Canada) will soon follow. In addition to deca-BDE, the program will cover HBCD and tetrabromo bisphenol-A (TBBPA). HBCD and deca-BDE are used in backcoatings for upholstered furniture fabrics.

VECAP[TM] participants should be well prepared to comply with the European Union's Registration, Evaluation, and Authorization of Chemicals (REACH) legislation.[53,54] REACH requires that manufacturers of chemicals that are made in or imported into the European Union in large volumes, such as brominated flame retardants, register those chemicals and provide information about how they can be safely used. REACH is a complex piece of legislation that will have an impact on a vast spectrum of substances contained within materials, including electronics products, which are made, sold, used, and disposed of across the EU. Under the proposed new system, enterprises that manufacture or import more than one metric ton of a chemical substance per year would be required to register it with the European Chemicals Agency by providing a dossier outlining the properties, uses and safe handling of the substance in question. In a similar manner to the RoHS directive (EU RoHS), firms face the dilemma of adapting their products to meet the EU regulation. REACH provisions will be phased-in over 11 years. Companies can find explanations of REACH in the *guidance documents* and can address themselves to a number of *helpdesks*.[54]

That means the manufacturer must understand in-depth how customers use its products and prepare a report describing how to work safely with the chemical.

8.7.2 Risk reduction strategies for decabromodiphenyl ether

A technical assessment of flame retardant alternatives to decaBDE/antimony oxide, the preferred flame resistant system available for furniture textile applications, was performed for the Swedish Chemicals Inspectorate.[55] This assessment suggested that intumescent systems and phosphorus chemistry were among the flame retardant systems most likely to be suitable to replace antimony-decaBDE based on research publish by Horrocks.[56] These systems included, phosphorus compounds, aluminum and zinc borate, swelling (intumescent) systems, new synergistic combinations, surface-active fiber systems, and systems of graft polymers.[56] In addition, inherently flame resistant fibers combined with combustible fibers may be used in some applications.

Risk reduction strategies should be based on a solid understanding of what works and how effectively substitution can be a direct replacement. However, some of the 'risk reduction' measures proposed from the EU Risk Assessment Process appear to be based on 'avoidance' rather than sound science. For example, with phosphorus chemistry, the behavior of manufactured or synthetic fibers cannot easily be compared to the performance of natural fibers and cotton, in particular. It must be recognized that some risk reduction approaches could be unintentionally directing what type of fibers should be used.

8.8 Future trends

Worldwide textile fiber utilization in 2004[57] was: cotton ~42%, polyester ~40%, nylon ~6.6%, acrylic 4.5%, all others (including rayon, polypropylene, wool) ~6.9%. Most of the emphasis for developing new flame retardant technology for textile products will probably be focused on the major fibers. The inherently flame resistant fibers and other specialty fibers will be important to meeting flammability standards for apparel and soft furnishings.

Because of current and potential restrictions on the use of chlorinated and brominated flame retardants,[10,13,14] non-halogen compounds will most likely be the main chemical systems used for the various textile applications that are required to meet flammability regulations in all areas except furnishings where backcoating technologies are preferred. It is possible that decaBDE and HBCD in combination with antimony trioxide could continue to be used, where acceptable, due to their high efficiency and cost-effectiveness on almost all fibers and blends. Phosphorus (phosphates, phosphoniums, and alkyl phosphonates) and phosphorus/nitrogen chemistry will most likely be the main source of new chemical systems. Increased use of phosphorus chemistry could lead to an increased use of natural fibers. The flame resistance of wool and other polyamide fibers can be enhanced by depositing in the fibers a complexed zirconium compound formed with an organic chelating agent or a halide.[58] From 0.5 to 5% of Zr (as ZrO) is preferred. Particularly preferred complexes include fluorozirconates, chlorozirconates and oxalic and citric acid complexes. Other zirconium compounds (zirconium hexafluoride complex and zirconium silicates) are also being investigated.[55] For some textile systems inherently flame resistant fibers combined with combustible fibers will likely be used, dependent upon the level of flame retardancy required and the test specifications defined. Nanocomposite technology, using various nanoparticles (organoclay/ phylosilicate clays, polyhedral oligomeric silsesquioxanes (POSS) and carbon nanotubes (CNT)) have been extensively studied with polypropylene and polyacrylonitrile fibers and as coatings on fabrics as well as with bulk polymers.[59–62] Nanocomposites show some promise as part of a system but appears not to be the silver bullet that was expected from initial studies with bulk polymers.[59–61] Low flammability of nanocomposites is only achieved in terms of low heat release rate (HRR) but they fail in terms of limiting oxygen index (LOI).[62] To help overcome this problem, nanoparticles are combined with traditional flame retardants (intumescents) or with plasma treatment using various chemicals for fabric coatings/finishes.[62] The nanofillers act as synergists and offer a way for making fire safe polymeric materials. Encapsulated flame retardants that are added to fibers prior to spinning have potential for improving durabilities of otherwise water-soluble systems. Durable systems that react with the fibers or that form insoluble polymers that do not cause exposure to humans and the environment will also be a research goal. New technology based on intumescent systems may be developed but fundamental issues with finding a truly insoluble phosphorus source and effectiveness at lower fiber loadings need to be resolved. 'Green chemistry' is pushed as a solution[63,64] but those pushing the concept of green chemistry at this point only 'think' they will be able develop new non-toxic systems. They have not developed acceptable commercial technology yet.

Cigarettes (i.e., smoulder ignition) are the major ignition source in mattress and upholstered furniture fires but open-flame ignition is also a major consideration. Fire-safe ('reduced ignition propensity' (RIP)) cigarettes are considered by many to be a proven, practical, and effective way to reduce/eliminate the risk of cigarette-ignited fires.[65] Many US states (>81% of US population) and all of Canada now have fire-safe cigarette requirements and other US states have introduced legislation.[65] However, even though RIP cigarettes do reduce the risk of smouldering ignition, the risk is not eliminated. The European Union is also considering making all cigarettes sold in EU countries self-extinguishing.[65] R J Reynolds Tobacco Co, producer of 35% of the USA cigarette market has announced that it will phase in RIP cigarettes for all its brands within the next two years.[66] A fire-safe cigarette has a reduced propensity to burn when left unattended. The most common fire-safe technology used by cigarette manufacturers is to wrap cigarettes with two or three thin bands of less-porous paper that act as 'speed bumps' to slow down a burning cigarette.[65] If a fire-safe cigarette is left unattended, the burning tobacco will reach one of these 'speed bumps' and self-extinguish. Fire-safe cigarettes meet an established cigarette fire safety performance standard (based on ASTM E 2187-04[67]). Estimates from New York State, the first USA state to pass this legislation, suggest RIP cigarettes could lead to a one-half to two-thirds reduction in fire deaths.[66] By 2009 most of the USA and possibly the EU will require RIP cigarettes, which could lead to less flame retardant chemical use in soft furnishings where cigarette ignition is of concern.

The concern for potential hazard to the environment of flame retardant chemicals[64,68] and chemicals in general[63] is part of the current trend toward the ecological marketing of textiles. It was reported at the first ever RITE (Reducing the Impact of Textiles on the Environment) Group Conference on sustainable textiles and clothing 10 October 2007 in the UK (London) that companies throughout the supply chain are under growing pressure to execute a 'green' business model.[69] Much is based on perception and geared toward reducing synthetic chemicals in the environment, recycling, reduction in energy and water use, and understanding the carbon footprint of textiles. Some of the emerging concerns about flame retardant chemicals are whether they are endocrine disrupters[64,68,70] and/or affect children and pregnant women. US EPA[71] is currently developing and validating screening and testing assays for their Endocrine Disruptor Screening program. Concern about human and ecological effects of chemical treatments most likely will increasingly affect all textile chemicals used. Whatever new chemicals are used will have to be fully evaluated for human and environmental effects.

8.9   Acknowledgements

Peter J. Wragg, UK Sales manager for Schill + Seilacher AG and Struktol Co. of America, and Dr Robert Barker, American Fiber Manufacturers Association, provided advice and input that was very helpful to me in preparing this chapter.

8.10   References

1. Wakelyn P J, Adair P K, and Barker R H, Do Open Flame Ignition Resistance Treatments for Cellulosic and Cellulosic Blend Fabrics Also Reduce Cigarette Ignitions? *Fire and Materials* **29**, 15–26 (2005).
2. Horrocks A R, Flame retardant finishes. *Rev. Prog. in Colouration*, **16**, 62–101 (1986).
3. Kandola B K, Horrocks A R, Price D and Coleman G V, Flame-retardant treatments of cellulose and their influence on the mechanism of cellulose pyrolysis. *J. Macromol. Sc., Rev. Macromol. Chem. Phys.*, **C36**, 721–794 (1996).
4. Horrocks A R, Developments in flame retardants for heat and fire resistant textiles – the role of char formation and intumescence, *Polym. Degrad. Stability*, **54**, 143–154 (1996).
5. Horrocks A R, Textiles, in *Fire Retardant Material*, Horrocks A R and Price D, Woodhead Publishing, Cambridge (2001), pp 128–181.
6. Horrocks A R, Flame retardant finishes and finishing, in *Textile Finishing*, Ed. Heywood D, Society of Dyers and Colourists, Bradford (2003), pp. 214–250.
7. Horrocks A R, Kandola B K, Davies P J, Zhang S and Padbury S A, Developments in flame retardant textiles – A review. *Polym. Degrad. Stab.*, **88** (1) 3–12 (2005).
8. Hearle J W S, *High Performance Fibres*, Woodhead Publishing, Cambridge (2001).

9. Wakelyn P J, Rearick W A and Turner J, Cotton and flammability – overview of new developments. *Am. Dyestuff Reporter* **87**(2), 13–21 (1998).
10. US EPA report, *Furniture Flame Retardancy Partnership: Environmental Profiles of Chemical Flame-Retardant Alternatives for Low Density Polyurethane Foam* Vol 1, EPA 742-R-05-002A, Sep 2005. http://www.epa.gov/dfe/pubs/projects/flameret/index.htm [Design for the environment, www.epa.gov/dfe].
11. NAS, *Toxicological risks of selected flame-retardant chemicals. Sub-committee on Flame-retardant Chemicals of the United States*, National Research Council, Washington, DC; National Academy Press, Washington (2000).
12. *Brominated Flame Retardants, Environmental Transport and Fate, Atmospheric Transport and Fate, Proceedings Dioxin 2003*, Boston, MA, Aug. 24–29, 2003; and Studies Show Flame Retardants Breaks Down, Data Said to Refute Previous Industry Studies. *BNA Daily Report for Executives*, 11-24-03, p. 24 (2003).
13. *Fire retardants: toxic effects*, California Assembly Bill 706 (AB 706), 2007. http://www.leginfo.ca.gov/pub/07-08/bill/asm/ab_0701-0750/ab_706_bill_20070827_amended_sen_v93.pdf.
14. Formulating environmentally friendly flame retardants (2007). ES&T Online News, Technology News – Sep 26, 2007 [http://pubs.acs.org/subscribe/journals/esthag-w/2007/sept/tech/kb_flameretard.html].
15. State of Maine, Maine to phase out flame retardant. Portland Press Hearld May 25, 2007. http://pressherald.mainetoday.com/story_pf.php?id=107956&ac=PHnws.
16. Washington State, Act phasing out polybrominated diphenyl ethers; New Chapter Title 70 RCW, 2007 (effective July 22, 2007). http://www.leg.wa.gov/pub/billinfo/2007-08/Pdf/Bills/Session%20Law%202007/1024-S.SL.pdf.
17. EUROPEAN UPDATE on DecaBDE (deca-brominated diphenyl ether): Current Status and History of DecaBDE Initiatives in Europe. 2006. http://safer-products.org/downloads/Deca%20History%20Nov06.doc.
18. Green Supply Line 2007, http://www.greensupplyline.com/showArticle.jhtml;?articleID=192300632.
19. BSEF, EU Risk Assessment of BFR's, 2007, http://www.bsef.com/regulation/eu_risk_assessm/.
20. VECAP^{TM}, Product Stewardship Program, http://www.bsef.com/product_stew/vecap_us.
21. US CPSC. 16 CFR 1610. Standard for the Flammability of clothing textiles.
22. US CPSC. 16 CFR 1615. Standard for the Flammability of children's sleepwear: Sizes 0- 6X.
23. US CPSC. 16 CFR 1616. Standard for the Flammability of children's sleepwear: Sizes 7-14.
24. Rearick W A, Wallace M L, Martin V B, and Wakelyn P J, Flammability Considerations for Raised Surface Apparel. *AATCC Review* **2**(2), 12–15 (2002).
25. Barker R H and Drews M J, Development of Flame Retardants for Polyester/Cotton Blends. NBS-GCR-ETIP 76-22, Experimental Technology Incentives Program, National Bureau of Standards (now National Institute of Science and Technology), Washington, DC, USA, p28 (1976).
26. Yeh K, Valente J A, Smith B F, Drews M J and Barker R H, Calorimetric Study of Polyester/Cotton Blend Fabrics. Part IV. Effect of Bromine Containing Retardants, *J Fire Retardant Chemistry*, **7**, 87 (1980).
27. Yeh K, Drews M J and Barker R H, Calorimetric Study of Polyester/Cotton Blend Fabrics. Part V. Characterization of Flame Retardant Action, *J Fire Retardant Chemistry*, **7**, 99 (1980).

28. Kruse W, *Proc Study Conf Textile Flammability and Consumer Safety*, Gottlieb-Dutweiller, Inst Econ Social Studies, Ruschlikon-Zurich, Switzerland, p 137 (1959).
29. Wakelyn P J, Bertoniere N R, French A D, Thibodeaux D, et al., *Cotton Fibre Chemistry and Technology*, Series: International Fibre Science and Technology, CRC Press (Taylor and Francis Group) (2007).
30. US CPSC Recall Alert #07-563, July 26, 2007, Coldwater Creek Recalls Sweaters Due to Burn Hazard.
31. Rearick W A, Wallace M L, and Wakelyn P J, Fire Retardants for Cotton: Market Opportunities, Technical requirements and Regulations. *Recent Advances in Flame Retardancy of Polymeric Materials* Vol.11. (Proc. Eleventh BCC Conf. on Flame Retardancy of Polymeric Materials, Stamford, CN, May 2000). Lewin M, Ed. Business Communications Co. (BCC). Norwalk, CN, pp. 222–230 (2000).
32. Martin V B, Cotton Flammability. *Proc 2005 Beltwide Cotton Research Conferences* (Nonwoven Symposium: Flammability). New Orleans, LA. Jan. 7, 2005.
33. *Proposed Airbourne Toxic Control Measure (ATCM) to Reduce Formaldehyde Emissions from Composite Wood Products*, California Air Resources Board (2007) http://www.arb.ca.gov/toxics/compwood/factsheet.pdf.
34. Yang C Q, Wu W, and Stowell J, New Development in Flame Retardant Finishing of Cotton Textiles. *Recent Advances in Flame Retardancy of Polymeric Materials*. Vol.14. (Proc. Fourteenth BCC Conf. on Flame Retardancy of Polymeric Materials, Stamford, CN, May 2003). M. Lewin, Ed. Business Communications Co. (BCC). Norwalk, CN (2003).
35. Stowell, J and Yang C Q, A Durable Low-Formaldehyde Flame Retardant Finish for Cotton Fabrics, *AATCC Review* **3**(2), 17–20 (2003).
36. Sawhney A P S, Ruppenicker G F, and Price J, A polyethylene staple-core/cotton wrap duck fabric for military tentage. *Proc 1997 Beltwide Cotton Conferences*, National Cotton Council, Memphis, TN, pp. 734–736 (1997).
37. Ruppenicker G F, Sawhney A P S, Kimmel L B, and Calamari T A, Flame-retardant cotton fabrics for the military. *Proc 2003 Beltwide Cotton Conferences*, National Cotton Council, Memphis, TN, pp. 2202–2206 (2003).
38. Wakelyn, P J, Adair P K, and Wolf S, Cotton and Cotton Modacrylic Blended Batting FireBlocking Barriers for Soft Furnishings to Meet Federal and State Flammability Standards. *Proc. 2004 Beltwide Cotton Conferences*. National Cotton Council, Memphis, TN. 2004, pp. 2829–2842 (2004).
39. Wolf S, Wakelyn P J and Adair P K, Cotton and Cotton Blended Fire Blocking Barriers for Soft Furnishings to Meet Federal and State Flammability Standards. *AATCC Book of Papers 2004 International Conf. and Exhibition*. American Association of Chemists and Colorists, Research Triangle Park, NC (2004).
40. Tenney A, Developing a Federal Flammability Standard for Mattresses. *2005 Beltwide Cotton Research Conferences* (Nonwven Symposium: Flammability). New Orleans, LA. Jan. 7, 2005.
41. Wakelyn P J, Wolf S, and Oliver K, 2003. Cotton Batting Barriers for Soft Furnishings. *Recent Advances in Flame Retardancy of Polymeric Materials*, Vol. 14 (Proc. Fourteenth Annual Conf. on Flame Retardancy of Polymeric Materials June 2–4, 2003). M. Lewin, Ed. Business Communications Co. (BCC). Norwalk, CN, pp. 250–264 (2003).
42. United States Patent 6,790,795. Fire Blocking Fabric. Sep 14, 2004.
43. United States Patent. 20050208855 Modacrylic/cotton/aramid fiber blends for arc and flame protection Sep 22, 2005. http://www.freepatentsonline.com/20050208855.html.

44. Stevens G C, Ghanem R, Thomas J L, Horrocks A R and Kandola B, Understanding flame retardant release to the environment, *Proc of Flame Retardants 2004*, Interscience Communications, London, pp. 37–50 (2004).
45. Horrocks A R, Wang M Y, Hall M E, Sunmonu F and Pearson J S, Flame retardant textile back-coatings. Part 2: Effectiveness of phosphorus-containing retardants in textile back-coating formulations, *Polym. Int.*, **49**, 1079–1091 (2000).
46. Horrocks A R, Davies P J, Kandola B K and Alderson A, The potential for volatile phosphorus-containing flame retardants in textile back-coatings, *J Fire Sciences*, **25**, 523–540 (2007).
47. California Technical Bulletin 117 (CA TB 117) http://www.bhfti.ca.gov/techbulletin/117.pdf; California. (Draft 2/2002) *Proposed Update of Upholstered Furniture Flammability Standard*. Technical Bulletin 117 Requirements, Test Procedures and Apparatus for Testing the Flame and Smolder Resistance of Upholstered Furniture. California Bureau of Home Furnishings and Thermal Insulation (2002).
48. California Technical Bulletin 604 (CA TB 604), Test Procedures and Apparatus for the Open Flame Resistance of Filled Bedclothing, October 2007. http://www.bhfti.ca.gov/techbulletin/tb604_final_draft.pdf.
49. Lowell Center for Sustainable Products, *Decabromodiphenylether: An Investigation of Non-Halogenated Substitutes in Electronic Enclosures and Textile Applications*. U of Massachusetts Lowell (2005). http://www.sustainableproduction.org/downloads/DecaBDESubstitutesFinal4-15-05.pdf.
50. Gabara V, Hartzler J D, Lee K-S, Rodini D J and Yang H H, Aramid Fibers. Chapter 13. *Handbook of Fiber Chemistry* (3rd edition, revised and expanded). Series: International Fiber Science and Technology, M. Lewin, Ed., CRC Press (Taylor & Francis Group), pp. 975–1029 (2007).
51. VECAP^{TM} *Guidance* (http://www.bsef.com/product_stew/vecap_us.
52. VECAP^{TM} Code of Practices for Textiles ('Controling Emission of Persistent Chemicals by Proactive Commitment to Good Practices'; http://www.bsef.com/product_stew/vecap_us/Textiles%20NA%20Final%2006.pdf).
53. REACH (2007). http://ec.europa.eu/environment/chemicals/reach/reach_intro.htm.
54. REACH, Guidance Document (2007). http://reach.jrc.it/guidance_en.htm.
55. Posner S, *Survey and technical assessment of alternates to decabromodiphenyl ether (decaBDE) in textile applications*. Swedish Chemicals Inspectorate, PM Nr 5/04 KEMI, Order No.510792 (2004).
56. Horrocks A R, Textile flame retardant challenges for the 21st century, In conf *Proc; Flame Retardants 2000*, Interscience Publications, London (2000), p147ff.
57. Anon., *Fiber Organon* **76**(7) (2005).
58. United States Patent 4,160,051 Zirconium flame-resist treatments July 3, 1979.
59. Gillman J W, Recent Advances in Nanocomposites. *11th Euopean Meeting on Fire Retardant Polymers*, July 4–6, Manchester, UK (2007).
60. Gilman J W, Kashiwagi T, and Lichtenhan J D, Nanocomposites: A Revolutionary New Flame retardant approach. *Macromolecular Symposia* Vol **233** No 1, *Special Issue: Fillers, Filled Polymers and Polymer Blends*. Edited by Dubois P., Groeninckx G, Jérôme R, Legras R (2006).
61. Bourbigot S, Heat and Fire Resistance of Textile Materials: A Review. *11th Euopean Meeting on Fire Retardant Polymers*, July 4–6, Manchester, UK (2007).
62. Bourbigot S, Duquesne S, and Jama C, Polymer Nanocomposites: How to Reach Low flammability. *Macromolecular Symposia* Vol **233** No 1, *Special Issue: Fillers, Filled Polymers and Polymer Blends*. Edited by Dubois P, Groeninckx G, Jérôme R,

Legras R (2006).
63. Wilson M P, Chia D A and Ehlers, B C, *Green Chemistry in California: A Framework for Leadership in Chemical Policy and Innovation.* California Policy Research Center, University of California (2006).
64. Blum A, The Fire Retardant Dilemma. *Science* 2007 **318**, 194.
65. Coalition for Fire-Safe Cigarettes, http://www.firesafecigarettes.org/categoryList.asp?categoryID=9&URL=Home%20-%20The%20Coalition%20for%20Fire%20Safe%20Cigarettes.
66. 2nd International Conf on Fire 'Safer' cigarettes, Harvard School of Public Health, Dec 11–12, 2006. http://www.hsph.harvard.edu/tobacco/agenda.html.
67. ASTM E 2187-04, Standard Test Method for Measuring the Ignition Strength of Cigarettes; http://firesafecigarettes.org/assets/files/NISTstandard.pdf.
68. Birnbaum L S and Cohen E A, Brominated flame-retardants: Cause for Concern?, *Environmental Health Perspectives*, **112**(1), 9–17 (2004).
69. RITE Group 2007 http://www.ritegroup.org/; Crowds pack first RITE Group event *Ecotextile News*, Nov 2007.
70. *Global assessment of the state-of-the-science of endocrine disruptors.* International Programme on Chemical Safety, World Health Organization (2002).
71. US Environmental protection Agency, Endocrine Disruptor Screening Program (EDSP), 2007. http://www.epa.gov/endo/pubs/assayvalidation/index.htm.

9
Recycling and disposal of flame retarded materials

A C A S T R O V I N C I, University of Applied Sciences of Southern Switzerland, SUPSI, Switzerland, M L A V A S E L L I, Centro di Cultura per l'Ingegneria delle Materie Plastiche, Italy and G C A M I N O, Politecnico di Torino, Italy

Abstract: Legislations, coupled with an increasing environmental sensitivity, have enhanced the interest in plastic recycling with the intent of reducing the waste stream. This chapter provides the state of the art of flame retardancy of polymeric materials, the approaches and technologies available for plastic recycling as well as the related legislations and European Directives. The chapter then analyses how the presence of flame retardants affects recycling of the plastic waste stream.

Key words: recycling, polymeric materials, plastic waste, flame retardants.

9.1 Introduction on flame retardants

The use of synthetic polymers has been continuously growing from the early decades of the 20th century. Ever since, polymeric materials have been replacing other materials in a number of applications such as transport (automotive, trains, aircraft, ships, etc.), buildings (e.g., thermal insulation), furniture (e.g., upholstery, etc.), electrical and electronic devices (e.g., cables, external cases for computers and notebooks, etc.).[1]

Common plastics are carbon-based polymeric materials which will burn under suitable conditions. They have become a large and growing fraction of the fire load in homes, commercial environments, and transports.[2]

The flammability of common polymeric materials, i.e. the tendency to ignite and/or propagate flames, represents one of the most important limitations in their use, because of the related fire hazard and risks.[3] Many efforts to reduce plastics flammability, decreasing the correlated fire hazard and risks, have involved the use of additives or of chemical modifications of polymers, generally called fire/flame retardants (FR). The goal of flame retardants is primarily to decrease the ease of ignition of the polymers and/or their rate of flame propagation possibly reducing the smoke production rate and the release of toxic gases when the fire inevitably occurs. The essential role of flame retardants is to increase the time to 'flash over', that is the moment in which the fire becomes generalised and the combustion can be confined but not interrupted,[4] to allow escape of people involved in the fire and intervention of extinguishing means. Flame retardants are chemical structures introduced in the polymer matrix either by copolymerisation, by grafting or by melt blending.[5-10] The fire retardant loading varies with its efficiency and the requested grade of flame retardancy. The use of flame retardants should not interfere with other additives, e.g. antioxidants, and should be environmentally friendly and easily processable.

Flame retardants can act in the gas-phase, where the combustion of volatile products from polymer thermal degradation takes place, or in the condensed phase, where polymer pyrolysis occurs.[11-14] Flame retardants may have a chemical and/or physical action. For example, in the gas phase, flame retardants can trap free radicals responsible for propagation of the combustion reactions (chemical action), e.g. halides, or can dilute flammable gases (physical action), e.g. metal hydroxides.[11-16] In the condensed phase, charring or surface protection can be promoted by chemical reaction, e.g. nanofillers,[17] intumescent systems[18] and/or by physical interaction between the additive and the polymer, e.g. metal hydroxide,12,19–21 borates,[11,12] expandable graphite.[22-25]

Usually flame retardancy is not provided just by a single mechanism, e.g. melamine compounds,[11,26-28] metal hydroxide,[12,19-21] nanofillers[17] and phosphorus compounds,[11,12,29] but it is the result of a combination of mechanisms that depends on the type of polymer and flame retardant used.[11-14,17]

9.2 Recycling

9.2.1 Recycling processes

The realisation of a wide variety of plastic products, particularly those with a short service life, has generated a rising production of waste. Legislation, coupled with an increasing environmental sensitivity, have enhanced interest in plastic recycling with the intent of reducing this waste stream.

Recycling methods are summarised in Fig. 9.1:[30]

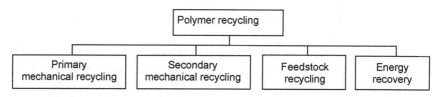

9.1 Plastic recycling categories.

Mechanical recycling

Mechanical recycling consists of the recovery and reuse by mechanical processes of plastics from industrial scraps or from post-consumer goods otherwise designated for disposal.[30]

Primary mechanical recycling

Primary mechanical recycling consists in the re-utilisation of uncontaminated homogeneous plastics, for example scraps derived from processing. Primary recycling generally takes place in the same plant that generates the recycled scraps; therefore the recovered material is homogeneous with a very small amount of contaminants.[30]

Secondary mechanical recycling

The recyclability of polymers from post-consumer plastic goods depends on a combination of intrinsic factors, due to the specific nature of the polymers, and on extrinsic factors such as the presence of additives, dyestuff, impurities and the extent of the degradation occurred during service life.[31,32]

Heterogeneous mixtures of polymers with noticeable differences in terms of processing conditions may prevent re-processing. For example, small amounts of polyethylene terephthalate (PET) dispersed in a high density polyethylene (HDPE) matrix can lead to the plugging up of the injection nozzle, because of its higher melting temperature.[32]

Also mixtures of virgin and used polymers of the same nature can be incompatible when compounded together. This is an important issue for mechanical recycling, where a percentage of reprocessed polymers is usually compounded with virgin polymers to obtain the so called 'homopolymer' or 'monopolymer' blends.[33]

Polymers subject to environmental ageing, e.g. photo-oxidative degradation due to UV and oxygen exposure, may undergo modification of the polymer chain chemical structure, crosslinking and branching, formation of oxygen-containing structures, changes of crystallinity[34–36] with subsequent depletion of properties.[37] Additionally, transformation processes may lead to a further decay of recycled material properties because of high temperature and mechanical stresses.[38–40]

The negative effects of service life and reprocessing can be so detrimental for material properties[41] that the recovered plastics can be useless for any application. To preserve recycled material properties, thermal and light stabilisers are added during the re-extrusion process, i.e. re-stabilisation.[42–47]

Mechanical recycling procedures can be described as a sequence of steps:[30,48]

- reduction of material dimensions (shredding)
- sorting to classify different type of plastics and to separate plastics from other materials (in the case of secondary recycling)
- washing to remove contaminants
- drying
- extrusion and granulation.

Reduction of material dimensions is achieved with different shredding technologies mainly based on cutting, impact, shearing and combination of impact and pressure[49] depending on the polymer nature.

Sorting technologies are based on density,[30,50–52] wettability,[53,54] spectroscopy systems,[30,55] selective dissolution,[30] electrostatic methods[56–58] and molecular markers.[30]

Feedstock recycling

Feedstock recycling, also known as chemical recycling or tertiary recycling, allows converting plastic waste into monomers and/or oligomers or other chemical products by physical or chemical processes.

Different approaches to feedstock recycling currently exist:[30,59,60] depolymerisation,[61] partial oxidation[62] and cracking.[63]

Depolymerisation

Depolymerisation leads to recovery of the starting monomer or oligomers from polymers with functional groups in the main chain by the treatment of polymer scraps in the presence of a chemical reagent (chemical depolymerisation). Depending on the reagent, different depolymerisation reactions take place: hydrolysis,[64] aminolysis,[65] glycolysis,[66] methanolysis.[67]

Chemical depolymerisation gives good results in the case of polyamides and polyesters,[59] whereas depolymerisation of polyurethanes has to be further developed.[68]

A few addition polymers such as polymethylmethacrylate (PMMA) and polystyrene (PS),[69–72] depolymerise on heating (thermal depolymerisation).

Partial oxidation

Partial oxidation, also known as gasification, consists of the oxidation of plastic waste to a hydrocarbons, hydrogen and carbon dioxide (CO_2) mixture using oxygen or high temperature steam (1200–1500 °C) as the oxidation medium.[59,60,73,74] An emerging technology consists of the use of supercritical water as the oxidative medium.[62]

Cracking

Cracking of macromolecules can be obtained by three different approaches: thermal-, catalytic- and hydro-cracking.[59]

Thermal cracking, also known as pyrolysis, consists of heating the polymer in an inert atmosphere, promoting macromolecules thermal bond scission to a variety of low molecular weight hydrocarbons, i.e. a liquid fraction that includes paraffins, olefins, naphtenes and aromatics, a volatile fraction and a solid residue.[63,75] Depending on the polymer chain characteristics (presence of defects, halogens and other hetero-atoms, aromatic degree and bond dissociation energy) different cracking mechanisms take place.[63,76]

As for thermal cracking, catalytic cracking involves heating the polymer in an inert atmosphere in the presence of catalysts that allow operating at lower temperatures, saving energy and costs with respect to thermal cracking. Reactions, due to higher rates, take place in reactors with smaller dimensions with shorter residence times as compared to non-catalysed cracking. Reaction products can be tailored with better selectivity and higher productivity through the selection of the best reaction conditions and catalysts.[63,77–79] Limits of catalytic cracking derive from the heterogeneity of the plastic stream designed for the recycling: the presence of certain hetero-atoms such as sulphur or nitrogen can poison the catalyst. Moreover toxic and corrosive compounds can be formed because of the presence of chlorine, therefore separation or de-chlorination is needed.[79]

Hydrocracking is conducted using catalysts, as in the case of catalytic cracking, and it is based on the reaction between polymer and hydrogen introduced at low pressure (3^{-10} MPa).[59,80–82]

Plants operating, in pilot or industrial scale, are summarised in Table 9.1. The reader can find a more detailed description of these processes in the reported references.[76,83–90]

Energy recovery

Polymeric materials are characterised by high values of heat of combustion, in some cases higher than oils. Polymer heats of combustion depend on the macromolecule chemical structure and on the nature and concentration of additives.[91] For example, polyolefins generally combust to carbon dioxide and water with a heat of combustion over 40 MJ/kg whereas the oxygen and nitrogen containing polymers have heats of combustion between 25 and 40 MJ/kg. The presence of halogen atoms in the polymer strongly influences the burning behaviour, leading to heats of combustion lower than 25 MJ/kg. Energy recovery is usually conducted following two possible approaches:

- monocombustion, when fuels are burned alone
- co-combustion, when the waste derived fuels are combusted together with conventional fossil fuels.

Depending on the composition and level of waste sorting, four types of stream can be identified and treated:[92]

- municipal solid waste (MSW), composed of unsorted household waste with an average energy value of 10 MJ/kg;
- refuse derived fuel (RDF), which derives from the elimination of non-combustible fractions from MSW, with an energy value of 15–17 MJ/kg;

Table 9.1 Chemical recycling operating plants

Process	Description	Products
Thermofuel	• Main reactor (inert atmosphere) • Catalytic column (cracking and reforming) • Selective condensation	• High quality diesel with lower content of sulphur and lower emission of SO_x, NO_x and particulate compared with commercial diesel
Smuda	• Catalytic pyrolysis with nitrogen purging (layered silicate catalysts) • Condensation • Distillation column	• 85% transportation grade diesel with very good low temperature properties • 15% gasoline used for the production of electric energy employed for the process
Polymer engineering	• Catalytic depolymerisation (ion exchange catalyst) • Oil used as pyrolysis medium • Distillation column	• High-grade stabilised diesel
Royco	• Low temperature thermal cracking • Distillation tower	• Output used for electricity production
Reentech	• First dehydrogenation with nickel-based catalysts • Fluid catalytic cracking • Reforming of the gasoline fraction	• Gasoline, kerosene and diesel
Hitachi	• Pyrolysis • Two condensers	• Kerosene and gasoline from the two condensers • Uncondensed gases used for the furnace
Chiyoda	• Dechlorination unit • Thermal cracking • Distillation column	• Fuel oil and feedstock for petrochemical plants
Blowdec	• Fluidised bed reactor	• Diesel with low sulphur content
Conrad	• Pyrolisis in a rotary kiln reactor • Recovery system (water spray system, gas-filter and cyclone)	• Liquid petroleum • Non-condensable gases • Carbonaceous material
BP	• Cracking in fluidised bed reactor • Capture of fine particles by cyclone • Neutralisation of HCl by lime absorber	• Hydrocarbons • Distillate feedstock
Akzo	• Pyrolysis in fluidised bed reactor with steam • Quenching (recovery of HCl) • Burning of the residual tar in a fluidised bed combustor	• Fuel gas • HCl
Ebara TwinRec	• Fluidised bed reactor • Separation of metals • Combustion in cyclonic chamber • Heat boiler unit	• Metals • Fuel gas for the production of power in the plant

Table 9.1 Continued

Process	Description	Products
Veba	• First depolymerisation and dechlorination • Hydrogenation in a liquid phase reactor	• Synthetic crude oil • HCl • Hydrogenated solid residue for the production of coke • Off-gas
Haloclean	• Rotary kiln reactor	• Phenolic-based oil • HBr • Residue
Niigata	• Dehydrochlorination • Pyrolysis • Distillation column	• Oil used as fuel
Sapporo	• Dehydrochlorination • Pyrolysis in rotary kiln reactor • Distillation column	• Oils used for the energy recovery and feedstock for refinery plants
Mikasa	• Dehydrochlorination • First cracking • Second cracking by catalysts • Condensation	• Gas oil • Naphta and non-condensed gases used in the plant
Zadgaonkar's	• Dechlorination • Catalytic cracking • Condensation • Distillation	• Liquid fuels • Gases • Coke
Fuji	• First cracking reactor • Reforming reactor	• Gasoline, kerosene, diesel
BASF	• Liquefaction in agitated tank (with absorption of HCl) • Cracking in tubular reactor • Separation in distillation column • Conversion of naphta in a steam cracker	• HCl • Naphta • Monomers • High-boiling oils
Hamburg University	• Fluidised bed reactor (the fluidising gas can be varied) • Separation of solids by a cyclone • Washing cooler with styrene • Quenching in packed column • Two distillation columns	• Oil and gas: the composition of the products depends on the composition of the plastic
Hunan University	• Catalytic cracking in a fluidised bed reactor • Column for catalytic reforming • Fractionation through a rectifying tower • Cooling	• Heavy oil • Diesel • Gasoline
United Carbon	• Extrusion • Thermal craking • Collection of the products	• Wax
Likun	• Cracking in the first reactor • Catalytic reforming in the catalytic tower	• Gasoline • Diesel oil

- packaging derived fuel (PDF), composed by paper and plastic from packaging, with an energy value of around 20 MJ/kg;
- polymer fuel derived from the collection and processing of plastics with an energy value of 30–40 MJ/kg.

Mono-combustion and co-combustion of these fuels are conducted through the application of different technologies.[92]

The main concern about burning of plastic materials is the formation of toxic substances in the combustion exhaust products. Actually, it has been observed that the presence of a plastic stream in pilot- and full-scale combustors generally improves various burning parameters and gives a more stable combustion, with a lower CO emission and a lower amount of unburned carbonaceous residue can be observed.[93] The presence of plastics in the feed composition seems to have no noticeable effect on the release of heavy metals.[93]

The directive 2002/95/CE 'Restriction of certain Hazardous Substances' (RoHS)[94] conceived with the objective to protect human health, restricts to few parts per million the presence of several substances in electrical and electronic equipment. Among these substances two classes of halogenated flame retardants emerge: polybrominated biphenyls (PBB) and polybrominated diphenyls ethers (PBDE). Decabromodiphenylether (DecaBDE) was exempted in October 2005 by the European Commission[95] on the basis of the conclusions of a 10-year European Union Risk Assessment Report.[96]

Polymeric materials produced before the RoHS was in place (July 2006) may contain some of the banned substances that could be released during recycling, for example penta and octa bromodiphenyl ether flame retardants are nowadays considered toxic. Furthermore, during recycling processes, e.g. mechanical recycling, toxic products may be formed due to thermal and/or mechanical degradation of non-toxic chemicals.[97,98] The possibility of toxic compounds being released during recycling of flame retarded polymers has to be considered because of its potential risks for operators and environmental impact.

9.3 Directives on recycling of flame retarded materials

9.3.1 European and local directives and ordinances

Waste Electrical and Electronic Equipment (WEEE) Directive – European Community

In Europe, post-consumer treatment of materials used in the electrical and electronic fields is regulated by the directive 2002/96/EC on Waste Electrical and Electronic Equipment (WEEE), effective since 31st December 2006.[99] This European Directive establishes different recovery and recycle targets depending on the type of electrical and electronic application. Small household appliances must be recovered at a minimum of 70% by an average weight per appliance and a minimum of 50% of component materials and substances must be recycled. For information and technology (I&T) devices a minimum of 75% by an average weight per appliance must be recovered and 65% must be recycled; for large household appliance these percentages are 80% and 75% respectively. Note that on the basis of the WEEE Directive the term *recycling* does not include energy recovery.

The WEEE Directive (Annex II) establishes that plastics containing brominated flame retardants have to be separated from a stream of waste of electrical and electronic equipment before it undergoes recycling and selectively treated. Treatments must avoid the dispersion of pollutants into the recycled material or the waste stream.

The Chemikalienverbotsverordnung (ChemVerbotsV) German ordinance

The German Chemicals Prohibition Decree Chemikalienverbotsverordnung (ChemVerbotsV)[97] sets limits to the presence of determined polybrominated dibenzo-p-dioxins and furans cogeners (PBDD/F) in goods, which can be dangerous for human health. The sum of the first four congeners reported in Table 9.2 is set in 1 $\mu g/kg$ while the total sum of the eight congeners must not exceed 5 $\mu g/kg$.

Since thermo-mechanical stress during secondary mechanical recycling can promote the formation of PBDD/F, this regional ordinance has to be taken into account in mechanical recycling of plastics containing the DecaBDE flame retardant.

9.4 Recyclability of flame retarded polymers

The aim of this section is to investigate the influence of flame retardant additives on recycling of polymers.

Table 9.2 PBDD/F involved in the ChemVerbotsV

N	Congener
1	2,3,7,8 – Tetra brominated dibenzo dioxine (TeBDD)
2	1,2,3,7,8 – Penta brominated dibenzo dioxine (PeBDD)
3	2,3,7,8 – Tetra brominated dibenzo furane (TeBDF)
4	2,3,4,7,8 – Penta brominated dibenzo furane (PeBDF)
5	1,2,3,4,7,8 – Hexa brominated dibenzo dioxine (HxBDD)
6	1,2,3,7,8,9 – Hexa brominated dibenzo dioxine (HxBDD)
7	1,2,3,6,7,8 – Hexa brominated dibenzo dioxine (HxBDD)
8	1,2,3,7,8 – Penta brominated dibenzo furane (PeBDF)

9.4.1 Mechanical recycling of flame retarded polymers

Since heterogeneous mixtures of polymers containing heterogeneous flame retardant chemicals can be affected in different ways by mechanical recycling, it is rather impossible to define general routines capable of processing the vast set of flame retarded plastic materials available for recycling nowadays. Focus will be on the effects of mechanical recycling on material properties retention. The nature of the flame retardant and its chemical interaction with the polymer matrix, strongly affects the recyclability of flame retarded polymers[98] in terms of retention of mechanical properties and fire performances.[100–102] For example, nitrogen-based flame retardants, due to their high thermal stability, appear resistant to mechanical recycling.[103]

ABS, HIPS, PS, PPO and their blends are often used in electrical and electronic equipment which is the subject of the WEEE directive.[104,105] Secondary recycling of HIPS containing brominated flame retardants (ethane 1,2-*bis* (pentabromophenyl) (EBP) and ethylene 1,2- *bis* (tetrabromophtalate) (EBTBP))[106] does not significantly affect the mechanical, thermal and flammability properties and so is technically feasible. However, as for the WEEE Directive these plastics have to be separated from the waste plastic stream to be selectively treated.[99]

The retention of properties after ageing and secondary recycling of PC/ABS blends flame retarded with phosphorous based chemicals (resorcinol diphenyl phosphate (RDP) and bisphenol A diphenyl phosphate (BPADP)) depends on the chemical structure of the flame retardant.[107,108]

A limiting aspect of secondary recycling of flame retarded polymers, particularly those containing DecaBDE, is the formation of polyhalogenated dibenzo-*p*-dioxines and furanes (PXDD/F) owing to the thermo-mechanical stress involved in the recycling process. The production and presence of these congeners in the recycled flame retarded plastics is a much studied matter. Particular attention is paid to compliancy to the regulatory requirements posed by the above mentioned German Chemicals Banning Ordinance.[97] For example, HIPS flame retarded with decabromodiphenylether (DecaBDE) and antimony trioxide[109] showed a low or non-detectable production of PBDD/F after multiple extrusions. A general rule for DecaBDE flame retarded plastics is not available, and tests have to be carefully carried out before establishment of an industrial scale secondary recycling process. Separation processes have been suggested, to eliminate unwanted halide compounds from polymers to be mechanically recycled.[105]

9.4.2 Feedstock recycling of flame retarded polymers

In order to fulfil the WEEE directive, industrial plants able to process heterogeneous mixtures of plastics separating unwanted compounds from valuable chemical products have been developed.[83,110–113] Thus no technological issues limit the application of feedstock recycling to flame retarded polymers. For example, the Haloclean process technology,[83] developed within the European Project Haloclean (Contract No: G1RD-CT-1999-00082), consists of a thermochemical process that uses a low temperature pyrolysis rotary kiln to separate noble metals from plastics contained in electric and electronic wastes.

9.4.3 Energy recovery of recycled polymers

There are many concerns about the supposed hazard of burning flame retarded polymers, particularly halide-containing polymers, because of the release of toxic and corrosive halogen compounds in the exhaust gas flow. The correlation between the feed composition and the evolution of polychlorodibenzodioxines and furanes (PCDD/F) is still under debate.[93] This is despite results that show that the emission of PCDD/F is not altered by the presence of flame retarded plastics in co-combustion plants where a relatively small fraction of bromine containing plastics are burned together with municipal solid waste (MSW).[110,114,115]

However, state of the art incinerator technology allows burning of flame retarded plastics avoiding the evolution of halide compounds in the exhaust gas flow, e.g. by using appropriate technologies involving post-combustion catalysis and appropriate scrubbing.[114]

9.5 Conclusions

Amongst the different polymer recovery techniques available nowadays, mechanical recycling of post-consumer plastics goods is the more economical one, but one which gives a positive balance only in the case of primary recycling of scraps. Being the less economically favourable and environmentally desirable, energy recovery is generally only considered when there is no other option available to avoid the use of landfill. Mechanical recycling of heterogeneous plastics waste containing unwanted flame retardants, requires the setting up of complex and expensive sorting/separation procedures in order to comply with European and Regional Directives, ensuring both process feasibility and material properties preservation. From this point of view, the most recent feedstock recycling plants which are able to process a wide range of plastics, even containing flame retardants, without making worse their environmental performances, appears a valuable alternative to mechanical recycling.

From a general point of view, all the recycling options could be successfully developed for fire retarded plastics; however, economical and political issues will establish which is the most suitable approach.

9.6 References

1. Nelson G L, 'The changing nature of fire retardancy of polymers', in Grant A F and Wilkie C A, *Fire retardancy of polymeric materials*, New York, Marcel Dekker, 2000, 1–26.
2. Bourbigot S, Le Bras M, Troitzsch J, 'Introduction', in Trotizsch J, *Plastics Flammability Handbook*, Munich, Carl Hanser Verlag, 2004, 3–7.
3. Hull J R, Bukowsky R W, 'Fire-hazard and fire-risk assessment of fire-retardant polymers', in Grant A F and Wilkie C A, *Fire retardancy of polymeric material*, New York, Marcel Dekker, 2000, 533–566.
4. Drysdale D, *An Introduction to Fire Dynamics*, New York, Wiley and Sons, 2002.
5. Ebdon J R, Hunt B J, Joseph P, Konkel C S, 'Flame retarding thermoplastics: additive versus reactive approach', in Al-Malayka S, Golovoy A and Wilkie C A, *Specialty Polymer Additives*, Oxford, Blackwell Science, 2001, 235–258.
6. Shi L, 'An approach to the flame retardation and smoke suppression of ethylene–vinyl acetate copolymer by plasma grafting of acrylamide', *React & Functl Polyms*, 2000 **45** (2) 85–93.
7. Armitage P, Ebdon J R, Hunt B J, Jones M S, Thorpe F G, 'Chemical modification of polymers to improve flame retardance – I. The influence of boron-containing groups', *Polym Degrad Stabil*, 1996 **54** (2–3) 387–393.
8. Ebdon J R, Hunt B J, Jones M S, Thorpe F G, 'Chemical modification of polymers to improve flame retardance – II. The influence of silicon-containing groups', *Polym Degrad Stabil*, 1996 **54** (2–3) 395–400.
9. Chiang W Y, Hu C H, 'Approaches of interfacial modification for flame retardant polymeric materials', *Compos Part A–Appl S*, 2001 **32** (3–4) 517–524.
10. Tsafack M J, Levalois-Grützmacher J, 'Plasma-induced graft-polymerization of flame retardant monomers onto PAN fabrics', *Surf Coat Tech*, 2006 **200** (11) 3503–3510.
11. Levchik S V, 'Introduction to flame retardancy and polymer flammability', in Morgan A B and Wilkie C A, *Flame retardant polymer nanocomposites*, Hoboken, John Wiley & Sons, 2007, 1–29.
12. Lewin M, Weil E D, 'Mechanisms and modes of action in flame retardancy of polymers', in Horrocks A R and Price D, *Fire retardant materials*, Cambridge, Woodhead Publishing Limited, 2001, 31–68.
13. Georlette P, Simons J, Costa L, 'Halogen-containing fire-retardant compounds', in Grant A F and Wilkie C A, Fire Retardancy of Polymeric Materials, New York, Marcel Dekker, 2000, 245–284.
14. Lyons J W, *The chemistry and uses of fire retardants*, USA, John Wiley & Sons, 1970.
15. Horn jr. W E, 'Inorganic hydroxides and hydrocarbonates: their function and use as flame-retardant additives', in Grant A F and Wilkie C A, *Fire retardancy of polymeric materials*, New York, Marcel Dekker, 2000, 285–352.
16. Hornsby P R, Rothon R N, 'Fire retardant fillers for polymers', in Le Bras M, Wilkie C A, Bourbigot S, Duquesne S and Jama C, *Fire retardancy of polymers – New applications of mineral fillers*, Cambridge, The Royal Society of Chemistry, 2005, 19–41.
17. Castrovinci A, Camino G, 'Fire retardant mechanisms in polymer nano-composite materials', Duquesne S, Magniez C, Camino G, *Mutifunctional Barriers for Flexible Structure: Textile, Paper and Leather*, Berlin, Springer-Verlag Publisher, 2007, 87–105.

18. Lewin M, 'Physical and chemical mechanisms of flame retarding polymers', in Le Bras M, Camino G, Bourbigot S, Delobel R, *Fire Retardancy of polymers: the use of intumescence*, Cambridge, The Royal Society of Chemistry, 1998, 3–32.
19. Camino G, Riva A, Vizzini D, Castrovinci A, Amigouët P, Bras Pereira P, 'Effect of hydroxides on fire retardance mechanism of intumescent EVA composition', in Le Bras M, Wilkie C, Duquesne S, Jama C and Bourbigot S, *Fire retardancy of polymers: New applications of mineral fillers*, Cambridge, The Royal Society of Chemistry, 2004, 248–263.
20. Castrovinci A, Camino G, Drevelle C, Duquesne S, Magniez C, Vouters M, 'Ammonium polyphosphate–aluminum trihydroxide antagonism in fire retarded butadiene–styrene block copolymer', *Eur Polym J* 2005 **41** (9) 2023–2033.
21. Lomakin S M, Zaikov G E, *New concepts in polymer science – Ecological aspects of polymer flame retardancy*, Utrecht, VSP, 1999.
22. Xie R, Qu B, 'Synergistic effects of expandable graphite with some halogen-free flame retardants in polyolefin blends', *Polym Degrad Stabil*, 2001 **71** (3) 375–380.
23. Li Z, Qu B, 'Flammability characterization and synergistic effects of expandable graphite with magnesium hydroxide in halogen-free flame-retardant EVA blends', *Polym Degrad Stabil*, 2003 **81** (3) 401–408.
24. Modesti M, Lorenzetti A, Simioni F, Camino G, 'Expandable graphite as an intumescent flame retardant in polyisocyanurate–polyurethane foams', *Polym Degrad Stabil*, 2002 **77** (2) 195–202.
25. Modesti M, Lorenzetti A, 'Halogen-free flame retardants for polymeric foams', *Polym Deg and Stab*, 2002 **78** (1) 167–173.
26. Bertelli G, Camino G, Costa L, Locatelli R, 'Fire retardant systems based on melamine hydrobromide: Part I – fire retardant behaviour', *Polym Degrad Stabil*, 1987 **18** (3) 225–236.
27. Bertelli G, Busi P, Costa L, Camino G, Locatelli R, 'Fire retardant systems based on melamine hydrobromide: Part II – overall thermal degradation', *Polym Degrad Stabil*, 1987 **18** (4) 307–319.
28. Levchik S V, Levchik G F, Balabanovich A I, Camino G, Costa L, 'Mechanistic study of combustion performance and thermal decomposition behaviour of nylon 6 with added halogen-free fire retardants', *Polym Deg and Stab*, 1996 **54** (2–3) 217–222.
29. Lu S-Y, Hamerton I, 'Recent developments in the chemistry of halogen-free flame retardant polymers', *Prog Polym Sci*, 2002 **27** (8) 1661–1612.
30. Lundquiest L, Leterrier Y, Sunderland P, Månson J-A E, *Life Cycle Engineering of Plastics – Technology, Economy and the Environment*, Oxford, Elsevier, 2000.
31. McEwan I, Arrighi V, Cowie J M G, 'Structure and Properties of Commonly Recycled Polymers', in La Mantia F, *Handbook of Plastics Recycling*, Shrewsbury, Rapra Technology Limited, 2002, 1–22.
32. Scheirs J, Camino G, 'Effect of Contamination on the Recycling of Polymers', in La Mantia F P, *Recycling of PVC and mixed Plastic Waste*, Toronto, ChemTec Publishing, 1996, 167–183.
33. Scaffaro R, La Mantia F P, 'Virgin/recycled homopolymer blends', La Mantia F, *Handbook of Plastics Recycling*, Shrewsbury, Rapra Technology Limited, 2002, 195–219.
34. Angulo-Sanchez J L, Ortega-Ortiz H, Sanchez-Valdes S, 'Photodegradation of polyethylene films formulated with a titanium-based photosensitizer and titanium dioxide pigment', *J Appl Polym Sci*, 1994 **53** (7) 847–856.
35. Sebaa M, Servens C, Pouyet J, 'Natural and artificial weathering of low-density

polyethylene (LDPE): Calorimetric analysis', *J Appl Polym Sci,* 1993 **47** (11) 1897–1903.
36. Tidjani A, Arnaud R, Dasilva A, 'Natural and accelerated photoaging of linear low-density polyethylene: Changes of the elongation at break', *J Appl Polym Sci,* 1993 **47** (2) 211–216.
37. Vilaplana F, Ribes-Greus A, Karlsson S, 'Degradation of High Ipact Polystyrene. Simulation by reprocessing and thermo-oxidation', *Polym Degrad Stabil,* 2006 **91** (9) 2163–2170.
38. La Mantia F P, Gaztelumendi M, Eguiazábal J I, Nazábal J, 'Properties – Reprocessing Behaviour of Recycled Plastics', in La Mantia F, *Handbook of Plastics Recycling,* Shrewsbury, Rapra Technology 2002, 127–194.
39. Wypych G, *Handbook of Material Weathering,* Toronto, ChemTech Publishing, 2003.
40. Hinsken H, Moss S, Pauquet J R, Zweifel H, 'Degradation of polyolefins during melt processing', *Polym Degrad Stabil,* 1991 **34** (1–3) 279–293.
41. Tzankova Dintcheva N, La Mantia F P, Acierno D, Di Maio L, Camino G, Trotta F, Luda M P, Paci M, 'Characterization and processing of greenhouse films', *Polym Degrad Stabil,* 2001 **72** (1) 141–146.
42. Kartakis C N, Papaspyrides C D, Pfaendner R, Hoffmann K, Herbst H, 'Mechanical recycling of postused high-density polyethylene crates using the restabilization technique. I. Influence of re processing', *J Appl Polym Sci,*1999 **73** (9) 1775–1785.
43. Kartalis C N, Papaspyrides C D, Pfaendner R, 'Recycling of post-used PE packaging film using the restabilization technique', *Polym Degrad Stabil,* 2000 **70** (2) 189–197.
44. Kartakis C N, Papaspyrides C D, Pfaendner R, Hoffmann K, Herbst H, 'Mechanical recycling of post-used HDPE crates using the restabilization technique. II: influence of artificial weathering', *J Appl Polym Sci,* 2000 **77** (5) 1118–1127.
45. Boldizar A, Jansson A, Gevert T, Möller K, 'Simulated recycling of post-consumer high density polyethylene material', *Polym Degrad Stabil,* 2000 **68** (3) 317–319.
46. Martins M H, De Paoli M A, 'Polypropylene compounding with recycled material I. Statistical response surface analysis', *Polym Degrad Stabil,* 2001 **71** (2) 293–298.
47. Braun D, 'Recycling of PVC', *Prog Polym Sci,* 2002 **27** (10) 2171–2195.
48. Santos A S F, Teixeira B A N, Agnelli J A M, Manrich S, 'Characterization of effluents through a typical plastic recycling process: An evaluation of cleaning performance and environmental pollution', *Resour Conserv Recy,* 2005 **45** (2) 159–171.
49. Bledzki A K, Sperber V E, Wolff S, 'Methods of Pretreatment', in La Mantia F, *Handbook of Plastics Recycling,* Shrewsbury, Rapra Technology Limited, 2002, 89–125.
50. Ferrara G, Meloy T P, 'Low dense media process: a new process for low-density solid separation', *Powder Technol,* 1999 **103** (2) 151–155.
51. Altland B L, Cox D, Enick R M, Beckman E J, 'Optimization of the high-pressure, near-critical liquid-based microsortation of recyclable post-consumer plastics', *Resour Conserv Recy,* 1995**15** (3–4) 203–217.
52. Pascoe R D, 'Investigation of hydrocyclones for the separation of shredded fridge plastics', *Waste Manage,* 2006 **26** (10) 1126–1132.
53. Takoungsakdakun T, Pongstabodee S, 'Separation of mixed post-consumer PET–POM–PVC plastic waste using selective flotation', *Sep Purif Technol,* 2007 **54** (2) 248–252.
54. Shent H, Pugh R J, Forssberg E, 'A review of plastics waste recycling and the flotation of plastics', *Resour Conserv Recy* 1999 **25** (2) 85–109.

55. Leitner R, Mairer H, Kercek A, 'Real-time classification of polymers with NIR spectral imaging and blob analysis', *Real-Time Imaging,* 2003 **9** (4) 245–251.
56. Hearn G L, Ballard J R, 'The use of electrostatic techniques for the identification and sorting of waste packaging materials', *Resour Conserv Recy,* 2005 **44** (1) 91–98.
57. Lungu M, 'Electrical separation of plastic materials using the triboelectric effect', *Minerals Engineering,* 2004 **17** (1) 69–75.
58. Yanar D K, Kwetkus B A, 'Electrostatic separation of polymer powders', *J Electrostat,* 1995 **35** (2–3) 257–266.
59. Garforth A A, Ali S, Hernández-Martínez J, Akah A, 'Feedstock recycling of polymer wastes', *Curr Opin Solid St M,* 2004 **8** (6) 419–425.
60. Pilati F, Toselli M, 'Chemical Recycling', in La Mantia F, *Handbook of Plastics Recycling*, Shrewsbury, Rapra Technology Limited, 2002, 297–336.
61. Newborough M, Highgate D, Vaughan P, 'Thermal depolymerisation of scrap polymers', *Appl Therm Eng,* 2002 **22** (17) 1875–1883.
62. Lilac W D, Lee S, 'Kinetics and mechanisms of styrene monomer recovery from waste polystyrene by supercritical water partial oxidation', *Adv Environ Res,* 2001 **6** (1) 9–16.
63. Buekens A G, Huang H, 'Catalytic plastics cracking for recovery of gasoline-range hydrocarbons from municipal plastic wastes', *Resour Conserv Recy,* 1998 **23** (3) 163–181.
64. de Carvalho G M, Muniz E C, Rubira A F, 'Hydrolysis of post-consume poly(ethylene terephthalate) with sulfuric acid and product characterization by WAXD, 13C NMR and DSC', *Polym Degrad Stabil,* 2006 **91** (6) 1326–1332.
65. Shukla S R, Harad A M, 'Aminolysis of polyethylene terephthalate waste', *Polym Degrad Stabil,* 2006 **91** (8) 1850–1854.
66. Ghaemy M, Mossadegh K, 'Depolymerisation of poly(ethylene terephthalate) fibre wastes using ethylene glycol', *Polym Degrad Stabil,* 2005 **90** (3) 570–576.
67. Kurokawa H, Ohshima M, Sugiyama M, Miura H, 'Methanolysis of polyethylene terephthalate (PET) in the presence of aluminium tiisopropoxide catalyst to form dimethyl terephthalate and ethylene glycol', *Polym Degrad Stabil,* 2003 **79** (3) 529–533.
68. Mahmood Zia K, Nawaz Bhatti H, Ahmad Bhatti I, 'Methods for polyurethane and polyurethane composites, recycling and recovery: A review', *React Funct Polym,* 2007 **67** (8) 675–692.
69. Newborough M, Highgate D, 'Thermal depolymerisation of poly-methyl-methacrylate using mechanically fluidised beds', Matcham J, *Appl Therm Eng,* 2003 **23** (6) 721–731.
70. Kaminsky W, Eger C, 'Pyrolysis of filled PMMA for monomer recovery', *J Anal Appl Pyrol,* 2001 **58–59,** 781–787.
71. Kaminsky W, Predel M, Sadiki A, 'Feedstock recycling of polymers by pyrolysis in a fluidised bed', *Polym Degrad Stabil,* 2004 **85** (3) 1045–1050.
72. Smolders K, Baeyens J, 'Thermal degradation of PMMA in fluidised beds', *Waste Manage,* 2004 **24** (8) 849–857.
73. Wallman P H, Thorsness C B, Winter J D, 'Hydrogen production from wastes', *Energy,* 1998 **23** (4) 271–278.
74. Borgianni C, De Filippis P, Pochetti F, Paolucci M, 'Gasification process of wastes containing PVC', *Fuel,* 2002 **81** (14) 1827–1833.
75. Walendziewsky J, 'Continuous flow cracking of waste plastics', *Fuel Process Technol,* 2005 **86** (12–13) 1265–1278.

76. Buekens A, 'Introduction to feedstock recycling of plastics', in Scheirs J and Kaminsky W, *Feedstock Recycling and Pyrolysis of Waste Plastics: Converting Waste Plastics into Diesel and Other Fuels*, Chichester, John Wiley & Sons, 2006, 3–41.
77. Garforth A A, Lin P Y H, Sharratt N, Dwyer J, 'Production of hydrocarbons by catalytic degradation of high density polyethylene in a laboratory fluidised-bed reactor', *Appl Catals A – Gen*, 1998 **169** (2) 331–342.
78. van Grieken R, Serrano D P, Aguado J, García R, Rojo C, 'Thermal and catalytic cracking of polyethylene under mild conditions', *J Anal Appl Pyrol*, 2001 **58–59** 127–142.
79. Aguado J, Serrano D P, Escola J M, 'Catalytic Upgrading of Plastic Wastes', in Scheirs J and Kaminsky W, *Feedstock Recycling and Pyrolysis of Waste Plastics: Converting Waste Plastics into Diesel and Other Fuels*, Chichester, John Wiley & Sons, 2006, 73–110.
80. Ding W, Liang J, Anderson L L, 'Hydrocracking and Hydroisomerization of High-Density Polyethylene and Waste Plastic over Zeolite and Silica-Alumina-Supported Ni and Ni-Mo Sulfides', *Energ Fuel*, 1997 **11** (6) 1219–1224.
81. Karayildirim T, Yanik J, Uçar S, Saglam M, Yüksel M, 'Conversion of plastics/HVGO mixtures to fuels by two-step processing', *Fuel Process Technol*, 2001 **73** (1) 23–35.
82. Luo M, Curtis C W, 'Effect of reaction parameters and catalyst type on waste plastics liquefaction and coprocessing with coal', *Fuel Process Technol*, 1996 **49** (1–3) 177–196.
83. Hornung A, Seifert H, 'Rotary kiln pyrolysis of polymers containig heteroatoms', in Scheirs J and Kaminsky W, *Feedstock Recycling and Pyrolysis of Waste Plastics: Converting Waste Plastics into Diesel and Other Fuels*, Chichester, John Wiley & Sons, 2006, 549–564.
84. Behzadi S, Farid M, 'Liquid fuel from plastic wastes using extrusion-rotary kiln reactors', in Scheirs J and Kaminsky W, *Feedstock Recycling and Pyrolysis of Waste Plastics: Converting Waste Plastics into Diesel and Other Fuels*, Chichester, John Wiley & Sons, 2006, 531–547.
85. Scheirs J, 'Overview of commercial pyrolysis processes for waste plastics', in Scheirs J and Kaminsky W, *Feedstock Recycling and Pyrolysis of Waste Plastics: Converting Waste Plastics into Diesel and Other Fuels*, Chichester, John Wiley & Sons, 2006, 383–431.
86. Arena U, Mastellone M L, 'Fluidized bed pyrolysis of plastic wastes', in Scheirs J and Kaminsky W, *Feedstock Recycling and Pyrolysis of Waste Plastics: Converting Waste Plastics into Diesel and Other Fuels*, Chichester, John Wiley & Sons, 2006, 435–471.
87. Zadgaonkar A, 'Process and equipment for conversion of waste plastics into fuels', in Scheirs J and Kaminsky W, *Feedstock Recycling and Pyrolysis of Waste Plastics: Converting Waste Plastics into Diesel and Other Fuels*, Chichester, John Wiley & Sons, 2006, 709–728.
88. Kaminsky W, 'The Hamburg fluidized-bed pyrolysis process to recycle Polymer wastes and tires', in Scheirs J and Kaminsky W, *Feedstock Recycling and Pyrolysis of Waste Plastics: Converting Waste Plastics into Diesel and Other Fuels*, Chichester, John Wiley & Sons, 2006, 475–490.
89. Okuwaki A, Yoshioka T, Asai M, Tachibana H, Wakai K, Tada K, 'The liquefaction of plastic containers and packaging in Japan', in Scheirs J and Kaminsky W, *Feedstock Recycling and Pyrolysis of Waste Plastics: Converting*

Waste Plastics into Diesel and Other Fuels, Chichester, John Wiley & Sons, 2006, 665–708.
90. Xingzhong Y, 'Converting waste plastics into liquid fuel by pyrolysis: developments in China', in Scheirs J and Kaminsky W, *Feedstock Recycling and Pyrolysis of Waste Plastics: Converting Waste Plastics into Diesel and Other Fuels*, Chichester, John Wiley & Sons, 2006, 729–750.
91. Curlee T R, Das S, *Plastic Waste – Management, Control, Recycling and Disposal*, eds. US Environmental Protection Agency, Park Ridge, William Andrew Publishing, 1991.
92. Barrales-Rienda J M, 'Energy Recovery from Plastic Materials', in La Mantia F, *Handbook of Plastics Recycling*, Shrewsbury, Rapra Technology Limited, 2002, 337–410.
93. Mirza R, 'A review of the role of plastics in energy recovery', *Chemosphere*, 1999 **38**(1) 207–231.
94. Directive 2002/95/EC of the European Parliament and of the Council of 27 January 2003 on the restriction of the use of certain hazardous substances in electrical and electronic equipment (RoHS).
95. Commission Decision 2005/717/EC, document number C(2005) 3754.
96. European Union Risk Assessment Report, CAS No: 1163-19-5, EINECS No: 214-604-9, available on the European Chemical Bureau website: http://ecb.jrc.it/.
97. Chemikalienverbotsverordnung (ChemVerbotsV) German ordinance.
98. Ebert J, Bahadir M, 'Formation of PBDD/F from flame-retarded plastic materials under thermal stress', *Environ Int*, 2003 **29** (6) 711–716.
99. Directive 2002/96/EC of the European Parliament and of the Council of 27 January 2003 on waste electrical and electronic equipment (WEEE).
100. Imai T, Hamm S, Rothenbacher K P, 'Comparison of the Recyclability of Flame-Retarded Plastics', *Environ Sci Technol*, 2003 **37** (3) 652–656.
101. Wanzke W, Goihl A, Nass B, 'Intumescent Systems for polyolefins – performance profile in electrical applications', *Flame retardants '98*, London, Interscience Communications Limited, 1998.
102. Rankin T N, Papazoglou E, 'Property Changes in Recycled Flame Retarded Polypropylene', in *ANTEC97: Plastics Saving Planet Earth, Volume 3: Special Areas*, Society of Plastics Engineers, 1997.
103. Horacek H, Garbner R, 'Advantages of flame retardants based on nitrogen compounds', *Polym Degrad Stabil*, **54** (2–3) 205–215.
104. Riess M, Ernst T, Popp R, Mueller B, Thoma H, Vierle O, Wolf M, Van Eldik R, 'Analysis of flame retarded polymers and recycling materials', *Chemosphere*, 2000 **40** (9–11) 937–941.
105. Schlummer M, Mäurer A, Leitner T, Spruzina W, 'Report: Recycling of flame-retarded plastics from waste electric and electronic equipment (WEEE)', *Waste Manage Res*, 2006 **24** (6), 573–583.
106. Dawson R B, Hardy M H, Landry S D, Yamada H, 'Recyclability of Flame Retardant Plastics', in *Proceedings EcoDesign 2001: Second International Symposium on Environmentally Conscious Design and Inverse Manufacturing IEEE 2001*, 107–115.
107. Dawson R B, Landry S D, 'End-of-Life Regulatory Issues For Flame Retardant Plastics Used in Electrical and Electronic Equipment Applications', in *Proceedings of the 2004 IEEE International Symposium on Electronics and the Environment, 2004 IEEE*, 46–50.
108. Dawson R B, Landry S D, 'Recyclability of Flame Retardant HIPS, PC/ABS, and

PPO/HIPS used in Electrical and Electronic Equipment', in *Proceedings of the 2005 IEEE International Symposium on Electronics and the Environment IEEE 2005*, 77–82.

109. Hamm S, Strikkeling M, Ranken P F, Rothenbacher K P, 'Determination of polybrominated diphenyl ethers and PBDD/Fs during the recycling of high impact polystyrene containing decabromodiphenyl ether and antimony oxide', *Chemosphere*, 2001 **44** (6) 1353–1360.

110. Tange L, Drohmann D, 'Waste electrical and electronic equipment plastics with brominated flame retardants – from legislation to separate treatment – thermal processes', *Polym Degrad Stabil*, 2005 **88** (1), 35–40.

111. Vasile C, Brebu M A, Karayildirim T, Yanik J, Darie H, 'Feedstock recycling from plastics and thermosets fractions of used computers. II. Pyrolysis oil upgrading', *Fuel*, 2007 **86** (4), 477–485.

112. Uddin M A, Bhaskar T, Kaneko J, Muto A, Sakat Y, Matsui T, 'Dehydrohalogenation during pyrolysis of brominated flame retardant containing high impact polystyrene (HIPS-Br) mixed with polyvinylchloride (PVC)', *Fuel*, 2002 **81** (14), 1819–1825.

113. Yamawaki T, 'The gasification recycling technology of plastics WEEE containing brominated flame retardants', *Fire Mater*, 2003 **27** (6), 315–319.

114. Fischer M M, Mark F E, Kingsbury T, Vehlow J, Yamawaky T, 'Energy recovery in the sustainable recycling of plastics from end-of-life electrical and electronic products', in *Proceedings of the 2005 IEEE International Symposium on Electronics and the Environment, 2005*, 83–92.

115. Tange L, Drohmann D, 'Environmental issues related to end-of-life options of plastics containing brominated flame retardants', *Fire Mater*, 2004 **28** (5), 403–410.

Part II

Testing, regulation and assessing the benefits of fire retardant materials

10
Challenges in fire testing: a tester's viewpoint

M L J A N S S E N S, Southwest Research Institute, USA

Abstract: This chapter deals with experimental methods for measuring the flammability of a material, i.e., its ease of ignition, burning intensity once ignited, propensity to spread flame over its surface and the rate at which it generates smoke and toxic products of combustion. A distinction is made between two types of experimental methods. The first type consists of tests that assess the ease of ignition of a material when exposed for a short time to a small heat source. Experimental methods of the second type measure how the material responds to a thermal exposure that is representative of pre-flashover fire conditions. Examples are given of both types of tests. The effects of test conditions and specimen details on the results and their meaning in terms of real fire performance are discussed. Challenges in assessing material flammability are identified and uses and limitations of flammability tests are reviewed. The chapter concludes with a section on future trends and recommendations for further reading.

Key words: calorimetry, cone calorimeter, ignition, heat release rate, material flammability, FMVSS 302, reaction to fire, room/corner test, smoke production, surface flame spread, toxic potency, UL 94.

10.1 Introduction

ASTM E 176 Terminology of Fire Standards defines the term 'flammable' as 'subject to easy ignition and rapid flaming combustion'. This definition implies that a flammable material is likely to perform rather poorly in a fire. Identifying materials that are very easy to ignite and that burn with high intensity once ignited has merit because restricting the use of these materials significantly improves fire safety. However, cost-effective design of fire-safe buildings and transportation vehicles requires the ability to discriminate between materials over a much broader range of fire performance. The fire performance of a material can be quantified on the basis of observations and measurements of its response in laboratory tests. It is therefore more useful to define 'flammability' in terms of the characteristics that are measured in these tests. The following section is taken from the chapter on 'understanding material flammability' in a recently-published book on flammability testing of materials used in construction, transport and mining.[1]

> With this in mind, a better yet still imperfect definition of flammability is: the ease with which a material ignites, the intensity with which its burns and releases heat once ignited, its propensity to spread fire, and the rate at which

it generates smoke and toxic combustion products during gasification and burning. A comprehensive evaluation of a material's overall flammability may require data from several laboratory tests, perhaps combined with some form of analysis or modeling to interpret the results properly.

This is the definition that is used in this book.

A common application of flammability tests is to obtain data for regulatory compliance, as will be discussed in Section 10.6. Fire safety codes and regulations are generally based on two strategies. The first strategy involves preventing, or at least minimizing the likelihood of ignition. Since in practice it is not possible to completely eliminate ignition, the second strategy involves managing the impact of a subsequent fire. There are two distinct types of flammability tests, each associated primarily with one of the two strategies.

In Type I flammability tests a specimen (linear dimensions of the order of centimeters) of the material is exposed to a small heat source (Bunsen burner type flame, hot wire, etc.) for a short duration (seconds). Pass/fail criteria are based on ignition of the specimen during exposure, formation of flaming droplets and/or sustained flaming or smoldering after removal of the heat source.

Type II flammability tests characterize the behavior of materials under more severe thermal exposure conditions that are representative of the growing pre-flashover stages of a compartment fire. These tests essentially determine how a material responds to the temperatures and heat fluxes in a growing fire and are therefore also referred to as 'reaction-to-fire' tests. The fire conditions are simulated with radiant panels in bench-scale tests (Type II.A) or large gas burner flames in intermediate to large-scale tests (Type II.B).

10.2 Type I flammability tests

10.2.1 General concepts

A large number of flammability tests have been developed to evaluate the ignition propensity of a wide range of materials exposed, usually for a short duration, to a small heat source. For example, Table 10.1 lists the 30 small-flame exposure ignition tests that are included in the most recent compilation of ASTM fire standards.[2] When adding methods that use an electric arc, cigarette, hot wire, hot surface, etc., instead of a flame and those developed by other standards organizations (NFPA, UL, ISO, IEC, etc.), the total number of tests is mind-boggling. Conceptually these tests are all very similar. A relatively small specimen of the material is exposed to the heat source, usually for a short duration (seconds). Performance is measured on the basis of the duration of flaming and/or glowing combustion after removal of the heat source and the extent or rate of flame propagation over the surface of the specimen. Two commonly used small-flame ignition tests are described in more detail in the next two sections.

Table 10.1 Standard ASTM small-flame exposure ignition tests

Designation	Product	Sample size	Orientation	Duration
C 1166	Elastomeric gaskets	25 × 460 mm	Vertical	5–15 min
D 229	Rigid electrical insulation	13 × 13 mm	Vertical	2 × 10 s
D 350	Electrical sleeving	≥ 180 mm	Vertical	Ignition
D 378	Flat rubber belting	13 × 150 mm	45°	60 s
D 470	Wire & cable insulation	250 mm	Horizontal	30 s
D 635	Self-supporting plastics	13 × 125 mm	Horizontal	30 s
D 777	Paper & paperboard	70 × 210 mm	Vertical	12 s
D 876	PVC tubing	560 mm	70°	15 s
D 1000	Electrical insulation tape	19 × 375 mm	Horizontal	30 s
D 1230	Apparel textiles	50 × 150 mm	45°	Varies[a]
D 1360	Paints[b]	150 × 305 mm	45°	Unknown[c]
D 2633	Wire & cable insulation	560 mm	Vertical	5 × 15 s
D 2671	Heat-shrinkable tubing	560 mm	70°	5 × 15 s
D 2859	Textile floor coverings	230 × 230 mm	Horizontal	Unknown[d]
D 2863	Plastics	W: 7–52 mm L: 70–200 mm	Vertical	6 × 5 s
D 2939	Emulsified bitumens	100 × 100 mm[e]	Vertical	2 × 10 s
D 3014	Thermoset foam plastics	19 × 254 mm	Vertical	10 s
D 3032	Hookup wire insulation	600/800 mm[f]	Vertical/60°[f]	5/1 × 15 s[f]
D 3801	Solid plastics	13 × 125 mm	Vertical	2 × 10 s
D 4151	Blankets	70 × 70 mm	Horizontal	1 s
D 4804	Nonrigid solid plastics	50 × 200 mm	Vertical	2 × 3 s
D 4982	Waste	5 g in aluminum boat		2–3 s
D 4986	Cellular polymers	50 × 150 mm	Horizontal	60 s
D 5048	Solid plastics	13 × 125 mm	Vertical	5 × 5 s
		150 × 150 mm	Horizontal	
D 5132	Flexible foams & rubber	100 × 300 mm	Horizontal	15 s
D 6413	Textiles	76 × 300 mm	Vertical	12 s
D 6545	Children's sleepwear	89 × 254 mm	Vertical	3 s
E 69	Treated wood	10 × 19 mm	Vertical	4 min
F 1358	Protective clothing	75 × 400 mm	Vertical	3 s & 12 s
F 1955	Sleeping bags	300 × 360 mm	Horizontal	30 s

[a] The ignition burner is removed when flame spreads to the top of the specimen
[b] Wet film thickness applied to 6 mm thick wood substrate
[c] Ignition source consists of a cup filled with 5 mL of pure ethanol
[d] The ignition source consist of a methanamine tablet placed on the specimen
[e] Emulsified bitumen applied to a 0.3–0.4 mm thick 150 × 150 mm metal plate
[f] The standard specifies two methods (A and B) with different specimen length and orientation

10.2.2 Example 1: UL 94 20-mm vertical burning test

The UL 94 standard provides procedures for bench-scale tests to determine the acceptability of plastic materials for use in appliances or other devices with respect to flammability under controlled laboratory conditions. The standard includes several test methods that are employed depending upon the intended

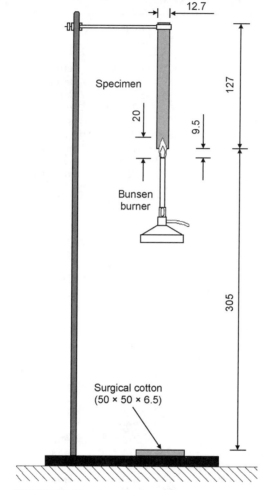

10.1 UL 94 20-mm vertical burning test.

end-use of the material and its orientation in the device. The standard outlines two horizontal burning tests, three vertical burning tests and a radiant panel flame spread test. The most commonly used method is summarized below.

The first vertical burning test described in the UL 94 standard is the '20-mm vertical burning test; V-0, V-1, or V-2'. The method is also described in ASTM D 3801 (see Table 10.1). A schematic of the test setup is shown in Fig. 10.1. The V-0, V-1, or V-2 classification is based on the duration of flaming or glowing following the removal of the burner flame, as well as the ignition of cotton by dripping particles from the test specimen.

Specimens measuring 127 mm in length by 12.7 mm wide are suspended vertically and clamped at the top end. A thin layer of cotton is positioned

Table 10.2 UL 94 20-mm vertical burning test classification

Observation	Classification		
	V-0	V-1	V-2
t_1 or t_2 for any specimen	$\leq$10 s	$\leq$30 s	$\leq$30 s
$t_1 + t_2$ for 5 specimens	$\leq$50 s	$\leq$250 s	$\leq$250 s
$t_2 + t_3$ for any specimen	$\leq$30 s	$\leq$60 s	$\leq$60 s
Flame propagation or glowing up to clamp, any specimen	No	No	No
Ignition of cotton	No	No	Yes

305 mm below the test specimen to catch any molten material that may drop from the specimen. A 20 mm long flame from a methane burner is applied to the center point on the bottom end of the specimen. The burner is positioned such that the burner barrel is located 9.5 mm below the bottom end of the material specimen. If the material melts, the burner barrel is held at an angle to avoid catching burning droplets that might ignite the cotton.

The flame is maintained for 10 seconds, and then removed to a distance of at least 150 mm. Upon flame removal, the specimen is observed for flaming and its duration time recorded (t_1). As soon as the flame ceases, the burner flame is reapplied for an additional 10 seconds, then removed again. Duration of flaming (t_2) and/or glowing (t_3) after the second flame application are noted. Based on the results for five specimens tested, the material is classified as either V-0, V-1, or V-2 based on the criteria outlined in Table 10.2.

10.2.3 Example 2: FMVSS 302 test

Federal Motor Vehicle Safety Standard (FMVSS) 302 describes a small-scale horizontal flame spread test to evaluate the flammability of automotive materials. In the US, components within 12.7 mm of the exposed surface inside the passenger compartment of motor vehicles have to be tested according to this method and need to have a burn rate that does not exceed 102 mm/min. The test is used throughout the world and also forms the subject of several international standards.

Test specimens are prepared to dimensions of 102 × 356 mm, with a maximum thickness of 12.7 mm. The specimen is mounted in a U-shaped frame facing down in the direction that provides the most adverse test results (see Fig. 10.2). Specimens that are less than 51 mm wide are supported in a special frame with wire supports. The frame is placed in a ventilated 203 × 381 × 356 mm chamber to protect against drafts.

The gas flow to a Bunsen burner is adjusted to provide a 38 mm flame, and the burner is placed 19 mm below the center of the open end of the frame. The specimen is exposed to the flame for 15 seconds, and the time is recorded when

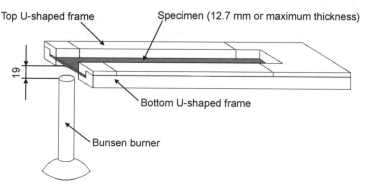

10.2 FMVSS 302 test.

the flame front reaches a point 38 mm from the exposed end. The time for the flame to travel along the underside of the specimen, from a point 38 mm from the exposed end of the frame to a point 38 mm from the clamped end of the specimen is also recorded. The burn rate is equal to the distance between the two marks (254 mm) divided by this time. The burn rate must not exceed 102 mm/min for any of the five specimens tested.

10.2.4 Heat source effects

The severity of the tests depends in part on the intensity of the ignition source (size of the flame, for example) and the duration of application. However, it also depends on the orientation of the specimen and the pass/fail or classification criteria. For example, the FMVSS 302 uses a larger flame than the UL 94 20-mm vertical burning test, but the total application time is shorter and the distance between the burner barrel and the specimen is greater. The FMVSS 302 test uses a horizontal specimen and allows sustained burning and flame propagation over the specimen surface after the burner has been removed, while the UL 94 test applies the burner flame at the bottom of a vertical specimen and only a short duration of sustained flaming and smoldering after removal of the burner is permitted. The combined effect is that it is much easier to pass the FMVSS 302 test than to obtain a UL 94 rating.

In any Type I flammability test it is essential to follow carefully the instructions in the standard, in particular those that pertain to the size, location and duration of application of the heat source. A recent study has shown that the results of the UL 94 20-mm vertical burning test are very sensitive to the exact duration of flame application and to the technique that is used to apply and remove the flame.[3] The magnitude of the effect depends on the type of material that is being tested. Loosely following the test procedure might result in an unwarranted V-0 classification.

10.2.5 Specimen effects

Specimen thickness

The width and length of Type I flammability test specimens is usually fixed, but the thickness is often variable. The thickness of the specimen affects performance in two ways. Thick specimens are generally more difficult to ignite as there is more bulk material acting as a heat sink. Once ignited, thick specimens have the potential to burn for a longer time. The first effect is usually dominant in Type I flammability tests. For example, in a recent study specimens from 18 plastics were evaluated in the UL 94 20-mm vertical burning test.[4] The materials were tested in two thicknesses, 1.6 and 3.2 mm. Specimen thickness did not affect the rating for most materials but for some the rating changed from V-0 to V-2 (two materials), from V-0 to not rated (one material) and from V-1 to not rated (one material). Type I flammability test standards therefore nearly always require that at least the minimum thickness be tested.

Specimen orientation

The orientation of the specimen has a significant effect on its propensity for flame spread. The following configurations are ordered according to their likelihood of supporting flame propagation over the surface:

1. Vertical specimen ignited at the bottom edge;
2. Horizontal specimen facing down ignited at the underside;
3. Horizontal specimen facing up ignited on the top side;
4. Lateral specimen ignited at one end; and
5. Vertical specimen ignited at the top edge.

This means, for example, that it is much easier to sustain upward flame spread over the surface of a vertical specimen ignited at the bottom edge than it is to support flame spread over the surface of the same specimen exposed to the same heat source from the top down. The effect of specimen orientation partly explains why it is much easier to pass the UL 94 HB test (configuration #4) than to obtain a rating in the UL 94 20-mm vertical burning test (configuration #1).

Specimens in Type I flammability tests are usually not backed by a substrate and the effect on the test results of the type of substrate used is not an issue.

Specimen mounting

Variations in specimen mounting methods may have a significant effect on the test results. For example, FMVSS 302 allows the use of specimens that are pieced together for materials that are not planar or automotive components that are smaller than the dimensions of FMVSS 302 specimens. The discontinuities between the different pieces affect the flame spread rate and might have a significant effect on the outcome of the test. FMVSS 302 also allows the use of support wires when the width of the pieces is less than the gap of the specimen holder. Wires have a similar effect on flame spread as discontinuities. The fact that wire may be used for certain types of materials essentially penalizes those that can be tested without support.

10.3 Type II.A flammability tests

10.3.1 General concepts

Type II.A flammability tests characterize the behaviour of materials under thermal exposure conditions that are representative of the growing pre-flashover stages of a compartment fire. The fire conditions are simulated in these bench-scale reaction-to-fire tests by exposing the specimen to the heat flux from a gas-fired radiant panel or an electrical heater. The range of heat fluxes that can be obtained in a particular test apparatus depends on the specimen-heater geometry and the type of radiant heat source that is used. Between different Type II.A flammability tests, heat fluxes can vary over a broad range, from approximately $1\,kW/m^2$ to more than $100\,kW/m^2$. The heat flux to the specimen is specified in some test standards and is variable in others. Tests for measuring flame spread characteristics typically expose the specimen to a heat flux that varies in the direction of its largest dimension. A pilot flame, spark plug or hot wire is used to ignite the gases and vapors that are generated by the pyrolysis of the heated specimen. When a pilot flame is used, it is either located in the gas phase or impinging on the specimen surface. The latter is less desirable because it locally increases the heat flux to the specimen by an unknown amount.

The purpose of Type II.A tests is to measure the flammability characteristics of materials, i.e., ease of ignition, flame spread propensity, heat release and production of smoke and toxic combustion products. Some tests are designed to measure only one of these characteristics. Other tests are more sophisticated and can be used to measure several characteristics at the same time.

An important example of such tests is the cone calorimeter which is discussed in detail in Chapter 11. This was developed at the National Institute of Standards and Technology or NIST by Dr. Vytenis 'Vyto' Babrauskas in the early 1980s.[5] It is a bench-scale fire test apparatus used primarily to measure the heat release rate of materials on the basis of the oxygen consumption method.[6] It is presently the most commonly used small-scale calorimeter. A bibliography compiled by the inventor indicates that over one thousand papers on cone calorimeter studies had been published at the end of 2002.[7] The cone calorimeter is standardized internationally as ISO 5660 Part 1 (heat release) and Part 2 (smoke production), in Australia and New Zealand as AS/NZS 3837 and in North America as ASTM E 1354 and NFPA 271.

10.3.2 Heat source effects

Radiant heat source

The discussion in this section pertains to Type II.A material flammability tests that are used primarily to obtain ignition and heat release characteristics. To be meaningful, these characteristics must be measured at heat fluxes that remain constant during the test and that are relatively uniform over the surface of the specimen. This is most easily accomplished by using a heat source that consists of a radiant panel remote from the specimen. Porous gas panels as well as electrical heating elements are suitable for this purpose.

The radiant heat flux can be adjusted by changing the power of the heater or by changing the distance between heater and specimen. If the second method is used, there are practical upper and lower limits to the range of radiant heat flux levels that can be created. If the heater is too close to the specimen, convective heat transfer becomes significant. Therefore, the upper limit corresponds to the minimum distance that has to be maintained in order to ensure predominantly radiative heat transfer. The lower limit is determined by the uniformity of the incident radiant heat flux, which drops with increasing distance between heater and specimen. The exact limits depend on the geometrical configuration, power of the heater, and the degree of non-uniformity of the incident heat flux profile that is deemed acceptable.

It is important that the radiant heat flux to the specimen be maintained at a constant level during the test. If the heater is operated at a constant power level, incident radiant heat flux changes during testing. At the start of a test, a cold specimen is inserted. The specimen acts as a heat sink, resulting in a decrease of the heater temperature, and consequently a decrease of the incident radiant heat flux. After ignition, the heat released by the specimen results in an increase of the heater temperature and incident radiant heat flux. To maintain the incident radiant heat flux during a test, it is therefore necessary to keep the temperature of the heater constant. This is not easy for a gas panel, but relatively straightforward for electrical heating elements. With the oxygen consumption method, another drawback of using a gas panel is that its products of combustion result in an oxygen depletion that is usually much larger than the oxygen consumed for combustion of the specimen. Thus, small fluctuations in panel flow can result in significant error of the measured heat release rate. This 'baseline' problem can be avoided by using a separate exhaust system for the heater.

Two types of electrical heaters are used in Type II.A flammability tests: low and high temperature. To obtain heat fluxes that are representative of real fire exposure conditions, resistance heating elements typically operate at temperatures between 800 and 1200 K. This is comparable to the temperatures that are observed during the pre-flashover stages of compartment fires. The cone calorimeter uses this type of low-temperature heating element. Babrauskas concluded from correlations between heater temperature and radiant heat flux that the electric heater in the cone calorimeter behaves as a grey body with an emissivity close to unity.[5]

Tungsten filament lamps are commonly used heaters that operate at high temperatures (typically around 2600 K). According to Wien's displacement law, thermal radiation from these lamps peaks at a much shorter wavelength than the radiation from flames and hot gases in fires. Piloted ignition studies on plastics and wood have shown that these materials absorb much less radiation in the visible and near-infrared range, than at higher wavelengths.[8,9] As a result, it was found that the ignition time at a specified incident heat flux is consistently longer and that the minimum heat flux for ignition is higher for tungsten filament lamps than for resistance heaters. This problem can be eliminated by applying a thin black coating to the surface of specimens that are tested in an apparatus with high temperature heating lamps. This procedure is specified in ASTM E 2058 (Fire Propagation Apparatus).

Ignition pilot

Type II.A flammability tests are usually conducted with an ignition pilot. The use of a pilot reduces the variation in time to sustained flaming between multiple tests conducted under identical test conditions. Because the duration of the preheat period prior to ignition affects burning rate after ignition, use of a pilot also improves repeatability of heat release rate measurements. Furthermore, piloted ignition is used because it is representative of most real fires, and conservative in other cases.

The ignition pilot in Type II.A flammability tests consists of a small gas burner flame, a glowing wire, or an electric spark. An impinging flame should not be used because it locally enhances the incident heat flux to the specimen by an unknown amount. A potential problem with pilot flames is that it is sometimes extinguished by fire retardants or halogens in the fuel volatiles. A glowing wire is not an efficient method for igniting fuel volatiles, sometimes resulting in poor repeatability. An electric spark remains stable when fire retardants or halogens are present. Several studies have shown that the exact location of a pilot flame may be critical to ensure consistency in the ignition times that are measured, at least under some conditions.[10–12] Since a spark plug occupies a smaller volume, positioning is even more critical than with pilot flames.

10.3.3 Specimen effects

Specimen area

Increasing the specimen area appears to reduce the ignition time at a specified heat flux. Östman *et al.* compared piloted ignition times from cone calorimeter tests on 13 materials for 200 × 200 mm and 100 × 100 mm specimens.[13] It was found that quadrupling the size decreased ignition time by 20% or less. A subsequent study comparing ignition times measured in the cone calorimeter and the ISO ignitability test (ISO 5657) concluded that the latter were slightly shorter (on average 10%) due to the larger specimen size.[14]

Comparable piloted ignition times were found in the cone calorimeter and the intermediate-scale calorimeter (ASTM E 1623), in spite of the fact that the specimen area is a factor of 100 larger in the latter.[15] However, the effect of specimen size may be offset by other differences between the two test apparatuses.

To predict real scale burning rates from bench-scale data, differences in flame heat transfer, and up to a lesser extent, heat transfer at the edges have to be accounted for. The heat flux from the flame increases and the effect of heat losses at the edges decreases with increasing fire size. The heat release rate per unit area is therefore higher for large specimens than for small specimens exposed to the same external heat flux.

The aforementioned ignition study by Östman and Nussbaum also involved heat release rate measurements. As expected, the increase in specimen size from 100×100 mm to 200×200 mm resulted in a 12% increase of the average heat release rate per unit area over the first minute after ignition at heat flux levels exceeding $25\,kW/m^2$. Larger increases were observed at the $25\,kW/m^2$ exposure level, and for peak heat release rate.

Janssens and Urbas presented a comparison of heat release rate data for nine wood products obtained in a cone calorimeter and an intermediate-scale calorimeter.[15] A 100-fold increase in specimen size only resulted in a 10% increase of the heat release rate. This modest effect can be explained by the fact that the heat feedback from the flame is relatively insensitive to specimen area for testing in the vertical orientation, in particular for materials that do not produce very luminous flames such as wood, because the flame has the shape of a thin sheet in front of the specimen instead of a large, roughly conical volume above the surface of specimens tested in the horizontal orientation.

Specimen thickness

Specimens in Type II.A flammability tests are usually backed by a substrate. The nature of the substrate does not affect the time to ignition if it occurs before the thermal wave hits the back surface. Under these conditions the specimen is said to behave as a thermally thick solid. Whether a specimen is thermally thick depends on the physical thickness of the specimen, the thermal properties of the material and the exposure time. Specimens of a specific material with a given thickness can behave as a thermally thick solid at high heat fluxes (short ignition times) or as a thermally intermediate or thin solid at low heat fluxes (long ignition times). Thermally thin solids have a negligible temperature gradient across the thickness. The gradient is not negligible for thermally intermediate

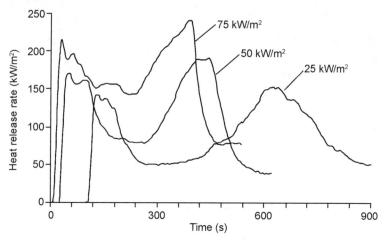

10.3 Heat release rate vs. time of plywood at different heat flux levels.

solids. In both cases the surface temperature, and consequently the time to ignition, is affected by the nature of the substrate. Since one does not know *a priori* whether a specimen will deviate from thermally thick behavior, it is important to test materials on the same substrate as used in practice.

Since calorimeter tests generally take much longer than ignition tests, heat release rate measurements are nearly always affected by the nature of the substrate. For example, standard cone calorimeter specimens are backed with ceramic fiber blanket. This explains the bi-modal shape of the heat release rate curve measured for charring materials (see Fig. 10.3). The heat release rate for charring materials reaches a maximum shortly after ignition and then decreases as the char layer builds up. Toward the end of the test the burning rate exhibits a second peak due to the reduced heat losses from the back face of the specimen. A similar phenomenon is observed for thermoplastics, although in this case the burning rate gradually increases as the specimen is consumed and a single peak is reached toward the end of the test (see Fig. 10.4). To obtain cone calorimeter data that are representative of a material's behavior under end-use conditions, it is essential to perform tests on specimens that are backed by the substrate that is used in practice.

Specimen orientation

Specimen orientation (horizontal facing up vs. vertical) generally does not appear to have a significant effect on the time to ignition at a specified heat flux, although in one case shorter ignition times were found for specimens tested in the horizontal orientation.[16,17] Researchers at the University of Edinburgh measured the time to piloted ignition and the corresponding temperature for a 6-mm thick slab of PMMA for a range of radiant heat fluxes applied perpendicular

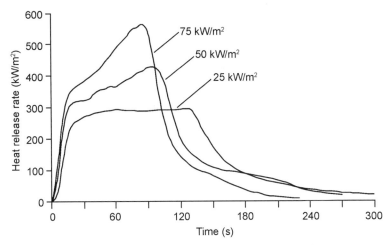

10.4 Heat release rate vs. time of GRP at different heat flux levels.

to the surface at a number of orientations.[18] Tests were conducted with the specimen surface inclined at 0°, 30°, 45°, 60° and 90°. The most surprising finding in this study was that the shortest time to ignition was measured at an inclination of 30° regardless of the incident heat flux.

Products do not necessarily have to be tested in the same orientation as they are used. For practical reasons, the preferred orientation for small-scale testing is horizontal facing upward. The vertical orientation might be preferable for collecting specialized data for research purposes.

Specimen mounting

Edge effects have been studied extensively in the cone calorimeter. ASTM and ISO standards of the cone calorimeter prescribe that, except for calibrations with PMMA, the specimen is to be wrapped with aluminum foil on the sides and bottom. The main purpose of the foil is to eliminate mass transfer along all boundaries except the exposed face of the specimen. Furthermore, the ISO standard requires all tests be conducted in the horizontal orientation with the stainless steel retainer frame.

The retainer frame is a relatively large mass of steel that acts as a heat sink, reducing the energy transferred to the specimen. As a result the ignition time at a specified heat flux is shorter when the retainer frame is not used.[19,20]

Several investigators have studied the effect of the retainer frame on heat release rate. The standard retainer frame reduces the actual exposed area from 100 × 100 mm to 94 × 94 mm, or from 0.01 m^2 to 0.0088 m^2 and the heat release rate at a specified heat flux is therefore expected to decrease by at least 12% when the retainer frame is used. An extensive study of the effects of specimen edge conditions on heat release rate measured was conducted in the cone calorimeter at NIST.[21] This study concluded that the heat sink effect of the retainer frame further reduces heat release rate values by approximately 8%. Researchers in Sweden found the same reduction for the average heat release over the first three minutes following ignition, but the peak heat release rate was reduced by as much as 25%.[22] Urbas reported smaller differences between heat release rates per unit area measured with and without the retainer frame at 75 kW/m^2 and even found that the retainer frame slightly increased the heat release rate of fire-retardant treated cellulosic insulation.[23]

Urbas and Sand designed an alternative retainer frame, comprising an insulating collar made of medium-density or high-density refractory material.[24] They concluded that the best edge conditions are obtained using the insulating frame with insulation material that most closely resembles the specimen in thermal properties. Researchers at FM Global recently proposed a similar approach to address edge effects in the ASTM E 2058 fire propagation apparatus.[25] The aforementioned NIST study confirmed that the use of an insulated frame as proposed by Urbas, Sand and deRis gives heat release rate values that are slightly closer to the expected true values. However, the insulated frame makes the test procedure significantly more complicated and it was therefore not recommended for routine testing.

It can be concluded from these studies that the specimen holder configuration in a Type II.A flammability test may have a significant effect on the measurements, and that this should be accounted for if the test data are used to predict performance in real fires.

10.3.4 Observations and measurements

The fact that ignition times are generally determined on the basis of visual observations introduces a significant element of uncertainty, in particular for fire-retardant treated materials that often exhibit flashes and transitory flaming before sustained ignition. To alleviate this problem, researchers at FM Global proposed a definition of ignition based on the inflection point in the mass loss rate curve.[26] This obviously requires that the mass of the specimen be measured during the test. A heat release rate criterion has been used to define ignition in the cone calorimeter.[27] Based on this criterion the specimen is considered to ignite when the heat release rate first exceeds 25 kW/m^2.

10.4 Type II.B flammability tests

10.4.1 Example: room/corner test

Room/corner tests are by far the most frequently conducted large-scale fire experiments throughout the world. Several standard room/corner test protocols are now available and are specified in codes and regulations for qualifying interior finishes. For example, US model building codes require that textile wall coverings for use in unsprinklered compartments meet specific performance requirements when tested according to NFPA 265 (previously known as UBC 8-2). The principal requirement of these tests is that flashover does not occur. The same codes also require that all other interior wall and ceiling finish materials comply with requirements based on NFPA 286, including a limit on the total smoke released.

The Safety Of Life At Sea (SOLAS) convention promulgated by the International Maritime Organization (IMO) permits the use of combustible bulkhead and ceiling linings on high speed craft, provided they meet stringent fire performance requirements based on assessment according to ISO 9705. ASTM E 2257 is the American version of ISO 9705.

The test apparatus and instrumentation described in the NFPA and ISO room/corner test standards are very similar (see Fig. 10.5). However, there are some significant differences in terms of specimen configuration and ignition source.

The apparatus consists of a room measuring 3.6 m deep × 2.4 m wide × 2.4 m high, with a single ventilation opening (open doorway) measuring approximately 0.8 m wide × 2 m high in the front wall. Walls and ceiling are lined for tests according to ISO 9705. For tests according to the NFPA standards, the interior surfaces of all walls (except the front wall) are covered with the test material. NFPA 286 is also suitable for evaluating ceiling finishes (see below).

The test material is exposed to a propane burner ignition source, located on the floor in one of the rear corners of the room opposite the doorway. The burner is placed directly against (ISO 9705 and NFPA 286) or at a distance of 50 mm (NFPA 265) from the walls. The ISO burner consists of a steel sandbox measuring 0.17 × 0.17 × 0.145 m. Propane is supplied to the burner at a specified rate such that a net heat release rate of 100 kW is achieved for the first 10 minutes of the test, followed by 300 kW for the remaining 10 minutes (20-minute test

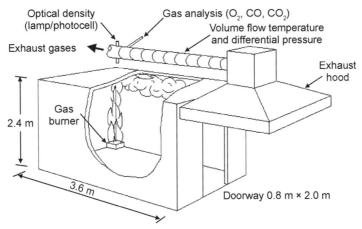

10.5 Room/corner test apparatus.

duration, unless terminated when flashover occurs). The NFPA burner consists of a steel sandbox measuring 0.305 × 0.305 × 0.152 m, with the top surface positioned 0.305 m above the floor of the room. Propane is supplied at a specified rate so that a net heat release rate of 40 kW is achieved for the first five minutes of the test, followed by 150 kW (NFPA 265) or 160 kW (NFPA 286) for the remaining 10 minutes (15-minute test duration unless terminated when flashover occurs). A fundamental difference between NFPA 265 and NFPA 286 is the fact that the flame from the burner alone just touches the ceiling in NFPA 286. This makes it suitable for assessing the fire performance of an interior ceiling finish, an application for which NFPA 265 is unsuitable. This effect is partly due to the higher energy release rate of the NFPA 286 burner, but primarily because of the burner being in direct contact with the walls, thereby reducing the area over which the flames can entrain air and increasing the overall flame height.

All combustion products emerging from the room through the open doorway are collected in a hood. Instrumentation is provided in the exhaust duct for measuring heat release rate based on the oxygen consumed (ISO and NFPA standards) and smoke production rate (ISO 9705 and NFPA 286 only). The room contains a single heat flux meter located in the center of the floor. The NFPA standards also specify that seven thermocouples be installed in the upper part of the room and doorway to measure the temperature of hot gases that accumulate beneath the ceiling and exit through the doorway. In addition to quantitative heat release and smoke production rate measurements, time to flashover (if it occurs) is one of the main results of a room/corner test. Different criteria are commonly used to define flashover, e.g. upper layer temperature of 600 °C, flames emerging through the doorway, heat flux to the floor of 20 kW/m, heat release rate of 1 MW, etc.

10.4.2 Ignition source and specimen configuration effects

An extensive comparison of room/corner test data obtained in seven studies was made at the Forest Products Laboratory in Madison, WI in the late 1990s.[28] A total of 65 tests on 24 materials (primarily wood products) were reviewed. Four different test protocols were evaluated. The four test protocols were the combinations of two burner scenarios and two specimen configurations. The two burner scenarios were (1) 40 kW for 5 min followed by a change to 160 kW for 10 min (NFPA 286 burner) and (2) 100 kW for 10 min followed by a change to 300 kW for 10 min (ISO 9705 burner). The second variable in the test protocols was whether or not the ceiling was lined with the test materials. It was found that the 40/160 kW protocol with the ceiling unlined was always the least severe (longest flashover times) and that the 100/300 kW protocol with the ceiling lined was always the most severe (shortest flashover times). The order of the remaining two protocols in terms of severity varied depending on the product that was tested.

In tests of untreated wood products, good agreement was found between the following flashover criteria: flames out the door and 20 kW/m² to the floor. However, there were inconsistencies between these criteria for fire-retardant treated wood products and treated polyurethane foam. These inconsistencies were observed in some 100/300 kW tests with lined and unlined ceilings.

10.5   Challenges in assessing material flammability

Materials that melt or sag when heated are often difficult to test. The problem can be avoided by conducting tests in the horizontal orientation. The question then is whether the test results are meaningful in terms of real fire performance. It might not be possible to evaluate these materials in a representative manner in a bench-scale test, and full-scale testing is often needed.

Layered products, in particular panels consisting of a plastic foam core sandwiched in between two sheets of a non-combustible material, are difficult to test. The weakest spot of this type of construction is at a joint or in a corner which are difficult to include in a small specimen. In addition, the thickness of these panels usually exceeds the maximum thickness that can be accommodated in bench-scale tests. So, again a full-scale test is usually needed.

The majority of flammability test standards give detailed and specific instructions on how to prepare the specimens, conduct the test and report and possibly interpret the results. In some cases, however, it is left to the user to select the test conditions and decide which results to use. For example, cone calorimeter standards do not specify the heat flux to be used. Tests can be run in the horizontal or vertical orientation, with or without the retainer frame and with or without the spark igniter. Various end-of-test criteria can be used and different approaches are available to test intumescing specimens. In addition, the user has to decide which subset of the vast amount of test data to focus on. Unless the objective is to meet a regulation or specification that spells out the details, there are a lot decisions that need to be made and it is not easy to make the best choice.

10.6   Use of material flammability tests

There are three distinct reasons for conducting fire tests:

1. Compliance with fire safety codes, regulations and specifications;
2. Quality assurance and product development research; and
3. Obtaining data for fire safety engineering design and analysis.

Codes and regulations have been developed in many countries to ensure a level of public fire safety that is acceptable to society. The focus hereby is primarily on buildings and transportation vehicles, where the potential exists for large life loss in case of fire.

Codes and regulations provide a minimum level of safety based on loss objectives that are acceptable to society. Other 'stakeholders' often specify additional requirements. For example, an insurance company might have more stringent requirements to reduce the risk of incurring excessive losses in the event of a fire in a building. Building owners and operators might require a level of safety that exceeds the code. Manufacturers of transportation vehicles might issue specifications for their product suppliers that go beyond the regulations. The documents that describe the additional requirements are commonly referred to as technical specifications.

Fire tests are also used in support of production control. The objective of a production control program is to ensure that a product that is sold in the marketplace performs at least as well as the specimen that was originally tested for regulatory compliance. Often it is not practical or economically feasible to use the qualification test(s) in a follow-up production control program and a simple test is usually adequate to verify consistency of the product. Relatively simple fire tests are therefore most often used for the purpose of quality assurance. The same tests are then also used in research and development of new products. They are often referred to as 'screening tests'.

Fire safety codes and regulations are largely prescriptive. This means that they contain acceptance criteria for materials, products and assemblies based on performance in one or several tests. The overall level of safety that is accomplished is not explicitly stated. Sometimes it is not possible to demonstrate that a material, product or assembly meets the specified acceptance criteria. For example, it might not be possible to obtain a valid test result for a product that melts and shrinks away from the heat source in a fire test. Fire safety codes and regulations generally contain a clause that gives the enforcing official the authority to accept a variance from the specified acceptance criteria. To persuade the enforcing official, it is necessary to demonstrate that the material, product or assembly does not compromise the level of safety that is implicit in the code or regulation. This usually requires an engineering analysis and tests to obtain material properties that are needed in support of the analysis.

In recent years, performance-based codes and regulations have become increasingly popular. Performance-based codes and regulations explicitly state the minimum level of performance of the system (building, railcar, etc.) in case of fire. Whether a particular design meets the specified level of performance is again based on an engineering analysis. Fire tests are needed to obtain material properties in support of the analysis. Large-scale fire tests might be required to validate mathematical fire models that are used in the analysis. Fire experts who investigate and reconstruct actual fire incidents now often rely on similar engineering analyses and have the same needs for fire test data.[29]

10.7 Limitations of material flammability tests

The characteristics that are measured in a fire test are indicative of the performance of the material under the conditions of the test and are therefore referred to as fire-test-response characteristics. They may or may not be indicative of the hazard of a material under actual fire conditions.

Type I flammability tests can serve a useful purpose. For example, it has been demonstrated that the lower number of fatalities in fires involving TV sets in the US versus Europe can be attributed to the UL 94 V-0 requirement for the plastic TV housing in the US.[30] In addition, many flammability tests of this type are suitable for quality assurance and can be used in support of research and development of new fire-retardant-treated products. The equipment is inexpensive, only a small quantity of material is needed, the results are usually reasonably repeatable and reproducible and a qualified laboratory technician can run many tests in a short time.

However, these tests do not provide a complete and quantitative assessment of real fire performance. For example, it is logical to assume that a V-0 rated material is unlikely to ignite when subjected in a real fire to a heat source similar to that in the test. But what would happen if the real source is more severe or persists beyond the exposure time in the test? The results could be dramatically different, i.e., ignition might occur and flames might subsequently propagate over the surface and quickly result in a catastrophic fire. There are numerous examples of materials that pass the test with flying colors, but perform miserably under slightly more stringent real fire conditions.

In 1974 the US Federal Trade Commission issued a consent order that resulted in the withdrawal of ASTM D 1692. The order was motivated by the fact that some ASTM flammability test standards used terms such as 'non-burning' and 'self-extinguishing' to label materials that perform well in the test. The Commission determined that these terms give the public a false sense of confidence in the actual fire performance of these materials. ASTM subsequently drafted a policy to ensure that the limitations of using test results for assessing the fire hazard and risk of materials, products and assemblies be clearly stated in its fire test standards. The policy is no longer in effect, but the issue is addressed by the requirement that the following caveat be included in the scope section of every ASTM fire test standard:

> This standard is used to measure and describe the response of materials, products, or assemblies to heat and flame under controlled conditions, but does not by itself incorporate all factors required for fire hazard or fire risk assessment of the materials, products, or assemblies under actual fire conditions.

The same limitations apply to Type II.A and up to a lesser extent to Type II.B flammability tests. For example, static smoke chamber methods have major limitations in terms of being indicative of the fire hazard due to smoke toxicity of products and materials in actual fires. The most commonly used static smoke chamber method is described in ASTM E 662. The test apparatus, often referred to as the NBS smoke chamber, consists of a 0.914 × 0.610 × 0.914 m enclosure. A radiant heater with a diameter of 76 mm is used to provide a constant radiant heat flux of $25\,kW/m^2$ to the specimen surface. The specimen measures 76 × 76 mm and is oriented vertically. Tests are conducted in two modes. A six-tube premixed pilot burner is used in the flaming mode. The burner is removed to conduct tests in the non-flaming mode. Smoke density is measured based on the attenuation of a light beam by the smoke accumulating in the closed chamber. A white light source is located at the bottom of the enclosure, and a photo-multiplier tube is mounted at the top. As combustion products accumulate in the chamber during a test, the burning behavior of the test specimen might have a significant effect on the level of vitiation (oxygen concentration) and temperature rise in the chamber. Consequently, rather than simulating a specific fire scenario, conditions in smoke chamber tests vary over time and are not well-defined.

10.8 Future trends in material flammability assessment

There are hundreds of flammability tests. Each test is unique and provides information on how the material that is tested responds to the specific fire conditions simulated by the test. A material that passes a test with flying colors might perform very poorly in a real fire if, for example, the thermal exposure conditions in the fire are more severe than those in the test. There is a trend to move away from tests that simulate a specific fire scenario and to promote tests that measure fundamental properties that can be used in conjunction with mathematical models to predict the performance of a material in a range of scenarios. Tremendous progress has been made in the past few decades in our understanding of the physics and chemistry of fire, mathematical modeling of fire phenomena and measurement techniques so that this approach is now feasible. However, there will always be materials that exhibit a behavior that cannot be captured in bench-scale tests and computer models. The fire performance of those materials can only be determined in full-scale tests.

Flammability tests have traditionally been simple in concept. Many tests that have been used for a long time do not require any sophisticated measurements. For example, the size of a flame is often specified on the basis of its length and not in terms of the flow rate of the gas. This is changing. With the introduction of increasingly sophisticated test methods, there is a growing awareness of the importance of measurement uncertainty. In 2007 ASTM published the first standard on the subject of measurement uncertainty in fire tests. Reports from testing laboratories will soon, as a matter of course, include uncertainty estimates for the test results obtained.

10.9 Sources of further information and advice

Comprehensive compilations and an extensive discussion of various uses of flammability test methods can be found in the following two books:

Troitzsch, J., ed. (2004). *Plastics Flammability Handbook – Principles, Regulations, Testing, and Approval*, 3rd Edition, Hanser Publisher, Munich.

Apte, V. (2006). *Flammability Testing of Materials Used in Construction, Transport and Mining*, Woodhead Publishing Limited, Cambridge, England.

The most complete collection of fire test standards is published by ASTM international:

ASTM Fire Standards and Related Technical Material, 7th Edition, ASTM International, West Conshohocken, PA, 2007. Available from *http://www.astm.org*.

An extensive discussion of the principles of fire physics and chemistry and the behavior of materials in fire can be found in:

DiNenno, P., ed. (2002). *The SFPE Handbook of Fire Protection Engineering*, 3rd Edition, Society of Fire Protection Engineers, Bethesda, MD.

Finally, a tremendous amount of useful information can be found on the Internet. For example the web site of the Building and Fire Research Laboratory at the National Institute of Standards and Technology (URL: https://www.bfrl.nist.gov) is a great resource. Instead of providing a list of web sites in this chapter, the reader is referred to the site of Dr. Vyto Babrauskas' company, Fire Science and Technology (URL: http://www.doctorfire.com). Not only is there a lot of useful information on this site, but you will also find what is, to the author of this chapter's knowledge, the longest list to fire-related web site links in the world (URL: http://www.doctorfire.com/links.html).

10.10 References

1. Lautenberger, C., Torero, J., and Fernandez-Pello, C., 'Chapter 1: Understanding Material Flammability', *Flammability Testing of Materials Used in Construction, Transport and Mining*, V. Apte, ed., Woodhead Publishing Limited, Cambridge, England, 1–22, 2006.
2. *ASTM Fire Standards and Related Technical Material*, 7th Edition, ASTM International, West Conshohocken, PA, 2007.
3. Blaszkiewicz, M., Bowman, P., and Masciantonio, M., 'Understanding the Repeatability and Reproducibility of UL 94 Testing', *18th BCC Conference on Flame Retardancy*, Stamford, CT, 2007.
4. Morgan, A., and Bundy, M., 'Cone Calorimeter Analysis of UL 94 V-Rated Plastics', *Fire and Materials*, 31, 257–283, 2007.
5. Babrauskas, V., 'Development of the Cone Calorimeter – A Bench-Scale Heat Release Rate Apparatus Based on O_2 Consumption', *Fire and Materials*, 8, 81–95, 1984.
6. Janssens, M., 'Measuring Rate of Heat Release by Oxygen Consumption', *Fire Technology*, 27, 234–249, 1991.
7. Babrauskas, V., *Cone Calorimeter Annotated Bibliography*, Fire Science Publishers, Issaquah, WA 2003.

8. Hallman, J., 'Ignition of Polymers by Radiant Energy', PhD Thesis, University of Oklahoma, Norman, OK, 1971.
9. Koohyar, A., 'Ignition of Wood by Flame Radiation', PhD Thesis, University of Oklahoma, Norman, OK, 1967.
10. Moysey, E., and Muir, W., 'Pilot Ignition of Building Materials by Radiation', *Fire and Materials*, 4, 46–50, 1968.
11. Simms, D., 'On the Pilot Ignition of Wood by Radiation', *Combustion & Flame*, 7, 253–261, 1963.
12. Tzeng, L., Atreya, A., and Wichman, I., 'A One-Dimensional Model of Piloted Ignition', *Combustion & Flame*, 80, 94–107, 1990.
13. Östman, B., and Nussbaum, R., 'Larger Specimens for Determining Rate of Heat Release in the Cone Calorimeter', *Fire and Materials*, 10, 151–160, 1986.
14. Östman, B., and Tsantaridis, L., 'Ignitability in the Cone Calorimeter and the ISO Ignitability Test', *5th Interflam Conference*, Canterbury, England, 175–182, 1990.
15. Janssens, M., and Urbas, J., 'Comparison of Small and Intermediate Scale Heat Release Rate Data', *7th Interflam Conference*, Cambridge, England, 285–294, 1996.
16. Atreya, A., Carpentier, C., and Harkleroad, M., 'Effect of Sample Orientation on Piloted Ignition and Flame Spread', *First International Symposium on Fire Safety Science*, Gaithersburg, MD, 97–109, 1985.
17. Babrauskas, V., and Parker, W., 'Ignitability Measurements with the Cone Calorimeter', *Fire and Materials*, 11, 31–43, 1987.
18. Thomson, H., and Drysdale, D., 'Effect of Sample Orientation on the Piloted Ignition of PMMA', *5th Interflam Conference*, Canterbury, England, 35–42, 1990.
19. Toal, B., Shields, T., and Silcock, G., 'Observations on the Cone Calorimeter', *Fire and Materials*, 14, 73–76, 1989.
20. Toal, B., Shields, T., and Silcock, G., 'Suitability and Preparation of Samples for the Cone Calorimeter', *Fire Safety Journal*, 16, 85–88, 1990.
21. Babrauskas, V., Twilley, W., and Parker, W., 'The Effect of Specimen Edge Conditions on Heat Release Rate', *Fire and Materials*, 17, 51–63, 1993.
22. Östman, B., and Tsantaridis, L., 'Communication: Retainer Frame Effects on Cone Calorimeter Results for Building Products', *Fire and Materials*, 17, 43–46, 1993.
23. Urbas, J., 'Effects of Retainer Frame, Irradiance Level and Specimen Thickness on Cone Calorimeter Test Results', *Fire and Materials*, 29, 1–13, 2005.
24. Urbas, J., and Sand, H., 'Some Investigations on Ignition and Heat Release of Building Materials Using the Cone Calorimeter', *5th Interflam Conference*, Canterbury, England, 183–192, 1990.
25. deRis, J., 'Sample Holder for Determining Material Propeties', *Fire and Materials*, 24, 219–226, 2000.
26. Khan, M., and DeRis, J., 'Operator Independent Ignition Measurements', *Eighth International Symposium on Fire Safety Science*, Beijing, China, 163–174, 2005.
27. Hansen, A., and Hovde, J., 'Prediction of Time to Flashover in the ISO 9705 Room Corner Test Based on Cone Calorimeter Test Results', *Fire and Materials*, 26, 77–86, 2002.
28. White, R., Dietenberger, M., Tran, H., Grexa, O., Richardson, L., Sumathipala, K., and Janssens, M., 'Comparison of Test Protocols for the Standard Room/Corner Test', *Fire and Materials*, 23, 139–146, 1999.
29. Icove, D., and DeHaan, J., *Forensic Fire Scene Reconstruction*, Pearson Education Inc., Upper Saddle River, NJ, 2004.
30. Roed, J., 'Low Voltage Directive', *FIRESEL 2003*, Borås, Sweden, 2003.

11
Challenges in fire testing: reaction to fire tests and assessment of fire toxicity

T R HULL, University of Central Lancashire, UK

Abstract: Fire testing includes the fire resistance of structural components, and the reaction to fire of flammable components. It can be applied to materials and products. Fire scenarios vary greatly, and the fire test should ideally represent the scenarios of the application, although this is not generally the case. Instead, standardised approaches are used, both for material development and product classification. These approaches generally focus on only some aspects of fire behaviour, such as ignitability, flame spread, heat release rate, or toxicity of fire effluent. The simplest methods include the limiting oxygen index (LOI or OI) (ISO 4589), and Bunsen burner tests such as the UL94 classifications (IEC 60695-11-10). Larger bench scale methods, which are more suited to providing data for prediction and modelling of fire behaviour, include the cone calorimeter (ISO 5660), the lateral ignition and flamespread test (IMO LIFT) (ISO 5658), the fire propagation apparatus (ASTM E 2058), the microscale combustion calorimeter (ASTM D 7309), and the steady state tube furnace (Purser furnace) (ISO 19700) for fire assessment of fire toxicity. The merits and limitations of various tests are reviewed from the perspective of the fire retardant material developer.

Key words: fire, flammability, ignition, toxicity, fire toxicity, cone calorimeter, heat release rate.

11.1 Introduction

The dependence of burning behaviour on the fire scenario, and the desire to quantify such behaviour within a single parameter, has resulted in the concept of 'flammability', which, like 'intelligence', is readily understood, but difficult to define. The difficulty arises from the different responses of a material in different fire scenarios. This has been discussed in some detail in Chapter 10.

Fire testing may be subdivided into fire resistance testing, designed to ensure the structural integrity of building components, such as steel beams or doors designed as barriers to fire spread, and reaction to fire, such as flammability and fire toxicity testing, see Fig. 11.1. There are relatively few tests for materials which yield fundamental information relating to flammability parameters. In contrast, there are a large number of industry-specific, empirical tests used to ensure the fire safety of a wide variety of environments, often by testing products, especially in more hazardous applications such as mass transport, upholstered furnishings, and electrical products. These are discussed in detail in

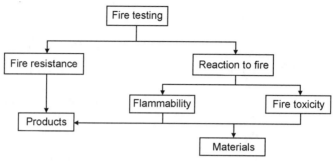

11.1 Areas of fire testing.

the previous chapter. Reaction to fire is concerned with the flammability and ignitability of products, or how they will contribute to fire growth. Reaction to fire applies to a wide range of products not just those with fire retardant properties. UK building regulations define the reaction to fire performance required for building products on their end-use application,[1] using classifications based on test results.

Running parallel to flammability testing is the quantification of fire effluent toxicity. It has long been recognised that the major cause of death in fires is the inhalation of toxic and incapacitating gases, but outside the mass transport sector, only China and Japan have embraced fire toxicity testing as part of their construction product requirements. A large body of research was undertaken in fire toxicity from 1975 to 1995 in recognition of this problem, but the absence of a suitable bench-scale device capable of replicating the toxicity of the most dangerous developed (under-ventilated) fires, coupled with the scientific and widespread public resistance to the use of animal subjects, left this area underdeveloped until recently. The recent development of the steady state tube furnace (ISO TS 19700)[2] able to replicate individual stages of a fire, and the advent of affordable gas phase infrared analysis capable of quantifying individual toxicants in fire effluents, has fuelled a renaissance in the assessment of fire toxicity.

Recent developments in flammability testing have emerged, aimed at addressing the ultimate goal of predicting large-scale fire behaviour from small-scale tests, or even measurements of material properties coupled to models of full-scale fire behaviour. These include the use of enriched oxygen atmospheres to replicate large-scale fire behaviour, and the development of the micro-scale calorimeters capable of determining rates of heat release.

11.1.1 Conditions of each fire stage

Fire tests focussing on particular fire stages should address the prevailing conditions appropriately. Most fires start from small beginnings. There may be an induction period (involving smouldering) before flaming ignition takes place,

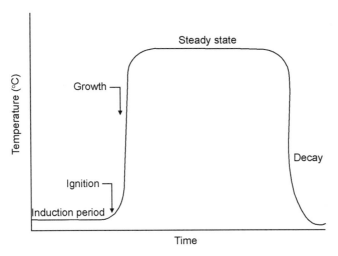

11.2 Stages in a fire.

then a rise in temperature until ventilation controlled burning takes place (usually 800–1000 °C) then a decay as fuel is consumed. This behaviour is shown schematically in Fig. 11.2.

In order to simulate the effects of fire, materials and products must therefore be heated under realistic conditions.

- **Ignition:** piloted ignition is the onset of flaming combustion, characterised by an ignition source (flame, cigarette, glow wire, etc.), small sample size (1–10 cm), with the sample surface temperature around the ignition temperatures (300–400 °C), and well-ventilated.
- **Developing fire:** the continuation of flaming combustion during fire growth is characterised by an external heat flux (around 20–60 kW m^{-2}), requiring larger sample sizes (10 cm–1 m), ambient temperatures above the ignition temperature (400–600 °C), and still well-ventilated.
- **Fully developed fire:** the penultimate stage of fire growth is characterised by a high external heat flux (>50 kW m^{-2}), large sample sizes (1–5 m), ambient temperatures above the spontaneous ignition temperatures (> 600 °C), and low ventilation.

11.1.2 Chemical and physical processes

The chemical composition of the polymer together with the presence of fire retardants, additives, etc., are all important in determining the degree to which flammable products will be released as the decomposing polymer temperature increases. Untreated natural materials, such as wood, cotton and paper tend to release flammable products and ignite at relatively low temperatures in comparison with synthetic materials (polyethylene, PVC, etc). However, the physical nature of the material also plays an important part, sometimes more so than the chemistry, in determining whether a material will reach decomposition temperatures.

A material's *thermal inertia* ($\kappa \rho c$) is the product of its thermal conductivity (κ), density (ρ) and specific heat capacity (c). It dictates the time for the surface temperature to reach the ignition temperature, describing the characteristics of materials according to their heat insulation or heat sink properties. A block of wood is much more difficult to ignite with a small ignition source than wood shavings, cellular polymers of inherently combustible compositions (such as polyurethane foam) will burn very rapidly in comparison to their solid counterparts because their heat insulation properties cause heat to be retained at the surface. The thermal inertia is low for insulating materials, and high for heat conducting materials.

Ultimately most fire science and hence most fire testing is focussed on specific protection goals, for good reasons. Common protection goals include preventing sustained ignition, limiting the contribution to fire propagation, or acting as a fire barrier. Most of the better established fire tests try to simulate a specific, realistic fire scenario and monitor a specific fire risk or hazard from a specific specimen within that scenario, rather than to determine the material's properties. Furthermore, the way a specimen responds in a fire, or in a fire test, may make a significant contribution to the overall fire scenario. Hence, three general remarks can be made:

- Comparing the fire behaviour in different fire tests is difficult. Exact predictions often fail because different material properties determine the performance in different scenarios. However, rough correlations or correlations limited to specific classes of materials have been successful.
- Scaling up and down is a key challenge in fire science, since the sample size plays such a major role. Typically, empirical approaches fail to predict fire behaviour satisfactorily; particularly attempts to span multiple orders of magnitude. Advanced predictive models have been developed which are moving towards reliable predictions of fire behaviour.
- The interactions between properties of components and 'intrinsic' material properties are complex and variable.

11.1.3 Defining the stages of a fire

ISO have identified a number of different fire stages, Table 11.1. While some real-life fires may be represented by a single fire stage, other fires may pass through several different stages.[3] The ISO fire stages, from non-flaming to well-ventilated flaming to under-ventilated flaming, have been classified in terms of heat flux, temperature, oxygen availability, and CO_2 to CO ratio, equivalence ratio ϕ and combustion efficiency. Although all fires may be regarded as unique,

Table 11.1 ISO classification of fire stages, based on ISO 19706

Fire stage	Heat (kW m^{-2})	Max temp (°C)		Oxygen (%)		Equivalence ratio ϕ	$\dfrac{V_{CO}}{V_{CO_2}}$	Combustion efficiency (%)
		Fuel	Smoke	In	Out			
Non-flaming								
1a. Self-sustained smouldering	n.a.	450–800	25–85	20	0–20	—	0.1–1	50–90
1b. Oxidative, external radiation	—	300–600		20	20	—		
1c. Anaerobic external radiation	—	100–500		0	0	—		
Well-ventilated flaming								
2. Well-ventilated flaming	0–60	350–650	50–500	~20	0–20	<1	<0.05	>95
Under-ventilated flaming								
3a. Low vent. room fire	0–30	300–600	50–500	15–20	5–10	>1	0.2–0.4	70–80
3b. Post flashover	50–150	350–650	>600	<15	<5	>1	0.1–0.4	70–90

burning behaviour and toxic product yields depend most strongly on a few factors. Amongst them material composition, temperature and oxygen concentration are normally the most important. The generalised stages in the development of a fire have been recognised and are used to classify fire growth into a number of stages,[3] from smouldering combustion and early well-ventilated flaming, through to fully-developed under-ventilated flaming.

The formation of CO, often considered to be the most toxicologically significant fire gas, is favoured by a range of conditions from smouldering to developed flaming. Although CO yield or CO_2/CO ratio can be indicative of fire conditions, it must be used with care. CO results from incomplete combustion, which can arise from:

- Insufficient heat in the gas phase (e.g., during smouldering).
- Quenching of the flame reactions (e.g., when halogens are present in the flame, or excessive ventilation cools the flame).
- The presence of stable molecules, such as aromatics, which survive longer in the flame zone, giving high CO yields in well-ventilated conditions, but lower than expected yields in under-ventilated conditions.[4]
- Insufficient oxygen (e.g., in under-ventilated fires, large radiant heat fluxes pyrolyse the fuel even though there is not enough oxygen to complete the reaction).

The high yields of CO from under-ventilated fires are held responsible for most of the deaths through inhalation of smoke and toxic gases, but this under-ventilated burning is the most difficult to create on a bench-scale. Research predicting the carbon monoxide evolution from flames of simple hydrocarbons, reviewed by Pitts,[5] has shown the importance of the equivalence ratio ϕ.

$$\phi = \frac{\text{actual fuel to air ratio}}{\text{stoichiometric fuel to air ratio}}$$

	Typical CO yield g/g
$\phi < 1$ fuel lean flames	0.01
$\phi = 1$ stoichiometric flames	0.05
$\phi > 1$ fuel rich flames	0.2

In a fully developed fire, with low ventilation, ϕ can be as large as 5. For many hydrocarbon polymers, the CO yield increases rapidly with increase in ϕ, almost independent of polymer.[4] In addition, a close correlation between CO formation and HCN formation has been established in full-scale fire studies,[6] as the formation of both species appear to be favourable under the same poorly ventilated fire conditions.

For most materials the yields of toxic species have been shown to depend critically on the fire conditions. Figure 11.3 illustrates the generalised change in toxic product yields during the growth of a fire from non-flaming through well-ventilated flaming to restricted ventilation. Although the toxic product yields are

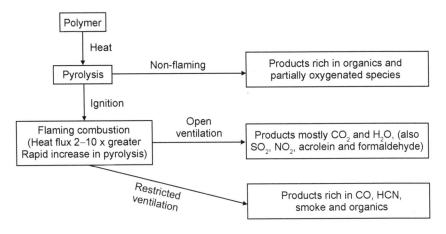

11.3 Effect of fire stage on toxic gas production.

often highest for non-flaming combustion, the rates of burning and the rate of fire growth are much slower, so under-ventilated flaming is generally considered the most toxic fire stage.

Data from large scale fires[7,8] shows much higher levels of the two asphyxiant gases (CO and HCN) under conditions of reduced ventilation. It is therefore essential to the assessment of toxic hazard from fire that these different fire stages can be adequately replicated, and preferably the individual fire stages treated separately. The drive for internationally harmonised methods for assessment of combustion toxicity, through adoption of international standards, such as those of ISO, provides the framework for meaningful and appropriate use of toxic potency data in the assessment of fire hazard. As structures and means of transportation become larger and more complex, there is movement away from the more traditional methods of ensuring fire safety by prescriptive codes, towards fire risk assessments and engineering solutions. Reliable rate of heat release, fire effluent toxicity and smoke generation data are all essential elements of such an assessment.

11.1.4 Factors affecting ignitability and fire development

The following factors affect polymer combustion in real fires, and should therefore influence the outcome of a suitably designed test.

- **Fuel production** – when the fuel in the gas phase reaches a critical concentration, ignition and flaming can occur. While the fuel production rate during heating is essentially a material property, the air flow around the sample may profoundly alter the ignition temperature.
- **Presence of inhibitors or diluents** – Cl· or Br· or PO· are stable radicals which will reduce the critical concentration of active radicals such as H and

OH, in the flame zone. The effect is most pronounced at ignition, and least evident under developed fire conditions.

- **Rheology of decomposing polymer** – some polymers depolymerise during decomposition, reducing their viscosity. This allows better dispersion of heat, and material flow away from the source of heat. This can result in harmless dripping away from the flame zone, or flaming drips allowing flaming to spread downwards. Some additives (e.g., high surface area fillers, such as nanofillers) will increase the viscosity, reducing dripping, resulting in a more rapid increase in the surface temperature. This will reduce the time to ignition. In some cases free radical initiators are added purely to promote dripping[9] to remove the fuel from the source of heat.
- **Char formation** – the formation of a char on the surface of the polymer will reduce the flow of heat to the sample. Intumescent chars bubble up and provide a more effective barrier. However, in a typical fire test, the direction of swelling is often towards the heat source, increasing the radiant flux to the sample.
- **Orientation of sample** – as flames rise, flame spread is easiest from below (going upwards) and hardest from above (going downwards). Because of flow of molten material and ultimately dripping, it is very difficult to correlate vertical burning behaviour with horizontal burning behaviour.
- **Absorption of radiation** – radiation from flames or a radiant panel must be absorbed by the polymer. The presence of absorbing centres (conjugated double bonds, or black pigments) can increase the localisation of the heating. Conversely, a highly reflective surface can significantly lengthen the time to ignition in certain tests.
- **Smoke formation** – smoke can act as both the source of radiation (a sooty yellow diffusion flame radiates much more than a blue premixed flame) or block radiation from the flame back to the polymer.

11.2 Fire testing – principles and problems

Quantifying flammability behaviour necessitates numerical measurement through some kind of test. It is to be hoped that such tests can be correlated to the real fires that we are working to avoid because large-scale fire tests are extremely expensive. Most tests are small-scale, because experimental materials are only available in small quantities. Small-scale fire tests are more reproducible and much more easily replicated than full-scale tests. However, many tests have an inherent deficiency in that they fail to reproduce the massive effect of heat present in a large-scale fire, giving results which could be misleading, if applied out of context.

Fire science is divided between two schools of thought. One group believe that the key to preventing fire growth is to minimise the peak heat release rate,[10] since this parameter appears to control the initial flame spread from the first object ignited to other objects, while the other believe that ignition resistance (quantified in terms of higher piloted ignition temperature or longer delay time before ignition) controls the fire growth. The time to ignition is proportional to the square of the ignition temperature for thermally thick materials, or the ignition temperature for thermally thin materials,[11] (Section 11.4.2), and typical flame spread occurs over a two-dimensional surface, so the fire growth rate is proportional to the time following ignition squared, giving the strongest dependence on the ignition time or temperature. Exactly which effect predominates will depend on the fire scenario.

11.3 Fire resistance tests

Fire resistance tests are intended to assess the performance of elements of construction for their load-bearing or fire separating properties – usually termed their fire resistance – for their regulated use in buildings. Fire resistance of beams, columns, doors, wall sections, etc., is determined by their performance in large furnaces (capable of holding specimens of 3 m long or of area 3 m × 3 m) against a standard temperature-time curve, typically rising from ambient to 850 °C over 20 minutes. These tests are designed for concrete, brick or steel (protected) elements and fire resistance is usually measured in terms of ½, 1, 2, 4, 8 hours integrity. Since fire resistance is mostly required in static structures such as buildings, which are regulated by local jurisdictions, and in many cases also by local building practices, there are a large number of test specifications, and little international agreement. Across Europe, standardised testing is being introduced through the Construction Products Directive.[12] The other main requirement for fire resistance is in mass transportation, where typically each industry has its own suite of tests. In the aerospace industry, where metal is gradually being replaced by polymer composites, new test methods are required, dealing with both structural integrity and flammability, see Chapter 15. Thermoplastic materials do not have fire resistance.

11.4 Reaction to fire tests

A large variety of industry standard tests are used to qualify a product's suitability for a particular application, and ultimately all products with specified fire performance must meet these criteria. However, the design of fire retardant materials requires simple quantifications of fire behaviour, usually while also meeting many other physical specifications. To achieve this, a relatively small number of material fire tests are employed by fire retardant formulators and developers in order to measure progress towards their fire retardant goal. The most common of these are the limiting oxygen index (LOI) and the Underwriters' Laboratory UL 94 test, both ease of extinction tests, and the cone calorimeter, which measures time to ignition and rate of heat release.

11.4.1 Ignitability and flame spread

As explained in Section 11.1.4, ignition is a very important parameter controlling flame spread and fire growth. However, the source of ignition necessarily impacts on the result, and, therefore ignition temperature is found to depend on the design of the test used to measure it. The characteristics of several common ignition sources have been reported[13] and are shown in Table 11.2. Tests which use sparks, electric arcs, hot surfaces or open flames for less than 30 seconds will not deliver more than 100 kJ, and represent low severity tests. Crumpled or folded paper can deliver between 200 and 4000 kJ in 1 to 8 minutes, representing a medium severity exposure, while burning bedding can deliver 130 000 kJ in 20 minutes, representing a high severity exposure. Many standard tests use a gas or liquid fuel ignition source, of specified energy or power, corresponding to one of the 'unwanted fire' ignition sources.

However, from a fire retardant formulator's perspective, measurement of the

Table 11.2 Characteristics of some common ignition sources[13]

Source of ignition	Duration of source (s)	Total heat (kJ)	Maximum heat flux (kW m^{-2})
Match flames	2–35	6	18–20
Cigarette lighter	30	24	16–24
Diffusion flame, small	30	8	18–32
Diffusion flame, large	30	15	6–37
Premixed flame, small	30	50	58
Premixed flame, large	30		120
Electric spark		<100 mJ	
Electric arc	1	0.4	
Electric arc	5	15	
Electric bulb, 60 W	30	3	
Electric bulb, 100 W	30	8	
Electric hot plate, 1 kW	30	30	
Electric radiator	30	90	20–25
Crumpled paper			
1/2 sheet	85	175	7–10
1 sheet	152	340	7–22
2 sheets	223	680	7–21
3 sheets	333	1020	5–22
4 sheets	335	1600	6–23
Folded paper			
5 sheets	380	1680	14
10 sheets	420	3500	15
Wastepaper basket	360	3400	10–40
	1600	5000	10–40
Small stuffed toy	330	9500	20–39
Scatter cushion	513	11000	17–28
Bedding	1200	130000	26

ignition delay time is probably just as important as the rate of heat release, and relative performance can readily be assessed using a cone calorimeter.

11.4.2 Ignitability in the cone calorimeter

Ignition is a complex subject, which has been addressed by several authors[14–18] and comprehensively described and summarised elsewhere.[19] Ignition occurs when the oxidising volatiles feed enough heat back to the polymer to volatilise a similar concentration under the conditions of the test. Thus, the fraction of the heat of combustion passed back to the polymer for a given mass of fuel must be greater than its heat of gasification. This critical condition can be described by the mass loss rate at ignition. For the cone calorimeter the critical mass loss rate is around 1–6 g s^{-1} m^{-2}, and the resulting heat release rate at ignition (HRR$_{ig}$) is around 20–100 kW m^{-2}.[20–22] Ignition does not directly or necessarily correspond to 'flammability' measured by LOI or UL 94, since both of these are ease of extinction tests, and correspond better to the minimum mass loss rate needed for sustained burning or fire propagation.[23]

The critical mass loss rate for ignition in the cone calorimeter occurs when the material's surface temperature equals its ignition temperature. This ignition temperature (T_{ig}) behaves like an intrinsic material property, since it is independent of the applied heat flux. The time to ignition (t_{ig}) is thus the time required for the surface to reach ignition temperature.[29] For thermally thick samples it can be shown that:

$$t_{ig} = \frac{\pi}{4} \kappa \rho c \left[\frac{T_{ig} - T_0}{\dot{q}_{ext} - CHF} \right]^2$$

where κ is thermal conductivity; c is heat capacity; ρ is density; T_0 is ambient/starting temperature; CHF = critical heat flux for ignition; and q_{ext} is applied heat flux.

Thus there is a linear relationship between external heat flux and $t_{ig}^{-0.5}$, and there is a critical heat flux necessary to reach the ignition temperature. In the cone calorimeter, the critical heat flux is characteristic for each material. Therefore, as well as the time to ignition, and possibly the ignition temperature, the critical heat flux can be used to describe the ignitability of a material. However, these are not truly intrinsic properties, since their values change significantly if the cone calorimeter set-up is modified, such as by using a pilot flame instead of a spark igniter or changing the distance between the cone heater and the sample surface.

11.4.3 IMO lateral ignition and flame spread test/spread of flame apparatus ISO 5658[24]

The LIFT or Spread of Flame apparatus measures the lateral spread of flame along the surface of a material or product, vertically orientated, in response to radiative heat in the presence of a pilot flame. It provides data suitable for comparing the performance of essentially flat materials, composites or assemblies, which are used primarily as the exposed surfaces of walls. It is one of the best methods for the determination of fundamental fire properties relating to flame spread. Some profiled products (such as pipes) can also be tested under specified mounting and fixing conditions. The complete test apparatus consists essentially of three main components, a radiant panel support framework and a specimen support framework which are linked together to bring the test specimen into the required configuration in relation to the radiant panel, and the specimen holder, which carries the test specimen.

Following ignition, any flame front which develops is noted and the progression of the flame front horizontally along the length of the specimen is recorded. The results are expressed in terms of the flame spread distance/time history, the flame front velocity versus heat flux, the critical heat flux at extinguishment and the average heat for sustained burning. The results of this test method are potentially useful to predict the time to ignition, t_{ig}, and the velocity of lateral flame spread on a vertical surface under a specified external flux without forced lateral airflow. Data are reported for convenient use in current fire growth models.

11.4.4 Ease of extinction tests

UL 94 'Bunsen burner' test IEC 60695-11-10[25]

This is a small-scale laboratory screening procedure for comparing the relative burning behaviour of vertically or horizontally oriented specimens made from plastic and other non-metallic materials, exposed to a small-flame ignition source of nominal 50 W power.

The method determines the linear burning rate and the afterflame/afterglow times, as well as the damaged length of specimens, and is applicable to solid and cellular materials with density of at least $0.25\,g\,cm^{-3}$, provided they do not shrink away from the applied flame without igniting. The test method described provides a classification system, which may be used for quality assurance, or the pre-selection of component materials of products, provided that the test sample thickness is the thinnest to be used in the application. The Underwriter's Laboratory designed this standard to indicate a plastic's flammability for use as part of an electrical appliance, rather than the hazards of a material under actual fire conditions. UL 94 flammability testing is the first step toward obtaining a plastic recognition and subsequent listing in the 'Plastics Recognised Component Directory' (formerly known as 'Yellow Cards').

The 94V test describes the Vertical Burn test (Fig. 11.4), which is a more stringent test than the Horizontal Burn method 94HB (Fig. 11.5). Another version, the 94VTM test, not detailed here, is for very thin materials, such as

Challenges in fire testing 267

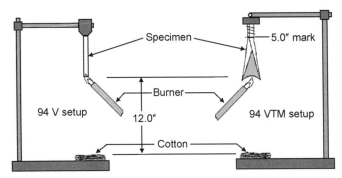

11.4 Vertical burning test for 94V and 94 VTM classifications.

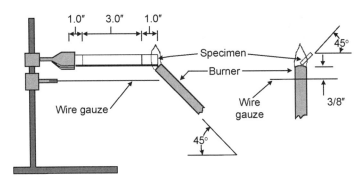

11.5 Horizontal burning test for 94 HB classification.

films or fabrics. The set up uses a very small Bunsen flame with a manometer and needle valve to control the gas flow. The criteria for each classification are shown in Table 11.3. While the test is crude, it is a realistic ignition scenario, and lets the user see what is happening during the test. It is easy to set up a small

Table 11.3 UL 94 classifications

V-0 Vertical Burn	Burning stops within 10 seconds after two applications of ten seconds each of a flame to a test bar. NO flaming drips are allowed.
V-1 Vertical Burn	Burning stops within 60 seconds after two applications of ten seconds each of a flame to a test bar. NO flaming drips are allowed.
V-2 Vertical Burn	Burning stops within 60 seconds after two applications of ten seconds each of a flame to a test bar. Flaming drips ARE allowed.
H-B Horizontal Burn	Slow horizontal burning on a 3mm thick specimen, with a burning rate less than 3″ per minute or stops burning before the 5″ mark. H-B rated materials are considered 'self-extinguishing'. This is the lowest (least fire retarded) UL 94 rating.

test burner with a 15 mm blue flame in a typical formulation laboratory, providing the formulator with direct and immediate feedback on their latest recipe.

Limiting oxygen index[26]

This test relates to the minimum concentration of oxygen that will just support flaming combustion in a flowing mixture of oxygen and nitrogen, Fig. 11.6. A specimen is positioned vertically in a transparent borosilicate glass test column and a mixture of oxygen and nitrogen is forced upwards through the column. The specimen is ignited at the top. If the flame remains for 3 minutes, or propagates down the length of the sample, the test is repeated at lower oxygen concentrations. If it self-extinguishes, the test is repeated at higher oxygen concentration. The oxygen concentration is adjusted in this manner until the specimen just supports combustion. The oxygen concentration reported is the volume percent, with repeatability often as good as $\pm 0.1\%$ O_2.

Downward flame spread may be regarded as a best case scenario, and while a material with limiting oxygen index (LOI) < 21% should be considered to support downward flame spread, materials with LOI $\gg$ 21% should still be considered flammable. Indeed, in a developed fire most materials will burn readily, while the oxygen concentration has fallen to a few percent. Particular problems arise with materials with a high dripping propensity, since ignition will only occur under extreme circumstances. Very thin materials often have

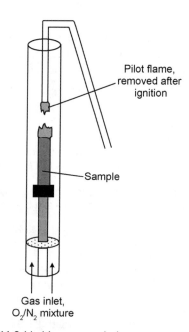

11.6 Limiting oxygen index test.

insufficient heat release per unit area to support combustion, while thicker materials conduct too much heat away from the flame zone. Thus there is a 'most flammable thickness' for many materials around 1.6 mm. For non-charring materials, the criteria for ignition, i.e. heat transfer from the flame > heat of gasification per unit mass, are replicated in the criteria for extinction. However, while ignition requires a source whose energy input will affect the result, extinction has no such dependence. The dilution of the flame by nitrogen causes the flame to swell, reducing the amount of heat fed back to the sample below the flame. As a rule of thumb, there is generally some correlation between the LOI value and time to ignition in the cone calorimeter but none between LOI and heat release rate.

11.5 Measurement of heat release

11.5.1 Bench-scale measurement of heat release

The cone calorimeter

The cone calorimeter (Fig. 11.7) was developed specifically to determine the rate of heat release and effective heat of combustion of building materials (ISO

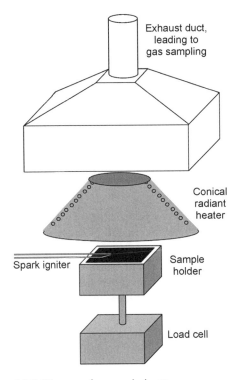

11.7 Diagram of cone calorimeter.

5660-1).[27] It was subsequently modified to determine smoke generation (ISO 5660-2)[28] and later applied to furniture.

A horizontal specimen, 100 mm square, typically 3–6 mm, but up to 50 mm, thick is mounted under a steel frame, such that only the surfaces, but not the edges are exposed to a conical radiator pre-set to between 10 and 100 kW m^{-2} mounted beneath an instrumented hood and duct. A spark ignition is used and the specimen is mounted on a load cell. Heat release is quantified by oxygen depletion calorimetry. Measurement of heat release from real fires by oxygen depletion calorimetry is well established, and gives sensible values which relate to the extent of burning. Provided the effluent flow through the exhaust is carefully controlled, the heat release will be proportional to the oxygen depletion. A sample of the effluent is cooled to remove water and analysed using a paramagnetic analyser and non-dispersive infrared CO and CO_2 analysers. However, depending on how the data analysis software has been configured, this may result in errors accompanying large yields of CO which are typical of materials incorporating flame inhibitors such as halogens or phosphorus. It does not take into account the reduction in heat release due to the endothermic decomposition of metal hydroxide fire retardants, such as aluminium hydroxide (ATH), although this can be compensated for separately, e.g. for PMMA containing 60% ATH this would result in an overestimation of total heat release by ~8%. A lower value of heat release will be observed in the presence of water soluble gases such as HCl or HBr, since these will be removed with water from the effluent stream by the cooler. For rigid PVC, this would incur an underestimation of the total heat release of ~5%.

A detailed description of the use and interpretation of data from the cone calorimeter for fire retardant materials development has recently been published.[29] The fire model of the cone calorimeter is always well-ventilated, corresponding to ISO stage 1b for non-flaming tests and to stage 2 for flaming tests. The standard requires the continuous measurement of mass loss and effluent gas concentrations and yields (CO and CO_2), oxygen depletion, smoke obscuration, and exhaust gas temperatures. The rate of heat release, effective heat of combustion, smoke generation, and gas yields are reported.

The cone calorimeter monitors a comprehensive set of fire properties in a well-defined fire scenario. The results can be used to evaluate material specific properties, setting it apart from many of the established fire tests which are designed to monitor the fire response of a certain specimen.

The cone calorimeter covers ignition followed by essentially penetrative flaming combustion, where the flame front moves through the bulk of the sample. The ignition parameter measured in the cone calorimeter is the time to ignition, which depends on the thermal inertia, critical heat flux and critical mass loss for ignition, or alternatively the critical surface temperature for ignition. Fire response properties more typical of fully developed or post flashover fire scenarios are not replicated in the cone calorimeter.

There are three distinct uses of cone calorimeter data:

- To compare the fire response of materials; to assess their fire performance; to perform screening for materials development; to develop pyrolysis and burning models.
- To determine data for input to simulations or predictions of full-scale fire behaviour.
- To determine characteristic parameters such as the maximum HRR (peak heat release rate, PHRR), fire growth rate index (FIGRA), THR, etc., for regulatory purposes.

These applications of the cone calorimeter define different techniques and data evaluation. For regulatory purposes, its strengths are its well-defined conditions, reproducibility and unambiguous data evaluation of one or two characteristic values. The use of defined, and in some way ideal, burning behaviour is suitable for developing pyrolysis and burning models and for obtaining reasonable input values for the simulation of fires. However, as a fire scenario, it is not representative of most real fires since small fires are

- not usually initiated with radiation from above,
- piloted by a spark ignition source,
- surrounded by a frame which acts as a large heat sink, producing an unusual gas flow field around the flame zone
- where the effects of sample dripping are negligible.

The heat release rate (HRR) during a cone calorimeter experiment gives rise to characteristic heat release rate curves versus the time, Fig. 11.8.[20,29]

Fire propagation apparatus[30]

This method is similar in principle to the cone calorimeter but the fire zone is contained within a wide vertical silica tube allowing better control of the fire atmosphere, and keeping it out of contact with the heaters, which are outside the tube, Fig. 11.9. Horizontal test specimens, typically 100 mm square and up to 25 mm thick, or vertical specimens 100 mm wide, 305 mm high and up to 25 mm thick, are exposed to thermal radiation and a pilot flame. The effluent flows through an instrumented duct and the rate of heat release is determined from oxygen consumption. The ISO fire stages generated are 1b, 1c, 2, 3a and 3b, see Table 11.1.

The method requires the continuous measurement of mass loss and effluent gas concentrations and yields of CO_2, CO, smoke and exhaust gas vitiation. Data are presented as HRR, effective heat of combustion, smoke generation, and gas yields. The same comments about the errors inherent in oxygen depletion calorimetry also apply here. Apart from the separation of the fire effluent from the radiant heaters which avoids some unrepresentative high temperature

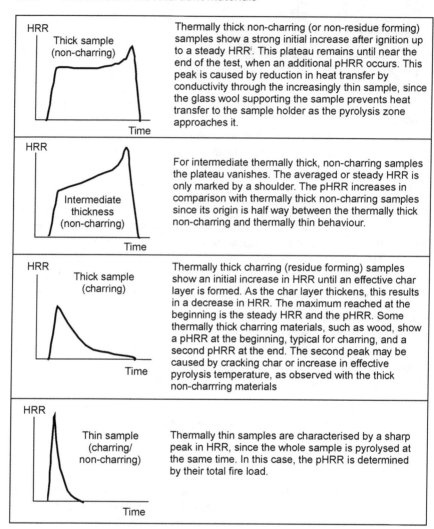

11.8 Types of heat release curves from cone calorimetry.

chemistry, the advantages and disadvantages of this method are generally similar to those of the cone calorimeter.

Correlations between the fire propagation apparatus, with the sample in a vertical orientation, and an intermediate scale fire test (testing upward flame spread in air between two parallel sheets 0.6 m × 1.2 m) were found to be improved by running the bench-scale fire propagation apparatus in 40% oxygen. This was ascribed to the ability to simulate the feedback from the flame, in terms of increased heat flux, compared to the fixed heat flux of the standard test.[31]

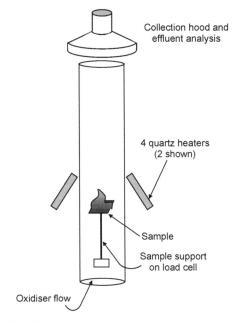

11.9 Fire propagation apparatus.

11.5.2 Micro-scale measurement of heat release

Pyrolysis-combustion flow calorimetry[32] (PCFC) evaluates the combustibility of milligram samples by separately reproducing the solid state and gas phase processes of flaming combustion by controlled pyrolysis of the sample in an inert gas stream, followed by high temperature oxidation of the volatile pyrolysis products. Oxygen consumption calorimetry is used to measure the heat of combustion of the pyrolysis products. The maximum amount of heat released per unit mass per degree of temperature ($J\,g^{-1}\,K^{-1}$) is a material property that appears to be a good predictor of 'flammability'. The heat release capacity (HRC) and total heat release (THR), obtained by PCFC, are related to the char yield and the heat of complete combustion of the volatiles. It takes no account of physical effects, such as dripping, wicking, and sample thickness; or chemical effects such as flame inhibition, because the conditions force pyrolysis and combustion to completion.[33] However, varying the combustion temperature or oxygen concentration results in incomplete combustion as occurring in real fires. The PCFC results have been correlated to:

- THR to LOI;
- HRC and char residue to LOI;
- HRC and THR with peak heat release rate (pHRR) in the cone calorimeter.

It has been used as a screening test for efficacy of flame retardant additives.[34]

11.5.3 Large-scale measurement of heat release

Tests for regulations and control of construction products

In the past, European countries had their own reaction to fire tests stipulated in regulations but differing in their approach, scale and exposure conditions. This led to uncertainties about how they relate to full-scale fire conditions and how results could be compared between them. Recently, the Construction Products Directive[12] has led to a harmonised system of testing and classification for construction products in the European Union, to remove potential barriers to trade.

All construction products (excluding floor coverings) will be classified into seven classes of increasing combustibility A1, A2, B, C, D, E and F based on their performance in the specified tests. F is reserved for products for which no data is available. Floor coverings will be classified separately.

The non-combustibility and bomb calorimeter are relatively small-scale tests for measuring the potential heat contribution (or lack of heat contribution) of a material in a fire situation, and are based on existing tests. The single burning item[35] (SBI) test is an important new development for the European classification, Fig. 11.10. It is an intermediate-scale test designed to measure the flammability of wall linings 2 m × 1 m and 2 m × 1.5 m formed into an internal corner, ignited by a simulated single burning item (gas burner) close to the corner of a room. It uses oxygen depletion calorimetry to quantify the HRR. It has been correlated with full-scale measurements of wall linings in a full room-corner test (ISO 9705 – the so-called reference scenario). The fire performance characteristics of the test are associated with the rate and quantity

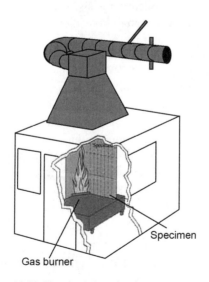

11.10 The single burning item test.

of heat released and, in particular, the likely relationship to the onset of flashover. However, the data have been reported to show an unacceptable level of repeatability in an assessment of composite panel flammability, due to the fact that small openings result in a wide variation of the classification result. The authors go as far as saying that it is dangerous to make a fire safety assessment of a sandwich panel based on small- or intermediate-scale tests.[36]

Tests for regulation and control of other products

Many tests exist for the selection and control of textiles, clothing, upholstery, etc., for use in public places dwelling, cars, etc. In most cases these tests are aimed at preventing ignition of the material from small sources such as cigarettes or small flames (matches), and inhibiting early growth of fire to enable occupants to escape before dangerous levels of heat, smoke and toxic gases are produced.

Many countries now have fire performance requirements for cables, since large quantities of communication cables with plastic insulating materials may be housed in buildings in vertical and horizontal runs often in voids where there is potential for rapid fire growth in the event of fire. Tests have been developed to study cables (including those with flame retarded insulation) under both small- and large-scale conditions to assess flammability, smoke and toxic gases (including halogen determination).

11.6 Fire toxicity

11.6.1 The need to quantify toxic gases as part of a hazard assessment

Analysis of fire hazards requires data describing the rate of burning of the material, and data describing the toxic product yield of the material. It has already been observed that these are not material properties, but are scenario dependent. The rate of burning will depend on the ignitability, heat release and rheological properties of the material, and on the material orientation (horizontal, vertical, etc.), proximity to a heat sink, thickness, fire conditions and so forth. While some materials clearly burn less easily than others, there is no consensus on how materials may be ordered in terms of increasing flammability. Fire toxicity is also scenario dependent, but a clear relationship has been demonstrated between the yield of toxic products (for example, in grams of toxicant per gram of polymer) and the fire condition, for many materials. Analysis of fire statistics shows that most fire deaths are caused by inhalation of toxic gases.[37] Assessment of toxic hazard is increasingly being recognised as an important factor in the assessment of fire hazard. Prediction of toxic fire hazard depends on two parameters:

1. Time-concentration profiles for major products. These depend on the fire growth curve and the yields of toxic products.
2. Toxic potency of the products, based on estimates of doses likely to impair escape efficiency, cause incapacitation, or death.

The replacement of prescriptive standards by performance-based fire codes requires a fire hazard assessment, which includes prediction of the toxic product distribution within the building from a fire.[38] The goal of any toxicity assessment is to generate reliable bench-scale toxicity data. Within the European Union, and other jurisdictions where routine animal testing is unacceptable[39] this effectively means reliable quantification of yields of toxic products. Toxic product yields depend on the material composition,[40] and the fire conditions.[41] The most significant differences arise between flaming and non-flaming combustion. For flaming combustion the most significant factor is the fuel/air ratio, although the oxygen concentration and the compartment temperature can also affect the yields. As an enclosure fire develops, the temperature increases and oxygen concentration decreases. This has been set out as series of characteristic fire types (Table 11.1), from smouldering, to post-flashover. Carbon monoxide is often considered to cause the greatest number of fire deaths, and the evolution of carbon monoxide is highly dependent on conditions, the most significant of which are difficult to create on a small scale. Further information is presented in a more detailed account of current protocols in toxicity testing.[42]

11.6.2 Toxic potency data

Toxic product data from chemical analysis may be expressed in various ways including, effluent gas concentrations, effluent gas yields, toxicity indices, Fractional Effective Dose (FED) values, Fractional Effective Concentration (FEC) values and Lethal Concentration to 50% of the exposed population (LC_{50}) and Incapacitation Concentration to 50% of the exposed population (IC_{50}) values. The results of animal-based tests are usually expressed as LC_{50} and IC_{50} values determined by direct observation.

11.6.3 Animal exposure methods

A comprehensive summary of results of animal-based tests has been published by Neviaser and Gann.[43] These data, as LC_{50} and IC_{50} values for exposed rats, were obtained for oxidative pyrolysis, well ventilated and vitiated flaming fire stages using the NIST cup furnace test, the NIST radiant furnace test (ASTM 1678) and a tube furnace test believed to be of the NF X 70-100 type. In general this shows that many materials have relatively low fire toxicity under well-ventilated conditions, but with large estimated uncertainties. Under poorly ventilated conditions, toxicity typically increases by a factor of 10, and for certain materials, even under well-ventilated conditions, the fire toxicity is up to six times greater than the norm. Thus it is important to test the toxicity of individual materials, and to ensure that this is done under appropriate conditions.

An interlaboratory evaluation of the DIN 53486 test involved the determination of CO, CO_2 and oxygen depletion with observations of the mortality of exposed rats and the carboxyhaemoglobin concentration in their blood.[44,45] Tests were carried out at 300, 400, 500 and 600 °C with softwood (*pinus sylvestris*) and showed good reproducibility. The tests also showed that the mortality of Wistar rats was higher than Sprague Dawley rats. This illustrates the difficultly in relating toxicity data for different animals, since in this case a noticeably different response occurred even with rats of different strains of the same species. In general, much greater differences are observed for different species, with insufficient trends to allow extrapolation to human responses.

The advantages claimed for animal based tests are that they determine toxicity directly, enable toxics effects to be observed and determine unusual toxicity. There are a number of disadvantages because although lethality is a well established end point, many of the techniques used to assess incapacitation have been criticised as being too close to lethality to be relevant to the human escape situation. The more relevant 'trained escape response' and 'maze escape' methods are rarely used for reasons of complexity and cost. It has been pointed out that difficulties exist when animal test data (rats, mice, dogs, monkeys and baboons have been used) is translated to humans because of different respiratory volume rates and airway dimensions, etc. In particular, that the rodent may not be a suitable model for evaluating the toxic effects in humans of smoke in which irritant gases are the principal toxicants.[46]

11.6.4 Chemical analysis methods

The general approach in generating toxic potency data from chemical analysis is to assume additive behaviour of individual toxicants, and to express the concentration of each as its fraction of the lethal concentration for 50% of the population for a 30-minute exposure (LC_{50}). Thus an FED equal to one indicates that the sum of concentrations of individual species will be lethal to 50% of the population over a 30-minute exposure.

11.6.5 Generation of fire effluents – general requirements

Guidance on assessment of physical fire models has been published in ISO 16312-1.[47] In all fire smoke toxicity tests, specimens are decomposed by exposure to heat, resulting in 'forced combustion' driven by an applied heat flux from a flame, radiant panel, etc. Some tests use a pilot flame or spark igniter to facilitate ignition, while others rely on self-ignition of the sample. When flaming combustion occurs, this will increase the radiant heat flux back to the sample typically between 2 and 10 times. This will have two significant effects on the fire effluent. First, the existence of flames will help to drive the combustion process to completion, by increasing the temperature and hence the reaction rates, which will tend to reduce the toxicity of the fire effluent (favouring CO_2 over CO and organic molecules). Secondly, the higher heat flux will pyrolyse more material at a greater rate, increasing the amount of material in the vapour phase, and reducing the concentration of oxygen, both of which will increase the toxicity of the fire effluent. Unfortunately, these effects are so large that, rather than cancelling each other out they can result in very large differences in the toxic product yield between different fire toxicity tests. Clearly, the presence or absence of flaming combustion is critical to the interpretation of the results from combustion toxicity assessments. In some conditions, specimens will either pyrolyse or self-ignite, but the scatter of results will be very large if flaming combustion is inconsistent. Once flaming is established, combustion will drive itself to completion (and hence the toxicity will be reduced), provided there is sufficient oxygen, and the flame is not quenched. If the flame is cooled rapidly, e.g. by excessive ventilation or a cool surface, the yield of toxic products will increase. Ultimately the value of the bench-scale toxicity assessment is dependent on its ability to predict large-scale burning behaviour, and therefore validation must involve comparison with large-scale test data. Unfortunately most large scale test data have been obtained under well-ventilated conditions, and when under-ventilated fire scenarios, such as the ISO 9705 Room test, are used the change of sample mass and the air flow to the fire during the test is not generally known.

Open tests

In just the same way that large-scale burning behaviour is not easily predicted from small-scale test data, so fire toxicity is highly dependent on the fire conditions, and particularly the sample size. Of the standard methods used for toxicity assessment, there are three general types: well-ventilated or open methods, closed box tests, and tube furnaces. Although most bench-scale fire tests, such as the cone calorimeter, are open, and run in well-ventilated conditions, they are generally unsuitable for estimation of toxic product yields because the high degree of ventilation coupled with the rapid quenching of fire gases, gives a high yield of products of incomplete combustion through premature flame quenching, rather than through vitiation.

The fire zone of the standard cone calorimeter apparatus is well-ventilated, Fig. 11.7. However, the apparatus can be modified for tests under oxygen-depleted, 'vitiated' conditions. This uses an enclosure around the specimen and radiator, and a controlled input flow of nitrogen and air, but has met with limited success. In some tests the effluent may continue to burn as it emerges from the chamber giving ultimately well-ventilated flaming. In others, under reduced oxygen concentrations, the fuel lifts from the surface, and ignition does not occur.[48] The CO yields in the cone calorimeter have been found to correlate with an equivalence ratio of 0.7 for a range of cable materials.[49] The relatively high dilution of fire gases in, and stainless steel construction of, the hood and duct, may lead to difficulties in detecting some effluent components. Fire gases pass through the conical heater which may modify their composition. Standardisation of the vitiated cone calorimeter is currently under discussion within ISO.

Closed chamber tests

Closed cabinet tests and their operation may be likened to a small fire burning in a closed room. The specimen is decomposed by a heat source and the resulting effluent accumulates within the cabinet. The decomposition system is either mounted within the cabinet as in the aircraft[50] and maritime tests[51] or may be outside, connected to the cabinet by a short duct, as in ASTM E 1678.[52]

A direct consequence of the closed cabinet is that the fire effluent accumulates within the cabinet and the fire gas concentrations therefore increase as the specimen burns and the gases will change with oxygen depletion. For laminated or layered specimens, the effluent will also change as flame burns through different layers.

As the specimen decomposes, the hot effluent rises to the upper part of the chamber where it may accumulate or circulate around the chamber due to natural convection. Thus the product concentration will depend on where the gas samples are taken from. The smoke density values will be unaffected provided a vertical light path is used. Although mixing fans are used in some smoke density tests, they are rarely used in toxicity tests. Both the aircraft and maritime tests require the smoke to be sampled at specified times (although burning may have proceeded at different rates) from gas sampling probes in the geometric centre of the cabinet.

If the effluent is stratified the gas sample is obviously unrepresentative, but if it is uniformly distributed, then the gas flowing into the fire zone will may be oxygen depleted and fire gases may be recycled through the fire zone. These latter effects will be greater with thicker specimens which would be expected to generate more smoke, due to under ventilation.

Therefore, the closed box tests give a complete product yield of burning from well-ventilated right through to fully vitiated conditions, but without giving any indication of how the yield varies with fire condition. Another potential source of error may occur as the fire effluent is heated and excess pressure is released or stickier components within the effluent, such as hydrogen chloride, are deposited onto the walls of the cabinet.

11.6.6 Tests based on the NBS smoke chamber ISO 5659-2[53]

The smoke chamber is a well-established piece of equipment, designed to monitor the smoke evolution from burning materials, in order to minimise visible obscuration of escape routes during a fire. Its widespread acceptance has led to its use in a number of industry specific toxicity tests, such as the Aircraft test[50] (prEN 2824, 5 and 6, uses the vertical radiator and test specimen of ASTM E 662[54]) and is specified for components for passenger aircraft cabins (Fig. 11.11). Airbus ABD 3 and Boeing BSS 7239[50] use the same apparatus but specify different gas analysis methods.

The IMO test[51] is based on ISO 5659-2 and is used to specify materials and products for large passenger ships and high speed surface craft. A reduced version of this test is used in the UK for railway vehicles[55] as BS 6853, B2. The draft European specification[56] (draft prEN 45545-2) uses the IMO toxicity test at $50kW/m^2$ without the pilot igniter and with FTIR analysis to determine the toxicity of railway vehicle components.

In the aircraft test, flaming conditions are generated by a series of small flames along the base of the vertical specimen, but in other tests it occurs when specimens are ignited by the pilot flame or self ignite. In all of these tests, the specimens, 75 mm square and up to 25 mm thick, are exposed to radiant heat with and without a pilot flame(s). Decomposition takes place inside a closed cabinet of $0.51\,m^3$. There is no control of the air flow or oxygen concentration through the fire zone and the effluent is mixed by natural convection as it accumulates within the closed cabinet. Gases are sampled using probes mounted in the centre of the cabinet.

Flaming tests result in some oxygen depletion which can vary with the thermal stability and thickness of the specimen and also decreases with increasing test duration. The flaming fire stage is difficult to assess but may be

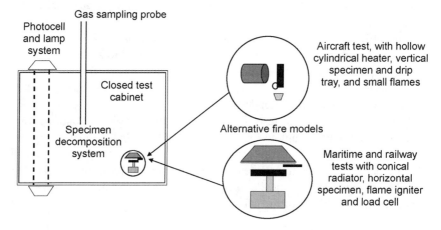

11.11 Diagram of fire smoke toxicity test based on NBS smoke chamber.

related to ISO stage 2, well-ventilated. The IMO tests at $50\,kW\,m^{-2}$ could possibly represent stage 3a, low ventilation flaming. The fire stage may change from 2 to 3a during a test.

These tests exclusively use chemical analysis methods to assess toxicity, and the concentrations of specified gases are determined using specified or approved analytical methods such as ion chromatography and ion specific electrodes, FTIR, etc. The aircraft tests specify the determination of CO, HCN, HBr, HCl, HF, NO_x and SO_2. Additionally, CO_2 and HBr are also determined in IMO and BS 6853 B2 tests.

The aircraft test[50] compares gas concentrations after 90 and 240 seconds (aircraft theoretical escape and flashover times respectively) with reference data. The IMO test[51] compares gas concentrations using different reference values determined during the 3 minutes around the time to maximum smoke density, without stating the rationale for doing so. The BS railway test[55] calculates an index using reference data related to immediate danger to life and health (IDLH) values for a 30 minutes exposure.

The advantages of these tests are that they use a widely available, standard smoke test apparatus, with the addition of simple gas sampling probes in the centre of the cabinet and relatively simple gas analysis systems to determine specified gases. The test specimen is heated from one side and the effects of surface protection layers can be determined. The principal limitations are that the air supply to the fire zone is not controlled, and testing can cause oxygen depletion, which will change the toxic product yield by an unknown amount while effluent may be recycled through the fire zone. Alternatively, the effluent may stratify and gas samples may not be representative of the effluent generated. Specimens which drip in the aircraft test may give erroneous results if the liquid falls to the cabinet floor of the cabinet and is not burned.

11.6.7 NIST Radiant Furnace method (ASTM E1678)[52]

This test method (Fig. 11.12) was developed to determine toxic potency data for materials and products used in the building and furnishing industries. Horizontal test specimens, up to 76 mm × 127 mm and up to 50 mm thick are exposed to a radiant heat flux of $50\,kW\,m^{-2}$ for 15 minutes with a spark igniter in a small chamber connected by three parallel, vertical ducts to the upper, closed chamber of 200 litres. The central slot acts as a chimney, while the outer slots replenish the air to the fire. Nose exposure animal ports are fitted to the upper chamber. Natural convection causes the effluent to move into the upper, closed cabinet where it accumulates and is mixed by natural convection. The specimen decomposition stage lasts for 15 minutes after which the exposures continue for a further 15 minutes, i.e. 30 minutes total.

The ISO fire stage is 1b, for samples that do not ignite. For specimens that ignite, it is probable that the fire zone corresponds to Stage 2, as the test

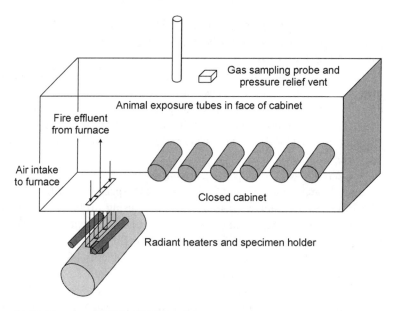

11.12 Diagram of ASTM E 1678 toxicity test.

specimen area is adjusted to give an FED value of 0.5 to 1.5. The underestimation of CO yield for developed and/or vitiated fires is addressed by numerically adjusting the results to give a CO yield of 0.2 g/g in underventilated conditions.[57] No compensation is made for other species whose yield increases with vitiation, such as hydrogen cyanide. The test is essentially an animal test with limited chemical analysis, though no corresponding correction is applied to the exposed animals. Rather tellingly, one of the test's developers reports that: 'the actual use of this test has been minimal. The reason is that it was mostly developed for precautionary reasons: If products should come onto the marketplace which produce toxic effects significantly in excess to what could be expected from their basic HRR (or mass loss) traits, then such a test could be used to quantify such effects.'[10]

11.6.8 Flow-through tests

In these methods the specimen is thermally decomposed with or without flaming in a furnace over flowing air which drives the effluent to the gas determining or sampling systems.

Simple tube furnace flow through test.

The NF X 70-100 method[58] (Fig. 11.13) was developed to estimate the toxicity of materials and products used in railway vehicles, initially in France. This is a small

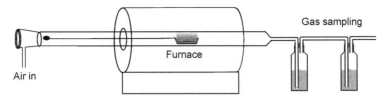

11.13 Schematic of NFX 70-100 test.

scale static tube furnace in which the test specimen, (typically 1 g or 0.1 g for low density materials), is pushed, in a crucible, into the middle of the furnace tube and thermally decomposed, using furnace temperatures of 400, 600 and 800 °C to represent smouldering, well-ventilated and under-ventilated conditions in flowing air at 2 litres min^{-1}, where they may pyrolyse and auto-ignite. At a temperature of 600 °C, the rate of burning may be fairly steady, and well-ventilated, whereas at 800 °C, the fire condition may be closer to under-ventilated as the rate of pyrolysis exceeds the air supply rate. The effluent is driven through gas detection systems, or bags or bubblers for subsequent analysis.

This method is easy to use, uses simple equipment with specified operating conditions of temperature and air flow. This method is increasingly used for fire toxicity testing of materials used for railway vehicles. The lack of requirement for flaming to be observed leaves the assignation of fire stages 3a or 3b to be assumed for most materials at 600 °C. A practical limitation is the number of replicate test runs needed to obtain sufficient samples for complete gas analysis.

Steady state tube furnace methods

Steady state tube furnace methods BS 7990 and ISO TS 19700 (known as the Purser furnace) allow the possibility of controlling the fire conditions during burning. These methods force combustion by feeding the sample into a furnace of increasing heat flux at a fixed rate, thus replicating each fire stage by steady state burning.

The Purser Furnace ISO TS 19700[2]

This is a test method for materials and products, whose results are intended to form part of the input to ISO 13344,[59] ISO 13571[38] and fire risk assessments, which are specifically related to the ISO fire stages. The test, shown in Fig. 11.14, uses the same apparatus as BS 7990[60] and IEC 60695-7-50 and -51,[61] with the air flow and temperature required to replicate each fire stage shown in Table 11.4.

Adjustment of temperature, air flow or specimen introduction rate may be required to simulate a specified ISO fire stage. A strip specimen or pieces are spread in a silica boat over a length of 800 mm at a loading density of

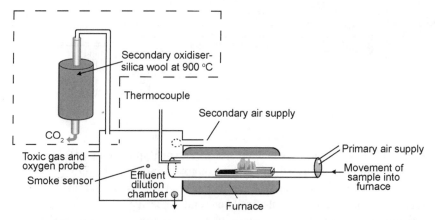

11.14 Diagram of apparatus of ISO TS 19700. The secondary oxidiser (inside dotted line) is for determination of total hydrocarbons in the ISO standard.

25 mg mm^{-1} and fed into a tube furnace at a rate of 1 g min^{-1} with flowing air. Secondary air is added in a mixing chamber to give a total gas flow of 50 l min^{-1} for analysis. The toxic potency of the effluent assessed during the steady state burn period.

This protocol enables the toxic potency of a material of unknown composition to be determined under known, steady state fire conditions (temperature and equivalence ratio) which relate directly to the end-use fire hazard. The use of a high secondary air flow usually permits the required gas samples to be taken during a single run. Smoke obscuration may also be determined. Unlike the closed box methods which may give toxic product data

Table 11.4 Furnace conditions corresponding to characteristic stages of burning behaviour

	Temperature (°C)	BS 7990 Primary air flow (l min^{-1})	ISO TS 19700 Primary air flow (l min^{-1})	IEC 60695-7-50 Primary air flow (l min^{-1})
1b Smouldering (non-flaming fires)	350	2	2	1.1
2 Well-ventilated flaming	650	10*	10*	22.6
3a Small under-ventilated flaming fires	650	Twice stoichiometric fuel/air ratio	Twice stoichiometric fuel/air ratio	--
3b Full developed under-ventilated fires	825	Twice stoichiometric fuel/air ratio	Twice stoichiometric fuel/air ratio	2.7

* subject to verification of ventilation condition

for a continuum of fire stages, in this method a separate run is required for each fire stage. In addition to analysis of the ISO gases (CO_2, CO, O_2, HCN, NO_x, HCl, HBr, HF, SO_2, acrolein and formaldehyde) there is a requirement to determine the total hydrocarbons. This may be achieved by passing part of the air-diluted test effluent through a secondary combustion furnace to allow the determination of the products of incomplete combustion. This also enables the equivalence ratio to be calculated directly. This protocol determines the equivalence ratio required for different fire stages for a material of unknown composition and thus enables the toxic potency of a material to be determined under known, steady state fire conditions (temperature and equivalence ratio) which relate directly to the end-use fire hazard.

Crucially, the method has been shown to replicate the toxic product yields from large-scale tests. Comparison of the yields of carbon monoxide from burning polypropylene (Fig. 11.15) and nylon 6.6 (Fig. 11.16) show a strong dependence on equivalence ratio and consistency between bench and large scale.[62]

Fire propagation apparatus[63]

This method is also suited to obtaining toxicity data (Fig. 11.9). Yield data have been published as a function of equivalence ratio and have been used to calculate FED and LC_{50} values. A significant advantage of this method is that the air flow and composition in the fire zone is controlled and consequently the apparatus could be used with pre-determined values of equivalence ratio (ϕ) to generate effluent yield data for the different ISO fire stages.[64,65]

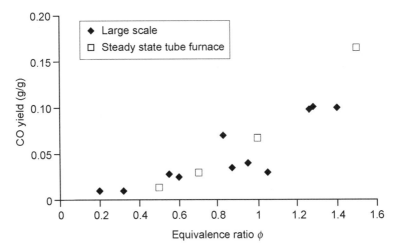

11.15 Comparison of tube furnace CO yields with large scale for polypropylene.

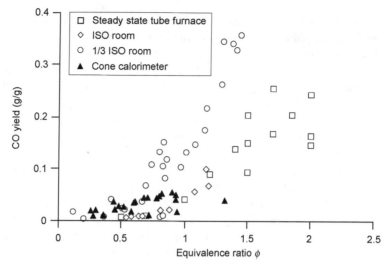

11.16 Comparison of CO yield for nylon 6.6 from steady state tube furnace with ISO Room and $\frac{1}{3}$ ISO room with controlled atmosphere cone calorimeter as a function of equivalence ratio ϕ.

11.7 Future trends: prediction of fire behaviour from material properties and models of fire growth

The International Forum of Fire Research Directors recently adopted a position paper in which they argued for a radical change in our approach to flammability assessment.[66] The current position, where a large range of tests, established by different authorities across the world is used to characterise the flammability of materials, ranging from small-scale tests with materials exposed to Bunsen burner flames to 15 m high simulations of fire scenarios was no longer appropriate. In particular, using the results to select general end-use applications for materials, from comparative tests that were specific to the test configuration and conditions, was not the best way to ensure fire safety. Although much more expensive, large-scale simulation tests at least provide results that approximately represent the end-use application. Indeed, current small- or large-scale flammability tests, cannot even be used to reliably predict the results of one another, let alone undefined unwanted fire scenarios. They propose that such tests be superseded by fundamentally based small-scale test methods for making material property measurements that can be used as input to validated, first principle, end-use computer models. Though more complex than most current ad hoc tests, they are well-demonstrated, well-documented and repeatable.

11.8 Sources of further information and advice

Manufacturers of fire testing equipment

Fire Testing Technology Ltd., Unit 19, Charlwoods Road, East Grinstead, West Sussex, RH19 2HL, UK Tel: + 44 (0)1342 323600; Fax: + 44 (0)1342 323608
Website: http://www.fire-testing.com
The Govmark Organization, Inc, 96D Allen Blvd, Farmingdale, NY 11735-5626 USA Tel: (631) 293-8944 Fax: (631) 293-8956
Website: http://www.govmark.com

Publications

Interpretation of cone calorimetry results
B. Schartel and T. R. Hull Application of cone calorimetry to the development of materials with improved fire performance, *Fire and Materials*, 31, 327–354 (2007).
Critical review of fire toxicity assessment
T Richard Hull and Keith T Paul, Bench-scale assessment of combustion toxicity – a critical analysis of current protocols, *Fire Safety Journal*, 42, 340–365 (2007).

11.9 References

1. *Approved Document B – Fire Safety Volume 1: Dwellinghouses; and Volume 2: Buildings other than Dwellinghouses*; TSO Publishing (2007).
2. ISO TS 19700:2006 Controlled Equivalence Ratio Method for the Determination of Hazardous Components of Fire Effluents.
3. ISO TS 19706:2004 Guidelines for assessing the fire threat to people.
4. Hull T R, Carman J M and Purser D A, Prediction of CO evolution from small-scale polymer fires, *Polymer International*, 49, 1259–1265 (2000).
5. Pitts W M, *Progress in Energy and Combustion Science*, 21, 197–237 (1995).
6. Purser D A, *Polymer International*, 49, 1232–1255, (2000).
7. Blomqvist P and Lonnermark A, *Fire and Materials*, 25, 71–81 (2001).
8. Andersson B, Markert F and Holmstedt G, *Fire Safety Journal*, 40, 439–465 (2005).
9. Cullis C F and Hischler M M, *The Combustion of Organic Polymers*, p. 324. Oxford Science Publications, Clarendon Press, Oxford (1981).
10. Babrauskas V, *Fire Test Methods for Evaluation of FR Efficacy in Fire Retardancy of Polymeric Materials* Ed. A F Grand and C A Wilkie, Marcel Dekker, New York (2000).
11. Drysdale D, *An Introduction to Fire Dynamics*, 2nd Edition, p. 218, John Wiley & Sons, Chichester (1998).
12. The Construction Products Directive, Council Directive 89/106/EEC.
13. Paul K T and Christian S D, *Journal of Fire Sciences*, 5(3), 178–211 (1987).
14. Hopkins Jr D and Quintiere J G, Materials fire properties and predictions for thermoplastics. *Fire Safety J*, 26, 241–268 (1996).
15. Rasbash D J, Drysdale D D and Deepak D, Critical heat and mass transfer at pilot

ignition and extinction of a material. *Fire Safety J*, 10, 1–10 (1986).
16. Thomson H E, Drysdale D D and Beyler C L, An experimental evaluation of critical surface temperatures as a criterion for piloted ignition of solid fuels. *Fire Safety J*, 13, 185–196 (1988).
17. Mikkola E and Wichman I S, On the thermal ignition of combustible materials. *Fire Mater*, 14, 87–96 (1989).
18. Kashiwagi T, Radiative ignition mechanism of solid fuels. *Fire Safety J*, 3, 185–200 (1981).
19. Babrauskas V, *Ignition Handbook*. Fire Science Publishers, Issaquah WA, USA and SFPE, USA (2003).
20. Lyon R E, Plastics and Rubber. In: *Handbook of Building Materials for Fire Protection*, Harper CA (ed). McGraw-Hill, chap 3: 3.1–3.51 (2004).
21. Lyon R E, Walters R N and Stoliarov S I, Thermal Analysis of Polymer Flammability. Presented at *228th ACS Meeting Philadelphia* (2004).
22. Tewarson A, Generation of Heat and Chemical Compounds in Fires. In *The SFPE Handbook of Fire Protection Engineering*, 3rd edition, DiNenno P J, Drysdale D, Beyler C L, Walton W D, Custer R L P, Hall Jr J R, Watts Jr J M (eds). National Fire Protection Association, Inc., chap. 3.4:3-82–3-161 (2002).
23. Sibulkin M and Little M W, Propagation and extinction of downward burning fires. *Combustion Flame*, 31, 197–208 (1978).
24. ISO TS 5658-1:2006, Reaction to fire tests – Spread of flame – Part 1: Guidance on flame spread; ISO 5658-2:2006 Part 2: Lateral spread on building and transport products in vertical configuration.
25. IEC 60695-11-10:1999 Fire hazard testing – Part 11-10: Test flames – 50 W horizontal and vertical flame test methods.
26. ISO 4589-2:1996 Plastics – Determination of burning behaviour by oxygen index – Part-2: Ambient temperature test.
27. ISO 5660-1:1993 Fire tests – Reaction to fire – Part 1: Rate of heat release from building products (cone calorimeter method).
28. ISO 5660-2: 2002, Reaction-to-fire tests – Heat release, smoke production and mass loss rate – Part 2: Smoke production rate (dynamic measurement).
29. Schartel B and Hull T R, Application of cone calorimetry to the development of materials with improved fire performance, *Fire and Materials*, 31, 327–354 (2007).
30. E 2058 – 02a Standard Test Methods for Measurement of Synthetic Polymer Material Flammability Using a Fire Propagation Apparatus (FPA) ASTM (2002).
31. Wu P K and Bill R G, Laboratory tests for flammability using enhanced oxygen, *Fire Safety Journal*, 38, 203–217 (2003).
32. Lyon R E and Walters R N, Pyrolysis combustion flow calorimetry, *J. Anal. Appl. Pyrolysis*, 71, 27–46 (2004).
33. Schartel B, Pawlowski K H and Lyon R E, Pyrolysis combustion flow calorimeter: A tool to assess flame retarded PC/ABS materials? *Thermochimica Acta*, 462, 1–14 (2007).
34. Lyon R E, Walters R N, Beach M and Schall F P, Flammability Screening of Plastics Containing Flame Retardant Additives, ADDITIVES 2007, 16th International Conference, San Antonio, TX (2007).
35. EN 13823:2002 Reaction to fire tests for building products. Building products excluding floorings exposed to the thermal attack by a single burning item.
36. Axelsson J and Van Hees P, New data for sandwich panels on the correlation between the SBI test method and the room corner reference scenario, *Fire and Materials*, 29, 53–59 (2004).

37. *Fire Statistics United Kingdom 2002*; Office of the Deputy Prime Minister: London, April (2004).
38. ISO 13571:2007, Life-threatening components of fire – Guidelines for the estimation of time available for escape using fire data.
39. Council Directive 86/609 EEC Article 3.
40. Hull T R, Quinn R E, Areri I G and Purser D A, *Polymer Degradation and Stability* 77, 235–242 (2002).
41. Stec A A, Hull T R, Lebek K, Purser J A and Purser D A, The Effect of Temperature and Ventilation Condition on the Toxic Product Yields from Burning Polymers, *Fire and Materials*, 32, 49–60 (2008).
42. Hull T R and Paul K T, Bench-Scale Assessment of Combustion Toxicity – A Critical Analysis of Current Protocols. *Fire Safety Journal*, 42, 340–365 (2007).
43. Nevaiser J L and Gann R G, Evaluation of toxic potency values for smoke from products and materials. *Fire Technology*, 40, 177–199 (2004).
44. Klimisch H J, Hollander H W and Thyssen J, Comparative measurements of the toxicity to laboratory animals of products of thermal decomposition generated by the method of DIN 53436. *J. Comb. Tox.*, 7, 209–230 (1980).
45. Pauluhn J, A retrospective analysis of predicted and observed smoke lethal toxic potency values, *Journal of Fire Sciences*, 11 (2), 109–130 (1993).
46. Kaplan H L, Effects of irritant gases on avoidance/escape performance and respiratory response of the baboon. *Toxicology*, 47, 165–79 (1987).
47. ISO 16312-1:2006 Guidance for assessing the validity of physical fire models for obtaining fire effluent toxicity data for fire hazard and risk assessment – Part 1: Criteria.
48. Christy M R, Petrella R V and Penkala J J, Controlled-atmosphere cone calorimeter. In *Fire and Polymers II*. ACS Symposium Series 599, 498–517 (1995).
49. Hull T R, Wills C L, Artingstall T, Price D, Milnes G J, Mechanisms of smoke and CO suppression from EVA composites. In *Fire Retardancy of Polymers New Application of Mineral Fillers*. Le Bras M, Wilkie C A, Bourbigot S, Duquesne S, Jama C (eds). Royal Society of Chemistry (2005).
50. prEN 2824. (Aerospace Series) Burning behaviour, determination of smoke density and gas components in the smoke of materials under the influence of radiating heat and flames – Test equipment apparatus and media.
prEN2825 – Determination of smoke density.
prEN 2826 – Determination of gas concentrations in the smoke.
ABD 0031 Fire-Smoke-Toxicity (FST) Test Specification (Airbus Industries).
Boeing BSS 7239, Test method for toxic gas generation by materials on combustion.
51. IMO MSC.41(64), Interim standard for measuring smoke and toxic products of combustion, International Maritime Organisation.
52. ASTM E1678, Standard method for measuring smoke toxicity for use in fire hazard analysis.
53. ISO 5659-2, Plastics – Smoke Generation – Part 2, Determination of Specific Optical Density.
54. ASTM E 662, Test for Specific Optical Density of Smoke Generated by Solid Materials.
55. BS 6853, Code of Practice for fire precautions in the design and construction of passenger carrying trains, Annex B.
56. pr EN 45545-2:2004 Railway applications – Fire protection on railway vehicles – Part 2: Requirements for fire behaviour of materials and components – Annex D Testing procedure for analysis of toxic gases.

57. Babrauskas V, Gann R G, Levin B C, Paabo M, Harris R H, Peacock R D and Yusa S, *Fire Safety Journal*, 31, 345–358 (1998), and detailed in ASTM E1678 (1998).
58. NFX 70-100, Analysis of pyrolysis and combustion gases. Tube furnace method. Part 1, Methods of analysis of gas generated by thermal degradation. Part 2, Method of thermal degradation using tube furnace.
59. ISO 13344:1996, Estimation of lethal toxic potency of fire effluents.
60. BS 7990:2003, Tube furnace method for the determination of toxic products yields in for effluents.
61. IEC 60695 Fire Hazard testing – Part 7-50: Toxicity of fire effluents – Estimation of toxic potency: Apparatus and test method.
    IEC 60695 Fire hazard testing – Part 7-51: Toxicity of fire effluent – Estimation of toxic potency: Calculation and interpretation of test results.
    PD IEC 60695-7-3:1998, Fire hazard testing, Part 7-3: Toxicity of fire effluent, Use and interpretation of test results.
62. Hull T R, Lebek K, Stec A A, Paul K T and Price D, Bench-scale assessment of fire toxicity, in *Advances in the Flame Retardancy of Polymeric Materials: Current perspectives presented at FRPM'05*, Editor B. Schartel pp. 235–248, Herstellung und Verlag, Norderstedt (2007).
63. E 2058 – 02a Standard Test Methods for Measurement of Synthetic Polymer Material Flammability Using a Fire Propagation Apparatus (FPA) ASTM (2002).
64. Tewarson A, Generation of heat and chemical compounds in fires, *SFPE handbook of fire protection engineering*, 3rd edn, pp. 3–82. Quincy MA (2002).
65. Tewarson A, Jiang F H and Morikawa T, Ventilation controlled combustion of polymers. *Combust Flame*, 95, 151–169 (1993).
66. Bill Jr. R G and Croce P A, The International FORUM of Fire Research Directors: A position paper on small-scale measurements for next generation standards, *Fire Safety Journal*, 41, 536–538 (2006).

12
New and potential flammability regulations

J T R O I T Z S C H, Fire and Environmental Protection Service, Germany

Abstract: The chapter shows the present and future fire safety regulations, classification systems and tests for products in building construction, electrical engineering and electronics (E&E), transportation, and furniture. The main topics are:

- Building. Construction Products Directive, Euroclasses, reaction to fire tests, CE marking and approval procedures.
- E&E. Main flammability tests and certification procedures.
- Transportation. New standardization rail vehicles (EN 45545); revision fire tests for ships.
- Upholstered furniture. Fire safety levels in the UK, France, and the USA.
- Trends influencing future use of products in building (smoke, acidity), E&E (glow wire, external ignition), transportation (higher fire safety in rolling stock and ships), and upholstered furniture (private homes).

Key words: Construction Products Directive, reaction to fire and flammability tests, CE-marking, external ignition sources in electrical engineering, fire tests in transportation, fire safety of furniture.

12.1 Introduction: overview of present fire safety regulations in Europe

The aim of fire protection is to minimize the risk of a fire thus protecting life and possessions. The state, as official custodian of public safety, ensures such protection via relevant legislation, which includes laws and statutory orders. Standards and codes of practice based on recognized technical principles are the means of putting the general requirements of fire protection defined in the legislation into practice. Materials, semi-finished and finished products are tested according to methods laid down in the standards and classified according to the test results. Such tests are carried out by officially recognized materials testing institutes. The certificate of the test result and classification provides a basis for the use of the material or product. A test mark is frequently required as evidence of the suitability of a material and its identity with the material tested. This implies a quality check on production either by the manufacturer or an outside body. In the latter case an agreement must be entered into with a state recognized institution.

Numerous public and private organizations are concerned with the rules and regulations of fire protection. In addition to governmental bodies, these include professional societies and industrial, commercial, technical and insurance associations.

In the European Union (EU), the harmonization of the legislative, regulatory and administrative provisions of the 27 Member States is the responsibility of the EC-Commission and aims at creating a single European market. The 'Construction Products Directive (CPD)' (89/106/EC) deals with types of products and only includes the main requirements. For this purpose, the Interpretative Document ID 2 (Essential Requirement 'Safety in Case of Fire') was issued. However, the fire safety regulations of the single Member States and the fire safety levels laid down therein are not part of European harmonization and remain the responsibility of the Member States.

European standards are issued by the European Committee for Standardization (CEN) (with members from 30 countries) and the European Committee for Electrotechnical Standards (CENELEC). These standards assist in eliminating technical barriers to trade between Member States as well as between these and other European countries.

The new EN standards are in the course of being transferred into national standards, e.g. BS EN (UK), NF EN (France) or DIN EN (Germany). After a definite time ('transition period') all national fire and product standards will be withdrawn in favour of the EN standards.

In other fields, regulations exist which are, in the main, internationally valid. This applies in transportation and in civil aviation in particular. The most important new technologies were developed in the USA, and thus the US Federal Aviation Regulations (FAR) of the Federal Aviation Administration (FAA) set the criteria for corresponding internationally accepted regulations. Evidence of airworthiness must be furnished with the aid of these regulations, which have been partially or totally adopted by most countries.

Sets of regulations in international use also exist in other areas of transportation. For example, the International Convention for the Safety of Life at Sea (SOLAS) and the Fire Test Procedures Code (FTP Code) is accepted in nearly all countries. The regulations, recommendations and conditions of supply relating to railways are being harmonized within the EU starting with the Interoperability of the Trans-European High-Speed Rail System Directive.

International harmonization of standards and mutual recognition of test results are ongoing in electrical engineering in order to remove trade barriers. International electrical standards are developed by the International Electrotechnical Commission (IEC) in order to achieve this aim. The Commission for Certification of Electrical Equipment (CEE) promotes mutual recognition of test results by all member countries on these standards.

While the IEC is responsible for international electrical standards, all other technical fields are covered by the International Organization for Standardization (ISO). This organization aims at promoting the worldwide development of standards in order to facilitate the exchange of goods and services and to encourage mutual cooperation in intellectual, scientific and economic activities. ISO member committees are the national standards organizations. A comprehensive overview of fire protection regulations and test procedures is given in Mitzlaff and Troitzsch.[1]

12.2 Building

The statutory regulations and provisions relating to fire protection are furthest advanced in the field of building, particularly in the industrialized countries, where comprehensive sets of regulations may differ significantly. Most industrialized countries belong to one or other economic grouping. The regulations are then affected by the attempts at harmonization within such groups in order to eliminate trade barriers. In Europe, the European Union (EU) now basically harmonizes legal and administrative regulations.

Regulations covering fire protection in building are more or less centralized depending on the particular form of government. Countries with centralized regulations include France, Belgium, Italy and Japan (where additional local regulations exist). In principle, the UK also belongs to this group. However, separate regulations apply in England and Wales, Scotland and Northern Ireland. The regulations, however, are gradually being brought into line with those in force in England and Wales.

In countries with a federal structure, each province or state usually has its own decentralized building regulations, which are generally based on a model building code and framework guidelines in order to maintain some uniformity. Germany, Austria, Switzerland, Canada, Australia and the USA all have a federal structure. The execution of regulations is the responsibility of the Laender, Cantons, Provinces or States. The situation is almost chaotic in the USA, where the individual states are theoretically responsible for building regulations. In practice, however, local authorities determine their application via some 20 000 local regulations.

In the following, the European harmonization of construction products and the related reaction to fire classifications and tests will be discussed.

12.2.1 Regulations, construction products, fire performance requirements, and tests in the European Union

The single European market

The single European market was established under the Single European Act,[2] and is the core of the process of European economic integration, involving the removal of obstacles to the free movement of goods, services, people, and capital between Member States of the EU. It covers, among other benefits, the elimination of customs barriers, the liberalization of capital movements, the opening of public procurement markets, and the mutual recognition of professional qualifications. It came into effect on 1 January 1993.

The Construction Products Directive

The European Construction Products Directive (CPD) is the Council Directive 89/106/EC of 21 December 1989 on the approximation of laws, regulations and administrative provisions to the Member States of the European Union relating to construction products.[3] At the beginning of the nineties, the CPD was transposed into the national laws of the Member States, and at the end of the nineties in the EFTA States Norway and Switzerland. However, the CPD only regulates the free exchange of goods in the European internal market and does not harmonize the fire safety levels of existing national building regulations.

The CPD regulates the establishment and functioning of the Internal Market in construction products, by means of technical harmonization to ensure:

- products are fit for their intended use
- the free circulation of goods, without reducing the existing and justified levels of protection in the Member States
- the contribution to competitiveness of enterprises.

Essential requirements and safety in case of fire

The CPD contains six essential requirements (ER) covering:

- Mechanical resistance and stability
- Safety in case of fire
- Hygiene, health and the environment
- Safety in use
- Protection against noise
- Energy, economy and heat retention.

Building products must comply with these ERs in order to obtain the CE mark, which allows their free circulation in Member States. Member States cannot refuse CE-marked products to be placed on their market. The CE mark gives evidence that the marked product complies with harmonized technical specification.

The ER 'safety in case of fire'[4] is based on uniform fire classes 'Euroclasses' and harmonized tests for reaction-to-fire and fire resistance valid throughout Europe. Only the Euroclasses and the new harmonized fire test methods and not fire safety levels have been integrated into the national building regulations. This means that the respective philosophies regarding the fire safety of building products applications remain the responsibility of the single Member States.

12.2.2 CE marking

CE marking is only possible for building products described in harmonized technical specifications (Product Standards (PS) or European Technical Approvals (ETA)) and meeting harmonized classifications and tests. The building product assessment leading to the CE mark is valid throughout the European Union.

The manufacturer is free to choose where to have his product assessed. The assessment in Member State X must be recognized by Member State Y.

An example of what the CE mark affixed to a product looks like is shown in Fig. 12.1.

CE marks cannot be granted for construction products by only meeting the reaction-to-fire tests and the required Euroclasses described in the following. Only the product standards, where the fire tests and classes to be met, the mounting and fixing procedures, as well as other properties like, for example, conductivity and flexural strength are specified, lead to a CE mark. Examples of European harmonized product standards, which after their publication and a co-existence period substitute existing national standards, are thermal insulation products for buildings:

- EN 13163 Factory made products of expanded polystyrene (EPS)
- EN 13164 Factory made products of extruded polystyrene foam (XPS)
- EN 13165 Factory made products of rigid polyurethane foam (PUR)

* No performance determined

12.1 CE mark.

12.2.3 Attestation of Conformity

In order to be CE marked, a construction product under the CPD is subjected to attestation provisions with Attestation of Conformity (AoC) levels defined by a mandate of the European Commission.[5]

The AoC level depends on the consequences of the failure of the construction product:

- risk of sudden failure of the product with no/very little warning, resulting in risk to life and limb, indicates a high level of attestation.

The AoC level is determined by the product characteristics:

- effect of variability on serviceability
- susceptibility to defects in manufacture
- nature of the product.

What are the different AoC levels? There are 4 (+2) main systems of AoC for CE marking:

Highest level 1: Manufacturer declares product conforms and full 3rd party (notified body) certification of product and factory production control (FPC).
level 2: Manufacturer declares product conforms and full 3rd party certification of product.
level 3: Manufacturer declares product conforms and FPC and 3rd party initial type testing.
Lowest level 4: Manufacturer declares product conforms and FPC (requires documented FPC).

A more detailed view of the AoC levels is shown in Table 12.1.

Basically, the producer is fully responsible for the AoC of his product, even if

Table 12.1 Attestation of Conformity levels

	AoC – CEC numbering system					
	1+	1	2+	2	3	4
Tasks for the MANUFACTURER						
1 Factory production control (FPC)	✓	✓	✓	✓	✓	✓
2 Regular and systematic sample testing	✓	✓	✓			
3 Initial type testing (ITT)	✓	✓		✓		
Tasks for the NOTIFIED BODY						
4 ITT	✓	✓			✓	
5 Certification of FPC		✓	✓	✓	✓	
6 Surveillance of FPC		✓	✓	✓		
7 Audit testing of samples	✓					

a third party, a notified body, is involved. However, in most AoC systems, notified bodies take over tasks and responsibilities from the producer.

For construction products having to satisfy reaction-to-fire requirements and classified to the Euroclasses (see below), Level 3 in the AoC – CEC numbering system is usually sufficient to comply with national requirements. In Germany, however, where surveillance of FPC and audit testing of samples are part of German building regulations (this is required for building materials of Class A2 and B1 to DIN 4102), the situation regarding the use of Level 1+ or 3 is still unclear. In fact, one still has to comply with the old German system.

12.2.4 Notified bodies

Member States shall notify the Commission through the appropriate national organization bodies to carry out attestation procedures (certification, inspection, testing) under the CPD. The legal basis for designating notified bodies is set out in Article 18 in Annex IV of the CPD.

There are four different kinds of notified bodies:

- Bodies performing product certification have to comply with
  EN 45011 'General requirements for certification bodies operating product certification systems'.
- Bodies performing Factory Production Control FPC certification have to comply with
  EN 45012 'General requirements for certification bodies operating assessment and certification oblige/registration of quality systems' and or EN 45011.
- Bodies performing FPC inspection have to comply with
  EN 45012 or EN 45004 'General criteria for the operation of various bodies performing inspection'.
- Testing laboratories have to comply with
  EN ISO/IEC 17025 'General requirements for the competence of testing and calibration laboratories'. Since 2003, it has replaced EN 45001 'General criteria for the operation of testing laboratories'.

12.2.5 Harmonised product standards

Harmonised product standards (hEN) enable products to be CE-marked and put on the European market. Product standards comply with the Essential Requirements of the Construction Products Directive. They will substitute existing national product standards after a period of co-existence: This is the time period where national and European technical specifications (fire and product standards) are both available.

Twenty-one months after the hEN availability, national product standards will be withdrawn. There are two periods:

- 9 months to test hEN applicability (meaning: is it applicable in real life?) followed by
- 12 months (for implementing it into national regulations) until the date of withdrawal of national standards.

Fire standards (fEN) were developed by CEN/TC127 'Fire Safety'. They allow the EC to define classification systems for construction products and are referred to in product standards. They cannot be used alone for CE-marking because they do not cover other Essential Requirements than Safety in Case of Fire.

Practice shows that it often takes 5 (to 10) years until national fire classifications and standards are withdrawn, because in many cases, national regulations are not adequately fulfilled by the harmonized product standards.

12.2.6 Classification of reaction to fire of construction products: the Euroclasses

The European Commission published the Euroclasses on February 8, 2000. Reaction to fire testing is done following a new concept compared to existing procedures in Europe. Seven main classes are introduced, the Euroclasses. These are A1, A2, B, C, D, E and F. A1 and A2 represent different degrees of limited combustibility. For linings, B–E represent products that may go to flashover in a room and at certain times. F means that no performance is determined. There are 7 classes for linings and 7 classes for floor coverings. Additional classes of smoke and any occurrence of burning droplets are also given. The classes for building products and floor coverings are shown in Tables 12.2 and 12.3.

Table 12.2 Classes of reaction to fire performance for construction products excluding floorings (*)

Class	Test method(s)	Classification criteria	Additional classification
A1	EN ISO 1182 (^{1}); And	$\Delta T \leq 30\,°C$; and $\Delta m \leq 50\%$; and $t_f = 0$ (i.e. no sustained flaming)	–
	EN ISO 1716	$PCS \leq 2.0\ MJ.kg^{-1}$ (^{1}); and $PCS \leq 2.0\ MJ.kg^{-1}$ (^{2}) (^{2a}); and $PCS \leq 1.4\ MJ.m^{-2}$ (^{3}); and $PCS \leq 2.0\ MJ.kg^{-1}$ (^{4})	–
A2	EN ISO 1182 (^{1}); Or EN ISO 1716; And	$\Delta T \leq 50\,°CC$; and $\Delta m \leq 50\%$; and t_f 20s $PCS \leq 3.0\ MJ.kg^{-1}$ (^{1}); and $PCS \leq 4.0\ MJ.m^{-2}$ (^{2}); and $PCS \leq 4.0\ MJ.m^{-2}$ (^{3}); and $PCS \leq 3.0\ MJ.kg^{-1}$ (^{4})	–
	EN 13823 (SBI)	$FIGRA \leq 120\ W.s^{-1}$; and $LFS <$ edge of specimen; and $THR_{600s} \leq 7.5\ MJ$	Smoke production(^{5}); and Flaming droplets/ particles (^{6})

Table 12.2 Continued

Class	Test method(s)	Classification criteria	Additional classification
B	EN 13823 (SBI); And	FIGRA $\leq$ 120 W.s^{-1}; and LFS < edge of specimen; and THR$_{600s}$ $\leq$ 7.5 MJ	Smoke production ([5]); and Flaming droplets/ particles ([6])
	EN ISO 11925-2 ([8]): Fs $\leq$ 150 mm within 60 s Exposure = 30 s		
C	EN 13823 (SBI); And	FIGRA $\leq$ 250 W.s^{-1}; and LFS < edge of specimen; and THR$_{600s}$ $\leq$ 15 MJ	Smoke production ([5]); and Flaming droplets/ particles ([6])
	EN ISO 11925-2 ([8]): Fs $\leq$ 150 mm within 60 s Exposure = 30 s		
D	EN 13823 (SBI); And	FIGRA $\leq$ 750 W.s^{-1}	Smoke production ([5]); and Flaming droplets/ particles ([6])
	EN ISO 11925-2 ([8]): Fs $\leq$ 150 mm within 60 s Exposure = 30 s		
E	EN ISO 11925-2([8]): Fs $\leq$ 150 mm within 20 s Exposure = 15 s		Flaming droplets/ particles ([7])
F	No performance determined		

(*) The treatment of some families of products, e.g. linear products (pipes, ducts, cables, etc.), is still under review and may necessitate an amendment to this decision.
([1]) For homogeneous products and substantial components of non-homogeneous products.
([2]) For any external non-substantial component of non-homogeneous products.
([2a]) Alternatively, any external non-substantial component having a PCS $\leq$ 2.0 MJ.m^{-2}, provided that the product satisfies the following criteria of EN 13823(SBI): FIGRA $\leq$ 20 W.s^{-1}; and LFS < edge of specimen; andTHR$_{600s}$ $\leq$ 4.0 MJ; and s1; and d0.
([3]) For any internal non-substantial component of non-homogeneous products.
([4]) For the product as a whole.
([5]) **s1** = SMOGRA $\leq$ 30m^2.s^{-2} andTSP$_{600s}$ $\leq$ 50m^2; **s2** = SMOGRA $\leq$ 180m^2.s^{-2} andTSP$_{600s}$ $\leq$ 200m^2; **s3** = not s1 or s2.
([6]) **d0** = No flaming droplets/ particles in EN13823 (SBI) within 600s; **d1** = No flaming droplets/ particles persisting longer than 10s in EN13823 (SBI) within 600s; **d2** = not d0 or d1; Ignition of the paper in EN ISO 11925-2 results in a d2 classification.
([7]) Pass = no ignition of the paper (no classification); Fail = ignition of the paper (d2 classification).
([8]) Under conditions of surface flame attack and, if appropriate to end-use application of product, edge flame attack.

Symbols. The characteristics are defined with respect to the appropriate test method.

ΔT	temperature rise
Δm	mass loss
t_f	duration of flaming
PCS	gross calorific potential
FIGRA	fire growth rate
THR$_{600s}$	total heat release
LFS	lateral flame spread
SMOGRA	smoke growth rate
TSP$_{600s}$	total smoke production
Fs	flame spread

Table 12.3 Classes of reaction to fire performance for floorings

Class	Test method(s)	Classification criteria	Additional classification
$A1_{FL}$	EN ISO 1182 ([1]); And	$\Delta T \leq 30\,°C$; and $\Delta m \leq 50\%$; and $t_f = 0$ (i.e. no sustained flaming)	–
	EN ISO 1716	$PCS \leq 2.0\ MJ.kg^{-1}$ ([1]); and $PCS \leq 2.0\ MJ.kg^{-1}$ ([2]); and $PCS \leq 1.4\ MJ.m^{-2}$ ([3]); and $PCS \leq 2.0\ MJ.kg^{-1}$ ([4])	–
$A2_{FL}$	EN ISO 1182 ([1]); Or	$\Delta T \leq 50\,°C$; and $\Delta m \leq 50\%$; and $t_f \leq 20s$	–
	EN ISO 1716; And	$PCS \leq 3.0\ MJ.kg^{-1}$ ([1]); and $PCS \leq 4.0\ MJ.m^{-2}$ ([2]); and $PCS \leq 4.0\ MJ.m^{-2}$ ([3]); and $PCS \leq 3.0\ MJ.kg^{-1}$ ([4])	–
	EN ISO 9239-1 ([5])	Critical flux ([6]) $\geq 8.0\ kW.m^{-2}$	Smoke production ([7])
B_{FL}	EN ISO 9239-1 ([5]) And EN ISO 11925-2([8]): Exposure = 15 s	Critical flux ([6]) $\geq 8.0\ kW.m^{-2}$ $Fs \leq 150\ mm$ within 20 s	Smoke production ([7])
C_{FL}	EN ISO 9239-1 ([5]) And EN ISO 11925-2([8]): Exposure = 15 s	Critical flux ([6]) $\geq 4.5\ kW.m^{-2}$ $Fs \leq 150\ mm$ within 20 s	Smoke production ([7])
D_{FL}	EN ISO 9239-1 ([5]) And EN ISO 11925-2([8]): Exposure = 15 s	Critical flux ([6]) $\geq 3.0\ kW.m^{-2}$ $Fs \leq 150\ mm$ within 20 s	Smoke production ([7])
E_{FL}	EN ISO 11925-2([8]): Exposure = 15 s	$Fs \leq 150\ mm$ within 20 s	
F_{FL}	No performance determined		

([1]) For homogeneous products and substantial components of non-homogeneous products.
([2]) For any external non-substantial component of non-homogeneous products.
([3]) For any internal non-substantial component of non-homogeneous products.
([4]) For the product as a whole.
([5]) Test duration = 30 minutes.
([6]) Critical flux is defined as the radiant flux at which the flame extinguishes or the radiant flux after a test period of 30 minutes, whichever is the lower (i.e., the flux corresponding with the furthest extent of spread of flame).
([7]) **s1** = Smoke $\leq$ 750%.min; **s2** = not s1.
([8]) Under conditions of surface flame attack and, if appropriate to the end-use application of the product, edge flame attack.

Symbols
See Table 12.2.

Definitions

- **Material**: A single basic substance or uniformly dispersed mixture of substances, e.g. metal, stone, timber, concrete, mineral wool with uniformly dispersed binder, polymers.
- **Homogeneous product**: A product consisting of a single material, of uniform density and composition throughout the product.
- **Non-homogeneous product**: A product that does not satisfy the requirements of a homogeneous product. It is a product composed of one or more components, substantial and/or non-substantial.
- **Substantial component**: A material that constitutes a significant part of a non-homogeneous product. A layer with a mass per unit area $\geq 1.0\,\text{kg/m}^2$ or a thickness $\geq 1.0\,\text{mm}$ is considered to be a substantial component.
- **Non-substantial component**: A material that does not constitute a significant part of a non-homogeneous product. A layer with a mass per unit area $<1.0\,\text{kg/m}^2$ and a thickness $<1.0\,\text{mm}$ is considered to be a non-substantial component.

Two or more non-substantial layers that are adjacent to each other (i.e., with no substantial component(s) in-between the layers) are regarded as one non-substantial component and, therefore, must altogether comply with the requirements for a layer being a non-substantial component.

For non-substantial components, distinction is made between internal non-substantial components and external non-substantial components, as follows:

- **Internal non-substantial component**: A non-substantial component that is covered on both sides by at least one substantial component.
- **External non-substantial component**: A non-substantial component that is not covered on one side by a substantial component.

A Euroclass may be determined as **Bs2d1**. **B** stands for the main class, **s2** stands for smoke class 2 and **d1** stands for droplets/particles class 1. This system gives basically a total of about 40 classes of linings and 11 classes of floor coverings to choose from (see Table 12.4). However, each country is expected only to use a very small fraction of the possible combinations.

12.3 Fire testing of construction products in the European Union

The main fire testing methods used to classify construction products according to the Euroclasses for surface products (linings) and floor coverings are summarized in Table 12.5. All these reaction-to-fire tests standards were published in 2002 and have been taken over later as national standards (DIN EN, BS EN, NF EN, etc.) by the European Union Member States. The national standards in use before the introduction of the new European standards are withdrawn after a co-existence period of 1 to 2 years, in which both standards

Table 12.4 Euroclasses for surface products (linings) and floor coverings

Surface products			Floor coverings	
A1			A1$_{fl}$	
A2s1d0	A2s1d1	A2s1d2	A2$_{fl}$s1	A2$_{fl}$s2
A2s2d0	A2s2d1	A2s2d2		
A2s3d0	A2s3d1	A2s3d2		
Bs1d0	Bs1d1	Bs1d2	B$_{fl}$s1	B$_{fl}$s2
Bs2d0	Bs3d1	Bs2d2		
Bs3d0	Bs4d1	Bs3d2		
Cs1d0	Cs1d1	Cs1d2	C$_{fl}$s1	C$_{fl}$s2
Cs2d0	Cs3d1	Cs2d2		
Cs3d0	Cs4d1	Cs3d2		
Ds1d0	Ds1d1	Ds1d2	D$_{fl}$s1	D$_{fl}$s2
Ds2d0	Ds3d1	Ds2d2		
Ds3d0	Ds4d1	Ds3d2		
E			E$_{fl}$	
Ed2				
F			F$_{fl}$	

were valid. As in some cases there are still problems regarding the suitability of the new European fire tests and classifications to implement fire safety levels specified in national regulations, the old national fire testing systems often continue to be used.

EN 13501 Fire classification and EN 13238 standard substrates

EN 13501[6] basically repeats the classification criteria given in the tables above. It also gives general requirements, provides a model for reporting and gives background information of the testing and classification system. Classification reports on products will be given based on EN 13501.

Table 12.5 Reaction-to-fire tests for linings and floor coverings

Test method	Standard
Fire classification	EN 13501
Standard substrates for product samples	EN 13238
Non-combustibility – Furnace test	EN ISO 1182
Calorific value (PCS) – Bomb calorimeter	EN ISO 1716
Room/Corner test (Reference scenario)	EN 14390 (ISO 9705)
Single Burning Item (SBI) test	EN 13823
Small flame test	EN ISO 11925-2
Radiant Panel (Flooring) test	EN ISO 9239-1

Prior to all testing, product samples shall be prepared, conditioned and mounted in accordance with the relevant test methods and product standards. If relevant, ageing and washing procedures are carried out in accordance with the actual product standard.

EN 13501 allows for two additional tests to increase accuracy of classification under certain circumstances. This rule applies to all of the tests described below.

EN 13238[7] recommends standard substrates on which the product sample can be attached before testing. The standard substrates represent various end use conditions. Thus the test results become more general and the amount of testing can be kept down.

EN ISO 1716 Calorific Potential

EN ISO 1716[8] determines the potential maximum total heat release of a product when completely burning, regardless of its end use. The test is relevant for the classes A1, A2, A1$_{fl}$ and A2$_{fl}$.

The calorific potential of a material is measured in a bomb calorimeter. The powdered material is completely burned under high pressure in a pure oxygen atmosphere.

EN ISO 1182 Non-combustibility

EN ISO 1182[9] identifies products that will not, or significantly not, contribute to a fire, regardless of their end use. The test is relevant for the classes A1, A2, A1$_{fl}$ and A2$_{fl}$.

EN ISO 1182 is a pure material test and a product cannot be tested in end-use conditions. Therefore only homogenous building products or homogenous components of a product are tested. EN ISO 1182 was first published by ISO during the 1970s and is well known. The EN ISO version is shown in Fig. 12.2 and test specifications in Table 12.6.

Table 12.6 EN ISO 1182 test specifications

Specimens	5 cylindrical samples, diameter 45 mm, height 50 mm.
Specimen position	Vertical in specimen holder in the centre of the furnace.
Heat source	Electrical cylindrical furnace at 750 °C (measured by the furnace thermocouple).
Test duration	Depends on temperature stabilization.
Conclusions	Classification is based on temperature rise as measured by the furnace thermocouple, duration of flaming and mass loss of the sample. Details are given in the Euroclasses tables.

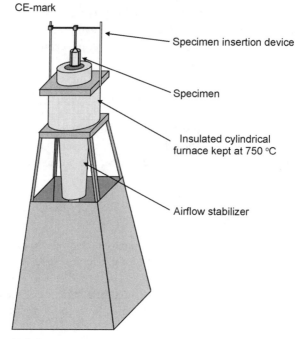

12.2 Schematic diagram of EN ISO 1182 test for non-combustibility.

EN 14390 (former ISO 9705) Room/Corner Test

In order to find limit values for the Euroclasses in a reference scenario, a room fire test was selected for the SBI. Test data from the reference scenario could then be used to determine the Euroclass of the product and to 'position' EN 13823, the SBI test procedure. The reference scenario chosen is the international standard Room/Corner Test, ISO 9705, which has been taken over as EN 14390.[10] The testing device and specifications are shown in Fig. 12.3 and Table 12.7.

The Room/Corner Test is a large-scale test method for measurement of the burning behaviour of building products (linings) in a room scenario. The product is mounted on three walls and on the ceiling of a small compartment. A door opening ventilates the room. The principal output is the occurrence and time to flashover. Also a direct measure of fire growth (Heat Release Rate, HRR) and light obscuring smoke (Smoke Production Rate, SPR) are results from the test.

The development of EN 13823, the SBI, included testing of 30 building products across Europe. The same products were tested according to the reference scenario, the Room/Corner Test. The subsequent analysis then resulted in a correlation between FIGRA (EN 14390) and FIGRA (SBI). The correlation between the tests is also relating flashover in EN 14390 to certain Euroclasses is shown in Table 12.8.

New and potential flammability regulations 305

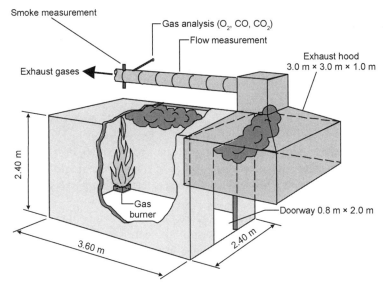

12.3 Schematic diagram of EN 14390 Room/Corner Test.

Table 12.7 EN 14390 Room/Corner Test specifications

Specimens	Sample material enough to cover three walls and the ceiling of the test room. The wall containing the doorway is not covered.
Specimen position	Forms a room lining.
Ignition source	Gas burner placed in one of the room corners. The burner heat output is 100 kW for the first ten minutes and then 300 kW for another ten minutes.
Test duration	20 min or until flashover.
Conclusions	A number of parameters relating to a room fire such as temperatures of the gas layers, flame spread and heat fluxes can be measured. However, the most important outputs are HRR, SPR and time to or occurrence of flashover.

Table 12.8 Description of the Euroclasses as the tendency of the actual product to reach flashover in the Room/Corner Test to EN 14390

Euroclass	Limit value FIGRA (SBI) (W/s)	Expected burning behaviour in the Room/Corner Test
A2	120	No flashover
B	120	No flashover
C	250	No flashover at 100 kW
D	750	No flashover before 2 min at 100 kW
E	>750	Flashover before 2 min

The role of the reference scenario to EN 14390 is not yet definitively decided. As it was used to identify the limit values in the SBI test for the Euroclasses, EN 14390 may be used in special cases for direct classification. Such needs might occur for products or product groups that for technical reasons cannot be tested in the SBI.

EN 13823 Single Burning Item Test (SBI)

EN 13823[11] SBI evaluates the potential contribution of a product to the development of a fire, under a fire situation simulating a single burning item in a room corner near to that product. The test is relevant for the classes A1, A2, B, C and D.

The SBI is the major test procedure for classification of linings, see Fig. 12.4 and Table 12.9. The SBI is of intermediate-scale size. Two test samples, 0.5 m × 1.5 m and 1.0 m × 1.5 m are mounted in a corner configuration where they are

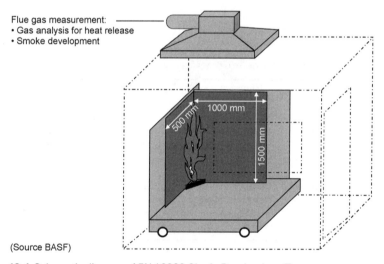

12.4 Schematic diagram of EN 13823 Single Burning Item Test.

Table 12.9 EN 13823 SBI Test specifications

Specimens	Samples for 3 tests. Each test requiring one sample of 0.5 × 1.5 m and one sample of 1.0 × 1.5 m.
Specimen position	Forms a vertical corner.
Ignition source	Gas burner of 30 kW heat output placed in corner.
Test duration	20 min.
Conclusions	Classification is based on FIGRA, THR_{600s} and maximum flame spread. Additional classification is based on SMOGRA, TSP_{600s} and droplets/particles. Details are given in the Euroclass tables.

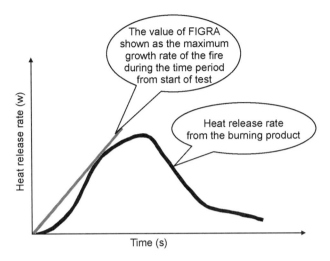

12.5 Graphical representation of the FIGRA index.

exposed to a gas flame ignition source. As for EN 14390, a direct measure of fire growth (Heat Release Rate, HRR) and light obscuring smoke (Smoke Production Rate, SPR) are principal results from a test. Other properties such as the occurrence of burning droplets/particles and maximum flame spread are observed.

The index FIGRA, i.e. FIre Growth RAte, is used to determine the Euroclass. The concept is to classify the product based on its tendency to support fire growth. Thus, FIGRA is a measure of the biggest growth rate of the fire during a SBI Test as seen from the test start. FIGRA is calculated as the maximum value of the function (heat release rate)/(elapsed test time), units are W/s. A graphical presentation is shown in Fig. 12.5.

The additional classification for smoke is based on the index SMOGRA, i.e. SMOke Growth RAte. This index is based on similar principles to FIGRA. SMOGRA is calculated as the maximum value of the function (smoke production rate)/(elapsed test time) multiplied by 10 000. The data on smoke production rate, SPR, is calculated as a 60 s running average to minimize noise. In addition, certain threshold values of SPR and integral values of SPR must first be reached before SMOGRA is calculated.

The detailed definitions of FIGRA and SMOGRA can be found in EN 13823 (SBI).

EN ISO 11925-2 Small flame test

EN ISO 11925-2[12] evaluates the ignitability of a product under exposure to a small flame. The test is relevant for the classes B, C, D, E, B_{fl}, C_{fl}, D_{fl} and E_{fl}.

The small flame test is quite similar to the DIN 4102 test used for

12.6 ISO EN 11925-2 Small flame test.

Table 12.10 ISO EN 11925-2 Test specifications

Specimens	250 mm long, 90 mm wide, thickness 60 mm.
Specimen position	Vertical.
Ignition source	Small burner. Flame inclined 45° and impinging either on the edge or the surface of the specimen.
Flame application	30 s for Euroclass B, C and D. 15 s for Euroclass E.
Conclusions	Classification is based on the time for flames to spread 150 mm and occurrence of droplets/particles. Details are given in the Euroclass tables.

determining German Class B2. Variants of this procedure are also found in other EU Member States' regulations. Test rig and specifications are shown in Fig. 12.6 and Table 12.10.

EN ISO 9239-1 Floor covering test

EN ISO 9239-1[13] evaluates the critical radiant flux below which flames no longer spread over a horizontal flooring surface. The test is relevant for the classes $A2_{fl}$, B_{fl}, C_{fl} and D_{fl}. Test apparatus and specifications are shown in Fig. 12.7 and Table 12.11.

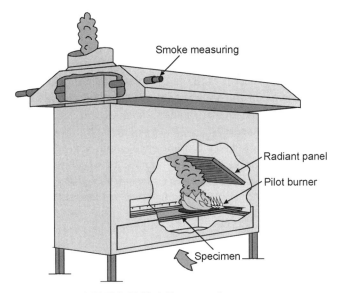

12.7 Sketch of EN ISO 9239-1 Floor covering test.

Table 12.11 EN ISO 9239-1 Test specifications

Specimens	1050 mm long × 230 mm wide.
Specimen position	Horizontal.
Ignition source	A gas fired radiant panel that gives a heat flux to the specimen. The maximum heat flux is 11 kW/m^2 that drops to 1 kW/m^2 at the end of the specimen. A pilot flame is impinging on the surface of the hot end of the specimen to initiate any flame spread.
Test duration	Until the flame extinguish or maximum 30 minutes.
Conclusions	Classification is based on the critical heat flux below which flame spread is not occurring. Additional classification is based on smoke density. Details are given in the Euroclass tables.

12.4 Fire safety requirements and tests for electrical engineering and electronics equipment

Fire regulations concerning electrical engineering and electronics (E&E) were basically drawn by professional bodies and taken over by the authorities. Classifications and fire tests are mainly international. This applies to technical parts, consumer and office electronics as well as to wire and cables. If E&E products are installed in a building or transportation, they are considered as building products or components of transportation vehicles and subjected to classification and tests mandatory in these applications.

12.4.1 Fire safety requirements

General safety requirements including fire safety are defined in European Directives like the Low Voltage (LVD)[14] and the Electromagnetic Compatibility (EMC)[15] Directives. Specific fire safety requirements and references to flammability tests are contained in national (like the Underwriters Laboratories UL in the USA) and international standards such as IEC, CENELEC and the corresponding national standards.

Plastics used in electrical and electronic applications very often have to meet fire performance requirements for materials and finished components. The main E&E applications covered by standards are consumer electronics (IEC 60065[16]), information technology (IT) in IEC 60950,[17] and appliances (IEC 60335-1[18]).

- Consumer electronics including TV-sets (UL 1410, UL 746, IEC 60065, EN 60065).
  The fire safety requirements for consumer electronics in these standards are UL 94 V1 in the USA with the UL tests, and UL 94 HB 40 and 75 in Europe with the EN tests. However, for achieving better fire safety, UL 94 V1 will also be used in Europe. External ignition will also be considered for housings and 'accessible parts' in the future with UL 94 V1 requirements.
- Office electronics including PCs, printers, copiers, fax machines (UL 1950, CSA 22.2, IEC 60950, EN 60950).
  Here UL 94 V1 has to be met worldwide, and as for consumer electronics, external ignition will be considered with the UL 94 V1 requirement.
- Appliances including household, dishwashers, washing machines, refrigerators (IEC 30335-1).
  In Europe, the glow wire tests are used to determine ignitability and flammability of materials and end products. Details of the requirements are given in the following under glow wire tests.
  - Technical parts like switches, connectors, capacitor housings (usually UL 94 V1).
  - Lighting including diffusers, housings, reflectors, lamp holders (usually UL 94 V1).
  - Printed circuit boards and laminates (usually UL 94 V1).

The flammability tests for plastics materials described in the UL 94 standards and their annexes were basically developed by Underwriters Laboratories to evaluate their performance with respect to resistance to ignition and flame propagation. They are mainly based on Bunsen/Tirrill burners and can be found in UL 94 'Tests for flammability of plastic materials for parts in devices and appliances' and its counterpart IEC 707. Depending on the fire safety requirements, materials have to meet horizontal burning tests (Class UL 94 HB) or the more stringent vertical burning tests (Class UL 94 V2, V1, V0 or 5V). These tests simulate ignition sources of low and medium intensity, which may occur in E&E equipment and impinge on plastics parts of electrical components. As can be seen in the following, these UL tests were taken over in IEC and EN test methods.

12.4.2 Fire safety standardization

The development of international standards is the task of such organizations as ISO and IEC. International electrical standards are established by the 'International Electrotechnical Commission' (IEC), which was founded more than 90 years ago.[19] The object of the IEC is to promote international cooperation on all questions of standardization and related matters in the fields of electrical and electronic engineering and thus promote international understanding.

The IEC is composed of 71 National Committees, (members and associate members) representing all the industrial countries in the world. The preparation of the Standards, in exceptional circumstance technical specifications, technical reports or guides is entrusted to technical committees. The IEC cooperates with numerous other international organizations, particularly with ISO, and also with the European Committee for Electrotechnical Standardization (CENELEC).[20]

The main IEC Technical Committees concerned with fire safety and related standards are:

- IEC/TC20 Electric Cables   IEC 60332
- IEC/TC61 Appliances   IEC 60335
- IEC/TC74 IT Equipment   IEC 60950
- IEC/TC89 Fire Hazard   IEC 60695
- IEC/TC92 Audio/Video (TVs)   IEC 60065
- IEC/TC108 IT + Audio   Founded in 2002

12.4.3 Fire safety requirements for electrical engineering and electronics materials and finished parts

The most important fire tests for E&E equipment are the flammability tests based on UL 94, taken over internationally by IEC and in Europe by EN as EN/IEC 60695-11-10 (HB, V2-V0) and EN/IEC 60695-11-20 (5VA and 5VB), the glow wire tests to EN/IEC 60695-2-10 to 13, and the needle flame test to EN/IEC 60695-11-5. They are discussed in more detail in the following. The main tests for end products and materials are as follows.

Flammability tests to IEC 60695-11-10 (also UL 94 tests HB and V)

The requirements for the flammability tests to IEC 60695-11-10,[21] also UL 94 Horizontal Burning (HB) and Vertical (V), are shown in Figs 12.8 and 12.9 and Tables 12.12 and 12.13.

Flammability test to IEC 60695-11-20 (also UL 94-5V)

The requirements for the flammability test to IEC 60695-11-20,[22] also UL 94-5V, have changed regarding the ignition source. Contrary to the past, where no

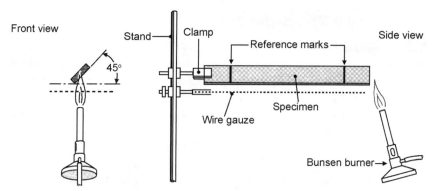

12.8 Schematic diagram of IEC 60695-11-10 (UL 94 HB): 50 W horizontal flame test.

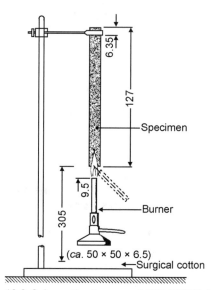

12.9 Schematic diagram of IEC 60695-11-10 (UL 94 V0, V1, V2): 50 W vertical flame test.

defined Bunsen or Tirril burner flame was used, the test is now operated with a defined flame ignition source of 500 W. The specifications are shown in Fig. 12.10 and Table 12.14.

Glow-wire tests to IEC 60695-2-10 to 13 and requirements to EN 60335-1

The glow-wire tests for end products and materials to IEC 60695-2-10 to 13 are required in Europe and increasingly in Asia, while they do not play a major role in the USA. The European standard EN 60335-1 'Household and similar electrical appliances – Safety – Part 1' contains safety requirements in Section

Table 12.12 IEC 60695-11-10 (UL 94 HB): 50 W horizontal flame test specifications

Specimen	Three bar-shaped test specimens (125 ± 5) mm × (13.0 ± 0.5) mm with two reference marks (25 ± 1) mm and (100 ± 1) mm from the end that is to be ignited. Maximum thickness 13 mm.
Specimen position	Longitudinal axis of specimen approximately horizontal, traverse axis inclined (45 ± 2) °C to horizontal.
Ignition source	Laboratory burner according to IEC 60695-11-4, (20 ± 2) mm non-luminous flame. The longitudinal axis of the burner is tilted at 45° towards the horizontal during application of flame.
Application of flame	30 s. If the flame front reaches the 25 mm on the specimen before 30 s have elapsed, the burner is removed.
Conclusion	Classification in class HB 40 if: a) there is no visible burning with a flame after removal of the flame b) the test specimen extinguishes before the 100 mm mark is reached c) the burning rate between the reference marks does not exceed 40 mm/min. Classification in class HB 75 if: the burning rate between the reference marks does not exceed 75 mm/min.

Table 12.13 IEC 60695-11-10 (UL 94 V0, V1, V2): 50 W vertical flame test specifications

Specimen	Two sets of 5 bar-shaped specimens (125 ± 5) mm × (13.0 ± 0.5) mm × max. thickness 13 mm.
Specimen position	Specimens suspended with longitudinal axis vertical. Surgical cotton below the lower edge of the specimen.
Ignition source	Laboratory burner according to IEC 60695-11-4, flames A, B or C, (20 ± 2) mm high non-luminous flame.
Application of flame	Twice (10 ± 0.5) s for each specimen. The second application starts as soon as the specimen, ignited by the first application, extinguishes
Conclusions	a) classification in class V-0 if: individual afterflame time is ≤ 10 s, total afterflame time does not exceed 50 s no ignition of the cotton indicator by flaming particles or drops samples do not burn up completely individual test specimen afterflame plus afterglow time <30 s after removal of flame. b) classification in class 94 V-1 if: afterflame time ≤ 30 s after removal of flame total afterflame time ≤ 250 s individual test specimen afterflame plus afterglow time ≤ 60 s after removal of the flame other criteria as for a). c) classification in class 94 V-2 if: ignition of surgical cotton by burning drops. Other criteria as for b).

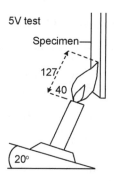

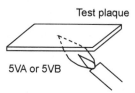

12.10 Schematic diagram of IEC 60695-11-20 (UL 94-5V): 500 W flame test.

Table 12.14 IEC 60695-11-20 (UL 94-5V): 500 W flame test specification

Specimen	Two sets of five bar-shaped specimens (125 ± 5) mm × (13.0 ± 0.5) mm × min. thickness 13 mm. or: two sets of three plates (150 ± 5) mm × (150 ± 5) mm × min. thickness normally supplied.
Specimen position	Procedure-bar test specimen: specimen suspended vertically, flame of the burner applied centrally to the lower front corner of the bar test specimen at an angle of 20° from the vertical procedure-plates: application of flame of the burner to the approximate centre of the bottom surface the horizontal plate at an angle of 20°.
Ignition source	Laboratory burner according to IEC 60695-11-3 (500 W) with axis inclined at 20° to vertical; flame length (125 ± 10) mm with (40 ± 2) mm blue inner cone.
Application of flame	5 times (5 ± 0.5) s with (5 ± 0.5) s intervals between:
Conclusions	a) burning category 5 VA: individual afterflame time is ≤ 60 s after the fifth flame application no ignition of the cotton indicator by flaming particles or drops. b) burning category 5 VB: the flame did penetrate through (burn through) any of the plates, other criteria as for 5 VA materials classified either 5 VA or 5 VB shall also conform to the criteria for materials classified either V-0, V-1 or V-2 in the same bar test specimen thickness.

30 'Resistance against fire and heat'. It refers to the glow wire tests described in IEC 606952-10 to 13:

- **IEC 60695-2-10** Ed. 1 (2000 10) Glow wire
  Apparatus and common test procedure
- **IEC 60695-2-11** Ed. 1 (2000 10) Glow wire
  Flammability test for end products **(GWT)**
- **IEC 60695-2-12** Ed. 1 (2000 10) Glow wire
  Flammability test for materials **(GWFI)**
- **IEC 60695-2-13** Ed. 1 (2000 10) Glow wire
  Ignitability test for materials **(GWIT)**

For non-attended appliances of >0.2 A, the requirements to EN 60335-1, Section 30 are:

- **IEC 60695-2-11** Flammability test for end products **(GWT)**
  750 °C $\leq$ 2 s
  if > **2 s Needle flame test** to IEC 60695-11-5 **or** Class **V0 or V1** to IEC 60695-11-10
- **IEC 60695-2-12** Flammability test for materials **(GWFI)**
  850 °C < 30 s
- **IEC 60695-2-13** Ignitability test for materials **(GWIT)**
  775 °C < 5 s

The glow wire test apparatus and the test for end products are shown in Figs 12.11 and 12.2.

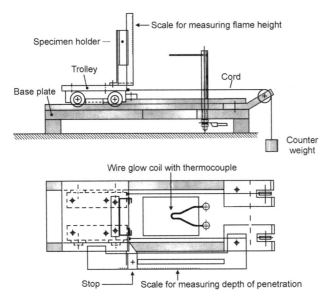

12.11 Schematic diagram of glow wire test apparatus.

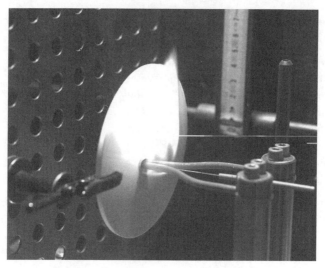

12.12 Ignitability test for materials.

Needle-flame test to IEC 60695-11-5

The needle-flame test specified in IEC 60695-11-5[23] is applicable to complete electrotechnical equipment, its sub-assemblies and components and to solid electrical insulating materials or other combustible materials. The test ensures that the test flame (which consists of a tube, for example prepared from a hypodermic needle) does not cause ignition of parts, or that a combustible part ignited by the test flame has a limited extent of burning (maximum 30 s after removal of the flame) without spreading fire by flames or burning or glowing particles falling from the specimen (Fig. 12.13).

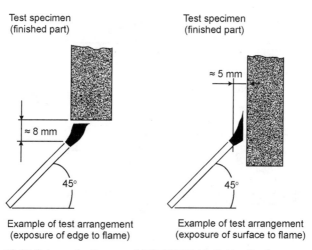

12.13 Schematic diagram of IEC 60695-11-5: Needle-flame test.

12.4.4 Certification

The Certified Body Scheme

Companies wishing to enter the global market have to overcome certain hurdles. The Certified Body (CB) Scheme[24] has been established by IEC and is based on the principle of mutual recognition by means of certification through the use of internationally accepted standards. By virtually eliminating duplicate testing, the CB-Scheme facilitates the international exchange and acceptance of product safety test results. The International Electrotechnical Committee for Conformity Testing to Standards for Electrical Equipment (IECEE) is a global network of National Certification Bodies (NCBs) that has agreed to mutual acceptance of CB test certificates and reports. The Scheme is based on the use of international (IEC) Standards. If some national standards of a member country adhering to the CB-Scheme are not yet completely harmonized to IEC-Standards, national differences are permitted, if clearly declared to all other members. Currently there are more than 35 member countries in the IECEE.

The CB-Scheme applies to many specific product categories established by the IECEE, for example IEC 60065: Audio, video and similar electronic apparatus – safety requirements, or IEC 60335: Safety for household and similar electrical appliances.

If a product is already Underwriters' Laboratories (UL)-Listed, Classified or Recognized or has a test Mark, like a VDE-Mark, the required testing for the CB-Scheme is mostly already completed and accepted. In some cases, additional testing is necessary to comply with national deviations required in countries to which the products are to be marketed.

Underwriters' Laboratories Certifications

Three types of UL[23] certifications exist: Listing, Classification and Recognition. Successful completion of a *Listing* Certification allows a finished product to carry the UL listing mark, which is similar to the German VDE test mark. In contrast, *Classification* covers specific characteristics of the finished product designated by the manufacturer. It results in a UL mark incorporating wording for a specific hazard, or if evaluated to a non-UL-standard, e.g. ASTM, OSHA, etc., *Recognition* is a pre-selection certification for materials and components, and results backwards in a UL mark. The UL categories of materials are very important in connection with the use of plastics in electrotechnical products, because specific ratings are often a pre-requirement of the Listing of a product by UL. By itself, neither a part nor a material can gain a UL mark, but the testing of a certified product is simplified if materials, which are already UL-Recognized, are used.

On behalf of UL, field representatives and authorized representatives pay periodical and unannounced visits to the manufacturers' facilities for inspecting and monitoring the manufacturers' production and verifying if the product is still in compliance with the UL-Report requirements. The field representative randomly selects production samples for Follow-Up testing at UL. The UL product directories help to find products that meet UL requirements. They contain the names of companies that manufacture products, components, devices, materials or systems in accordance with UL safety requirements.

These are products eligible to carry UL's Listing Mark, Recognized Component Mark or Classification Marking. The *Plastics Recognized Component Directory* contains the category of polymeric materials components *QMFZ2*. Recognition cards, also known as '*Yellow Cards*', are an excerpt of the Recognized Component Directory. They give the manufacturers the documentary evidence that his product meets the relevant UL requirements.

12.4.5 CE marking

CE marking is a declaration from the manufacturer that the product conforms to a specific Directive adopted in the EEA (European Economic Area). It further symbolizes the fact that the product has been subjected to an appropriate conformity assessment procedure contained in the directive. Directives on electrotechnical products with reference to fire safety are the Low Voltage, Electromagnetic Compatibility, and Medical Devices Directives.

12.4.6 German VDE approval procedures

Electrical safety requirements in Germany are laid down in the VDE (Association for Electrical, Electronic and Information Technologies),[26] which is a non-profit association for electrical science and technology. As a neutral and independent institution, the VDE Testing and Certification Institute carries out testing of electrotechnical products, components and systems and awards the VDE test mark, which is recognized worldwide. The VDE Institute is accredited on a national and international level. The VDE Mark indicates conformity with the VDE standards or European or internationally harmonized standards, respectively, and confirms compliance with protective requirements of the applicable EC Directive. The VDE Mark is a symbol for testing based on the assessment of electrical, mechanical, thermal, toxic, fire and other hazards.

12.4.7 New developments: external ignition sources

In E&E, the introduction of standards and requirements concerning fire hazards from external ignition sources is under development for consumer and office electronics.

IEC/TC 108 'Safety of electronic equipment within the field of audio/video, information technology and communication technology' was set up in 2002 and is a merger of the Technical Committees IEC/TC 74 (IT) and 92 (TV). In IEC/TC 108, the HBSDT (Hazard Based Standard Development Team) started to work and is currently developing the new IEC Standard 62368 'Audio/Video, Information and Communication Technology Equipment – Safety Requirements' and the Technical Specification IEC/TS 62441 'Accidentally caused candle flame ignition'.

IEC 62368 is **not** a merger of the IEC standards 60065 and 60950. The standard will implement a Hazard Based Concept in developing five single fragments:

- F 1: Principles, Scope, Definitions, Common Requirements
- F 2: Fire Hazard Requirements
- F 3: Electric Shock Hazard Requirements
- F 4: Mechanical, Chemical, Radiation Hazard Requirements
- F 5: Annexes

The Technical Specification IEC/TS 62441 requires the following criteria regarding 'resistance against external ignition':

- An individual item, if accidentally subjected to a candle flame, shall not release enough heat to ignite other items.
- An individual item having a candle flame accessible area (Fig. 12.14) is considered to comply is it fulfills external ignition requirements.
- The general requirement is UL 94 V1 or better rated materials for outer housings.
- Exemptions are:
  – The total mass of the outer housing in the accessible area does not exceed 300 g.
  – Combustible materials pass the Needle Flame Test with 3 min flame application, no sustained flaming (max. 30 s persistence of flames).
  – The mass of individual parts in the accessible area does not exceed 25 g and 300 g in total.

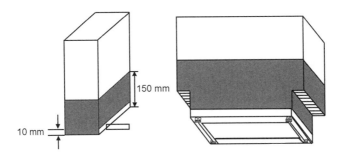

12.14 Sketch of accessible area to IEC/TS 62441.

12.5 Fire safety requirements and tests in transportation for rail vehicles and ships

In transportation, the fire safety requirements and tests are becoming more and more international. While motor vehicle and aircraft requirements basically originate from the USA, those for ships were internationally developed by the International Maritime Organisation IMO. Requirements for railways are still national in Europe, but will be harmonized shortly in the European Union. Fire safety requirements are prescribed for many applications and apply, of course, also for plastics. In many cases, the latter have to be flame retarded. Owing to very high new fire safety requirements, particularly in railways, in many cases, thermoplastics and elastomers, even flame retarded, may no longer be used in the future. Thermosets, however, will basically not be affected.

12.5.1 Rail vehicles

Technical legal provisions for rail vehicles vary from country to country. Generally the type of vehicle or the operation mode is taken into account. In addition, operators frequently stipulate that various specifications laid down in conditions for delivery must be met.

The most important national fire protection requirements and tests for rail vehicles in Europe come from the UK, France and Germany. For the UK, fire tests are defined in the BS 6853 Guidelines on rolling stock relating to fire. Various fire tests including a smoke test in a fire room are used. In France, all fire tests in use for railways are described in the standard NF F16-101. Besides various fire tests, smoke density is determined and the toxicity of pyrolytic gases analytically evaluated. The English BS 6853 and French NF F16-101 tests are required in the Channel Tunnel specifications for rolling stock. In Germany, DIN 5510 is dealing with fire precautions in railways. Part 2 describes classification, requirements and test methods for fire behaviour and fire side effects of materials and parts. Various fire tests are used depending on the materials and parts to be tested. Smoke development and dripping are also considered. Toxicity requirements are referred to in an Annex 1 to DIN 5510-2 published in 2007.

Fire safety of rail vehicles in the European Union: EU directives and standardization

Fire safety requirements are part of the European Directive on the interoperability of the trans-European high-speed rail system. A joint working group (JWG) compiled from members of CEN/TC 256 und CENELEC 9X has been working in the formulation of a seven part standard EN 45545 'Railway applications – Fire protection on railway vehicles' since 1991. In 1998, the EU Commission and industry financed the Firestarr Project to develop suitable fire tests and a classification system for materials and components in EN 45545-2. The project was completed in 2001 and the results taken over into the EN standard. However, there have been difficulties with the progress of the seven parts of the new standard. In order to speed up the standardisation procedure, it was decided to first publish the standard as a Technical Specification (foreseen for 2008), before releasing the final EN 45545 in 2010. The status of 2008 is shown in the following:

CEN/TS 45545 Railway applications – Fire protection on railway vehicles

	Status
Part 1 General	Rejected, revision
Part 2 Requirements for fire behaviour of materials and components	Under enquiry
Part 3 Fire resistance requirements for fire barriers and partitions	Rejected, revision
Part 4 Fire safety requirements for rolling stock design	Formal voting
Part 5 Fire safety requirements for electrical equipment	Adopted as TS
Part 6 Fire control and management systems	Formal voting
Part 7 Fire safety requirements for flammable liquid and flammable gas installations	Under enquiry

In Part 1, the objectives of the standard are given as minimization of the risks of a fire starting by accident, arson or by technical defects and allowing for the safety of staff and passengers if the first objectives fail. Four different design categories as well as four operational categories are defined. Fire hazard levels resulting from different dwell times for passengers and staff are related to combinations of the operation and design categories.

Part 2 describes requirements for the fire behaviour of materials and components. The main tests used are shown in Table 12.15.

In CEN/TS45545-2, heat fluxes of 25 and 50 kW/m^2 for materials > 0.25 m^2 or > 100 g are proposed for:

- flame propagation in radiant panel to ISO 5658-2[27] (Fig. 12.15)
- heat release in cone calorimeter to ISO 5660-1[28] (Fig. 12.16)
- smoke/toxicity in single chamber smoke box to ISO 5659-2[29] (Fig. 12.17).

12.5.2 Ships

The International Maritime Organization (IMO) develops international regulations and standards to improve the safety of sea vessels. In 1996, IMO developed the Fire Test Procedures Code (FTP Code), which contains fire testing methods for flammability, smoke and toxicity to meet fire safety

322 Advances in fire retardant materials

Table 12.15 Fire test methods in CEN/TS 45545

Test method	Reference
Reaction-to-fire tests – spread of flame Part 2: Lateral spread on building products in vertical configuration	ISO 5658-2
Reaction-to-fire tests – Horizontal surface spread of flame on floor-covering systems Part 1: Flame spread using a radiant ignition source	EN ISO 9239-1
Reaction to fire tests - Ignitability of building products subjected to direct impingement of flame Part 2: Single flame source test	EN ISO 11925-2
Fire tests – Reaction-to-fire Part 1: Heat release (cone calorimeter method)	ISO 5660-1
Plastics – Smoke Generation Part 2: Determination of optical density by a single chamber test	ISO 5659-2
Determination of burning behaviour by oxygen index Part 2: Ambient temperature test Furniture Calorimeter vandalised	EN ISO 4589-2 EN 45545-2 Annex C
Gas analysis in the smoke box ISO 5659-2, using FTIR technique	EN 45545-2 Annex D
Gas analysis for the 8 gases based on the French quartz tube test	NF X 70-100

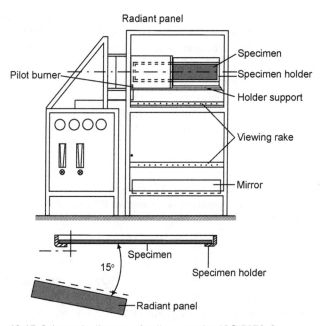

12.15 Schematic diagram of radiant panel to ISO 5658-2.

New and potential flammability regulations 323

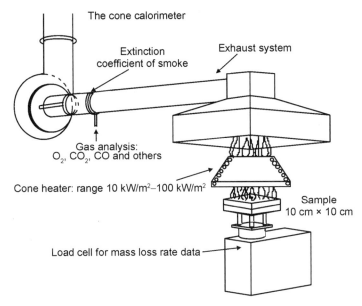

12.16 Schematic diagram of cone calorimeter, ISO 5660-1.

requirements for materials and components used on ships.[30] They are summarized in Table 12.16.

The fire tests for determining the fire performance of materials and components used in ships are basically ISO tests also used elsewhere (building and transportation).

Since 2006, the FTP Code has been under revision, and will be completed in 2008. The objective is to update the existing tests in the FTP Code and to add new fire test methods. It is foreseen to extend Part 2 Smoke and toxicity testing

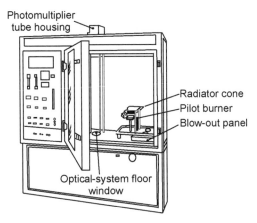

12.17 Sketch of smoke box, ISO 5659-2.

Table 12.16 Tests to the FTP Code

FTP code	Type of test	Referred test method	Similar test method
Part 1	Non-combustibility Test	ISO 1182:1990	–
Part 2	Smoke and Toxicity Test	ISO 5659-2	–
Part 3	Fire Resistance Test for Fire Resistant Divisions	IMO A.754(18)	ISO 834-1
Part 4	Fire Resistance Test for Fire Door Closing Mechanisms	–	–
Part 5	Surface Flammability Test	IMO A.653(16) IMO A.687(17)	ISO 5658-2
Part 6	Test for Primary Deck Coverings	IMO A.653(16)	ISO 5658-2
Part 7	Flammability Tests for Curtains and Vertically Suspended Textiles and Films	IMO A.471(XII) IMO A.563(14)	ISO 6940/41
Part 8	Test for Upholstered Furniture	IMO A.652(16)	BS 5852-1/-2
Part 9	Test for Bedding Components	IMO A.688(17)	EN 597-1/-2

to ISO 5659-2 by ISO 21489 for FTIR gas measurement and to use the Fractional Effective Dose Scenario described in ISO 13344.

In addition, it is planned to extend the spread of flame test to ISO 5658-2 by the heat release cone calorimeter test to ISO 5660-1 and the ISO 9705 full-scale room fire test for high speed craft.

12.6 Fire safety requirements and tests in furniture

Furniture and textiles may be subjected to building regulations, if they form part of the building, or to transportation rules if used there. If not an integral part of a building, upholstered furniture and textiles are subjected to specific fire safety regulations and tests if they are used in high risk areas like hospitals, prisons, hotels, homes for the aged, etc. In some countries (USA, California; UK) specific fire regulations and tests exist for upholstered furniture used in the private domain. In Europe, furniture flammability is controlled in the public building sector whether by regulation or by purchasing specification.

12.6.1 United Kingdom

Only the UK and Ireland have introduced regulations, which are aimed at the protection of the individual consumer, in that regulations are in force, which govern the type of private dwellings. Following a series of fires in dwellings which resulted in multiple deaths due to speed of fire growth, pressure was brought to introduce some level of control to limit the rate at which the upholstered furniture was allowed to burn.

As a result, under the Consumer Safety Act, the UK Furniture and

New and potential flammability regulations 325

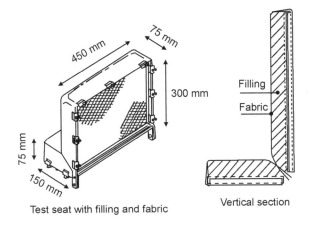

12.18 Sketch of BS 5852: Simulated chair.

Furnishings (Fire) (Safety) Regulations 1988, Statutory Instrument No. 1324, was published and amended in 1993.[31] This regulation essentially limits the mass of a small, simulated chair (Fig. 12.18), which is allowed to be lost due to combustion when exposed to defined fire sources (cigarette, simulated match and wood cribs). The test procedures are described in BS 5852 Parts 1[32] and 2[33] and have in the meantime been taken over as European standards EN 1021 Parts 1[34] and 2[35] (cigarette and match tests).

The BS 5852 tests ignition sources are:

- BS 5852 Part 1 Cigarette test
- BS 5852 Part 2 Simulated match (145 mm gas flame)
- BS 5852 Part 2 describes a further gas flame test (240 mm flame) and 4 wood crib tests (4 to 7 with 8.5, 17, 60, 126 g) (Fig. 12.19).

12.19 Woodcrib No. 5.

12.6.2 France

Furniture in French public buildings has to meet ERP Article AM 18 which, since 2006, requires a finished product testing (mock up for seat with filling and cover) to EN 1021 and no more the former classifications M2 for the covering and M4 for the filling. The seat frame, however, still has to meet M3. The ignition source is a simulated 20 g cushion with a smaller modified Belfagor burner and improved test criteria (i.e. mass loss 300 g instead of burnt surface distance) described in the new French standard NF D60-013.

The French GPEM (Groupement Permanent d'Etude des Marchés, which is a permanent group for studying markets) has issued an upholstered furniture and bedding guide for public procurement requiring tests with cigarette, match simulation and 250 mm flame. The cigarette test will be evaluated with a predictive grid for filling and covers described in the new French standard NF D60-015.

12.6.3 USA

In the USA, a flammability standard for mattresses (DOC FF 4-72) with a cigarette as ignition source is mandatory. California has additional regulations regarding the flammability of upholstered furniture, which were adopted by several other US states. The Consumer Product Safety Commission (CPSC) is pushing a flammability standard for upholstered furniture foreseen to be mandatory nationwide. In addition, legislation is under discussion that would make flame retarded furniture mandatory. At the end of 2007, however, CPSC adopted a 'proposed rulemaking' on furniture fire safety only requiring a cigarette test.

On 16 February 2006, CPSC adopted a federal regulation for strict new mattress flammability requirements, including fire safety against an open flame. Under the new mandatory federal rule, mattress sets must meet a performance standard. The CPSC does not specify how manufacturers are to design their mattresses to meet the standard. The new federal standard 16 CFR Part 1633[36] for mattresses went into effect on 1 July 2007. In the performance standard, the peak heat release rate is limited to 200 kW during a 30-minute test. The total heat release is limited to 15 MJ within the first 10 minutes of the test. Cigarette ignition is covered by a separate mandatory standard. That standard, 16 CFR Part 1632,[37] has been in place for more than 30 years during which deaths and injuries from mattress fires caused by smoking materials have fallen dramatically.

12.7 Future trends

12.7.1 Building

In the European Union, substantial progress has occurred in removing trade barriers by EU directives and CE-marking. In building, the Construction

Products Directive (CPD) was transposed into national laws in the beginning of the 1990s. However, the CPD will be revised, as the current version is unclear and has to be simplified. One objective is its transformation from a directive into a regulation. This would make it directly compulsory without the time consuming national implementation of the directive into the respective national regulations. Industry complains that the national authorities often do not accept the test reports from laboratories from other Member States. The reason is that the EU Euroclasses and standards may not always fit into the existing national regulations. National test methods may therefore remain valid in parallel for a not precisely defined period of time. This would not only require European testing, but, in addition, existing national fire testing to be met. This point will also be considered in the revised CPD.

The harmonization and publication of reaction to fire classifications and test methods has been completed in 2002 and taken over into national building regulations. The Euroclasses and the harmonized fire test methods basically differ from national systems. New voluntary classification criteria, which can be taken over into compulsory national regulations, are smoke, burning drips, and acidity for cables. In 2007, the revision of the reaction-to-fire test methods has started in order to refine the standards. The test procedures will not be changed, as this could impact on the already classified construction products and question the whole classification system.

12.7.2 Electrical engineering and electronics equipment

General safety requirements including fire safety are defined in European Directives such as the Low Voltage (LVD) and the Electromagnetic Compatibility (EMC) Directives. Specific fire safety requirements and references to flammability tests are contained in national (like Underwriters Laboratories UL in the USA) and international standards like IEC, CENELEC and the corresponding national standards.

The most important *flammability tests for E&E equipment* are the flammability tests based on UL 94, taken over internationally by IEC and in Europe by EN as IEC/EN 60695-11-10 (HB, V2-V0) and IEC/EN 60695-11-20 (5VA and 5VB), the glow wire tests to IEC/EN 60695-2-10 to 13 and the needle flame test to IEC/EN 60695-11-5.

The ongoing *problems with the glow wire*, particularly with the GWIT, in testing plastics and flame retarded plastics grades are under review. The test procedure and test conditions are currently being optimized:

- The flame after ignition has been defined (corona phenomenon are no more considered as a flame).
- Conditions to improve reproducibility and repeatability are addressed in round robins.

Certification of E&E equipment is provided by various systems like the CB scheme, the UL certifications, and approval procedures in Europe and Germany.

Since August 2003, the *CCC* (China Compulsory Certification) Mark is required for a wide range of manufactured electrical and non-electrical products before being exported to or sold in the Peoples Republic of China market.

External ignition requirements are foreseen to be introduced for TV/IT equipment worldwide and envisaged for appliances in Europe, where studies on the behaviour of appliances towards external ignition have already started.

12.7.3 Transportation

In transportation, the fire safety requirements and tests are becoming more and more international. While motor vehicle and aircraft requirements basically originate from the USA, those for ships were internationally developed by the International Maritime Organization (IMO). Requirements for railways are still national in Europe, but will be harmonized shortly and introduced in the European Union. Fire safety requirements are prescribed for many applications and apply, of course, also for plastics. In many cases, the latter have to be flame retarded. Owing to the new very high fire safety requirements, particularly in railways, in many cases, thermoplastics and elastomers, even if flame retarded, may no longer be used in the future. Thermosets, however, will basically not be affected.

In the European Union, a seven-part standard: EN 45545 'Railway applications – Fire protection on railway vehicles' has been under development since 1991, and there are still difficulties with the progress of the various parts of the new standard. In order to speed up the standardization procedure, it was decided to first publish the standard as a Technical Specification (foreseen for 2008), before releasing the final EN 45545 in 2010.

In CEN/TS 45545-2, heat fluxes of 25 and 50 kW/m^2 have been proposed for materials $>0.25\,m^2$ or >100 g. It has to be seen whether these very stringent requirements will virtually ban the use of products which have been shown over many years to fulfil high fire safety requirements. Smaller components mainly used in E&E will not be affected by these requirements and still allow the use of all flame retarded systems currently commercially used in rolling stock.

The IMO develops international regulations and standards to improve the safety of sea vessels. In 1996, IMO has developed the Fire Test Procedures Code (FTP Code), which contains flammability, smoke and toxicity fire testing methods to meet fire safety requirements for materials and components used on ships.

The revision of the FTP Code started in 2006 with the target of being completed in 2008. The objective is to update the existing tests in the FTP Code and to add new fire test methods. It is foreseen to extend Part 2 Smoke and toxicity testing to ISO 5659-2 by ISO 21489 for FTIR gas measurement and to use the Fractional Effective Dose Scenario described in ISO 13344. In addition, it is planned to extend the spread of flame test to ISO 5658-2 by the ISO 5660-1 cone calorimeter and the ISO 9705 full-scale room fire tests for high speed craft.

12.7.4 Furniture

If not an integral part of a building, upholstered furniture and textiles are subjected to specific fire safety regulations and tests if they are used in high risk areas. In some countries (USA, California; UK) specific fire regulations and tests exist for upholstered furniture used in the private domain. In the USA, a federal regulation for the fire safety of mattresses in private homes has become effective in July 2007. A proposed rulemaking for flame retarded upholstered furniture has been adopted since the end of 2007.

12.8 Sources of further information and advice

J. Troitzsch, *Plastics Flammability Handbook*, 3rd edn, Carl Hanser Publishers, Munich, 2004.

12.9 References

1. M. Mitzlaff, J. Troitzsch, 'Regulations and testing', p. 222, in *Plastics Flammability Handbook*, J. Troitzsch (ed.), 3rd edn, Carl Hanser Publishers, Munich, 2004.
2. http://europa.eu/scadplus/treaties/singleact_en.htm.
3. http://ec.europa.eu/enterprise/construction/internal/cpd/cpd.htm.
4. http://ec.europa.eu/enterprise/construction/internal/safire.htm.
5. http://ec.europa.eu/enterprise/construction/internal/guidpap/k.htm.
6. EN 13501-1:2007 Fire classification of construction products and building elements. Classification using data from reaction to fire tests.
7. EN 13238:2001 Reaction to fire tests for building products. Conditioning procedures and general rules for selection of substrates.
8. EN ISO 1716:2002 Reaction to fire tests for building products. Determination of the heat of combustion.
9. EN ISO 1182:2002 Reaction to fire tests for building products. Non-combustibility test.
10. EN 14390:2007 Fire test – Large-scale room reference test for surface products.
11. EN 13823:2002 Reaction to fire tests for building products. Building products excluding floorings exposed to the thermal attack by a single burning item.
12. EN ISO 11925-2:2002 Reaction to fire tests. Ignitability of building products subjected to direct impingement of flame. Single-flame source test.
13. EN ISO 9239-1:2002 Reaction to fire tests. Horizontal surface spread of flame on floor-covering systems. Determination of the burning behaviour using a radiant heat source.
14. http://ec.europa.eu/enterprise/electr_equipment/lv/index.htm.
15. http://ec.europa.eu/enterprise/electr_equipment/emc/index.htm.
16. IEC 60065 Edition 7.1 (2005-12) Audio, video and similar electronic apparatus – Safety requirements.

17. IEC 60950-1 Edition 2.0 (2005-12) – Information technology equipment – Safety – Part 1: General requirements.
18. IEC 60335-1 Consolidated Edition 4.2 (incl. am1+am2) (2006–09) Household and similar electrical appliances – Safety – Part 1: General requirements.
19. http://www.iec.ch.
20. http://www.cenelec.org/Cenelec/Homepage.htm.
21. IEC 60695-11-10 Consol. Ed. 1.1 (incl. am1) (2003-08) Fire hazard testing - Part 11-10: Test flames – 50 W horizontal and vertical flame test methods.
22. IEC 60695-11-20 Consol. Ed. 1.1 (incl. am1) (2003-08) Fire hazard testing - Part 11-20: Test flames – 500 W flame test methods.
23. IEC 60695-11-5 Ed. 1.0 (2004-12) Fire hazard testing – Part 11-5: Test flames – Needle-flame test method – Apparatus, confirmatory test arrangement and guidance.
24. http://www.iecee.org/cbscheme/default.htm.
25. http://www.ul.com/.
26. http://www.vde.com/vde_en/.
27. ISO 5658-2 (2006-09) Reaction to fire tests – Spread of flame – Part 2: Lateral spread on building and transport products in vertical configuration.
28. ISO 5660-1 (2002-12) Reaction-to-fire tests – Heat release, smoke production and mass loss rate – Part 1: Heat release rate (cone calorimeter method).
29. ISO 5659-2 (2006-12) Plastics – Smoke generation – Part 2: Determination of optical density by a single-chamber test.
30. http://www.imo.org/Safety/mainframe.asp?topic_id=777.
31. http://www.opsi.gov.uk/SI/si1988/Uksi_19881324_en_1.htm.
32. BS 5852-1:1979 Fire tests for furniture. Methods of test for the ignitability by smokers' materials of upholstered composites for seating.
33. BS 5852-2:1982 Fire tests for furniture. Methods of test for the ignitability of upholstered composites for seating by flaming sources.
34. EN 1021-1:2006 Furniture. Assessment of the ignitability of upholstered furniture. Ignition source smouldering cigarette.
35. EN 1021-2:2006 Furniture. Assessment of the ignitability of upholstered furniture. Ignition source match flame equivalent.
36. http://www.cpsc.gov/businfo/frnotices/fr06/mattsets.pdf.
37. http://www.access.gpo.gov/nara/cfr/waisidx_00/16cfr1632_00.html.

13
Life cycle assessment of consumer products with a focus on fire performance

M SIMONSON and P ANDERSSON,
SP Technical Research Institute of Sweden, Sweden

Abstract: Fire-LCA was invented in response to the need for a more holistic LCA of products containing flame retardants. This chapter describes how to conduct a Fire-LCA, illustrated using three product-based case studies: TV sets, furniture and cables. In the TV-set study a US TV was compared with a European TV, the furniture study compared a European sofa with a UK sofa, while two cables with similar fire performance were compared in the cable study. Finally future trends in holistic assessment of modern products are outlined, including a newly developed cost-benefit methodology.

Key words: fire, Fire-LCA, consumer products, fire statistics, flame retardant, environmental impact.

13.1 Introduction

Environmental issues are a vital part of our society and the ability to perform accurate estimates and evaluations of environmental parameters is an important tool in any work to improve the environment. Initially, environmental studies were mainly focussed on the various emissions sources, such as factory chimneys, exhaust gases from vehicles, effluents from factories, etc. However, in the 1980s it became apparent that a simple measurement of an emission did not provide a full picture of the environmental impact of a specific product or process. The emissions from a chimney, for example, only reflect one of several process steps in the production of a specific product. To fully describe the environmental impact of a product or activity, the entire process chain has to be described including raw material extraction, transport, energy and electric power production, production of the actual product, the waste handling of the product, etc. There was, therefore, an obvious need for a new methodology and an analytical tool able to encompass this new situation. The tool that was developed during this period (end of 1980s and 1990s) was Life Cycle Assessment (LCA).

However, the Life Cycle Assessment methodology also needs continuous improvement to incorporate new aspects and processes. An LCA typically describes a process during normal operation, and abnormal conditions such as accidents are left out of the analysis, usually due to lack of a consistent methodology or relevant data. For example, LCA data for power production usually assume normal conditions without any accidents. Provisions for certain accidents in the analysis of the life-cycle could be included, provided these could be specified in sufficient detail and occurred with sufficient regularity to make their inclusion relevant.

In traditional LCA models a higher fire performance is only included as a change in energy and material consumption and no account is taken of the positive effect of higher fire performance in the form of fewer and smaller fires. The emissions from fires contribute to the environmental impact from products and should be included in a more complete evaluation of the environmental impact of a product where the fire performance is an important parameter. In cases where the fire performance is not a critical product performance characteristic (e.g., underground piping) one need not include this in the product LCA.

This chapter describes a methodology for the incorporation of fires into a Life Cycle Assessment. Fires occur often enough for statistics to be developed, providing necessary information on material flows in the model. A model has been specifically developed to allow for this inclusion and will be referred to as the Fire-LCA model. The model itself is generally applicable, provided appropriate additions and changes are made whenever a new case is studied. The Fire-LCA method was originally developed by SP and IVL[1,2] and they have since applied the model to three different case studies: TV sets,[3-6] Cables,[7,8] and Furniture.[9-11] Furthermore, a set of Guidelines has been developed for the Fire-LCA model.[12] This chapter will describe both the methodology and present results from the case studies that have been developed to date.

13.2 Life Cycle Assessment

Life Cycle Assessment (LCA) is a versatile tool to investigate the environmental impact of a product, a process or an activity by identifying and quantifying energy and material flows for the system. The use of a product or a process involves much more than just the production of the product or use of the process. Every single industrial activity is actually a complex network of activities that involves many different parts of society. Therefore, the need for a system perspective rather than a single object perspective has become vital in environmental research. It is no longer enough to consider just a single step in the production. International standards for LCA methodology have been prepared by the International Organisation for Standardisation (ISO) which provide guidance on how to conduct an LCA.[13-16]

Generally the method can be divided into three basic steps with the methodology for the first two steps relatively well established, while the third step (Impact Assessment) is more difficult and controversial. The first two steps are usually referred to as the life cycle inventory (LCI) and can be applied separately without the following impact assessment. In addition to the different steps in the procedure there can also be an interpretation phase. The three basic steps are shown in Fig. 13.1.

Life cycle assessment of consumer products

Life Cycle Assessment framework

```
┌─────────────────┐         ┌─────────────────┐
│  Goal and scope │────────▶│                 │
│   definition    │         │                 │
└────────┬────────┘         │                 │
         ▲│                 │                 │
         │▼                 │                 │
┌─────────────────┐         │                 │
│    Inventory    │────────▶│  Interpretation │
│    analysis     │◀────────│                 │
└────────┬────────┘         │                 │
         ▲│                 │                 │
         │▼                 │                 │
┌─────────────────┐         │                 │
│     Impact      │────────▶│                 │
│   Assessment    │◀────────│                 │
└─────────────────┘         └─────────────────┘
```

13.1 The main phases of an LCA according to the ISO standard.[13]

The *goal and scope definition* consists of defining the study purpose, its scope, system boundaries, establishing the *functional unit*, and establishing a strategy for data collection and quality assurance of the study. Any product or service needs to be represented as a system in the inventory analysis methodology. A system is defined as a collection of materially and energetically connected processes (e.g., fuel extraction processes, manufacturing processes or transport processes) which perform some defined function. The system is separated from its surroundings by a *system boundary*. The entire region outside the boundary is known as the *system environment*.

The *functional unit* is the measure of performance, which the system delivers. The functional unit describes the main function(s) of the system(s) and is thus a relevant and well-defined measure of the system. The functional unit has to be clearly defined, measurable, and relevant to the input and output data. Examples of functional units are 'unit surface area covered by paint for a defined period of time', 'the packaging used to deliver a given volume of beverage', or 'the amount of detergents necessary for a standard household wash'. It is important that the functional unit contains measures for the efficiency of the product, durability or lifetime of the product and the quality/performance of the product. In comparative studies, it is essential that the systems are compared on the basis of equivalent functional unit.

In the *inventory analysis* the material and energy flows are quantified. The system consists of several processes or activities, e.g. crude material extraction, transport, production and waste handling. The different processes in the system are then quantified in terms of energy use, resource use, emissions, etc. Each sub-process has its own performance unit with several in- and outflows. The processes are then linked together to form the system to analyse. The final result of the model is the sum of all in- and out-flows calculated per functional unit for the entire system.

The most difficult and also the most controversial part of an LCA is the *Impact Assessment*. So far, no standard procedure exists for the implementation of an entire impact assessment. However, the ISO standard covers the so-called Life Cycle Impact Assessment (LCIA),[15] where different impact categories are used and recommendations for Life Cycle Interpretation.[16] Transparency of the LCA model is, however, important and inventory data must also be available in addition to aggregated data. Several methods/tools have been developed for impact assessment and the tools can usually be integrated with different LCA computer softwares. The modern tools today usually include a classification and characterisation step where the different parameters, e.g. emissions, are aggregated to different environmental classes such as acidification, climate change or eutrophication. There are of course also possibilities for direct evaluation/ interpretation of the different emissions or environmental classes.

13.2.1 The risk assessment approach

In a conventional LCA the risk factors for accidental spills are excluded. In the LCA data for the production of a chemical, for example, only factors during normal operation are considered. However, there can also be, for example, emissions during a catastrophic event such as an accident in the factory. Those emissions are very difficult to estimate due to a lack of statistical data and lack of emission data during accidents. The same type of discussion exists for electric power production in nuclear power plants.

In the case of the evaluation of normal household fires, the fire process can be treated as a commonly occurring activity in society. The frequency of fire occurrences is relatively high (i.e., high enough for statistical treatment) and statistics can be found in most countries. This implies that it is possible to calculate the different environmental effects of a fire if emission factors are available. Statistical fire models can be set up for other types of fires but the uncertainty in the statistical fire model will increase as the statistical data is more limited.

The fundamental function of better fire performance is to prevent a fire from occurring or to slow down fire development. Improving a product's fire performance will thus change the occurrence of fires and the fire behaviour. By evaluating the fire statistics available with and without different types of fire performance improvements the environmental effects can be calculated. The benefits of a higher fire performance must be weighed against the 'price' society has to pay for the production and handling of possible additives and/or other modes of production. The LCA methodology will be used to evaluate the application of higher fire performance in society. In this way a system perspective is applied.

13.2.2 The Fire-LCA system description

Schematically the LCA model proposed for a Fire-LCA can be illustrated as in Fig. 13.2. The model is essentially equivalent to a traditional LCA approach

Life cycle assessment of consumer products   335

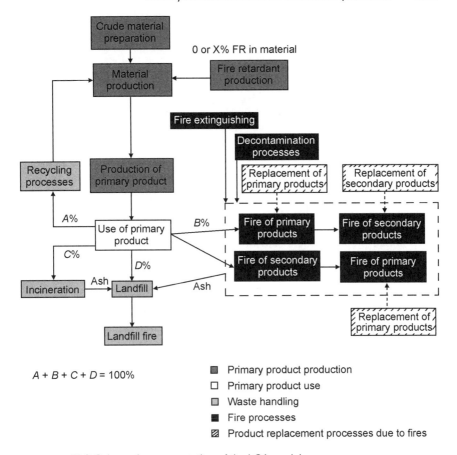

$A + B + C + D = 100\%$

- ■ Primary product production
- ☐ Primary product use
- ▦ Waste handling
- ■ Fire processes
- ▨ Product replacement processes due to fires

13.2 Schematic representation of the LCA model.

with the inclusion of emissions from fires being the only real modification. In this model a functional unit is characterised from the cradle to the grave with an effort made to incorporate the emissions associated with all phases in the unit's life cycle. Thus, the model includes production of material for the product to be analysed, as well as the production of the additives if applicable. If possible the model should be designed in such a way that the fire performance can be varied. Furthermore, the model should include production, use and waste handling of the product during its lifetime.

During the lifetime of the products to be analysed, some products will be involved in different types of fires. The Fire-LCA model will therefore include modules to describe the fire behaviour for the different types of fires. Fire statistics are used to quantify the amount of material involved in the different types of fires. In addition, the model should also include modules for handling the production of replacement materials that are needed due to the shortening of lifetime that the fires have caused. If possible the model should also include modules for the handling of the fire extinguishing process and the decontamination process.

Statistical fire model

A wealth of statistics is available concerning fires from a variety of sources (such as national fire brigades and insurance companies). Differences between countries and between different sources of data in the same country provide information concerning the frequency of fires and their size and cause.

The number of products that are involved in the different types of fires constitutes the statistical fire model to be used in a Fire-LCA model. The fire model should preferably be based on fire statistics but could also, if there are no statistics available, be based on some hypothesis and perhaps comparison to other similar products where statistics are available.

A *primary fire* is one where the functional unit is the ignition source (either by internal or external ignition). In this case both the functional unit and all other material that burns are included in primary fires. A *secondary fire* is where something other than the functional unit is the ignition source. In this case, it is anticipated that only emissions associated with the functional unit will be included.

In those cases where the functional unit is not the source of the fire we must assume that the functional unit will burn independently of whether it is flame retarded or not.

Both in the case of *primary fires* and *secondary fires* it is reasonable to assume that the functional unit and other products will be damaged and need to be replaced but the LCA should not necessarily be burdened with the cost of their replacement to the same degree. In the case of a *primary fire* the LCA will be burdened with the replacement both of the functional unit and other products that are included in the fire. In the case of *secondary fires* the LCA will be burdened with replacement of the functional unit only.

The fire in both the functional unit and in other products results in a shortening of the lifetime of these products. Assuming a normal distribution of the age for all products in a society, we can assume that the allocation of the environmental burden for replacement of the burnt products is 50% of the new production burden.

The system boundaries are important when defining what to include in the fire model. Fire extinguishing and decontamination processes may be included when relevant. A wide variety of extinguishment methods are available. The method employed to extinguish a fire depends largely on the size of the fire and the type of material involved. In the case of household fires it is possible that the occupant may extinguish a small fire without the aid of the fire brigade. In cases where the fire brigade is called to a fire, transport and deployment will be included as realistically as possible. Generally the mode of extinguishment is well defined in this case depending on the type of fire. Where possible and relevant the effect of the extinguishing medium should be included.

As in the case of extinguishment, the necessity of decontamination of a fire scene is largely dependent on the size of the fire and the material involved. In the case of household fires decontamination of the fire scene is seldom necessary in the sense of cleaning soil or ground water. Should an entire dwelling be burned down then an attempt should be made to include the effect of transport and deposition of any debris remaining after the fire. In the case of an industrial fire, however, decontamination will often be necessary. In principle transport and deposition of debris should also be included as realistically as possible even in the case of industrial fires.

Use of fire statistics

Fire statistics are compiled in most developed countries. These statistics vary in terms of details and in some cases definitions of what constitutes different types of fires varies. In general one can say that fire statistics from fire brigades tend to contain somewhat larger fires with consequences outside of the ignition source. In other words the fire brigade is seldom called to a very small fire. Statistics from insurance companies, however, cover a much broader range of fires. This can be explained since a consumer would be apt to report a small fire to an insurance company in order to make a claim on their insurance policy. Comparison of these statistics provides information on the total number of fires and also on the distribution according the size of the fire and how much material was actually burned.

Three different case studies using the Fire-LCA method have been conducted to date. These were concerned with television sets, cables and furniture and will now be described in detail.

13.3   Television case study

In the TV case study, a TV complying with the European regulations was compared to a TV complying with the regulations in US. The US TV contained high fire performance material (UL94 V0) in the outer enclosure while the European TV contained material that was easy to ignite and spread a fire readily (UL94 HB) in its outer enclosure. The high performance material was flame retarded high impact polystyrene (HIPS) with decabromodiphenyl ether (deca-BDE) while the low fire performance material was untreated HIPS.

Two scenarios were studied, one present day and one future scenario. In the present day scenario 1% of the TV sets was assumed to go to incineration, 2% to disassembly (for recycling) and the rest to landfill (with the exception of those involved in fires). In the future scenario 1% went to incineration, 89% to disassembly and the rest to landfill. The 'present day' end-of-life scenario was based on information from the major Swedish disassembly company while the future scenario was based on a possible future waste treatment system, designed for the year 2010, where no more than 10% would be allowed to go to landfill. It should be emphasised that this scenario is speculative. All TV enclosures from disassembly were assumed to go to incineration and all incineration was assumed to be run with energy recovery.

13.3.1 Statistical fire model

Based on the results of a previous investigation[17,18] the number of TV fires in Europe (normalised per million TV sets each year) was determined as: 100 TV fires due to internal ignition; 65 TV fires due to external ignition; and, 160 TV fires where the TV enclosure is not breached in the incident. This makes a total of 165 TV fires/million TVs where the TV enclosure is breached and 160 TV fires/million TVs where the TV enclosure is not breached, for a TV set with non-flame retarded enclosure material. A division of the TV fires where the enclosure is breached according to the probable size of the fire was made based on German statistics and is summarised in Table 13.1.

An investigation of US statistics revealed that TV fires are quite uncommon in the US. A total of 5 TV fires/million TVs occur each year in the US. These fires were classified as 'minor' in the LCA model as the US TV was found through experimentation to be extremely difficult to ignite. US statistics for TV fires when the enclosure is not breached had not been forthcoming and so the model assumed that the number of such fires is essentially the same in Europe and the US. It was estimated, both in Europe and the US that approximately one-third of these 160 TV fires/million TVs (where the enclosure is not breached) will be replaced while the remaining two-thirds will be repaired.

The full TV statistics model used as input to the Fire-LCA is summarised in Table 13.2. Products involved in a fire will generally need to be replaced in the Fire-LCA. The environmental impact of replacing the TV is included at only 50% of the full cost. This is to allow for the fact that the replacement products

Table 13.1 Division of European TV fires based on severity

Severity	Frequency %	#TVs in model	LCA category
Fire restricted to TV	35	58	Minor*
Fire spread beyond TV	53	88	Full TV
Major damage to room	5	8	Full room
Major damage to dwelling	5	8	Full house
Building destroyed	2	3	Full house
TOTAL	100	165	

*The 160 TV fires/million TVs each year in Europe where the enclosure is not breached are also classified as 'minor'.

Life cycle assessment of consumer products

Table 13.2 Fire-LCA input for European TV (HB enclosure) and US TV (V0 enclosure) where the TV enclosure is assumed breached

European TV (fires/million TV sets)	US TV (fires/million TV sets)
58 minor 88 TV only 8 full room 11 full house	5 minor

need only be stand-in for the destroyed product less that a full life cycle. The choice of 50% weighting is based on the assumption that a TV involved in a fire has on average only lived half of its full life cycle.

The fire emissions data for the 'TV only' and 'full room' categories were obtained through full-scale experiments. In these experiments a large variety of species were measured including HCl, HBr, HCN, NO_x, SO_2, CO, CO_2, PAH, dibenzodioxins and furans, TBBP A, deca-BDE, and PCB.[3]

13.3.2 Results and discussion

Large numbers of results are available due to the large number of input parameters present in the model and complex flows through the entire life cycle. It is therefore difficult to study them all. Instead one has to focus on certain key parameters and emissions, a few of these are presented here as examples.

The energy requirements throughout the whole life-cycle proved to come from many parts of the model with the majority of the energy requirement coming from the 'USE' module, see Fig. 13.3. This corresponds to the energy requirements for running the TV during its functional lifetime. The mixture of energy in the 'USE' module corresponds to the standard OECD mixture for the

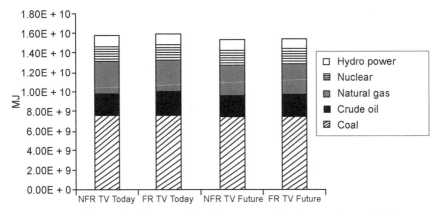

13.3 Comparison between the different scenarios shows only small differences in the energy use.

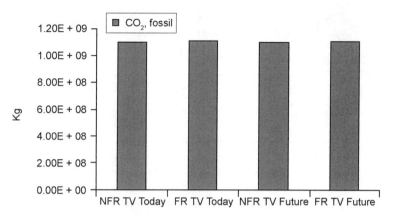

13.4 Comparison of CO_2 emission for 106 TV sets over their 10-year life cycles.

production of electricity. This is based on data available for 1997. The slight reduction in energy requirement in the future scenario is due to the provision of energy by incineration of waste and energy recovery.

Figure 13.4 shows that there is little difference in the total emissions of CO_2 from the various scenarios, with and without flame retardants. This is a reflection of the fact that fires represent a small part of the total emissions of carbon dioxide (CO_2) in the TV life cycle. The majority of the CO_2 emissions come from the production of energy during the TV USE and PRODUCTION modules.

In contrast, a significant decrease was seen in the polyaromatic hydrocarbon (PAH) emissions for the flame retarded (FR) TV relative to the non-flame retarded (NFR) TV (see Fig. 13.5), for both the present day and future scenarios. This is a direct effect of the fact that the NFR TV is involved in a greater number of fires.

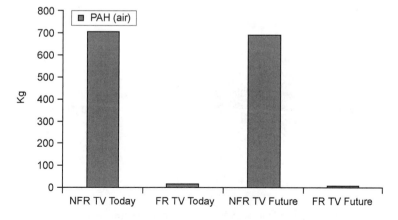

13.5 The emission of PAH to the air for 106 TV sets over their 10-year life cycle.

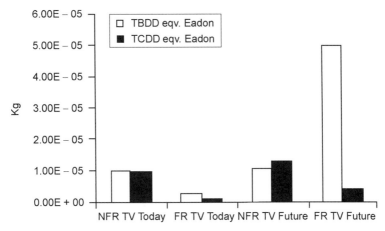

13.6 The emission of TCDD and TBDD equivalents to the air for 106 TV sets over their 10-year life cycle.

Similarly, a significant decrease was seen in the 2,3,7,8-tetrachlorodibenzo(p)dioxin-equivalents (TCDD-equivalents) for the FR TV relative to the NFR TV (see Fig. 13.6), for both the present day and future scenarios. The TCDD-equivalents were based on EADON factors as the majority of other input information from other sources throughout the life cycle cites EADON values. EADON refers to toxic equivalent factors (TEFs) according to Eadon *et al.* (1986).[19] Calculations based on the most recent WHO factors[20] gave only marginal differences with no change in the qualitative nature of the results.

There is presently no accepted international method of defining tetrabromodibenzo(p)dioxin-equivalents (TBDD-equivalents). For the sake of simplicity it was assumed that the same factors can be used to define the toxicity of the various members of the PBDD-equivalents as those used for the chlorinated variety. As in the case of TCDD-equivalents, a significant decrease was seen in the TBDD-equivalents for the FR TV relative to the NFR TV (see Fig. 13.6), for the present day scenario. This changed to an increase in the future scenario.

Again the result for the present day scenario is a direct effect of the fact that the NFR TV is involved in a greater number of fires than the flame retardant TV. The fact that the tetrabromodibenzo(p)dioxin-equivalent (TBDD-equivalent) emissions increase for the flame retardant TV in the future scenario is a direct result of the allocation of a small amount of TBDD-equivalent emissions to all flame retardant TV enclosures sent to incineration and energy recovery. The TBDD-equivalents are allocated according to bromine content with the assumption that the dioxin emission limit proposed in the waste directive, i.e., 0.1 ng TCDD eq./Nm^3, is met. The amount of TBDD-equivalent produced is then calculated in direct relation to the amount of bromine input to the energy recovery facility in the form of TV enclosures. In the future scenario the flame retardant TV enclosures are, therefore, allocated a large portion TBDD-equivalents from the energy recovery module.

One should note that the production of polyaromatic hydrocarbon is many times higher than the production of all types of dibenzodioxins and furans. A model is available for the comparison of different pollutants based on the assignment of 'Unit Risk Factors'.[21] Using this 'unit risk' model one can compare the risk that a person exposed to the same quantity of different substances over their lifetime would develop cancer. The application of the model requires that the PAH emissions be reduced to a single toxicity equivalence factor in essentially the same manner as that for dibenzodioxins and furans. In the case of PAH the species benzo(a)pyrene (BaP) has been defined as the most toxic species and is assigned a toxic equivalence factor of 1. All other species are then assigned toxic equivalence factors relative to BaP, allowing the calculation of BaP-equivalents.[22]

It was not possible to do this for the majority of PAH emissions, as information concerning the specific species emissions hidden behind the general term 'PAH' was missing in all cases except the fire emissions. A reduction of the amounts of the various PAH emitted from the fire experiments indicated that the BaP-equivalent of the fire gases was approximately 3% of all the PAH emissions. Thus, the BaP-equivalents for the LCA models were estimated as 3% of the total PAH production.

Based on the toxicity analysis presented in Table 13.3 it is clear that polyaromatic hydrocarbon emissions from the whole life cycle of one million TVs represents a far greater risk than that of dibenzodioxins and furans. This risk is significantly higher in the case of non-flame retarded TVs relative to flame retarded TVs, both in the present day scenario and the future scenario. This effect is largely due to the increased number of fires in the non-flame retarded TV models relative to the flame retarded TV scenarios. This confirms recently presented results concerning polyaromatic hydrocarbon and dibenzodioxin emissions from large fires in Germany.[23]

Table 13.3 Cancer risk of PAH relative to TCDD-equiv. for each scenario

Species	Unit risk factor (URF)	NFR today	FR today	NFR future	FR future
BaP-equiv. (kg)	$7 \times 10^{-2}\,\mu g/m^3$	1.46	3.05×10^{-2}	1.44	7.60×10^{-3}
TCDD-equiv. (kg)	$1.4\,\mu g/m^3$	1.36×10^{-5}	1.01×10^{-6}	1.82×10^{-5}	5.71×10^{-6}
Cancer risk factor*		108 000	30 200	79 200	1300

*(BaP-equiv. $\times$ URF$_{BaP}$)/(TCDD-equiv. $\times$ URF$_{TCDD}$), the values have been rounded to the nearest 100.

13.3.3 Conclusions

The conclusions from the TV-case study were:

- Results from a comparison between the LCA of a TV with a V_0-rated enclosure material to that with an HB-rated enclosure material indicate that in the case of a number of key emission species there is a markedly higher total emission over the whole life cycle from the non-flame retarded TV than from the flame retarded TV in the present day scenario, although the picture becomes more complicated in the future scenario.
- The emissions with the most marked difference between in flame retarded and non-flame retarded TV sets reflect those species that can be minimised from all controlled combustion sources. These include PAH and TXDD-equivalents (where 'X' can be bromine or chlorine). The emission of PAH dominates over the emission of TXDD-equivalents. Thus, the PAH emissions represent a far greater cancer potential that the TXDD-equivalent emissions.
- Fires are not, however, the dominant source of those species, such as CO, CO_2, NO_x, etc., that are given off in abundance from controlled combustion. In this case, the difference in total emission due to fires is marginal. This has been predicted previously[24] and is a confirmation of the fact that the emission of organic compounds such as PAH and dibenzodioxins and furans is perhaps the major environmental problem associated with fires.
- While a full analysis of the uncertainty of the data was not conducted, the relative uncertainties between the models should be low. Data sources were essentially equivalent in all cases. Further, greater emphasis was placed on obtaining detailed information for parts of the model that were found through a sensitivity analysis to be critical to the results.

Finally, in a risk analysis it is important to consider all potential risks. In this context it should be emphasized that according to European statistics at least 16 people die each year and 197 are injured as a direct result of TV fires. This number is a conservative estimate according to a study conducted by the UK Department of Trade and Industry.[25] A true estimate may be up to 10 times this number. In the US, however, there is no record of people dying as a result of a TV fire. From these results it is clear that there is a societal benefit from the use of flame retardants. Therefore, the risk to human lives of a fire should not be neglected in an overall assessment of the risk of using flame retardants.

13.4 Cable case study

The cables case study compared two indoor electrical cables with essentially the same fire performance. The two cables chosen in the comparison were a flexible PVC cable and a flame retarded polyolefin cable denoted CASICO. In both cases the material could be easily ignited and spread a flame readily in the IEC

60332-3 test (a large-scale test of cable fire performance). Thus the same fire model was used for both cables and the LCA became essentially a comparison of different material choices. In this context the end-of-life (EOL) scenarios were deemed to be very important to the outcome of the model. Four different EOL scenarios were selected for detailed study:

- Scenario 1: 100% landfill plastics and copper
- Scenario 2: 100% landfill plastics, 100% material recycling copper
- Scenario 3: 100% energy recovery plastics, 100% material recycling copper
- Scenario 4: 100% material recycling plastics, 100% material recycling copper

These scenarios were selected to focus on extreme situations. In real life one would expect a percentage of material to go to recycling in some form, which would be less than 100% but more than 0%.

In addition the lifetimes of the cables were varied. Based on literature information,[26,27] average lifetimes of 30 and 50 years were selected. Choosing a 50-year lifetime was found to have a marginal effect on the absolute results.

Finally an uncertainty analysis was performed in this case study which proved that the model is robust.

13.4.1 Statistical fire model

Primary fires are defined as those where the cable is the first item ignited. In Table 13.4, the primary fire categories are defined by the modules labelled: 'Cables Fire', 'Cable/Room Fire', and 'Cable/House Fire'. Secondary fires are defined as all fires where cables are involved but where cables are not the first item ignited. Using available European statistics, with a focus on UK and Swedish statistics due to their availability and detail, an approximate fire model was defined for use in the Cables Fire-LCA model.

In the primary fire model it was assumed that there are a total of 35 fires/million dwellings. Each dwelling contains on average 250 m of cable, which is assumed to correspond to approximately 50 m per room. This corresponds to 140 fires per million km cable.

Using statistics from the UK Home Office and from SRV Sweden concerning distribution of the size of the fire, the numbers of fires that are confined to the cable only, those that spread beyond the cable but are contained in the room of

Table 13.4 Summary of primary fires

Start object, 'Cables Fire'	Start room, 'Cable/room Fire'	Beyond start room, 'Cable/house fire'
60% 84 fires/million km cable	31% 43 fires/million km cable	9% 13 fires/million km cable

origin (so called 'Cable-room' category) and those that cause significant damage to the dwelling (so called 'Cable-house' fires) was estimated. The results of this division are summarised in Table 13.4. It is assumed that 50% of all the cables in the room in the 'Cables only' category are involved in the fire. All results for the primary fires represent average data based on statistics from the Netherlands,[28] the UK Home Office,[29] the Swedish Rescue Services Agency[30-33] and the Swedish National Electrical Safety Board.[34]

Based on UK Home Office statistics compiled by the University of Surrey[35] it was estimated that approximately 1400 serious fires occur per million dwellings. Assuming 250 m cable per dwelling this corresponds to 5600 fires per million km cable. Assuming that 10% of all the cables in each dwelling are consumed corresponds to 140 km cable emitting fire gases and needing to be replaced each year. This was used as a 'realistic estimate' of the contribution of secondary fires and was compared to a worst case scenario where 100% of all cables in the fires are burned. This corresponds to 1400 km cable emitting fire gases and needing to be replaced each year.

13.4.2 Results and discussion

Some results of the cable case study are presented below as examples of the information that can be obtained from the Fire-LCA analysis. Please note that in all figures 'Cable production' includes production of replacement cables while 'Repl. Prod.' includes production of all replacement material other than cables. In all cases, the categories shown in the bar charts have been selected such that they are visible on the diagram. In cases where the emission is too low to be visible, the category has not been included.

The energy use of the different scenarios decreases for each scenario as can be seen in Fig. 13.7, with the highest energy consumption for Scenario 1 and the lowest consumption for Scenario 4. The energy consumption of the CASICO cable is higher than the energy consumption of the PVC cable for Scenarios 1, 2 and 4. The majority of energy is used during the production of the cable in Scenario 1 through 3 for both the cables. When polymer recycling is undertaken in Scenario 4, the recycling stage is the most energy consuming stage. Recycling is also important in Scenarios 2 and 3, but polymer recycling in Scenario 4 becomes the dominant energy consumer in the life cycle of the cable.

The energy gain for PVC in Scenarios 1 and 2 comes from the production of CH_4 in landfills, where the CH_4 is collected and can be used as an energy source. The energy gain for the CASICO material in Scenarios 1 and 2 is, due to a small production of CH_4, much smaller than the gain from PVC. In Scenario 3, however, the energy gain is larger for the CASICO cable than for the PVC cable, due to the higher amount of energy recovered through incineration, as the energy content of the CASICO material is higher than that of the PVC material.

The total production of hydrocarbons, HCs, is dominated by emissions from

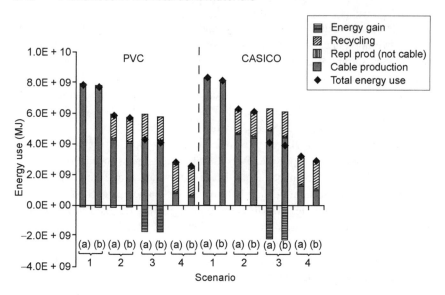

13.7 Use of energy resources for PVC and CASICO cables, 30-year life cycle. The division is based on the part of the model the energy is required or produced. The results are shown for (a) 100% secondary fires and (b) 10% secondary fires for each scenario.

the production of the cables as seen in Fig. 13.8. Approximately 70% of the total production of HCs from PVC in Scenarios 1 through 3 comes from the cable production, with 40% of the total cable production coming from the production of the PVC polymer. Owing to the importance of the production of PVC, the production of HCs for Scenarios 1, 2 and 3 are similar. In Scenario 4, however, the production is decreased by about 60%, due to the recycling of the PVC plastics.

The production of HCs from the CASICO material is approximately twice the total emissions of HCs from the PVC cable. The cable production is very important in this case, too, being responsible for about 85% of the total amount produced. Specifically, the production of Ethylene Butyl Acrylate copolymer (EBA) is responsible for about 80% of the total production of HCs. As seen in Fig. 13.8, the production of hydrocarbons (HCs) from Scenario 4 is similar for both CASICO and PVC cables. This is due to the significance of the material recycling in the PVC life-cycle in the cable production stage.

The importance of the production of HCs from fires in Scenarios 1 through 3 differs between the two materials. In Scenarios 1 through 3, fires are responsible for about 10% of the total amount of HCs produced by the PVC cable, and for about 5% of the total amount of HCs produced by the CASICO cable. In Scenario 4, however, the importance is much larger (approximately 26% for both the materials). If the emissions of HCs from fires and from the production of replacement products (not cable) in Scenario 4 are added together, fires are

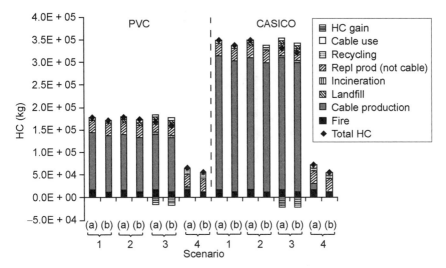

13.8 Hydrocarbon (HC) emission to air for PVC and CASICO cables, 30-year life cycle. The bars are coded to show the amount from the various parts of the model. The results are shown for (a) 100% secondary fires and (b) 10% secondary fires for each scenario.

responsible for about 60–70% of the total emissions of HCs. In the other scenarios this production, i.e. fire + production of replacements, corresponds to about 25% of the production from the PVC cable and about 13% of the production from the CASICO cable. The production of HCs from fires, however, is equally large in all of the scenarios although it differs between the two systems. The importance of fires increases as the production of HCs from other sources decreases.

Fire is the largest producer of dioxins for both of the materials as shown in Fig. 13.9. This is partly due to fires being one of the few areas in the model where detailed information concerning the emission of dioxins is available. This is probably a sound reflection of the most important sources of dioxins, however, as most other sources are bound to be minor (emission from energy production and energy recovery modules are included). Dioxins from the fire experiments with the CASICO material were due to a small amount of chlorine present in the plastic material. The chlorine was present in such a small quantity that it probably is a contaminant in, for example, the filler material. Its presence has, however, been confirmed through elemental analysis of the cable material.

For both of the materials the largest production of dioxin is from Scenario 3, due to the production from the incineration process. The increased production of dioxins from the CASICO material in Scenario 3 is, however, almost negligible being only 0.6% of the total production in Scenario 1, 2 and 4. In Scenarios 1, 2 and 4, fires are the only source of dioxins for the CASICO material, since no dioxins are emitted during the production of the cables. The total production of

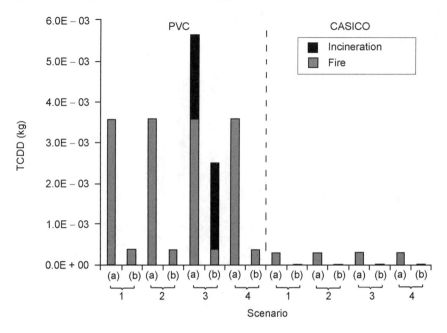

13.9 2,3,7,8-tetrachlorodibenzo(p)dioxin-equivalent (TCDD-equivalent) emission to air for PVC and CASICO cables, 30-year life cycle. The bars are coded to show the amount from the various parts of the model. The results are shown for (a) 100% secondary fires and (b) 10% secondary fires for each scenario.

dioxins from the CASICO material is approximately 1/10 of the dioxins produced from the PVC material, except for Scenario 3, where the total production is only 6% of the total production from the PVC material. Very small amounts of dioxins, in fact near 0.0% of the total production, are emitted during the production of the PVC material. This production decreases in Scenario 4, due to decreased production of PVC plastics. Because of the large production from fires, however, except for Scenario 3 the difference in the total production of dioxins between the scenarios is negligible. The only producer of dioxins for the CASICO material in Scenario 1, 2 and 4 is fire, thus no difference can be seen between these scenarios. In Scenario 3, the increase of emissions of 2,3,7,8-tetrachlorodibenzo(p)dioxin-equivalents (TCDD-equivalents) due to the incineration is very small, resulting in a total emission of TCDDs that is almost equal with the emissions from the other scenarios that have been examined. Thus, no significant difference can be seen between the scenarios studied with the CASICO cable.

13.4.3 Conclusions

The conclusions drawn from the cable case study were:

- For most species studied, one sees a decrease in emissions for each scenario from Scenario 1 to Scenario 4 with the exception of Scenario 3. In Scenario 3 one assumes energy recycling of the plastic material, which leads to increased emissions from the incineration part of the model. The effect of this is most marked for those species with very low emissions from other sources such as TCDD-equivalents and HCl. In these cases the treatment of allocation of TCDD and HCl emissions from incineration to the chlorine content in the waste leads to a significant increase of these emissions in Scenario 3. Similarly, a significant increase is seen in the emissions of CO and CO_2 as these species are produced in large quantities from incineration.
- When considering the gases that have been discussed above TCDD-equivalents, HCl, CO and HC were found to be influenced significantly by the incorporation of fires in the LCA model.

The uncertainty analysis showed that the model is stable and the species presented in detail are not affected to any great degree by changes in key parameters. This implies that the model is robust and the conclusions sound.

The results of the comparison indicate that the Fire-LCA model is able to include the impact of fire emissions on the overall life cycle of two products with equivalent fire performance and not just products of differing fire performance as was the case with the TV set case study. It is, however, noted that the model is perhaps most useful when comparing products with differing fire performance.

13.5 Furniture case study

The furniture case study compared two different high fire performance UK sofas with a low fire performance mainland European sofa. The high fire performance sofas contained a phosphorous-based flame retardant in the foam and two different flame retardants in the material used on the sofa decorative cover (one Br-containing and one without Br). The mainland European sofa did not contain flame retardant treated material or foam.

13.5.1 Statistical fire model

The statistics presently available in the UK are based on a sofa population that consists of both pre-1988 and post-1988 furniture. To define the effect of the presence of ignition resistant sofas in the 1999 statistics, two scenarios were considered. In the first case the model assumed that a sofa has a lifetime that is exponentially distributed with the expectation value of 10 years which results in a population with 63% of the sofas containing combustion-modified materials in 1999. In the second case a mean lifetime of 15 years was assumed, which results in a population with 49% of the sofas containing combustion modified material in 1999. In addition it was assumed that each household has two sofas and the sofas are placed in the living room.

According to the DTLR statistics,[36] in the late 1990s, on average 5500 fires per year spread beyond the room of origin but were confined to the building. Since 1989 there is an increasing trend for 'confined to item' and a decreasing trend for 'confined to room'. In 1989, 32 500 fires were *confined to room* (50%) and 26 300 *confined to item* (41%). In 1999, 29 200 fires were *confined to room* (41%) while 34 800 *confined to item* (50%). During this time period there has been a change in the way fires are reported, which makes it more difficult to interpret the data. This change in reporting data is to some extent taken into account by not including the fires reported as 'No fire damage' when calculating the percentages. Assuming that the change in *confined to item* and *confined to room* depends solely on the increasing number of flame retardant (FR) treated sofas results in a model where, if all sofas were flame retarded, 36% of the fires would be *confined to room* and 55% *confined to item*. Assuming that the change in *confined to item* depends solely on the change in the Fire Regulations is probably not correct but provides a starting point for the model.

In 1999 around 10 fires started in combustion modified upholstery and 500 in other upholstery.[36] The number of fires per year is then calculated using:

$$\# fires_{sofaA}/year = \frac{\# fires_{sofaA}}{\chi_{sofaA} \cdot X \cdot B}$$

where $\#fires_{sofaA}/year$ is the total number of fires in a particular type of sofa (i.e., combustion modified or otherwise) each year, χ_{sofaA} is the percentage of fires in this type of sofa each year, X is the number of sofas per household (i.e. 2) and B is the total number of households (assumed to be 23.9 million in the UK in 1999[37]).

Assuming that all 10 fires that started in combustion modified upholstery were fires in sofas. These would result in 0.33 fires per million flame retarded sofas and 28.3 fires per million non-flame retardant treated sofas. The DTI studies[38] indicate that the number of fires in flame retarded furniture may actually be higher based on an extrapolation of the data post-1988 that has not been included in this model.

According to statistics from the University of Surrey[39] the number of fires starting in the living room in UK was constant until 1986 (= 0.5 fires/1000 households) when the number suddenly decreased to 0.45 fires/1000 households in 1987 and has continued to decrease slightly since then. In 1999 the number of fires starting in the living room in UK was 8600. Subtracting the number of fires starting in sofas, i.e. 500 + 10, results in 8090 fires. Of these, 41% were *confined to room* but beyond the starting item itself assuming that *confined to room* is independent of the starting item, i.e. 3317 fires, which is equivalent to 69 fires/million sofas. The 5500 fires that were *confined to building*, resulted in 115 fires per million sofas.

Table 13.5 Number of fires assuming a 10-year half-life and that the change in 'confined to item' and 'confined to room' depends solely on flame retardant treatment of the sofas

	UK, fires/million sofas	Europe, fires/million sofas
Primary fires		
Small fires	215 fires/million sofas	187 fires/million sofas
Fires starting in sofa	0.33 fires/million sofas	28 fires/million sofas
Fires confined to sofa	0.55 × 0.33 = 0.18	0.41 × 28 = 12
Fires starting in sofa confined to room	0.36 × 0.33 = 0.12	0.5 × 28 = 14
Fires starting in sofa confined to building	0.09 × 0.33 = 0.030	0.09 × 28.3 = 2.5
Secondary fires		
Fires confined to living room not starting in sofa	69	69
Fires confined to building	115	115

The results for the LCA model for UK and mainland European fires in sofas, assuming that 'confined to item/room' is independent of room and starting item, are summarised in Table 13.5.

The number of sofas that are replaced when the fire is too small to be reported to the fire brigade is not available in the DTLR statistics. Swedish[30–33,40] and UK[41] statistics indicate that only about 13% of all fires are reported in the fire statistics. Assuming that the same figure applies to fires in sofas, then it can be estimated that the number of fires not reported to the fire brigade is 2.2 (i.e., 0.33/0.13 − 0.33) in the UK and 187 (i.e. 28/0.13 − 28) in the EU per million sofas each year. In this model it has been assumed that the same number of sofas is ignited independent of the presence of flame retardants but that a more limited number result in a fire that is reported to the fire brigade in the UK. Thus 215 fires (i.e. 28/0.13 − 0.33) in the UK and 187 fires in the EU are confined to the sofa of origin and result in the replacement of the sofa but do not have fire emissions included as LCA input.

In the LCA model it is assumed that on average half the mass of the sofas are consumed during a primary sofa fire, while 90% of the mass is consumed in the secondary fires. All the sofas that take part in fires are assumed to be involved in a fire after 50% of their lifetime (on average). This also means that 50% of the sofas that have been involved in a fire must be replaced with new sofas.

13.5.2 Results and discussion

A selection of the results achieved by using the 10-year lifetime fire model on the non-flame retardant and the two flame retardant treated sofas are presented in Figs 13.10–13.12. The chlorinated and brominated dioxins and furans are

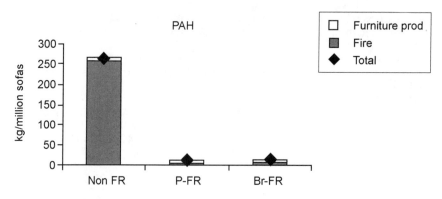

13.10 PAH for the different modules and cases of the LCA model.

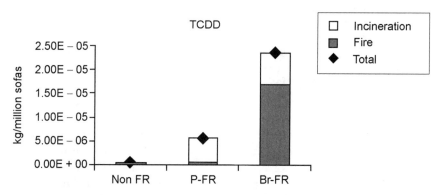

13.11 The TCDD-equivalent for the different modules and cases of the LCA model.

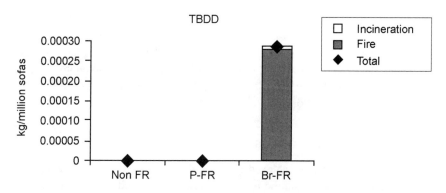

13.12 The TBDD-equivalent for the different modules of the LCA model.

presented as tetrachlorodibenzo(p)dioxin-equivalents (TCDD-equivalents) and tetrabromodibenzo(p)dioxin-equivalents (TBDD-equivalents). The TCDD-equivalents are calculated according to International Toxicity Equivalent ITEQ using the EADON method defined previously. The TBDD-equivalent is calculated based on contributions from both dibenzodioxins and furans and are based on the TCDD factors for the chlorinated equivalent.

The polycyclic aromatic hydrocarbons (PAH) emissions are much higher for the non-flame retardant treated case as can be seen in Fig. 13.10, and this is due to the higher number of fires that the non-flame retardant treated sofas cause. For the PAH emissions, the emission from the fires and furniture production are of the same order of magnitude for the flame retardant treated cases, while the emissions from the fires dominates the emissions in the non-FR treated case.

The TCDD- and TBDD- equivalent emissions are higher from the flame retardant treated sofas than from the non-flame retardant treated sofa as can be seen in Figs 13.11 and 13.12. Most of the TCDD-equivalent is due to the incineration for the case with a phosphorus-based flame retardant while the emissions from the fires contribute more for the case with a bromine-based flame retardant. Incineration causes some TBDD-equivalent emissions in the case with a bromine-based flame retardant, the major emissions originate, however, from the fires.

The emission level of dioxins from the sofas is low compared to the background level. The TCDD-equivalent emissions for the bromine-flame retardant case considered here are 23.4 mg per million sofas during their life cycle of 10 years. This corresponds to 2.3 mg per million sofas per year. This should be compared with the amount of chlorinated dibenzodioxins from all sources in UK which was 3870 g in 1989.[42] With 23.9 million households and two sofas per household the emissions from the sofas amount to 112 mg, which equals 0.003% of the total emissions.

One should note that all results are specific to the materials in these sofas and would change if the materials were changed. For example, an alternative choice of filling material or sofa construction would have repercussions for the sofa production module and potentially for the fire emissions. Further, definition of scope and boundary conditions for the model would impact on the final results.

When evaluating the two choices, non-flame retardant versus flame retardant treated one, the entire range of emissions including toxic environmental pollutants should be taken into account. There is presently no evaluation method for LCA analyses available that covers the whole range of species monitored in this study, i.e. PAH, dioxins, furans, etc. Therefore, the same comparison as in the TV case study was conducted, i.e a comparison between PAH and TCDD- and TBDD-equivalents using the Unit Risk Factor (URF) model. Similar to the TV case study the benzo(a) pyrene (BaP) equivalent was set to 3% of the PAH emissions.

Based on Table 13.6 it is clear that the PAH emissions from the whole life cycle of one million sofas represent a considerably greater cancer risk than that

Table 13.6 Cancer risk due to benzo(a)pyrene equivalents (BaP equivalents) and tetrachlorodibenzo(p)dioxin-equivalents (TCDD-equivalents)/fire emissions for the 10-year model. No units are included as the comparison is relative rather than absolute

Species	Non-FR	P-FR	Br-FR
BaP-equivalent	8.1	0.35	0.45
TCDD-equivalent	5.5×10^{-7}	55.7×10^{-7}	234×10^{-7}
Cancer risk BaP[1]	0.57	0.024	0.031
Cancer risk TCDD[1]	7.7×10^{-7}	78×10^{-7}	328×10^{-7}
Cancer risk factor[2]	734 000	3140	951
Sum cancer risk[3]	0.57	0.024	0.031

[1] Unit risk factor × equivalent
[2] Cancer risk BaP/Cancer risk TCDD
[3] Cancer risk BaP + Cancer risk TCDD

of dioxins and furans. This is in agreement with the data presented for the Fire-LCA TV case study.

13.5.3 Conclusions

The major conclusions from the study were:

- The major part of the NO_x, CO and CO_2 emissions was due to furniture production, while the fires were responsible for the major part of the HCN, PAH, PCDD/F and PBDD/F emissions.
- The emission levels were about the same for the CO and NO_x for the non-flame retardant and flame retardant treated cases. The CO emissions were 1–5% lower for the flame retardant treated cases than for the non-flame retardant treated case, due to the smaller number of fires in which these sofas are involved while the NO_x emissions were 5–10% higher for the flame retardant treated cases due to the flame retardant production and incineration.
- The energy consumption and CO_2 emissions were slightly larger for the flame retardant treated cases. The CO_2 emissions were 10–15% higher for the flame retardant treated cases due to the energy consumption during the flame retardant production.
- The non-flame retardant treated case gave higher HCN and PAH emissions due to the larger number of fires that these sofas are involved in. The HCN emission for the non-flame retardant treated case was 4–80 times the emissions from the flame retardant treated cases. For PAH, the emissions from the non-flame retardant treated case were 10–25 times the amount of PAH emissions from the flame retardant treated cases.
- The TCDD- and TBDD-equivalents were higher for the flame retardant treated cases. The TCDD emissions for the flame retardant treated cases were

8–50 times the TCDD emissions from the non-flame retardant treated case. Further, the HCl emissions were 1.5–4 times higher for the flame retardant treated cases than for the non-flame retardant treated case.
- Applying the 'Unit Risk Model' enables an indication of the relative importance of these two classes of toxic environmental pollutants to be obtained. This model compares the potential cancer risks due to chlorinated dioxins and furans to that from PAHs. Using this model one can see that the PAH emissions pose a far greater cancer risk than the PCDD/F emissions for all the three different sofa constructions considered and there is a significantly greater cancer risk due to PAH and TCDD emissions from the non-flame retardant sofas than from either of the flame retardant treated sofas.
- Changing the fire model had little influence on the results, thus demonstrating the robustness of the fire-LCA model.

A true estimate of the environmental impact of the adoption of high fire performance material, in a product that is often involved in a fire, cannot be made without the inclusion of the emissions associated with a fire. In the case of certain key species (e.g., PAH, dioxins) fire emissions are a significant part of the total environmental impact. Ideally, when evaluating the environmental and societal impact of the choices made to adopt a high level of fire safety, one should include all the emissions presented in this report. Currently no agreed evaluation method to take into account and compare all the different emissions using the Fire-LCA Model exists.

The number of lives saved due to the better fire performance of the UK flame retardant treated sofas is not included in the LCA model. The DTI investigation of the effect of the 1988 furniture legislation indicates that the number of lives saved in UK since the introduction of the stricter Fire Regulations was between 970 and 1860 by 1997. They estimate that the number of lives saved annually will be between 10 and 15 persons per million people, when all furniture has been replaced with flame retardant treated furniture.[35] Similarly, the same study estimated the number of injuries saved annually to be 110 per million of the population. In a complete analysis of the risks and benefits associated with any method used to create a high level of fire safety this should also be taken into account.

13.6 Conclusions

Fire-LCA is an LCA method that incorporates fires as one possible end-of-life scenario. It was developed by SP and IVL in order to be able to assess life cycle aspects of the fire performance of a product.

A great deal of input data is needed in order to conduct a Fire-LCA study. However, very little fire emission data is reported in the literature and only recently has detailed characterisation of fire emissions been conducted in some laboratories. Much of this data is confidential. However, as the number of fire-LCA studies and research on fire emissions increase, such data will become more readily available.

Application of the Fire-LCA model clearly shows that a holistic approach is necessary to adequately evaluate the environmental impact of, for example, a high level of fire safety. Merely including a flame retardant additive in a traditional LCA evaluation fails to take into account the function of the additive and the potential environmental benefits associated with its production and use.

The Fire-LCA has up to now been used in three different case studies. These case studies have confirmed the robustness of the model and demonstrated its applicability to different situations. It has been used in evaluating products with different fire performance, investigating the impact of different legislation or, with similar fire performance, investigating different solutions to comply with these requirements. Fire-LCA has proved to be useful to find which stages in the life cycle contribute most to each emission as the traditional LCA does.

While the Fire-LCA tool provides a good starting point for a more holistic interpretation of a realistic life cycle of a product, including information concerning the probability that the product may be involved in a fire, it does not provide information concerning, for example, the effect of the toxicity of chemicals used in the product, number of lives saved, costs associated with the different cases or the societal effect of manufacturing practice. The Fire-LCA concept would be a much more powerful tool if these aspects could be included. This requires that a multivariate analysis method be developed which would potentially assist decision-makers to evaluate fully all consequences of a change in regulations, the introduction of a new production method, a new product, etc. Full application of such a model would also require a significant amount of research into the toxicology of many of the emissions analysed within each model application.

13.7 Future trends

In recent years there has been an increased focus on sustainable development. Sustainable development includes environmental stewardship, economic development and the well-being of people, not just for today but for generations to come. The Fire-LCA methodology has been developed to investigate sustainable fire safety but it has the shortcoming of not being able to incorporate financial costs or toxicity aspects directly. In order to do this more work must be done on cost-benefit analysis and on the development of multi-parameter models.

In particular, sustainable development can only occur within a sustainable regulatory framework. For the past several decades, regulation of the environment has been covered by environmental protection agencies worldwide.

Development of sustainable regulations, however, requires tools that can evaluate the costs and benefits of a given regulation. Perhaps one of the most important lessons that has been learned from modern experience of environmental regulation is that regulations have significant costs, not just benefits and that analysis of both the cost and benefit of proposed legislation is imperative. To this end a specialised cost-benefit analysis model, the Fire-CBA, has been developed and applied to a specific case study, i.e. a TV set. The case study compares a TV set with low fire performance with another of high fire performance in the same way as the Fire-LCA TV case study described previously in this chapter. The results of the CBA calculation clearly indicate that in all cases investigated, the benefits of a high level of fire performance in a TV set far outweigh the costs associated with obtaining that high level of fire safety. The net benefit is a function of the choices made in the various scenarios but ranges from $657 million per year to $1380 million per year. Full details of this study can be found elsewhere.[43]

Clearly, future evaluations of the environmental cost of compounds such as flame retardants must incorporate both an assessment of their direct environmental impact through the use of a Fire-LCA model and an assessment of the societal cost-benefit of regulations to control their use through the Fire-CBA model. Reference to the precautionary principle when proposing bans may be convenient, but it is neither scientific nor defensible. Instead, an objective evaluation of toxicity and ecotoxicity issues together with the environmental impact and the costs associated with the different alternatives, should be required as the basic foundation of all our efforts to promote sustainable development. Without such an approach we risk an uncertain, potentially unsustainable, future.

13.8 References

1. Simonson, M., Boldizar, A., Tullin, C., Stripple, H. and Sundqvist, J.O., 'The Incorporation of Fire Considerations in the Life-Cycle Assessment of Polymeric Composite Materials: A Preparatory Study'. *SP Report* 1998:**25**, 1998.
2. Simonson, M., Stripple, H., 'The Incorporation of Fire Considerations in the Life-Cycle Assessment of Polymeric Composite Materials: A preparatory study', *Interflam*, 1999, pp. 885–895.
3. Simonson, M., Blomqvist, P., Boldizar, A., Möller, K., Rosell, L., Tullin, C., Stripple, H. and Sundqvist, J.O., 'Fire-LCA Model: TV Case Study'. *SP Report* 2000:**13**, 2000.
4. Simonson, M., Stripple, H., 'LCA Study of TV Sets with V0 and HB Enclosure Material', *Proceedings of the IEEE International Symposium on Electronics and the Environment*, 2000.
5. Simonson, M., and Stripple, H., 'LCA Study of Flame Retardants in TV Enclosures', *Flame Retardants 2000*, 2000, pp 159–170.
6. Simonson, M., Tullin, C., and Stripple, H., 'Fire-LCA study of TV sets with V0 and HB enclosure material', *Chemosphere*, **46**: 737–744 (2002).
7. Simonson, M., Andersson, P., Rosell, L., Emanuelsson, V. and Stripple, H., 'Fire-

LCA Model: Cables Case Study', *SP Report* 2001:2 available at http://www.sp.se/ fire/br_reports.HTM (2001).
8. Simonson, M., Andersson, P., Emanuelsson, V., and Stripple, H., 'A life-cycle assessment (LCA) model for cables based on the fire-LCA model', *Fire and Materials*, **27**:71–89 (2003).
9. Andersson, P., Simonson, M., Rosell, L., Blomqvist, P., and Stripple, H., 'Fire-LCA Model: Furniture Case Study', *SP Report* 2003:**22** (2003).
10. Andersson, P., Simonson, M., Blomqvist, P., Stripple, H., 'Fire-LCA Model: Furniture Case Study', *Flame Retardants 2004*, pp 15–26 (2004).
11. Andersson, P., Blomqvist, P., Rosell, L., Simonson, M. And Stripple, H., 'The environmental effect of furniture', *Interflam*, 2004, pp. 1467–1478.
12. Andersson, P., Simonson, M.,Tullin, C., Stripple, H., Sundqvist, J.O., Paloposki, T., 'Fire-LCA Guidelines', NICe project 04053, *SP Report* 2004:**43** (2005).
13. Environmental management – Life cycle assessment – Principles and framework. ISO 14040:1997.
14. Environmental management – Life cycle assessment – Goal and scope definition and inventory analysis. ISO 14041:1998.
15. Environmental management – Life cycle assessment – Life cycle impact assessment. ISO 14042:2000.
16. Environmental management – Life cycle assessment – Life cycle impact interpretation. ISO 14043:2000.
17. Simonson, M. and De Poortere, M., 'The Fire Safety of TV Set Enclosure Material', Fire Retardant Polymers, 7th European Conference (1999).
18. De Poortere, M., Schonbach, C. and Simonson, M., 'The Fire Safety of TV Set Enclosure Materials, A Survey of European Statistics', *Fire and Materials*, **24**, pp. 53–60 (2000).
19. Eadon, G., Kaminsky, L., Silkworth, J., Aldous, K., Hilker, D., O'Keefe, P., Smith, R., Gierthy, J., Hawley, J., Kim, N., DeCaprio, A., 'Calculation of 2,3,7,8-TCDD equivalent concentrations of complex environmental contaminant mixtures.', *Environ. Health Perspect.*, **70**, pp. 221–227 (1986).
20. Van den Berg, M., Peterson, R.E. and Schrenk, S, 'Human risk assessment and TEFs', *Food Addititives and Contaminants*, **17**(4), pp. 347–358 (2000)
21. Spindler, E.-J., 'Brandrusse – eine Risikoabschätzung', *Chemische Technik*, **49**(4), pp. 193–196 (1997), in German.
22. Nisbeth, I.C.T. and LaGoy, P.K., 'Toxic Equivalency Factors (TEFs) for Polycyclic Aromatic Hydrocarbons (PAHs)', *Reg. Toxicol. Pharmacol.*, **16**, pp. 290–300 (1992).
23. Troitzsch, J., 'Fire gas toxicity and pollutants in fires. The role of flame retardants', Proceedings of the Flame Retardants 2000 Conference, pp. 177–185, Interscience Communication Ltd, London (2000).
24. Persson, B., and Simonson, M., 'Fire Emissions into the Atmosphere', *Fire Technology*, **34**(3), pp. 266–279 (1998).
25. Sambrook Research International. *TV Fires (Europe)*, Department of Trade and Industry (UK), 14 March (1996).
26. Dinelli, G., Viti., N., Miola, G., Fara, A., de Nigris, M., Gagliardi, E., End-of-Life Management of Power Distribution Cables: Improvement Options Derived from an Analytical Approach, MEIE '96, Versaille (1996).
27. de Nigris, M., Use of LCA Approach to Evaluate the Environmental Impact of Distribution Transformers and Implementation of Improvement Options, MEIE 2000, Paris (2000).

28. 'Brandweerstatistiek 1998', Central Bureau voor de Statistiek, Nederland, ISBN 90-357-2627-8, in Dutch (1998).
29. Tabulated data from UK Home Office ordered specifically for this study (1996).
30. *Räddningsverket insatser* (The Swedish Rescue Service Agency Activities), ISBN 91-88891-18-6 (1996), in Swedish.
31. *Räddningsverket i siffor* (The Swedish Rescue Service Agency in Numbers), ISBN 91-88891-46-1 (1997), in Swedish.
32. *Räddningsverket i siffor* (The Swedish Rescue Service Agency in Numbers), ISBN 91-7253-039-1 (1998), in Swedish.
33. *Räddningsverket i siffor* (The Swedish Rescue Service Agency in Numbers), ISBN 91-7253-076-6 (1999), in Swedish.
34. Enqvist, I. (ed.), 'Electrical Fires – Statistics and Reality. Final report from the "Vällingby project"', Swedish National Electrical Safety Board ('Elsäkerhetsverket') (1997), in Swedish.
35. Stevens, G. (ed.), 'Effectiveness of the Furniture and Furnishings Fire Safety Regulations 1988', UK Department of Trade and Industry (DTI) URN 00/783 (2000).
36. http://www.safety.odpm.gov.uk/fire/rds/index.htm.
37. http://www.statistics.gov.uk/.
38. Emsley, A.M., Lim, L. and Stevens, G.C., 'International Fire Statistics and the Potential Benefits of Fire Counter-Measures, Proceeding from FR2002, London, February 2002, in pp 23–32 (2002).
39. Statistics specially prepared for this project supplied by Polymer Research Centre, University of Surrey, Guildford, UK.
40. www.forsakringsforbundet.com.
41. British crime survey, http://www.homeoffice.gov.uk/rds/pdfs/hosb1301.pdf.
42. Bartelds, H., Broker, G., Ham, J. van and Vicard, J.-F., 'Expertise on the measurement and control of dioxins', Directorate General XI, EC, Society for Clean Air in the Netherlands/Vereniging Lucht, Brussels/Delft, 1991.
43. Simonson, M., Andersson, P., and van den Berg, M., 'Cost Benefit Analysis Model for Fire Safety Methodology and TV (DecaBDE) Case Study', *SP Report* 2006:**28** (2006).

Part III

Applications of fire retardant materials

14
The risks and benefits of flame retardants in consumer products

A M EMSLEY and G C STEVENS,
University of Surrey, UK

Abstract: This chapter examines the risks and benefits of flame retardants used in consumer products and how they serve to reduce the hazards and risks associated with residential fires. This is put into context by examining residential fire statistics in the UK and the rest of the world, and the benefits of the 1988 furniture fire regulations in the UK. The role of flame retardants is discussed in combating fire and compared to active measures of protection such as smoke alarms. The human exposure risks associated with flame retardants is also discussed by reference to risk assessments carried out in Europe and the US for a number of common flame retardants. It is shown that the balance of risk for death and injury favours the use of flame retardants for those compounds whose risks have been assessed. Controlling and eliminating flame retardant risks to the environment and in regard to human exposure is also discussed in relation to the development of new flame retardant systems and improved industry and supply chain management of their use and release.

Key words: flame retardants, hazards and risks, fire statistics, residential fires, socio-economic benefits, consumer product safety.

14.1 Background

Fire kills upwards of 500 people a year in their homes in the UK, almost 3000 in Europe and a similar number in the USA, and injures many tens of thousands of others to varying degrees. Property damage and loss is also significant with residential fires accounting for over €12 billion and an average loss of typically 0.17% GDP in many European countries.

The introduction of legislation in the UK in 1988 to regulate the performance of upholstered furniture in fire was an attempt at improving fire safety in the home by giving house occupants adequate time to escape from a serious fire. This legislation did not prescribe the means to be adopted to achieve improved reaction to fire. However, in the case of furniture foams and textile coverings, the most commonly adopted technique was the use of flame retardants.

Similar measures have been taken to improve the fire performance of other consumer products, such as televisions, by the use of flame retardants in key components and in the casing materials, albeit in this case and in Europe, measures were undertaken voluntarily by the television manufacturers. Flame retardants are also commonly used to improve the fire performance of public places and mass transportation systems, where both legislative and non-legislative standards exist to ensure product safety.

However, some flame retardants that seek to protect consumer products, by their chemical nature, bring with them human exposure hazards and a degree of risk to both people and the environment and these risks must be balanced against their benefits in a fire situation.[1] This requires that fire impact risks be clearly understood as well as the potential risks from human exposure to individual flame retardants. It is estimated that there are more than 400 individual flame retardant compounds[1] and it is therefore important that each be considered on its own merits.

In this chapter, we discuss the hazards and risks of fire, and the risks to human health associated with fire retardants. We also consider the historic and potential future benefits arising from their use in furnishing materials and the potential development of conflicting directives in Europe that could undermine the current benefits of improved fire performance products.

14.2 The importance of understanding risks and benefits

14.2.1 Fire hazards and risks in residential fires

Principal human hazards from fires and the associated risks of death and injury arise from exposure to the toxic products in fire atmospheres and to a lesser extent from the thermal effects of fire (radiative and convective heat transfer), which cause death due to burns. Of these, the former is the more frequent cause of death.[1] The hazards of reduced visibility due to smoke, with the consequent increased risk of people being trapped by the fire or having their escape impeded, should also be noted and these may be directly related to the rate of spread of fire and the effective heat release rate of the materials involved in the fire.

Survivors of fires may experience post-exposure lung complications which may be an inflammatory response to irritants and may lead to delayed death.[2] There are possible long-term chronic effects from exposure to carbon monoxide, to other narcotic gases and to compounds such as carcinogens (e.g., polyaromatic hydrocarbons and certain dioxins) in fire atmospheres.

The evolution of toxic gases and particles depends on:

- stage and rate of fire development, e.g. whether it is smouldering, flaming or at the flashover stage
- type of material being burned, particularly in the early stages of fire development
- supply of oxygen to the fire and the effective ventilation condition.

The risks and benefits of flame retardants in consumer products 365

These factors and the associated hazards are strongly affected by the kind of enclosure in which the fire occurs. The location of people within the enclosure also affects their degree of exposure to fire effluents. So the dose-response curve for toxic products will be determined both by the fire dynamics and location, and the physiological characteristics of the individual. The exposure risks will be determined by the susceptibility of the individual to the received dose and their actions during the exposure experience. So when considering fire-hazard counter-measures and regulatory policies, we should be mindful of differences within the population and of the response of the most susceptible groups to toxic fire atmospheres.

It is now widely accepted that collective fire hazards and risks, and particularly those due to the fire atmosphere, are directly related to the rate of heat release and rate of flame spread. For occupants in enclosed spaces or in a typical room in a small domestic dwelling where a fire occurs, the pre-flashover stages of the fire, which lead to untenable fire gas concentrations, are likely to present the greatest hazards and risks. For occupants remote from the room containing the fire and in larger dwellings and buildings, the hazards and risks are likely to be greatest following flashover.

The toxicity of fire atmospheres is a function of the rate of production of compounds as well as their concentration and their toxic potency. This is determined by the mass burning rate and relates to the overall dynamics of individual fires. It must also be recognised that in some acute exposure situations irritant effects may be as important as systemic poisoning effects in preventing the victim from escaping from the fire. From this perspective it is clear that the principal components of fire atmospheres can be prioritised according to the acute hazards and risks they present.[1,3] These are:

> carbon monoxide > hydrogen cyanide > carbon dioxide > oxygen depletion > irritant gases > solid and liquid particulates > oxides of nitrogen > carcinogens > dioxins and products of extreme or unusual toxicity

If we consider chronic hazards and risks, the order of the above will be quite different.[3]

14.2.2 Fire risks in the UK and the rest of the world

Fire statistics are collected and reported in both the UK (Fire Statistics, United Kingdom) and the US (http://www.usfa.dhs.gov/statistics/national/residential.shtm), so it is relatively easy to make comparisons between dwelling fires in the two countries. In Fig. 14.1 the base of each triangle represents the potential for experiencing fire with the top culminating in reported loss of life. The figures in parentheses, in the diagram and throughout this section, represent 1995 data and those without parentheses, 2004 data.

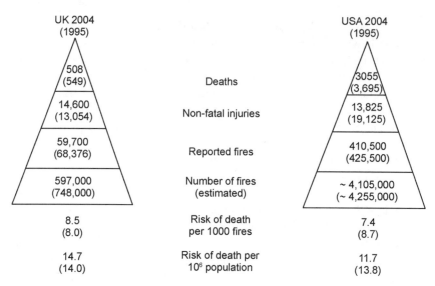

14.1 Human impact of fire for UK and US 2004 compared to 1995 (in brackets) residential/dwelling fires.

This comparison serves to illustrate the enormous potential for fire events in the home environment and the extent to which this can lead to potentially serious fire incidents causing injury and death. At the top of the triangle, the UK experienced 14 600 non-fatal injuries in 2004 (13 054 in 1995) and 508 (549) fatalities in dwellings. This equates to 8.5 (8) deaths per 1000 reported fires and a UK annual average population risk of death by fire of 14.7 (14) per million.

The major cause of death is gas and smoke inhalation with around 37–42% of fatalities being attributed to being overcome by gas and smoke alone and a further 16–19% being attributed to burns plus being overcome by gas and smoke, as reported in the UK fire statistics for 1995 and 1996. The equivalent figures for 2004 were 53% overcome by gas alone and 21% deaths from burns. We can see: (a) a close correspondence between the UK and US figures and (b) that the figures have not changed dramatically between 1995 and 2004, although the US numbers have gone down.

These figures are typical of developed countries, as can be seen from Table 14.1 for fire death statistics for European countries,[1,4] from data prior to the expansion of the European Union. If, we compare the causes of fire incidence, death and injury in the UK in 1995 (Table 14.2) and 2004 (Table 14.3), we can see that most of the numbers have not changed significantly. However, we can see that, while fires in upholstery decreased by about a third, the number of deaths halved and the number of injuries decreased by about 40%. The reasons for this are probably two-fold, as we shall show later:

1. an increased volume of flame retarded products in general use (1995 was only 7 years after UK furnishing regulations were introduced to require manufacturers to use improved fire performance materials), and
2. improvements in the efficiency (and number) of smoke alarms in use in UK homes.

Table 14.1 Estimated European EU-15 residential fire impacts from WFSC data

European EU-15 countries	Estimated residential fire impacts in Europe			
	Deaths pmp* pa (2000–2001)	Population, M (2000–2001)	Total deaths	Cost (MEuro)
B (Belgium)	10.2	10.25	105	450
DK (Denmark)	12.4	5.34	66	285
D (Germany)	7.0	82.18	575	2474
EL (Greece)	11.0	10.55	116	499
E (Spain)	5.1	39.46	201	865
F (France)	7.8	59.34	463	1990
IRL (Ireland)	16.7	3.80	63	273
I (Italy)	6.2	57.77	358	1540
L (Luxembourg)	(7.8)†	0.44	3	15
NL (Netherlands)	5.4	15.92	86	370
A (Austria)	6.1	8.16	50	214
P (Portugal)	(6.2)†	10.0	62	267
FIN (Finland)	15.8	5.18	82	352
S (Sweden)	13.0	8.87	115	496
UK (United Kingdom)	9.7	59.73	579	2491
Total loss			2926‡	12580‡
Loss (death pmp pa), %GDP			7.8	0.17

* pmp = per million persons.
† assumed value – as no data is available.
‡ equivalent to an additional 351 deaths and 15,100 MEuro in the EU-25 Member States.

14.2.3 Product and cause-related fire incidence

The UK fire statistics break down the causes of fire by product and cause of fire and Table 14.2 and Table 14.3 show how the major product items in general use in the UK affected fire statistics in 1995 and 2004 respectively. It is clear that the main causes of domestic fire are cooking appliances, and kitchen fires are also the largest cause of injury, due to burns and smoke inhalation, and the figures have changed very little in 10 years.

The main causes of death, however, are smokers' materials and fires in upholstery, textiles and furnishings, due in large part to the high toxicity of the

Table 14.2 UK 1995 product and cause related fire incidence, death and injury

Product	Numbers of fires and proportion (%)		Number of fatalities and proportion (%)		Number of non-fatal injuries and proportion (%)	
Smoker's materials[a]	5,580	8.2%	152	27.7%	1,926	14.8%
Cooking appliances[a]	28,540	41.7%	61	11.1%	5,671	43.4%
Space heating appliances[a]	2,822	4.1%	71	12.9%	758	5.8%
Electrical distribution[a]	2,344	3.4%	15	2.7%	293	2.2%
Other electrical appliances[a]	6,374	9.3%	2	5.3%	879	6.7%
Upholstery and covers[b]	2,645	3.9%	107	19.5%	1,092	8.4%
Other textiles, clothing and furnishings[b]	6,362	9.3%	113	20.6%	1,932	14.8%
Total (all products and sources)	68,376		549		13,054	

[a] reported by source of ignition; [b] reported as material or item first ignited.

fire gases from foam filling. It is interesting to note that the incidence of death from upholstery fires has halved over the 10-year period. The number of deaths related to space heating appliances has also decreased significantly, but this is probably due to decreased use in the home as more and more houses become centrally heated.

Table 14.3 UK 2004 product and cause related fire incidence, death and injury

Product	Numbers of fires and proportion (%)		Number of fatalities and proportion (%)		Number of non-fatal injuries and proportion (%)	
Smoker's materials[a]	4,336	7%	114	30%	1,667	14%
Cooking appliances[a]	27,248	46%	52	14%	5,337	45%
Space heating appliances[a]	1,677	3%	13	3%	363	3%
Electrical distribution[a]	2,850	5%	9	2%	315	3%
Other electrical appliances[a]	5,525	9%	13	3%	885	7%
Upholstery and covers[b]	1,784	3%	53	14%	650	5%
Other textiles, clothing and furnishings[b]	7,152	12%	110	29%	2,300	19%
Total (all products and sources)	59,743		375		11,977	

[a] reported by source of ignition; [b] reported as material or item first ignited.
Note: the percentage values in this table do not and should not add up to 100% because the data are for both products and causes.

14.2.4 Role of flame retardants in reducing fire hazards and risks

The main effects of flame retardants are to delay the spread of fire over the burning product and to inhibit the development of the fire. The overall effect in real fires is to generally reduce the heat release rate, reduce the consumption of substrate and consequently to reduce the evolution of toxic gases. In addition, flame retardants reduce the exposure of personnel to toxic gases by increasing the time available to escape before flashover occurs or the atmosphere becomes totally incapacitating. There appear to be a number of special cases – mostly demonstrated in smaller-scale test systems – in which the influence of flame retardants is to increase the production of toxic gases. This influence has not been demonstrated in room-scale tests or in real fires. It is also clear that small- and large-scale tests on materials containing modern commercial flame retardants indicate that they do not produce effluents of extreme or unusual toxicity.

14.2.5 Use of active and passive fire countermeasures

Protection against fire can be achieved through the use of both active and passive fire countermeasures. Passive measures would include elements such as materials choice, product design and the use of flame retardants to impart a reduced rate of fire development and propagation, largely through reducing the total heat release. In contrast active measures would include smoke alarms, fire extinguishers and sprinkler systems that respond to the presence of the fire to provide a warning in the case of smoke alarms or an extinguishing of the fire. Some flame retardant systems may also be considered as active measures as many are activated by the presence and development of the fire – they may respond in one or more ways:

1. prevent chain branching reactions in the gas phase – gas phase retardancy
2. absorb heat through undergoing endothermic reactions – solid phase retardancy
3. form protective physical barriers by intumescence – barrier retardancy.

14.3 Primary risks of flame retardants in consumer products

14.3.1 The toxicology of flame retardants

In previous work, we selected a group of six flame retardants for hazard assessment which were representative of the more common types, where concern had been expressed on their toxicology and where toxicity information was available to support an assessment – a detailed discussion of findings and references can be found elsewhere.[5] In a more extensive study of oral, dermal and inhalation exposure routes a National Academy of Sciences (NAS) subcommittee on toxicology examined the toxicological risks of 16 selected flame retardants that could be more widely used in the US.[6] More recent studies have been carried out in the US by the Environmental Protection Agency[7] (EPA) and also the Consumer Product Safety Commission.[8] In Europe a number of flame retardants have been examined in detail in the last 10 years by the European Chemicals Bureau and its appointed competent authorities and decisions on their risks taken by the European Commission Scientific Committee on Health and Environmental Risks (SCHER).[9]

What have these various studies shown for some of the highest use flame retardants?

Alumina trihydrate

Alumina trihydrate (ATH) is broadly used in a number of consumer products and in cable and wiring applications. Alumina, aluminium hydroxide and aluminium compounds in general have very low levels of toxicity except when there are very high exposure levels or unusual routes of exposure. In view of the lack of reported adverse effects from the very extensive environmental exposure to aluminium compounds, including alumina, it is extremely unlikely that any adverse effects would ensue from the levels of exposure to ATH in the use of consumer products.

Antimony trioxide

Antimony trioxide (ATO) is commonly used as a co-synergist with halogenated flame retardants to enhance their effectiveness. Recent comprehensive genotoxicity studies and a critical review by the European Commission have indicated that, contrary to the indications of earlier less well authenticated studies, antimony trioxide is not a genotoxic carcinogen. No adverse health effects are expected from antimony trioxide, although there remains some uncertainty on a possible cancer hazard arising from inhalation of particles, where better data on particular exposure is required. However, in most cases, exposure is probably minor compared with exposure to antimony trioxide from other sources in the domestic and urban environment.

Decabromodiphenyl ether

Decabromodiphenyl ether (DecaBDE) is widely used for consumer product fire safety and has been the subject of over 10 years of research and one of the most exhaustive chemical risk assessments ever undertaken by the European Commission. In regard to both environmental and human exposure risk it has been found to be safe, although some questions have been raised on environmental exposure as it appears to act as a persistent organic compound in the environment. In May 2004 the EU Competent Authorities officially closed the scientific assessment of decabromodiphenyl ether (DecaBDE) with no restrictions due to the lack of risks identified for the use of this substance and in March 2005 SCHER stated that the risk assessment had been well done. A consolidated risk assessment text is to be published by the European Chemicals Bureau.

The detailed history of this assessment is provided on the Bromine Science and Education Forum's (BSEF) web site.[10] This same site and others identify that other members of the same group of brominated diphenyl ethers or BDEs, notably PentaBDE and OctaBDE, which were once used commercially as flame retardants in the US and Europe, are no longer used due to acknowledged risks and potential risks associated with these compounds.

Melamine

Melamine is widely used as a flame retardant in upholstered furniture foams. All of the available information indicates that melamine has low acute and chronic toxicity, so no adverse effects are envisaged from the level of exposure expected from the use of melamine as a flame retardant. Melamine has been available commercially since the late 1930s and no significant reports of adverse health effects in humans have been found. It may therefore be concluded that for the low levels of exposure expected from the use of melamine as a flame retardant, it can be considered toxicologically safe. Interestingly in recent years the Federal Drug Administration (FDA) and other agencies in the US have carried out risk assessments related to the eating of meat of animals given feed containing melamine.[11] This assessment has concluded that even in the most extreme risk-assessment scenario, the level of potential exposure was way below any level of public health concern.

Tetrabromobisphenol A

Tetrabromobisphenol A (TBBPA) is the primary flame retardant used in printed wiring boards (PWBs) to improve the fire safety of electrical products. Earlier reviews have shown that subacute studies of oral, dermal and inhalation routes of exposure indicate such low levels of toxicity that it would be surprising if there were appreciable effects from low levels of long-term exposure. TBBA is covalently bound in the PWB application but there is evidence that Bisphenol A has weak oestrogenic properties and these may extend to TBBA. TBBA is currently the subject of a European risk assessment and the UK government is the Member State rapporteur responsible for the risk assessment, which is yet to be completed. Most of the risks identified for additive application in the manufacture of this flame retardant are manageable through VECAP (**V**oluntary **E**missions **C**ontrol **A**ction **P**rogramme – see BSEF web site[10]) with 100% of TBBPA additive users in Europe committed to control and emissions reduction.

Hexabromocyclododecane

The EU has started the scientific assessment of hexabromocyclododecane (HBCD). This chemical is mainly used in expanded and extruded polystyrene for thermal insulation foams, in building and construction and is also applied in the back-coating of textiles, mainly for upholstered furniture. Although not classified in the EU as a dangerous substance, in 2000 the European Commission Regulation 2268/95 listed HBCD in the Priority List 2 and designated the Swedish government as the Member State rapporteur for the risk assessment of this compound – the outcome of this study is still awaited. In the interim Stevens and Horrocks and co-workers have carried out detailed but largely unpublished work[12,13] on the release of this compound from model back-coated textiles and it has been shown to be as effectively contained as DecaBDE.

Tris-(chloropropyl)-phosphate

Tris-(chloropropyl)-phosphate (TCPP) is widely used in polyurethane foams in domestic, transport and public seating. None of the available data give any indication that TCPP is likely to cause any toxic risk at the exposure levels envisaged from its use in consumer products. However, this flame retardant, in common with other organophosphates, does not have a well established toxicology and there are significant data gaps in regard to the release of these compounds to the environment and of human exposure that can take place through either oral, dermal of inhalation exposure.

As yet unpublished work by Stevens and Horrocks and coworkers on the release behaviour of TCPP from polyurethane foams shows that release is strongly temperature and geometry dependent. This and other data has been collated by the UK Environment Agency as part of their risk assessment of this flame retardant on behalf of the European Commission – the outcome of this study is expected soon.

14.3.2 Human exposure hazards and risks

As shown by the discussion above and in common with many other groups of chemical compounds used by the public (from those in coffee to those in household cleaning products), there is an incomplete understanding of all the potential hazards and risks associated with many flame retardants. As we have seen a number of flame retardants such as the polybrominated diphenyl ethers (especially DecaBDE) have had very extensive risk assessments undertaken on them and a number have been shown to be safe in normal use. Indeed, our common experience to date indicates that there are no significant practical risks from human exposure to most flame retardants in consumer products, although there have been some exceptions that are now banned or have been withdrawn such pentadecabromodiphenylether.

The risks and benefits of flame retardants in consumer products 373

There is no evidence that flame retardants have caused significant acute or chronic human health impacts, but there have been concerns raised over the potential environmental impacts of some compounds. This exists against a background where the formal regulated risk assessment methods in place in various parts of the world, provide safeguards to reduce and prevent exposure risks to both humans and the environment. An example is REACH, the new European Community Regulation on chemicals and their safe use (EC 1907/2006) which became law on 1 June 2007 – this deals with the *R*egistration, *E*valuation, *A*uthorisation and restriction of *Ch*emical substances. In addition, the brominated flame retardant industry has adopted increased voluntary actions on limiting emissions, such as VECAP, mentioned above. It was set up to manage, monitor and minimise industrial emissions of brominated flame retardants through partnership with the supply chain including small companies. VECAP is consistent with the principles of cooperation across the chemical user chain which will be required under REACH and it requires continuous improvement as part of a responsible supply chain approach to product stewardship.

14.4 Balance of risk

There are two sides to the risk-benefit equation – one side is risks associated with acute exposure to fire and fire atmospheres and the other associated with possible long-term exposure to flame retardants in the absence of fire. Neither side of this equation has been fully enumerated and both have been poorly communicated historically. In the absence of a developed approach to risk-benefit analysis for fires, we have attempted a preliminary and simplistic balance of risk assessment for flame retardants which could set the scene for more detailed analysis in the future.

This approach recognises that the risk of a fire event (the prime-cause risk) can be reduced (or balanced) by the acceptance of an intervention or counter-measure risk, in this case, the use of a flame retardant. This approach requires that the risks being compared relate to the same end-point. Suitable risk end-points for fire events could be (i) death, (ii) short-term injury, (iii) long-term injury and (iv) the risk of experiencing a particular economic level of loss. Using such risk metrics, the size of the benefit conferred by the flame retardant will be directly related to the reduction in risk due to an unrestrained fire plus any additional risk (related to the same risk measure) arising from the flame retardant itself.

For a particular event and risk end-point, the risk reduction benefit is:

P(risk reduction benefit) = P(unrestrained fire risk) − P(restrained fire risk) + P (flame retardant risk)

where the P's are total risk probabilities, summed over all contributing risk probabilities, for a particular common end-point. If the risk reduction probability

374   Advances in fire retardant materials

is positive, a net benefit results; if it is negative, no benefit exists and the use of the flame retardant should be discontinued.

In suggesting this approach we recognise that definitive data are not currently available to fully quantify each risk assessment element. However, going through the exercise illustrates the potential value of this approach and what information will be needed in the future to better quantify the risk-reduction benefits of any fire intervention or countermeasure.

14.4.1 Risk of death

The risk of death due to fires in domestic dwellings can be measured in a variety of ways. We can adopt a number of appropriate measures for input into the balance of risk equation; using UK dwelling fire statistics we can identify the following risks:

- ~8 fatalities per 1000 reported fires and ~40 per 1000 for higher risk upholstery and covers
- ~0.7 fatalities per 1000 household fire events and ~9.4 fire fatalities per 10^6 of the population per annum.

For high risk consumer products, such as upholstered furniture, US statistics indicate that the risk of death can be as high as 86 fatalities per 1000 serious fires when the fire spreads beyond the room of origin. In the UK it is 40 fatalities per 1000 serious fires for the same product (but where part of the population of upholstered furniture meets the UK 1988 furniture fire regulation performance standards).

The reduction in risk associated with the use of flame retardants is not known precisely. However we can make some estimates:

1. A Brominated Flame Retardants Industry Panel funded study suggested risk reductions of 9–14 per 1000 fires p.a. on an average of 27 per 1000 fires (for fires that spread beyond the room of origin) in the US, a reduction of around 30–50%.
2. UK upholstered furniture fire regulations, gross risk reduction of 6 per 10^6 of the population p.a.
3. Television fire fatality reductions (peak to plateau without account of growth in total numbers) 70–80%, from UK, European and US experience.
4. Based on a US National Bureau of Standards fire retardant study:[16]
    - a threefold reduction in toxic gases could translate to a 67% risk reduction or better, based on risk being directly proportional to toxic hazard,
    - a twofold reduction in material consumption could conservatively translate to a 50% risk reduction, based on toxic hazard being directly proportional to mass loss,
    - a fourfold reduction in heat release rate (HRR) could translate to 75% risk reduction if risk is directly related to HRR (this is very conservative), and

- a fifteenfold escape time improvement could translate to 93% risk reduction or better assuming that risk reduction is inversely proportional to escape time; however, this is conservative as the number of casualties increases exponentially with time taken to evacuate.

These potential risk reductions conservatively span the range 30 to over 90% for products that could be considered high risk. It may be different for lower risk products and fire situations. This serves to illustrate that we need to partition risk reduction benefits further according to product type and according to individual flame retardant systems.

In the case of upholstered furniture in the UK, we will show later that the reductions in risk of death are significant. In 1997 these can be equated to 1 in 4.2×10^5 p.a. risk reduction for furniture as the item first ignited and 1 in 1.6×10^5 p.a. for overall risk reduction in dwellings. These equate to risk reductions in relation to pre-1988 trends of 35% and 67% respectively with the prospect of further significant reductions as more post-1988 furniture replaces older furniture in the UK economy.

What is the size of the associated counter-risk due to flame retardants in comparison with the above levels of risk reduction? This is not known precisely. However, current assessments of the toxicology of the flame retardants considered here suggest that the population exposure risks are negligible (less than 1 in 10^7 per annum). Indeed, we are not aware of any deaths occurring as a result of exposure to flame retarded products. If this is correct, the lethality risk reduction benefits are significant and are sustained in the face of the counter-measure risks; in this case the 'benefit risk reduction' to 'adverse risk' ratio is greater than 100:1.

However, caution is required in applying this result to all flame retarded products. The context and performance of each flame retardant including the general use and abuse that products experience must be considered. Also, alternative countermeasures should also be examined that may have very lower intrinsic risk, for example the use of non-combustible products or design changes that could confer equivalent fire performance. Alternatively, a parallel countermeasure such as investment in effective sprinkler systems, may significantly reduce the fire exposure risk of death.

14.4.2 Risk of short-term injury

To calculate this risk, we have adopted a number of appropriate statistical metrics based upon UK dwelling fire statistics:

- 191 injuries per 1000 reported fires (compared with 413 per 1000 for higher risk upholstery and covers)
- 16 injuries per 1000 household fires
- 225 fire injuries per 10^6 of the population per annum.

The same reasoning may be used as in the risk of death case to show that the estimated risk reduction levels for short term injury may be in the range 30% to 90%, or greater, for higher risk products. This leads to a greater degree of overall risk reduction than in the case of death because of the higher risks of injury from fire – typically in the range is 60 to 200 injuries per 10^6 of the population per annum.

What is the size of the associated counter-risk due to flame retardants conferring injuries which are comparable in type and magnitude to those of non-fatal fire injuries? Again this cannot be answered precisely, but we are not aware of any evidence to suggest that this is likely to be any higher than 1 in 10^6 per annum. Clearly, the risk reduction benefits are significant for this risk measure and are sustained in the face of counter-measure risks due to flame retardants. The corresponding population risk ratio is probably in excess of 100:1 to 200:1.

14.4.3 Risk of long-term effects

It is not possible to assess this area because there are no statistics available on the long-term human health impacts of fire and of exposure to flame retarded consumer products. However, we have noted[1] that there are a number of possible longer-term risks arising from exposure to fire atmospheres which may cause health impairment or death (e.g., long-term sequelea (illness developing over a long period of time post exposure) arising from CO exposure, exposure to known carcinogens in smoke, etc.). We would expect flame retardants to reduce these risks using similar reasoning to that above. However, in this case it is difficult to know what the likely countermeasure risks will be for the same risk end-point. Our toxicity assessment suggests the risks may be small but in the absence of reliable exposure data we cannot be sure.

14.4.4 Exposure to flame retardants

Examination of the toxicology of the more common flame retardants used in consumer products indicates that in general they do not pose any significant threats to human life and the environment. Moreover any indication of toxic risk from flame retardants themselves in isolation will be exaggerated because their bioavailability is constrained when they are incorporated into a polymer matrix – sometimes by covalent bonding as in the case of TBBA in reactive resin matrices or by strong physical containment.

However, while little data are available on bioavailability, the US Consumer Product Safety Commission, using solvent extractability from textile furniture covering as the criterion, has concluded that the flame retardants examined have low bioavailability. This remains a key issue in determining the degree of possible human and environmental exposure and more work is required to assess the effectiveness of the containment of flame retardants in polymer matrices where the flame retardant is not covalently bonded to the host matrix.

Caution is therefore required in generalising conclusions obtained on a limited number of flame retardants, to all flame retardants. This reflects the fact that data on many flame retardants, in common with most other chemicals on the European existing chemicals list, are incomplete. The corollary is that a number of flame retardants are among some of the most extensively studied chemicals on the existing chemicals list and have been shown to be safe. Despite this, it is important that each flame retardant is considered individually and in the context of its incorporation into consumer products and that the appropriate human exposure and environmental life-cycle risks are assessed.

14.5 Benefits of flame retardants in consumer products

14.5.1 Evidence of fire protection

In principle the use of flame retardants to reduce the probability or time to ignition and to reduce flame spread and heat release rate is also expected to reduce toxic hazard and risk. Hence our focus must always be on hazard and risk assessment measures rather than simple toxic potency measures when comparing the fire performance of materials and real fire situations. Babrauskas[14] and Hirschler[15] reached the same conclusion by comparing data drawn from a variety of work on heat release, toxic potency and smoke production across a wide range of materials, both natural and synthetic. They concluded in particular that heat release rates can be reduced by factors of 10 or more by the use of flame retardants

Babrauskas *et al.*[16] also compared five consumer products, and the materials they were made of, with and without flame retardants. Their conclusions from small-scale tests were:

- in four out of five product categories the fire retarded (FR) products had reduced burning rates
- none of the products gave smoke of extreme or unusual toxicity although, in the cases of the flame retarded TV cabinet, insulated wire and circuit board, unidentified agents made a small contribution to the specimen's toxicity
- smoke from flame retarded and non-retarded products was of similar toxic potency and comparable with smoke from materials commonly found in buildings.

In contrast, more significant conclusions came from the room/corridor tests on all of the consumer products:

- the time to untenability (i.e., time to flashover or to the occurrence of an incapacitating atmosphere) was more than 15 times greater for flame retarded than for non-retarded products

- the amount of material consumed in the fire was twice as much in the non-retarded tests as in the flame retarded tests
- the amount of heat released was four times as much in the non-retarded tests as in the flame retarded tests
- the quantity of toxic gases in 'CO equivalents' was three times as much in the non-retarded tests as in the flame retarded
- smoke production was not significantly different in both sets of tests.

14.5.2 Benefits of the UK furniture fire regulations

In their original form the regulations sought to address the fire resistance of upholstered furniture,[17,18] but were extended to include indoor and outdoor furniture and covering and upholstery on bedding. The regulations came into force progressively:

- from 1 November 1988 all fabric and polyurethane (PU) foams used in the construction of furniture plus PU mattress fillings, were required to be of a fire resistant type;
- requirements on the fire resistance of other filling materials were applied from 1 March 1989;
- finally, second hand furniture for retail sale was required to meet the regulations on 1 March 1993.

Careful examination of time series trends for dwelling fires, deaths and injuries pre and post 1988 show that, while pre-1988 the trend was upwards, or, at best, level, the trends post 1988 are downwards, or, at least, at a lower rate of rise.[19] The data in Fig. 14.2 have been corrected for the influence of smoke alarms, which, as we will show later, only had a significant effect on the statistics post about 1993. All data are also corrected for demographic factors.

We have applied some basic statistics to the data in Fig. 14.2 to calculate the number of fires, deaths and injuries saved since 1988. This has been done by projecting the trend line prior to 1988 into the future, as shown in Fig. 14.2 and subtracting from it the values from a trend line through the data post 1988[2]. The results are shown in Fig. 14.3 for deaths and for injuries and summarised in more detail in Fig. 14.4. Both figures show the average trend line solid and the 95% confidence limits dashed.

14.5.3 Economic losses and benefits

It is difficult to estimate the economic savings arising from the implementation of the regulations, but some estimates can be made from a few simple assumptions, using data from the year 2000 and using furniture market information to assist the benefits assessment:[20,21]

The risks and benefits of flame retardants in consumer products 379

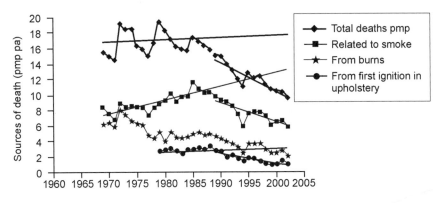

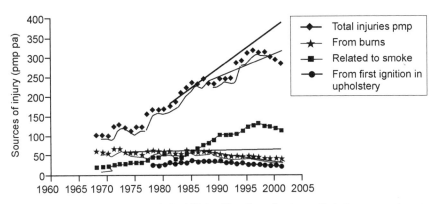

14.2 Time series trends for UK dwelling fires, deaths and injuries pre- and post-1988.

- the cost associated with the loss of a standard statistical life is £3m – others[22] have used a value of £1m, so we use this value to provide a lower estimate.
- the average insurance loss-adjusted cost per fire claim* was approximately £102,402 in 2000 based on a total claim figure of £8.2m.
- that loss-adjusted fire claims underestimate the total costs of fires and that, in reality, 10% of all reported dwelling fires would result in losses comparable with those of loss-adjusted fires.[3]

Our estimate of the total cost for 2000 is then £1,540m using £3m for the cost of a standard statistical life or £548m for a £1m value, in contrast to the £8.2m actually reported. In 1995 the loss would have been £351m according to this approach, in good agreement with the 1995 British Crime Survey,[23] which

---

* A loss-adjusted fire claim is defined as one in which either
(a) the total cost of a fire exceeded £50k and/or there was a fatality, or
(b) the total cost exceeded £25k and/or there was an injury (FPA data).

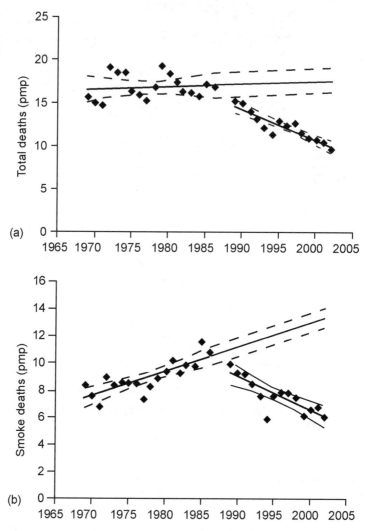

14.3 (a) Total deaths pmp 1969–1986 and 1988–2002 in UK dwelling fires in the UK pre- and post-1988; (b) Smoke-related deaths pmp 1969–1986 and 1988–2002 in UK dwelling fires in the UK pre- and post-1988.

estimated £355m for all home fires and the 2000 British Crime Survey,[24] which estimated it to be £375m in 1999.

Combining the actual loss adjusted costs with the fire and injury savings estimated above, we have calculated the effective cost saving benefits resulting from the regulations and some of the relevant figures are given in Table 14.4.

The figures for life costs have been estimated on the basis of £3m per life, with estimates at £1m per life in parentheses.

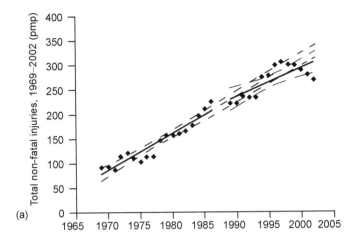

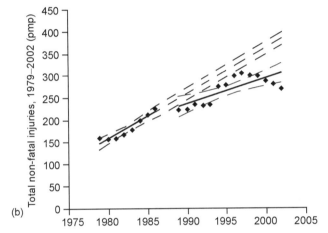

14.4 (a) Total non-fatal injuries pmp 1969–1986 and 1988–2002; (b) Smoke-related non-fatal injuries pmp 1979–1986 and 1988–2002.

14.6 Differentiating the effects of smoking and smoke alarm trends

14.6.1 Smoking trends

A factor that might influence the time trend statistics would be smoking, since the smoking-related figures in Tables 14.1 and 14.2 are quite significant. In reality though, the main change in smoking habits in the UK occurred prior to 1990, as shown in Fig. 14.5, when the main trends in Figs 14.2 to 14.4 were to increasing number of fires, death and injuries.

The effects of smoke alarms on fire statistics were first reported in the UK in

Table 14.4 Change in annual savings and cumulative benefits of the UK furnishing regulations up to 2002

Benefit measure	1994 Annual benefit	2002 Annual benefit	1988–2002 Cumulative benefit
Number of dwelling fires	2448	6188	44314
Total lives saved	269	468	4287
Lives saved for upholstery as item first ignited	69	139	1150
Total non-fatal injuries saved	2392	4578	39257
Injuries saved for upholstery as the first item ignited	896	1789	13442
Actual loss-adjusted cost saving, £m p.a.	9	52 (Year 2000)	182 (1988–2000)
Fatality cost saving, £m p.a. @ £3m per life (£1m in brackets)	806 (269)	1327 (442)	9900 (3300)
Total cost saving, £m p.a. @ £3m per life (£1m in brackets)	815 (275)	1399 (494)	10082 (3482)

1988, when <8% of households had working alarms. By 2000, the number of fires detected by alarms was still only around 1–2% of the total and about 10% of the number of dwelling fires and by 2004, the number of fires detected by smoke alarms was still only 32% of the total. The primary benefit of alarms is in decreasing the risk of death in fire – it is 4 per 1000 fires when fires are detected by alarms compared to 9 per 1000 fires in the absence of alarms.

So in summary, the number of people smoking in the UK decreased from over 40% in 1975 to around 30% in 1988. Thereafter the figure has fluctuated between 30% and 27%, i.e., it has been relatively constant post-1988.

From a national statistical perspective it appears that there is no connection between smoking and the prevalence of fire impacts. However, care is required as the rate of decline of smokers is not socio-economically uniform – it has been shown to be lowest in financially deprived households. Indeed, a statistical analysis of patterns of domestic fire risk (British Crime Survey 1995) shows a strong correlation between those groups most strongly linked to the occurrence of serious domestic fires and those that have the highest probability of not owning a smoke alarm. Complex social factors affect the patterns of fire risk and more work is required to investigate social factor links between smoking trends and fire risk trends.

In contrast, US fire statistics trends in the last few decades (provided by the Directorate for Economic Analysis, US Consumer Product Safety Commission in Washington, DC, US population and from US Government Census Office web pages) appear to be dominated by the influence of smoke alarms. In the period from 1976 the number of homes containing fire alarms increased to over 90% of the total (although not all may be working effectively) and over 50% of

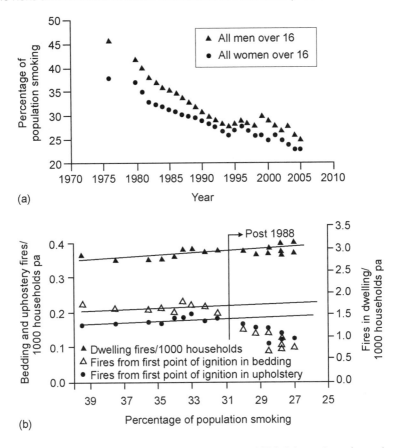

14.5 Smoking trends in the UK pre- and post-1988 (a) number of people smoking; (b) number of fires as function of the percentage of smokers.

domestic fires are now first detected by a smoke alarm. This has progressively contributed to reducing the total number of serious fires, deaths and injuries (although the number of smokers has also decreased, which may be a further contributory factor).

There is no evidence in the US for a change in statistical trends similar to those observed for the pre- and post-1988 UK dwelling fire statistics. We have carried out an equivalent detailed analysis of US fire statistics trends, taking into account demographic changes and the influence of smoke alarms, over the period 1980 to 1995 and we have found no evidence of any discontinuous change. However, interestingly we have found that overall dwelling fire risks are remarkably similar in the UK and the US; in 1995 both countries experienced a risk of death in household fires of 8 to 9 deaths per 1000 fires and 14 deaths per million of the population. While the figures were still comparable in 2004, the risk of death in the US had decreased a little to 11.7 per million of population, which we can probably attribute to improved efficiency of smoke alarms and perhaps an increased awareness of the need to keep them maintained.

14.6.2 Smoke alarms trends

In relation to smoke alarms trends in the UK shown in Fig. 14.6, up to 1988 the number and effectiveness of smoke alarms was very low. In 1988 it was estimated that around 10% of homes had alarms but these were judged by the UK Home Office to be less than 10% effective for raising an alarm when a fire occurred.

By 1993 the penetration had risen to 68% but alarms were still only considered to be 10% effective. In 1996 the penetration had increased further to around 70% but the effectiveness was judged to be 18%. In the next six years the penetration increased more slowly and has now levelled at around 80% and the effectiveness has been judged to be around 25% currently.

Figure 14.7 shows the trends in the effectiveness of the smoke alarms in the UK and USA as measured by three ratios (i) the number of fires per 1000 households, (ii) the number of deaths and (iii) the number of injuries, in all cases ratioed with respect to the percentage number of households with smoke alarms fitted.

Figure 14.7 shows that smoke alarm effectiveness in the UK has now probably converged with that in the USA which claims 90% penetration. In 1990/91, the UK was only as effective as the US was in 1980. The rapid catch-up in effectiveness appears to be related to UK Home Office campaigns to encourage better siting and maintenance of residential smoke alarms and the reducing cost of smoke alarms in real terms.

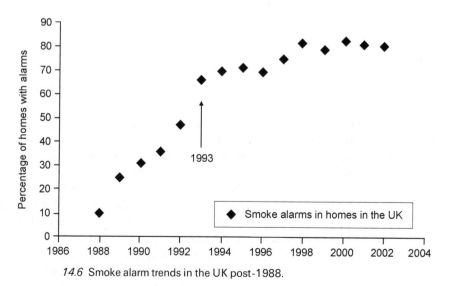

14.6 Smoke alarm trends in the UK post-1988.

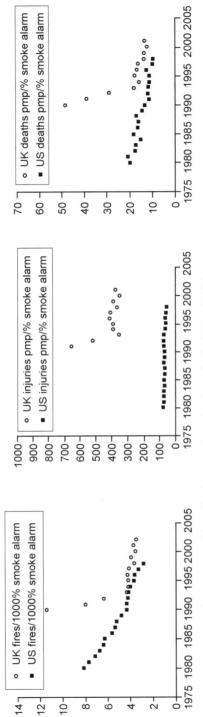

14.7 Trends in measures of the effectiveness of smoke alarms in the UK and USA.

In 1988 the penetration of smoke alarms in the UK was 10% and, if only 10% of these were effective, then only 1% at best of households would have had their risk of death and injury in fire affected by the presence of smoke alarms.

In 1996 when the penetration was 70% and the effectiveness was judged to be 18%, then at best 13% of households would have gained a benefit from reduced risk of death or injury in fire. In the 8-year period between 1988 and 1996 the reduction in the risk of death in residential fire in the UK dropped from ~17 per million persons or pmp pa to ~12 pmp pa, a 29% decrease is risk – of this we could suggest that no more than 13% was due to smoke alarms and the rest, at least 16%, was due to the 1988 furniture fire regulations.

Currently, the risk of residential death in fire in the UK is close to 10 pmp pa – a 41% reduction in risk compared with the pre-1988 position. Current smoke alarm penetration is close to 80% and the Home Office estimate the effectiveness to be ~25%, so at best 20% of households would gain a benefit from possible risk reduction due to the presence of their smoke alarm, leaving at least 21% of the risk reduction ascribed to the 1988 regulations.

Therefore, in 1988 practically no benefit was achieved by smoke alarms but since then benefit has grown as the penetration and effectiveness of alarms has increased. Today up to possibly 50% of the post-1988 annual benefit in reduced fire impacts could be ascribed to the presence of smoke alarms. If correct, then 3.5 deaths pmp pa are currently avoided by the presence of the 1988 furniture fire regulations and a further 3.5 deaths pmp pa avoided by the presence of more effective smoke alarms.

14.6.3 Benefits of passive and active measures working together

Since about 1993, we can identify further savings in UK fires, deaths and injuries, which we can attribute to increased efficiency/use of smoke alarms in domestic buildings, as shown in Fig. 14.8, where standard statistical methods have been used to fit linear lines to the data and calculate the 95% confidence limits (shown dashed). Total savings post-1993 must therefore be considered to be due to a combination of the 1988 regulations and improvements in smoke alarm efficiency and it is very difficult, if not impossible, to separate the two. Historically, however, research suggests that, in the absence of fire-inhibited furniture, which slows the rate of spread of fire, smoke alarms only alert a sleeping householder in time for them to wake up to die. The continuing, slow, downward drift in the number of people smoking must also have some effect, given that such a high proportion of fires is caused by smokers' materials. However, this effect is largely obscured by the increased prevalence of fire-retarded materials in the population, especially given the fact that most smoker-related fires start in soft furnishings.

The improvement in the figures for lives saved post-1993 can be attributed to

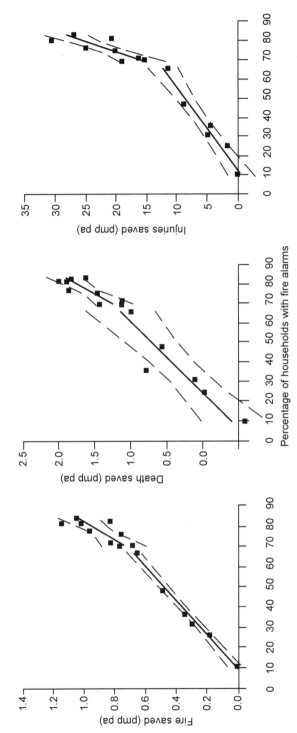

14.8 (a) Total fires saved pmp by smoke alarms pre- and post-1993; (b) Total deaths saved pmp; (c) Total injuries saved pmp.

a combination of increased use of smoke alarms in domestic property together with increased efficiency of the alarms themselves.

However, even in 2002 smoke alarms were absent in fires which accounted for 240 deaths and 6,881 non-fatal casualties. This clearly demonstrates that reliance on smoke alarms as a prime fire countermeasure is inappropriate and other measures are required such as fire-protected consumer products. The primary benefit of alarms is in decreasing the risk of death in fire; it is 3–4 deaths per 1000 fires when fires are detected by alarms compared to 7–9 deaths per 1000 fires in the absence of alarms.

14.6.4 Socio-economic factors

A statistical analysis of patterns of domestic fire risk was reported in the 2000 British Crime Survey[24] which gives an insight into the socio-economic factors that may influence domestic fire risks. A detailed bivariate and multivariate analysis was performed to try to understand the reasons for unequal fire risks across the population. This included an analysis of those groups most strongly linked to the occurrence of domestic fires and those same groups that have the highest probability of not owning a smoke alarm.

The groups which showed the highest risks of experiencing a domestic fire according to both bivariate and multivariate analysis were:

- the head of the household is between 16 and 24 years of age
- *the household is getting into financial difficulties*
- the respondent has a limiting disability
- *someone in the household smokes*
- the respondent has A-level educational qualifications or above
- the household contains one adult living alone with children
- the household has a low income (£2,500–£4,999 per annum).

The groups which showed the highest risks of not owning a smoke alarm according to both bivariate and multivariate analysis were:

- the household contains one adult living alone
- the property is privately owned
- the property is a purpose built flat
- *somebody in the household smokes*
- *the household is getting into financial difficulties*
- the respondent has no qualifications
- the respondent is Asian.

These results, and the others not cited here, indicate that complex social factors affect the patterns of fire risk. However, what is clear is that some common factors affect both risk areas, such as those highlighted in italics in the lists above. In such cases these groups will be at greatest risk because the individual risks compound. Under such circumstances the presence of passive fire protection methods, such as that achieved through the UK furniture fire regulations, in addition to encouraging active measures, such as working smoke alarms, is essential.

14.6.5 Future UK prospective trends and benefits

It would be useful to project the savings post-1988 into the future and estimate the total potential savings over an extended period of time. In order to do this, we first need a model for the penetration of new, fire-retarded furniture into the marketplace and fit it to existing data.

The percentage penetration (P_N) of the UK market by new upholstered furniture since 1988 has been estimated using an exponential penetration growth model of the form:

$$P_N = (1 - e^{-kt}) \cdot 100\% \qquad [1]$$

where k is the annual rate of penetration.

In the analyses that follow, we used annual furniture replacement rates of 4, 6 and 8%, which equate to product half-lives of 8, 11 and 16 years respectively (i.e. 50% of households change their furniture every 8 to 16 years). The fraction of new furniture in the population is calculated using the formula:

Current year's % population of new furniture =
$\{1 - P_N \cdot$ (previous-year % population of old furniture)$/100 \cdot 100\%$
$$N_{N,i} = \{1 - P_N \cdot N_{O,i-1}/100\} \cdot 100\% \qquad [2]$$

Post-1998 forward projections of the savings in the number of fires, deaths and injuries are dependent on making assumptions about:

- the rate of penetration of the market by new furniture (P_N),
- the effectiveness (E_I) of the regulations (and the measures that satisfy the regulations) in reducing the incidence of fire and
- its consequences (e.g., deaths, injuries, etc.).

In general, we can say that the number of savings (S) is a function (F) of the product $P_N \cdot E_I$, where P_N is also a function of time. Mathematically:

$$S = F[P_N(t), E_I] \qquad [3]$$

It is impossible to separate the two variables at this point in time from the data available. The best we can do is make assumptions about $P_N(t)$ and use the existing data to infer the effects of E_I.

Note that $P_N(t)$ and E_I must change in opposing senses if the savings product is to fit existing 1988–2000 data, i.e. the more rapid the assumed penetration, the less effective the regulations must be to give the same result. The corollary therefore is also true, that the scenario with the highest assumed rate of

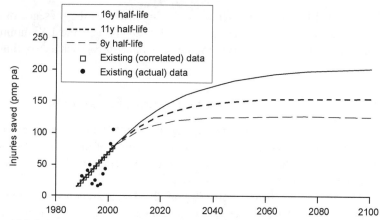

14.9 Estimated past and future lives saved for non-fire-inhibited UK furniture half-life scenarios of 8, 11 and 16 years (equivalent to replacement rates of 4%, 6% and 8% pa respectively).

penetration of new furniture will give the lowest predicted savings when $P_N(t)$ approaches 100%. The question to answer is 'when will the savings plateau out and at what level?'.

Plotting S against $P_N(t)$, as estimated above, gives a clue to the value of E. In our original studies,[19] we used a simple power law function of the form:

$$S = A \cdot P_N(t)^B \qquad [4]$$

However, using the latest set of UK fire statistics, we have recently refined the model to a parabolic equation of the form:

$$S = A \cdot P_N(t)^2 + B \cdot P_N(t) + C \qquad [5]$$

where A, B and C are constants which vary according to the penetration function chosen; between them the constants set the maximum value of S as $P(t)$ approaches 100%.

Table 14.5 Prospective annual and cumulative benefits of the regulations up to 2031 for an 11-year half-life (annual furniture replacement rate of 6%)

Benefit measure	2011 Annual	$\sum$ [a]	2031 Annual	$\sum$
Fires saved	9300	56,700	14,600	109,000
Lives saved	642 (577 : 658)[b]	9360 (8971 : 9432)	897	11912
Injuries saved	6,480 (5,160: 7,800)	90,400 (77,600: 104,500)	9,300	118,100

[a] The cumulative benefits are calculated from 1988.
[b] The values in parentheses are the 95% confidence limits.

We can use our three penetration scenarios as the best upper, lower and middle case options currently available, as shown in Fig. 14.9 for deaths saved.

Table 14.5 shows the projected estimated middle case scenario savings for an annual furniture replacement rate of 6%, for years up to 2031. The figures in parentheses indicate the upper and lower 95% confidence limits on the estimates.

14.7 Future trends

14.7.1 Maintaining knowledge of fire impacts in society

Our work has shown the importance of using reliable fire statistics and their longer-term trends, alongside laboratory- and fire-based tests of reaction to fire information of consumer products and simulated room fires, to quantify the impacts of fire for residential fires in society. This was only possible because of the effort expended in the gathering and consolidating of reliable fire statistics.

There is an unfortunate trend in Europe to stop the collection of fire statistics, whereas in Scandinavia the trend has been the opposite, with most countries in the region improving the quality and reliability of their fire data collection. Alternative sources of information on global performance are provided by the World Fire Statistics Centre which is hosted by the Geneva Association (www.genevaassociation.org/WFSC.htm) and has United Nations support. However, this operates through voluntary information contributions, which are helpful but do not have the same level of confidence that can be achieved with national fire reporting schemes.

With the increasing complexity of consumer products, the increasing use of plastics and natural materials with concomitant increases in fuel loads in domestic and public buildings, there are increasing risks of fire. These risks must be understood and actions taken to reduce and, where possible, eliminate them – the collection of reliable fire statistics linked to other policy measures is an effective way of achieving this.

14.7.2 Better understanding human exposure to flame retardants

Quantitative assessment of the societal benefits of flame retardant use in consumer products is illustrated well in the case of the UK regulations on the fire performance of upholstered furniture. However, with these benefits comes the need to better understand the potential risks associated with the production and use of flame retardants.

Both human exposure and environmental risk assessment require that releases during the manufacturing phase and the consumer use phase are well understood. Such understanding has been frustrated by a lack of research on

FR release behaviour and of the degree of human exposure that can occur during occupational and consumer exposure.

In 1999, when the NAS Subcommittee study[6] was undertaken, it was very difficult to obtain reliable exposure data – it was simply not available in the scientific and technical literature and only limited information was available from industry sources. At that time it was necessary to make crude estimates of what the release behaviour could be for the primary exposure routes of oral, dermal and inhalation exposure.

This position has improved since then with more systematic studies conducted on a limited number of flame retardants and consumer products. These data have informed more recent risk assessments but it is still apparent that gaps exist in our understanding of release behaviour, particularly during consumer product use and as a result of wear and ageing. Exploratory studies by the authors[12,13] suggest that ageing and wear play a major role in changing the release characteristics, in some cases beneficially but in others detrimentally. Similarly, there are no established and validated mathematical models to describe FR release and in the light of REACH this should be corrected.

14.7.3 Controlling and eliminating flame retardant exposure risks

Flame retardant exposure risks can be controlled in a number of ways. The most obvious way, during the manufacture of the compounds and their incorporation into consumer products, is by careful monitoring of losses to the environment and via waste streams, be they solid or liquid, and by careful occupational monitoring. The voluntary measures introduced by the flame retardant manufacturers both by individual companies, and by the extension of the VECAP approach to more FR sectors, are necessary steps to convince all stakeholders that exposure risks are being actively managed and effectively reduced. With the development of significant materials recycling activities, it is important that the lessons learnt and methods adopted by the primary manufacturers are applied in the waste management and recycling sectors and that the fate of historic FRs are managed well to avoid perpetuation of any residual risks. For some flame retardant-containing materials recycling for continued use may be a safe option but other FRs, which may not be approved for continued use, should be destroyed using appropriate incineration and other chemical destruction methods, such as thermal plasmas.

Controlling and eliminating consumer and environmental exposure risks in the use phase of consumer products is best achieved by adopting materials formulations or individual FRs that are intrinsically safe. Alternatively, if flame retardants are required which are found to present some risks, these should be strongly incorporated into the fabric of the materials by physical means or by chemical grafting or crosslinking, e.g., TBBA as mentioned earlier. In the case

The risks and benefits of flame retardants in consumer products 393

of TBBPA, the European risk assessment has very recently reported that after an eight-year study of its potential risk to human health and the environment, it is safe to use without restriction in the EU. TBBPA is used in more than 70% of printed circuit boards so this finding is of particular relief to FR4 board manufacturers. The EU approved a Risk Reduction Strategy which recommended an environmental permit to monitor the emissions at a particular plant in Europe and the European Commission's Scientific Committee has also confirmed the EU Risk Assessment conclusions.

In all cases it is necessary that the potential for release by fluid extraction, bulk physical diffusion and surface loss and by material ageing and wear processes be measured and the results used to assess the safety case and the need or otherwise for greater containment. Some protocols for this type of assessment have been put in place by CPSC and our group, GnoSys,[12,13] but they now need to be more widely and routinely applied to ensure a body of experience is established to handle the most general requirements. This need also exists to simplify and facilitate the operation of REACH in the future.

14.7.4 Maintaining fire safety in a world of competing and contradictory regulations

Consumer product safety in general and fire safety in particular are basic requirements of any developed society, but they do not at present appear explicitly on any agenda for sustainable development. In a world which is focused on challenges such as climate change, energy efficiency, carbon footprints, resource efficiency, and producer responsibility, to name only a few, many regulators and many non-governmental organisations may be failing to acknowledge the wider needs of public protection and safety.

Care is required on the part of regulators to ensure that fire safety is maintained and where possible improved and that measures intended to protect society are not compromised by contradictory regulation. An example is the introduction of REACH in Europe but the failure to introduce European collection of fire statistics or require Member States to maintain their own collection. Similarly, scientific evidence and SCHER (Scientific Committee on Health and Environmental Risks) decisions supporting the safety and continued use of certain brominated flame retardants in Europe may be circumvented by individual member states taking unilateral action to ban these compounds.

14.7.5 Targets for improvement in fire risk reduction

Our work on comparing fire safety trends in Europe and the USA[3] has shown that some of the better performing regions have been able to achieve death rates in the region of 4 to 8 deaths per million of the population per annum (pmp pa). In Europe more effective counter-measures could enable a target to be set of

6 pmp pa. which could be achieved in the absence of significant new investment in infrastructure particularly in the existing residential building stock. Such a target would lead to 850 fewer deaths in the current expanded European Union and save an estimated €3.6 billion pa.

This could be achieved by greater attention to pan-European consumer product safety, improved smoke detector effectiveness and improved education on the impacts of fire, particularly for the more vulnerable groups in society, who are known to experience casualty rates three times higher than those in better performing groups. Similar initiatives in the UK have led to a significant reduction in the residential fire death rate from close to 17 pmp p.a. to less than 10 pmp p.a. in less than 20 years. Similar initiatives in California have led to this State having one of the lowest death rates in the US at between 2 and 4 pmp pa.

Further improvements in residential fire safety in Europe can be achieved by continuing with these measures and investing in the installation of appropriate sprinkler systems in new and existing private and publicly owned residential buildings. Indeed, work by Ramachandran[25] has shown, on probabilistic grounds, that sprinklers may achieve a reduction in fire death rate to close to zero. However, to achieve this would require significant investment and take many years to fully accomplish. Consideration of the performance and reliability of different types of sprinklers systems would also be required.[26] Even this technology is not without its countervailing risks and these, too, would need to be better enumerated. Nevertheless the prospects of achieving very low fire death rates are extremely attractive, if the cost benefit case can be shown to exist and is politically acceptable.

14.8 Sources of information and advice

UK sources

UK fire statistics can be obtained from the Home Office, Fire Statistics Bulletins and individual reports back to 1966. UK population statistics were taken from the 1961, 1971, 1981, 1991 and 2001 decennial census reports. Inter-census estimates of population were provided by the Office for National Statistics.

Production and sales data of furniture and bedding were obtained from Business and Research Associates in their reports on The UK Market for Upholstered Furniture[20] and The UK Market for Beds and Bedding.[20]

Economic impacts were assessed using a variety of sources. Data on costs associated with insurance industry loss-adjusted fire claims can be obtained from the Fire Protection Association (FPA). These data extend back to 1986. More general data on patterns of fire risk in the home environment can be obtained from British Crime Survey results.[23,24] Account was also taken of the costs associated with loss of a statistical life; this was set at £3m, based on independent work for the DTI (now designated BERR) on this subject.[27] Use

The risks and benefits of flame retardants in consumer products 395

was also made of Home Office Research, Development and Statistics Directorate report on the 'The economic costs of fire'[28] which includes an appraisal of the whole costs of fire in the UK and in some other countries where such exercises have been carried out.

European

Additional economic data on international comparisons of fire economic impacts (including Europe) can be obtained from the International Association for the Study of Insurance Economics at the Geneva Association and its bulletins on World Fire Statistics (WFS).[29] However, the precise rationale that is used by individual country correspondents used to respond to the WFS questionnaire is not specified so the accuracy of the data is unknown.

Individual European county data can be obtained from:

- Ireland: http://www.environ.ie/general/fire.html#10
- The Federation of the European Union Fire Officers Associations: http://www.f-e-u.org/index.php?newlang=english
- WHO Statistics of Fire Deaths: http://www.genevaassociation.org/FiredeathsCentral%20Eastern%20Europe%20etc.pdf
- World Fire Statistics Centre: http://www.genevaassociation.org/WFSC.htm
- UK: http://www.homeoffice.gov.uk/rds/fire1.html
- Norway: Directorate for Fire and Explosion Prevention (DBE)
- Denmark: Raddningsverket

United States of America

USA sources of information can be obtained from a variety of direct sources and websites including the Consumer Products Safety Commission (CPSC) and its Statistical Office, the National Statistical Office and Census and their web sites, the National Fire Protection Association (NFPA) and its web site.

US fire statistics are also provided by the Directorate for Economic Analysis, US Consumer Product Safety Commission in Washington D.C. US population and household statistics were obtained from government census office web pages at http://www.census.gov/population/www/index.html.

Information can also be obtained from individual US States fire authorities and similarly in Europe and Scandinavia. These include:

- California: http://osfm.fire.ca.gov/
- Washington: http://www.wa.gov/wsp/fire/firemars.htm
- Texas: http://www.tdi.state.tx.us/fire/indexfm.html
- South Carolina: http://www.llr.state.sc.us/FMARSHAL/
- North Carolina: http://www.ncdoi.com/OSFM/default.asp
- Maryland: http://www.firemarshal.state.md.us/

- Connecticut: http://www.state.ct.us/dps/DFEBS/OSFM.htm
- Florida: http://www.doi.state.fl.us/SFM/indextest.html
- New York: http://www.dos.state.ny.us/fire/firewww.html
- Oregon: http://www.sfm.state.or.us/
- Alaska: http://www.dps.state.ak.us/fire/asp
- Colorado: http://www.state.co.us/gov_dir/cdps/dfs.htm
- Illinois: http://www.state.il.us/osfm
- Kansas: http://www.ink.org/public/ksfm
- Minnesota: http://www.dps.state.mn.us/fmarshal/fmarshal.html
- Oklahoma: http://www.oklaosf.state.ok.us/~firemar/
- Virginia: http://www.vdfp.state.va.us/

Further reading

Stevens G C, 'Countervailing Risks and Benefits in the Use of Flame Retardants'. *Flame Retardants 2000*, Interscience Publishers, London, 59-70, 2000, ISBN 0 9532312 4 0.

Hartzell G E and Emmons H W, 'The fractional effective dose model for assessment of hazards due to smoke from materials', *J. Fire Sciences*, **6**, 356–362, 1988; see also Hartzell G E, 'Overview of combustion toxicology', *Toxicology*, **115**, 7–23, 1996.

14.9 References

1. Stevens G C, Emsley A and Lim L, 'International comparison of fire statistics and the potential benefits of fire counter-measure', In *Flame Retardants 2002*, Interscience Communications, London, 2002, pp. 23–32.
2. Fardell P J, Murrell J and Lunt M G, 'Chemical analysis of fire gases'. *Chemistry in Britain*, 226–228, 1987.
3. Stevens G C and Mann A H, *Risks and Benefits in the Use of Flame Retardants in Consumer Products*, DTI ref. URN 98/1026, January 1999; and *Technical and Commercial Annexes* PRC07b/98/DTI, January 1999 (available for the Polymer Research Centre, University of Surrey).
4. Stevens G C, Emsley A, Lim L, and Williams P, 'The benefits of fire counter-measures in the UK and Europe from a consideration of UK and international fire statistics', In *Flame Retardants 2006*, Interscience Communications, London, 2006, pp. 235–246.
5. Stevens G C and Mann A H, 'Risks and benefits in the use of flame retardants in consumer products', In *Flame Retardants '98*, Interscience, London, 1998, pp. 59–70.
6. Toxicological Risks of Selected Flame-Retardant Chemicals, Subcommittee on Flame-Retardant Chemicals, National Research Council, National Academy of Sciences, Washington, 2000.
7. http://www.epa.gov/dfe/pubs/index.htm#ffr.
8. http://www.cpsc.gov/cgi-bin/pub.aspx.
9. European Chemicals Bureau http://ecb.jrc.it/ (search on Flame retardants) and SCHER http://ec.europa.eu/health/ph_risk/risk_en.htm.
10. BSEF web site http://www.bsef.com/fire_safety_benefits/.
11. FDA website May 7th, 2007 http://www.fda.gov/consumer/updates/

melamine051407.html.
12. Stevens G C and Horrocks A R, 'Textile back-coating environmental challenges for the UK furnishing fabrics industry: Release and exposure of flame retardant species', *J Ind.Text.*, **32**, 267–278 (2003).
13. Stevens G C, Ghanem R, Thomas J L, Horrocks A R and Kandola B, 'Understanding flame retardant release to the environment', . In proceedings of *Flame Retardants 2004*, Interscience Communications, London, 2004, pp. 37–50.
14. Babrauskas V, 'Toxic Hazards: Control by Limiting Toxic Potency or Control by Limiting Burning Rate?' In *Flame Retardants '94* Interscience Communications, London, pp. 239–250, 1994.
15. Hirschler M M, 'Fire Retardance, Smoke Toxicity and Fire Hazard'. In *Flame Retardants '94* Interscience Communications, London, pp. 225–237, 1994.
16. Babrauskas V et al., *Fire hazard comparison of fire-retarded and non-fire-retarded products*, NBS Special Publication 749, 1988.
17. HMG (1988), *The Furniture and Furnishings (Fire) (Safety) Regulations 1988*, Statutory Instruments 1988 No. 1324, Consumer Protection.
18. HMG (1989), *The Furniture and Furnishings (Fire) (Safety) (Amendment) Regulations 1988*, Statutory Instruments 1989 No. 2358, Consumer Protection, Public Health, England and Wales, Public Health Scotland.
19. HMG (2000), *Effectiveness of the Furniture and Furnishings (Fire) (Safety) Regulations 1988*, URN 00/783 – authors Emsley A and Stevens G C.
20. Business and Research Associates (1997a), *The UK Market for Beds and Bedding*, July 1997; obtainable from BRA, 9 Market Street, Disley, Stockport SK12 2AA.
21. Business and Research Associates (1997b), The UK *Market for Upholstered Furniture*, September 1997; obtainable from BRA, 9 Market Street, Disley, Stockport SK12 2AA.
22. Ball D J, Ives D P, Wilson I G and Postkle M (1998), *The Optimisation of Consumer Safety*, report for the Department of Trade and Industry, October 1997.
23. Home Office (1997b), *Fires in the Home in 1995: Results from the British Crime Survey*, Home Office Statistical Bulletin, Budd T and Mayhew P, 10 April 1997, Home Office.
24. Department of Transport, Local Government and Regions (DTLR), (2000), *Fires in the home: findings from the 2000 British Crime Survey*, Rebecca Aust, 2 August 2001.
25. Ramachandran G. – private communication and, 'Early detection of fire and life risk', *Fire Engineers J.*, 33–35, December 1993.
26. DCLG Research Report, *Effectiveness of sprinklers in residential premises – an evaluation of concealed and recessed pattern sprinkler products*, March 2006, Building Research Establishment.
27. Ball D J, Ives D P, Wilson I G and Postkle M (1998), *The Optimisation of Consumer Safety*, report for the Department of Trade and Industry, October 1997.
28. Weiner M, *The economic cost of fire*, Home Office Research Study 229, October 2001.
29. Wilmot T and Paish T, Information Bulletins of the World Fire Statistics Centre, WFSC.

15
Composites having improved fire resistance

B K KANDOLA and E KANDARE,
University of Bolton, UK

Abstract: For the use of fibre-reinforced polymeric composites as structural materials for different applications, the associated fire risks and their assessment are crucial. For critical applications especially, most products have to conform to certain specified fire performance requirements and regulations. Another important issue is the mechanical performance of these composites during heat/fire exposure and the residual strength. This chapter reviews the key issues and performance requirements for load-bearing fibre-reinforced polymer composites for various sectors of the industry. Several ways of improving their fire resistance are discussed and their effectiveness assessed by post-fire mechanical property evaluation.

Key words: fibre-reinforced polymeric composites, intumescent coatings, additive and reactive fire retardants, nanoparticulate fire retardants, post-fire mechanical performance.

15.1 Introduction

Technological advancement in aerospace, automotive and marine industries requires materials with superior properties and functionalities for new applications, something which is hard to achieve using monolithic materials such as metal, glass, polymer, etc. This has prompted research and development of innovative materials and structures for the next generation of aero-vehicles, automotives, boats, ships, etc. One of the major achievements of recent years has been the development of fibre-reinforced composite materials, which due to their high strength-to-weight ratio can replace conventional materials in load bearing applications. The basic concept of a composite is that when two or more different but compatible materials are brought together into one, the overall properties and functionalities of the resultant material, are different and improved compared to each constituent component.[1,2] Enhanced interfacial features of the ensuing composite component materials are crucial for the improved properties. Most composite materials are composed of a bulk binding material, the matrix, and a reinforcement component, usually in fibre form, serving the purpose of increasing the stiffness and hardness of the hybrid. The core materials such as polymer foams, honeycombs, wood, balsa and cedar are sometimes used to add volume to the composite. There are many different types of composite materials in use today; the following are brief accounts of the most common ones.

Polymer matrix composites

These composite materials utilise a polymer-based matrix reinforced with a variety of natural and synthetic fibres such as glass, aramid, carbon, boron, polyethylene, wood fibre, jute, sisal, flax, wheat straw and bamboo.[1,2] Whilst, polymeric resins such as epoxies, polyesters and phenolics, exhibit good chemical resistance and ease of processibility relative to metals or ceramics, their mechanical properties are very poor. On the other hand, fibres only have strong tensile properties along the fibre length making it impossible to use them in structural applications on their own. A combination of a polymer-based matrix with reinforcement fibres affords a hybrid structure with combined characteristic properties of the constituent components. These hybrid composite materials are chemically inert, abrasion and impact resistant, have high strength and stiffness and low densities, hence are more preferable than metals and metal composites. However, since the matrix in polymer matrix composites (PMCs) is polymer-based, these composites are prone to fire damage raising concerns over the safety of passengers in the construction and transport industries. They are also sensitive to deterioration of mechanical and bond properties.[3] Hence, before PMCs can be used in primary structures, they need to pass standard fire safety tests often regulated by international regulations and laws.

Metal matrix composites

Mostly used in the automotive industry, the matrix for metal matrix composites (MMCs) are usually light metals such as aluminium, magnesium and titanium reinforced with silicon carbide (SiC) and nickel or titanium boride coated carbon fibres. Compared to metals, MMCs offer increased specific strength, stiffness, improved wear resistance, damping capabilities, good corrosion resistance and low densities.[4]

Ceramic matrix composites

Ceramic matrix composites (CMCs) comprise a ceramic-based matrix in both oxide and non-oxide forms reinforced with treated short carbon fibres or whiskers. These materials are ideal for applications in components operating under severely harsh environmental conditions such as military aircraft and space shuttles. Unfortunately, the use of whiskers raises health and safety problems both during processing and in service. However, both MMCs and CMCs are inherently fire retardant as compared to PMCs.

Of all the different composite systems, fibre-reinforced polymeric composites (PMCs) are most commonly used and are ubiquitous in different applications such as land transport, marine, aviation and construction industries. While the mechanical properties exhibited by PMCs are ideal for their structural applications, their flammability needs to be assessed and resolved before their intended applications in various sectors. This chapter complements our earlier reviews,[1,2,5] where general overviews of composite flammability and methods of their fire retardation have been discussed. In this chapter the key fire issues pertaining to the use of fibre-reinforced, polymer matrix composite materials in aerospace, automotive and marine industries, fire retardant solutions, degradation of their mechanical properties at elevated temperatures and the overall performance of the composites under fire are discussed.

15.2 Composites and their constituents

There is a variety of fibre/resin combinations of composites used for different applications, and the most common ones are discussed here. In aerospace applications usually carbon/epoxy composites are used, mainly in the fuselage, wing and tail fin components, control surfaces and doors. A large percentage of the cabin interior of passenger aircraft is made out of polymer matrix composite, mostly of a glass/phenolic type. Phenolic composites are used as a single skin laminate or as a sandwich material that consists of thin glass/phenolic face skins over a polyaramid fibre (e.g., Nomex, DuPont) honeycomb core,[6] in ceiling panels, interior wall panels, partitions, cabinet walls, structural flooring and overhead stowage bins. Fibre reinforcement is usually of the unidirectional or woven roving form. In the marine industry use of glass/unsaturated polyester or glass/vinyl ester composites is more common, usually as thick laminates.[7] For sandwich structures, the core material is balsa or foam. Epoxy resin is sometimes used for smaller vessels, especially wooden boats that are often laminated from thin veneers saturated in epoxy resin. Wooden boats made of traditional planking materials are often coated in epoxy or sheathed with glass fabric wetted out in epoxy. For other applications such as glass-reinforced fibre polymer (GRP) hulls and small commercial vessels exclusively of GRP materials, polyester resin is often used. Mostly glass fibre is used, whereas carbon fibre is for specific uses only, e.g. high performance boats. For marine applications, mostly chopped fibres (mixed with resin and sprayed with a gun) woven roving or mats of non-woven random chopped or continuous strands of fibres are used. In the automotive industry, the external structures of racing cars are made out of carbon epoxy composites, whereas for the interiors of a majority of normal cars a variety of glass/thermoplastic and thermoset resin combinations, e.g. glass/epoxy, glass/nylon, etc., are used. For trains, glass/polyester or glass/vinyl ester composites are more common.

The properties of any composite depend upon the fibre and resin used for its fabrication. Here some generic resin and reinforcements are discussed in brief. For more detailed background information the reader is referred to our previous reviews.[1,2,5]

15.2.1 Resins

Polymeric resin systems constitute a substantial volume fraction of the composite material; hence a careful choice of resin for a specific application is essential. In general, resin systems must have good mechanical and adhesive properties whilst being able to resist environmental and service degradation. In composite materials and structures, adhesion between the resin and reinforcement is of paramount importance to allow efficient load transfer mechanisms while preventing debonding during stress cycles. Cracking and delamination are serious problems challenging the use of composite materials as load-bearing structures. Thus, the resin system selected for a specific application must be tough enough to resist crack propagation. Those most frequently used in structural composites are unsaturated polyesters, vinyl esters, epoxies and phenolics and are discussed in brief here. Our previous publications[1,2,5,8] present more detailed information about their chemical structures and thermal stabilities at higher temperatures.

- **Unsaturated polyester** resins are prepolymer mixtures containing unsaturated groups and styrene with the latter serving as both a diluent and cross-linking agent during the radical polymerisation process.[9] Unsaturated polyester resins are found in two forms based on either orthophthalic or isophthalic ester moieties with the latter preferred for conferring superior water resistance properties. Most unsaturated polyester resin formulations cure at room temperature; however, catalyst, accelerators and heat can be used to speed up the reaction.[10] Additives including fire retardants can easily be introduced during the processing stage. Easy processibility, good mechanical, water and corrosion resistance properties and relatively low cost make them ideal for the marine, automotive and construction applications.[11–13] However, the emission of styrene during the processing and thermal combustion of unsaturated polyester is a serious environmental problem and currently there is considerable research to lower styrene levels without compromising the integral properties of the polymer.
- **Vinyl esters resins** are similar in structure to unsaturated polyester resins differing only in the location of the primary vinyl reactive groups, which are positioned at the ends of the monomeric chains. Vinyl esters are extensively used in the marine industry[13–16] as a replacement for unsaturated polyesters as they offer better mechanical properties and are less prone to hydrolysis damage since they contain fewer ester linkages. Other applications include sewage pipelines, water and chemical storage tanks.
- **Epoxies** refer to a wide variety of cross-linked polymer chains based on polymer monomeric units containing an epoxide group. Epoxy resins are typically formed from the reaction between a diepoxide with a primary diamine. Reactants with higher reactive functionalities result in a highly cross-linked, stiff and tough epoxy network suitable for aerospace

applications.[17-19] Epoxy resins are also light, have increased adhesion, reduced degradation from water ingress and increased resistance to osmosis making them ideal for marine type applications.[20-22] Epoxy resins generally have superior functionalities and material properties when compared to unsaturated polyester resins or vinyl esters but their use is limited due to high cost which can be several times higher than the latter.

- **Phenolics** are obtained by the reaction of phenols with simple aldehydes and are used to make moulded products, coatings and adhesives. These resin systems are mostly used where high fire resistance is essential but they are brittle and do not possess high mechanical properties. They are found in aeronautical and aerospace vehicle construction in which, during atmospheric re-entry, the vehicle is subjected to severe aerodynamic heating[23,24] and the phenolic converts to an ablative char.

Besides these, other resin systems such as cyanate esters, polyimides and bismaleimides are also in use in the aerospace industry.[25]

15.2.2 Reinforcements

The role of reinforcements in composite materials is to increase the mechanical properties of the consolidating polymer resin matrix. Most reinforcements are based on uni-directional, chopped, braided, stitched and woven fibre rovings. Uni-directional fibres are preferred for high performance composites, e.g. aerospace. For high volume applications, such as in the construction industry, the use of chopped reinforcement is ideal, whilst in the rest of the composites, woven fibres (fabrics) are common. Of paramount importance in the choice of reinforcement for a specific composite application are the mechanical and interfacial (fibre-resin) properties, the volume fraction of the fibres and their orientation in the composite. Although different fibre types can be used, only those commonly used for structural composites are discussed herein.

- **Glass fibres** are found in various types depending on their physical properties such as E-glass (electrical), A-glass (alkali), C-glass (chemical) and S-glass (high strength). E-glass has good electrical properties, good tensile and compressive strength and stiffness and is cheaper than other types of glass reinforcement. C-glass is highly resistant to chemical attack; hence, it finds use mostly as the outer layer in laminates for applications in water pipes and chemical storage tanks. S-glass fibre has a higher tensile strength and modulus than E-glass but is expensive to produce. This type of glass reinforcement is usually used for aerospace and hard ballistic armour applications where cost is not as important as safety and protection.
- **Carbon fibres** are produced via thermal treatment of carbon-containing precursor fibres which involves oxidation, carbonisation and graphitisation. Fibres are made from cellulose, pitch and most commonly from

polyacrylonitrile precursors because of its superior fibre properties. Surface treatment is usually applied to improve matrix bonding followed by chemical sizing, which serves as protection during handling. Carbon fibres have the highest tensile strength and stiffness of all fibres used in composite fabrication and they are also highly resistant to corrosion, creep and fatigue. However, they are more expensive and have a low impact resistance relative to glass fibres.

- **Aramid fibres** normally comprise the para-aramid form (e.g., Kevlar®, DuPont) which are highly crystalline aromatic polyamides obtained by spinning the liquid crystalline polymer solutions into thin threads followed by stretching them to increase stiffness, which is a result of high molecular orientation in the longitudinal direction. Aramid fibres are highly anisotropic, have a high strength and low density hence, a very high specific strength. However, the fibres are not recommended for applications under compressive loads and they are considerably more expensive than glass type fibres on a weight basis.

15.2.3 Core materials

The flexural stiffness of a structural panel or beam is shown from the engineer's theory[26] to be proportional to the cube of its thickness. The use of lightweight core materials in composites is to increase the flexural stiffness while keeping the overall weight low. Usually thin laminate skins are glued on both sides of the core material capable of taking a compressive loading without premature failure, which otherwise might result with the outer skin wrinkling and buckling. Core materials are commonly found in the form of honeycomb structures of a variety of base materials including aluminium, aramid paper (Nomex®, DuPont) and phenolic resin impregnated fibreglass. Honeycombs can be processed into both flat and curved composites structures under mild conditions hence with minimum damage on the core material. The size of the cells, material thickness and web material strength determine the mechanical properties of the honeycombs. Aramid and aluminium honeycomb cores are extensively used in the aerospace industry, whereas for the marine industry their applications are limited due to the difficulty in bonding to complex geometrical shapes and also due to a potential for water absorption. The microstructure of wood is similar to the cellular hexagonal structure of synthetic honeycombs, thus wood can be described as nature's honeycomb. While wood can be used as a core material its use is limited by the service environment, as it can easily degrade in water and at elevated temperatures. Other types of wood structure-based, core materials are balsa and cedar, which are most commonly used in the marine industry.

Core materials can also be found in the form of synthetic polymer foams including polystyrene (PS), polyvinyl chloride (PVC), polyurethane (PU), polyetherimide (PEI), etc. Fire retardants can also be easily incorporated into polymeric foams during synthesis, thus improving the thermal stability of the composite on the whole.

15.3 Key issues and performance requirements for different sectors

For the inclusion of polymer-based composites in the transport and construction industries, it is essential to assess the fire risk associated with the application. All products have to conform to certain specified regulations for particular applications. International standards and codes used for fire-safety design of fibre-reinforced polymer composites which provide a summary of typical requirements needed for a specified application are discussed below.

15.3.1 Aerospace

In the manufacture of civilian aircraft, the use of composite materials has long been almost restricted to the secondary structures such as seats, passenger cabins and interior design; however, recently their use in primary structures has increased. The polymer-based composites used for the interior design of the aircraft must be self-extinguishing and exhibit low flame, smoke and toxicity (FST) characteristics and are required to comply with industry and international regulations such as FAR (the US Federal Aviation Regulations) which govern the requirements for materials used for such applications.[27] The suitability and durability of materials used for parts, the failure of which could adversely affect safety, must:

- be established on the basis of experience or tests;
- conform to approved specifications, such as industry and military specifications, or so-called Technical Standard Orders, that ensure their having the strength and other properties assumed in the design data; and
- take into account the effects of environmental conditions, such as temperature and humidity, expected in service. In general, composites used in likely fire zones must be constructed of fireproof material or be shielded so that they are capable of withstanding the effects of fire.

In the United States, the Federal Aviation Administration (FAA) controls the design and operation of aircraft under, the Federal Air Regulations. Various tests are documented and before a composite is used in a civilian or military aircraft, it must show compliance with the requirements of FAR. Some of the tests required are discussed below[28] while being described in more detail in Chapter 20:

- Flammability and flame propagation test (FAR 25.856 (a) Part VI Appendix F) – This is characteristic of thermal and acoustic insulation materials when they are exposed to either a radiant heat source or a naked flame;

- Heat Release Rate Test/Ohio State University calorimeter (FAR 25.853 Part IV Appendix F) – This test is designed to allow the determination of heat release rates for internal fuselage materials;
- Oil burner test (FAR 25.855) – This test evaluates the flame penetration resistance capabilities of a cargo compartment and weight loss characteristics of seat cushions when exposed to a high intensity open flame;
- Vertical Bunsen Burner Test (FAR 25.853, FAR 25.855) – The test determines the heat resistance of the composite to a Bunsen burner flame when tested in accordance which the above mentioned FAR requirements; and
- Smoke chamber test (FAR 25.853) – The method is used to determine the smoke generation characteristics of airplane passenger cabin interior materials to demonstrate compliance with the requirements of FAR 25.853.

15.3.2 Automotive and rail transport

For use of composite materials in railcars and automobiles, the rate of fire spread and toxicity of the fumes produced during their combustion are of vital importance, since in the former, a significant number of rail fires occur in underground tunnels where longitudinal flame spread is rapid (coinciding with escape routes) and ventilation is poor.[29,30] Smoke generation is also important as the smoke from the fire reduces visibility thereby impeding evacuation and causing panic, while the sub-lethal effects and toxicity cause irritation, incapacitation or even death.

In the US, for example, PMCs used in passenger cars and locomotive cabs are expected to pass the test procedures and performance criteria for the flammability and smoke emission characteristics of materials as regulated through the Federal Railroad Administration (FRA) and Department of Transportation (DOT).[31] Some of the tests required for specific applications are listed below:

- Flame spread index (ASTM 162-98, ASTM 3675-98) – This test determines flame spread defined as a product of the rate of flame propagation and the rate of heat release which is supposed to be under a defined level for different components such as cushions, mattresses, flexible cellular foams used for armrests and seat padding ($\leq 25\,kW/m^2$) and light diffusers and windows ($\leq 100\,kW/m^2$);
- Critical radiant flux (ASTM E 648-00) – This is a measure of the thermal behaviour of horizontally mounted flooring materials exposed to a flaming ignition source in a graded radiant heat energy environment in a test chamber. This fire-test-response standard measures the critical radiant flux at the point at which the flame advances the farthest. It provides a basis for estimating one aspect of fire exposure behaviour for exposed attic floor insulation; and
- Specific optical density (ASTM E 662-01) – This fire-test-response standard covers determination of the specific optical density of smoke generated by

solid materials accumulating within a closed chamber due to non-flaming pyrolytic decomposition and flaming combustion. The results are expressed in terms of specific optical density which is a measurement characteristic of the concentration of smoke.

Similarly, in United Kingdom (UK) and European Union (EU), there are various test procedures and performance criteria with respect to the flammability and smoke emission characteristics of materials generally and which are applied to automotive and rail transport. Tests include the well-established BS 5852: 2006 for upholstered furniture composites and/or complete pieces of furniture such as seats in rail compartments when they are subjected to a smouldering cigarette or to a flaming ignition source of thermal output ranging from a burning match to that approximating to the burning of four double sheets of full-size newspaper.[32] On the other hand, smoke production may be determined using ISO 5659-2:2006 based on the well known NBS smoke chamber method[33] and similar to ASTM E662.

15.3.3 Marine transport

The regulations governing materials specifications for use in marine type vessels are found in United States Code of Federal regulations (CFR) Title 46[34] and these are more fully described in Chapter 19. Some of the requirements are given below which have specific relevance to composites:

- Smoke tests (Subpart 164.006) – Composite materials used for deck coverings for merchant vessels need to be tested for smoke emission and the obtained values should be below a certain classification value for the material to be used;
- Structural insulations (Subpart 164.007) – The purpose of this specification is to set forth tests necessary to measure the insulation value of structural insulation specimens under fire exposure conditions; and
- Flame spread and smoke classification (Subpart 164.012) – Composites used for interior finishes for merchant vessels must qualify under defined specifications for flame spread and smoke generation.

As described in detail by Sorathia in Chapter 19, in US naval submarines the use of composites for the interior is governed by MIL-STD-2031 (SH), Fire and Toxicity Test Methods and Qualification Procedure for Composite Material Systems Used in Hull, Machinery and Structural Applications Inside Naval Submarines.[35] Two thermal characteristics required of a composite material used in a submarine are that; (i) it will be thermally resistant not to be the source of rapid flame spread and (ii) that time-to-ignition will be delayed to allow response to the primary source of fire. Several test methods are contained in MIL-STD-2031 such as the oxygen index test (ASTM D2863 modified), smoke obstruction (ASTM E662), flame spread (ASTM E162), cone calorimetry heat release rates (ASTM E1354), etc.

All commercial passenger and cargo ships in European countries have to comply with the fire performance requirements contained in the International Convention for the Safety of Life at Sea (SOLAS)[36] as IMO/HSC Code (Code of Safety for High Speed Craft of the International Maritime Organisation). There is no specific military regulation concerning the use of composite materials for manufacturing structural parts such as decks and bulkheads of surface ships in the UK and other European countries, hence they also comply to IMO/HSC codes. The fire tests to be carried out and the acceptance criteria are defined in the IMO/FTP code (International Code for Application of Fire Test Procedures), which have been mandatory since 1998.[37] This Code allows for use of non-conventional shipbuilding materials, defined as 'Fire restricting', materials which have low flame spread characteristics, limited rate of heat release and smoke emissions. Furthermore, for the areas of moderate and major fire hazard (e.g., machinery spaces, storerooms), the materials used should be 'Fire resisting', i.e., they should prevent the fire and smoke propagation to adjacent compartments during a defined period of time (60 min for high hazard and 30 min for low hazard areas). Composites used for load-bearing structures should be able to maintain their load-bearing capacity within the specified period of time (30 or 60 min). For fire-restricting material characterisation, specified tests include the room corner test, ISO 9705 and the cone calorimeter test, ISO-5660.

Apart from flame tests composite materials used for the external structure must also be corrosion resistant, have reduced degradation from water ingress and have increased resistance to osmosis.

15.4 Flammability of composites and their constituents

For high performance composites, e.g. for aerospace applications, there has historically been a popular misconception that composites do not burn. Such composites have resins with high temperature curing cycles, during which materials react and, cross-link which creates resistance to thermal degradation. So while it is true that they are not very flammable, it cannot be generalised that composites do not burn under any conditions. All types of resins and hence PMCs are prone to heat and fire damage. There are a number of factors that can affect the flammability of composites as discussed below.

15.4.1 Resin type

Most resins soften on heating and decompose into smaller products, which under heating either cross-link to produce non-combustible char or volatilise to combustible species, which burn in the presence of air producing smoke and in some cases toxic gases. This thermal softening and decomposition of the resin at elevated temperatures results in the degradation of mechanical properties of the composite laminate. As the matrix softens, the bonding strength with the fibres reduces, leading to debonding followed by delamination and frictional sliding at the fibre/matrix interface, which may result in catastrophic failure of load-bearing structures. Such losses in properties are exacerbated by parallel effects of chemical degradation which have been reviewed for different polymer matrices used in composites.[5,8] We have recently studied thermal degradation behaviour of a number of different types of phenolic, aerospace grade epoxy and unsaturated polyester resins by differential thermal analysis and thermo-gravimetric techniques.[8] The ranges of temperatures where the maxima of different reactions were observed are reported in Table 15.1. This table also reports the limiting oxygen index values of different thermoset resins.[38] From this and a number of other reviews it may be concluded that ranking of fire resistance of thermoset resin composite components is:

phenolic > polyimide > bismaleimide > epoxy > polyester and vinyl ester

The superior performance of phenolics is due to their non-thermoplasticity and char-forming ability which enables composites comprising them to retain mechanical strength for long times under fire conditions. In addition, it is observed that because such composites encapsulate themselves in char, they do not produce much smoke.[39] Epoxy and unsaturated polyesters, on the other hand, carbonise less than phenolics, produce more volatile fuels during pyrolysis and so continue to burn in a fire.[5] Furthermore, those containing aromatic structures such as styrenic moieties produce more smoke. Char formation is the

Table 15.1 Thermal behaviour of different resins[8] and limiting oxygen index values for resins and composites[38] at 23 °C

Resin	Temperature ranges of peak maxima from DTA-TGA analysis (°C)			LOI vol%	
	Curing	Decomposition	Char oxidation	Cast resin	[a]Composite: resin/glass fabric
Phenolic	140–160	440–520	>500	25	57
Epoxy	153–160	360–430	>500	23	27
Polyester	RT–120	215–400	>500	20–22	*
Vinyl ester	*	*	*	20–23	*
Polyaromatic	*	*	*	*	8
Amine				30	42
Bismaleimide	*	*	*	35	60

[a] Glass fabric reinforced polymer composites (40% resin : 60% glass reinforcement)
* = Test not performed.

key to achieving low flammability and good fire performance. This is because char is formed at the expense of possible flammable fuel formation.

15.4.2 Fibre type

Although the resin system primarily determines the overall fire behaviour of a composite, the reinforcement does have some influence on the burning characteristics of the material, though this is minor in comparison with that of resin. During a crash of an aircraft, small fibre pieces can be freed from the main body of the composite, which are sharp enough to puncture human skin, and small enough to be inhaled and carried down the trachea into the lungs. This can cause serious 'needle-stick' injuries to the rescue team. However, during fires, burnt carbon becomes porous, absorbing chemicals, dust and dirt from the immediate environment and the needle-stick wounds result in these chemicals being injected into skin or inhaled with carbon fibres.[40]

The fibre-reinforced composites are generally less flammable than cast neat resins of similar thicknesses as also seen from LOI results[38] in Table 15.1. This is due to lower resin contents in the former and also the fibres present act as fillers and thermal insulators. Thick laminates with more fabric layers burn slowly giving lower heat release than thin laminates with less fabric layers.[41] Amongst different composite types, there are different factors that can affect the flammability of the laminate and its residual mechanical property, namely: (i) fibre type (glass, carbon, aramid, etc.), (ii) volume content of fibre and (iii) reinforcement type (weave type, fibre/yarn arrangement, etc.). While glass fibres are non-combustible, exposure to high temperatures can result in softening and finally, melting, which effectively lowers their mechanical properties. On the other hand, other commonly used fibres such as carbon and aramid are non-thermoplastic but because they are organic in nature, will decompose and/or oxidise at elevated temperatures.

In our recent work we have studied the effect of different fibre types on the burning behaviour of epoxy composites.[42–44] The thermogravimetric (TGA) curves shown for glass, carbon and aramid fibres in air in Fig. 15.1(a) indicated that glass is inert, showing no major mass loss up to 800 °C, whereas, carbon fibres start losing mass at 280 °C probably by physical desorption of volatiles and are fully oxidised at 800 °C, leaving 5% charred residue. The aramid fibre starts losing mass earlier than carbon fibre. The curves for all fibre/resin combinations in 1:1 mass ratio shown in Fig. 15.1(b) indicate that apart from glass fibre, both resin and fibre decompose or lose mass. This shows that carbon and aramid fibres may add to the fuel content, which in case of carbon is due to solid state oxidation. However, the TGA technique is not truly representative of actual fire conditions. The burning behaviour of the laminates prepared from these fibres in terms of cone calorimetric investigation is given in Table 15.2 (samples 1–3) and Fig. 15.2. The carbon-containing sample shows lower time-

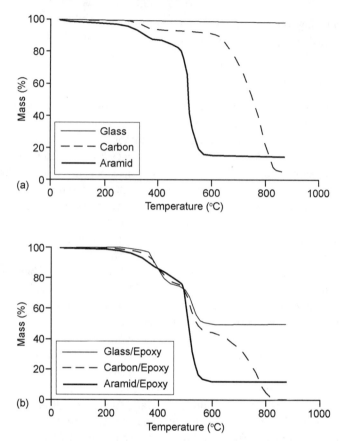

15.1 TGA curves in air of (a) glass, carbon and aramid fibres and (b) fibre/epoxy mixtures in 1:1 ratio.[42]

to-ignition (TTI) values and burns for a shorter period of time compared to the glass-containing sample. The difference between these two samples could be due to their difference in resin contents and fabric area densities. In Fig. 15.2, the sample containing aramid caught fire more slowly and burned for a longer time. The glass sample showed lower peak heat release rate (PHRR) and total heat release (THR) values compared to the carbon sample, but the resin content was lower in the former. Heat release values and curves in Fig. 15.2(a) show that the aramid-containing laminate has the lowest peak although their heat release curves are broader than other samples and they burn for longer times. Surprisingly, the aramid-containing composite has the highest THR value, although its PHRR value is one of the lowest. The higher THR is a consequence of its decomposition and fuel-generating property as shown by the TGA results. However, contrary to the TGA results, the carbon fibre did not show any significant degradation (see Fig. 15.2(b)), which is not surprising since carbon

Table 15.2 Physical and cone calorimetric results of composite laminates[43,44]

Sample (fibre/epoxy) composites	Physical properties			Cone results at 50 kW/m² heat flux			
	Density (g/cm³)	Average thickness (mm)	Fibre vol. (%)	TTI (s)	PHRR (kW/m²)	THR (MJ/m²)	Char after 300 s
1 Glass	1.26	2.7	65	34	373	20.7	71
2 Carbon	1.03	2.8	53	25	413	21.1	62
3 Aramid	0.89	3.3	47	31	295	40.6	37
4 Bi-directional hybrid	1.05	3.0	45	26	357	29.9	54
5 Hybrid 1 G/2C/2G/G	1.19	2.4	65	31	334	18.8	71
6 Hybrid 2 2C/4G/2C	1.18	2.6	61	33	284	20.7	67
7 Hybrid 3 G/2A/2G/2A/G	1.17	2.4	63	36	275	25.0	59
8 Hybrid 4 2A/4G/2A	0.92	3.4	57	27	217	30.4	55

G = E-glass fabric, 273 g/m², plain weave
C = Carbon fabric, 197 g/m², satin 4H weave
A = Aramid fabric, Kevlar®, 173 g/m², satin weave
Bi-directional hybrid: E-glass and carbon, 175 g/m², plain weave,
Hybrids 1 and 2: Glass and carbon fabrics with different lay-ups
Hybrids 3 and 4: Glass and aramid fabrics with different lay-ups

does not decompose and the mass loss in TGA is due to solid state oxidation. For detailed information about this work the reader is referred to our other publications, where these results are also analysed in terms of effect of resin content and fabric area density.[42–44]

Morrey[40] however, has reported that in a similar study the glass fibres present reduce TTI and produce both an earlier and a higher PHRR value compared to carbon fibres. Although in our study TTI and PHRR results are contrary to this (which is due to difference in resin contents, see Table 15.2), the PHRR value for glass sample occurs earlier than for the carbon-containing sample (see Fig. 15.2(a)). This is due to the thermal insulating property of glass causing the resin layer to rapidly reach the critical ignition temperature of the matrix, whereas carbon conducts the heat throughout the composite, prolonging the time needed to achieve this critical temperature.

Morrey[40] also reported the effect of reinforcement type for two glass-reinforced vinyl ester composites, with 30% weight fraction glass in chopped strand mat (CSM) and the other with woven roving. The CSM sample ignited earlier, burned longer but with a lower PHRR value. This was due to the resin-rich surface of this composite. Other researchers[45] also reported small differences between the fibre arrangements; however, the major variable was

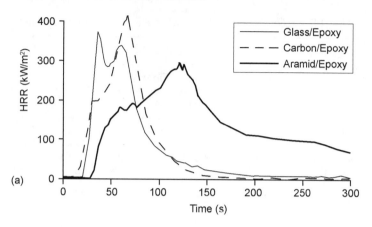

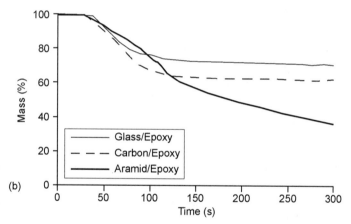

15.2 (a) Heat release rate and and (b) mass loss versus time curves for different fibre/epoxy composites at 50 kW/m² heat flux.[43]

fibre content. With fibre content less than 50% by weight, the composite easily burns in air.

15.4.3 Resin/fibre combinations

The flammability of a composite not only depends upon resin and fibre type, but also the combination of the two. A flame- or heat-resistant fibre can be effective in reducing flammability for one type of resin but not necessarily for another. This can be seen from Fig. 15.3(a), where the results reported are taken from Hshieh and Beeson's work on flame retarded epoxy (brominated epoxy resin) and phenolic composites reinforced with glass, aramid and graphite fibres, tested with the NASA upward flame test.[46] The phenolic/graphite combination has the

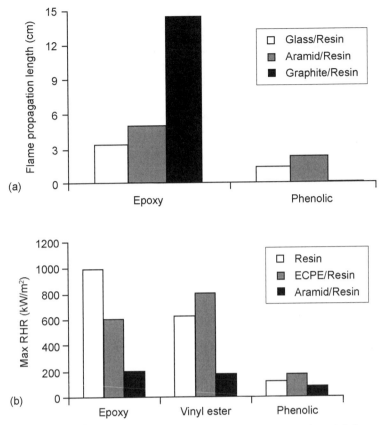

15.3 Effect of fibre type on flammability of different composites: (a) flame propagation lengths[46] and (b) maximum RHR values at 50 kW/m^2 cone irradiance.[47]

highest and epoxy/graphite composites the lowest flame resistance, indicating that overall fire performance is not simply the average of the components present. The most fire retardant graphite or carbon reinforcement has produced the most flammable composite with epoxy possibly because the carbon fibres prevent the liquid decomposition products from the resin from dripping away in the upward flame test – this is the so-called 'scaffolding effect', often seen in blends of thermoplastic and non-thermoplastic fibres in textiles.[5] Conversely, the presence of the char-forming phenolic will complement the carbon presence in the graphite reinforcement and so present an enhanced carbon shield to the flame.

Another interesting work in this context is the study of fire performance of extended-chain polyethylene (ECPE) and aramid fibre-reinforced epoxy, vinyl ester and phenolic composites by Brown and co-workers.[47] Various parameters were determined for ECPE and aramid fabrics only, matrix resins only and their composites and maximum rates of heat release (peak RHR) only are plotted in Fig. 15.3(b). ECPE reduced the flammability of epoxy but increased it for the vinyl ester matrix resins. Aramid, on the other hand, had little effect on time-to-ignition (compared to resin alone) except for the phenolic, but reduced RHR.

Hence, the overall burning behaviour of a composite depends on fibre type, resin type, their respective flame retardant mechanisms, and transferability and possible synergism or antagonism between component mechanism.

15.4.4 Thickness of laminate

In general, the composite thickness of a structure can affect the surface flammability characteristics up to a certain limiting value. Our cone calorimetric study on glass-fibre reinforced polyester composite samples[41] showed that as the thickness of the laminates increases, TTI and also duration of burning (flameout time) increases. When sample thickness becomes equivalent to being thermally thick (i.e., the heat wave penetration depth is less than the physical depth) stage, TTI is less affected and burning slows down. Mass loss curves in Fig. 15.4(a) show that thermally and physically thin samples burn quickly losing mass quickly, whereas for thicker samples mass loss rates are slower. Peak HRR values decrease with thickness (see Fig. 15.4(b)), but as thick samples burn for longer times, THR and average HRR values increases. In Fig. 15.5 the effect of thickness on polyester, epoxy and phenolic composites, based on data taken from a study by Gibson[48,49] is shown, where it can be seen that the effect of thickness is greater in polyester and vinyl ester compared to epoxy and in particular phenolic-containing composites. Phenolics being inherently flame retardant, do not burn vigorously even when are thermally and physically thin.

15.4.5 Intensity of fire

For a given composite of defined thickness, the flammability also depends on the intensity of the fire or more correctly, the incident heat flux. While many large-scale fire tests involve heat sources or 'simulated fires' having constant and defined fluxes, in real fires, heat fluxes may vary. For example, a domestic room filled with burning furniture at the point of flashover presents a heat flux of about $50\,kW/m^2$ to the containing wall and door surfaces; larger building fires present fluxes as high as $100\,kW/m^2$ and hydrocarbon fuel 'pool fires' may exceed $150\,kW/m^2$. Scudamore[50] has shown by cone calorimetric results that the thermally thin to thick effect decreases as the external heat flux increases. At 35 and $50\,kW/m^2$, thin samples (3 mm) ignited easily compared to thick samples (9.5 mm), but at 75 and $100\,kW/m^2$ there was not much difference and both sets of samples behaved as if they were 'thermally thin'.

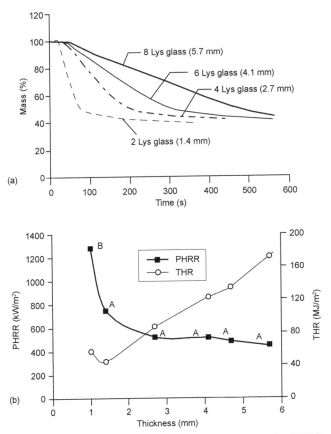

15.4 Effect of laminate thickness on (a) mass loss and (b) PHRR and THR at 50 kW/m^2 cone irradiance of glass/polyester composites (A samples are with random mat, B is with woven roving) (reprinted from reference 41 with permission from Elsevier.[41]

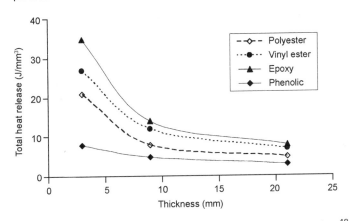

15.5 Effect of laminate thickness on total heat release of composites.[48,49]

15.4.6 Core materials

Core materials are usually made from paper, polymeric foams and wood which are all highly flammable. These core materials are fully encapsulated within the laminate, hence are not directly exposed to fire. However, when exposed to fire and/or heat, a foam will melt and burn, wood will burn and/or char depending upon the intensity of fire and an aramid will char. Aluminium cores, on the other hand, will be initially unaffected in fire, but will melt at very high temperatures and the mechanical properties of the sandwich structure will diminish as a function of time.

15.5 Fire retardant solutions

Although a variety of fire retardant solutions are available such as surface coatings,[51,52] resin modification,[53–56] additive/reactive fire retardants,[57–60] nanoclays,[61] nanotubes,[62–64] and nanofibres,[65] their use is very application specific. In most cases, imparting fire retardancy and smoke reduction to composites often results in reductions in their mechanical strength, which can be very detrimental if they are to be used in primary structures. Therefore, achieving a certain level of fire retardancy while maintaining other such properties is a major challenge. However, for some of the applications, e.g. cabin interiors of aircraft and ships, mechanical properties are not as important. Following our earlier reviews of the work done in this field,[1,2,5] here we extend these studies to include research work undertaken in our own and other laboratories since 2004.

15.5.1 Surface coating and thermally insulative layers

The most efficient way to protect materials against fire without modifying their intrinsic properties (e.g., mechanical properties) is the use of fire retardant coatings.[66] Ignition is most likely to occur at the heat-exposed surfaces, hence it is important to thermally insulate them.[67] This means of protecting flammable materials is called 'passive fire proofing' as it serves to decrease heat transfer from the fire to the structure. Most of the effective coatings are ceramic- or intumescent-based. Intumescence is defined as the swelling of a substance when exposed to heat typically forming a multi-cellular, carbonaceous or ceramic layer, which acts as a thermal barrier that effectively protects the substrate against rapid increase of temperature, thereby maintaining its structural integrity.

The ceramic coating consists of zirconia sometimes stabilised with yttria.[68] There are a number of coatings commercially available. Tewarson and Macaione[69] have evaluated the performance of intumescent and ceramic coatings when applied on glass/resin composites. The intumescent coatings were superior on the vinyl ester and phenolic resinated composites while the ceramic coating was best on the epoxy composites. However, a study by Sorathia and others[70] for potential use of intumescent coatings on the ships for the US Navy has shown that these coatings are not sufficient to protect shipboard spacings during a fire and are not equivalent when used alone as direct replacement for blanket-type fibrous fire insulation (mineral wool, StrutoGard®) installed aboard ships. However, if combined with blanket-type, fibrous fire insulation, they can be effective in meeting required fire resistance criteria. The alternative is to use ceramic fabrics, ceramic coatings, hybrids of ceramic and intumescent coatings, silicone foams or a phenolic skin, all of which showed good performance in a study carried out by Sorathia and co-workers using standard methods.[71] For less demanding applications, gelcoats, which are usually used to make composite surfaces attractive, can also be fire retarded. For example HexCoat® 01 epoxy gel coat is a flame-retardant coating system developed by Hexcel Composites for curing in combination with prepregs, wet lay-up systems and liquid resin infusion systems. Other resin manufacturers also have similar products.

Use of mineral and ceramic claddings is quite popular for naval applications to fireproof conventional composite hull, deck and bulkhead structures. These barriers function as insulators and reflect the radiant heat back towards the heat source, which delays the heat-up rate and reduces the overall temperature on the reverse side of the substrate. One commercial example of this sort of product is the Tecnofire® range of ceramic webs produced by Technical Fibres Ltd in the UK; these are available with a number of different inorganic fibres including glass and rock wool either with or without an associated exfoliated graphite present. They are designed to be compatible with whatever resin is used in composite production. In our recent work[72] we have studied the effect of different Tecnofire 60853A intumescent mats of thicknesses 1, 2 and 4 mm on glass-reinforced polyester composites. These mats contained vitreous (silicate) fibres and expandable graphite. A control composite laminate was prepared with eight layers of woven glass fibres, whereas in the other three samples the top glass layer was replaced with intumescent mats of different thicknesses. The cone calorimetric results at $50\,kW/m^2$ incident heat flux given in Fig. 15.6(a) show that with increasing thickness of the mat, the peak heat release decreases. In Fig. 15.6(b) the temperatures on the reverse side of the sample, measured by a thermocouple and plotted as function of time show the thermal insulative effect of these mats.

3M™ Nextel™ 312 woven fabrics from alumina-boria-silica fibres are also used as composite fire barriers. Phenolic foams are also very effective fire barriers, and particularly used for bulkhead structures in military ships, where a balsa core, sandwiched between glass fibre-reinforced vinyl ester laminate, is coated with an intumescent coating with a phenolic foam layer bonded to it.

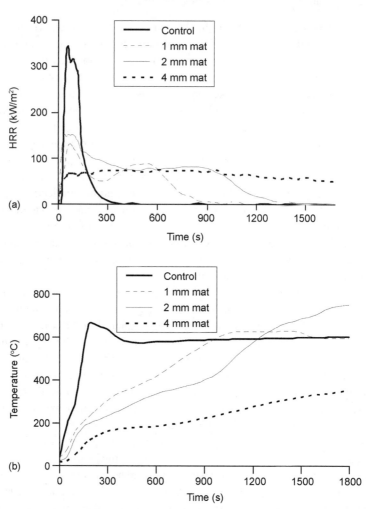

15.6 Effect of thermally insulative intumescent mats on (a) heat release rate and (b) temperature on reverse side of the composite laminates during cone exposure at 50 kW/m² heat flux.

15.5.2 Reinforcing fibre hybridisation

The hybridisation of reinforcing fibres is a common technique to fabricate composites with the optimum properties required for specific applications.[43,44] This can be achieved in two ways:

- by using the hybrid fabric (containing different fibre types in warp and weft directions)
- by using two different fabrics (of different fibre type) in a multi-layered laminate.

The most commonly used hybrid combinations are carbon/glass, carbon/aramid and aramid/glass. In carbon/glass hybrids, carbon fibre contributes high tensile, compressive strength, stiffness and reduces the density, while glass has a higher impact strength, better fire resistance, is an electrical insulator and is much cheaper. In carbon/aramid combinations, the high impact resistance, tensile strength and fire resistance of the aramid fibre combines with the high compressive and tensile strength of carbon. Both fibres have low densities but are relatively expensive. In aramid/glass hybrids the low density, high impact resistance and tensile strength of aramid fibre combines with the good compressive and tensile strength of glass, coupled with its low cost. Both fibre components are fire resistant making the resultant laminate a suitable candidate composite material for use at elevated temperatures.

Recent work in our own laboratories has investigated the effect of the relative positions of each fibre layer in glass/carbon and glass/aramid epoxy composite laminate on burning and burning-induced mechanical property degradation of the laminate. For both hybrid types two different lay-ups were prepared and are reported here in Table 15.2 (samples 4–8). Heat release rate versus time curves are shown in Fig. 15.7. The lay up sequence of hybrid samples affects the PHRR. For example in Hybrid 1, where glass fibre is on the outside, the PHRR value ($334\,kW/m^2$) is higher than for Hybrid 2, where carbon fibre is on the surface (PHRR = $284\,kW/m^2$). Similarly the PHRR value is reduced from $275\,kW/m^2$ in Hybrid 3 to $217\,kw/m^2$ in Hybrid 4 which has aramid on the surface. Hence, by changing the fibre arrangement in a laminate, the burning behaviour can be changed.

15.5.3 Additive and reactive fire retardants

The use of additive and reactive fire retardants in the resin matrix is a common method of fire retarding composites. However, if used in large quantities, the additives can have a detrimental effect on the mechanical properties of the composite laminate. To achieve a certain level of fire retardancy while maintaining high performance mechanical properties is a challenge.

Reduction in flammability is obtained by addition of various fillers and additives acting in different modes. While fillers such as talc and calcium carbonate undergo endothermic reactions at elevated temperatures leading to a reduction in the polymer flammability, nitrogen and phosphorus-containing additives, on the other hand, rely on promoting char formation to achieve their goal. Metal hydroxides have been shown to improve fire retardancy with common ones being aluminium hydroxide or alumina trihydrate (ATH) and magnesium hydroxide ($Mg(OH)_2$). The endothermic decomposition of the metal hydroxides into metal oxides reduces polymer temperature and also the released water into the vapour phase effectively dilutes the volatile species emanating from polymer degradation. However, fillers and metal hydroxides are effective

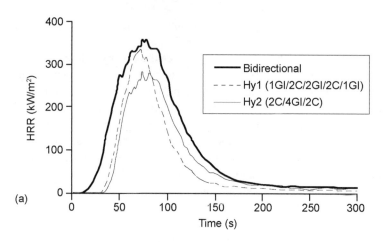
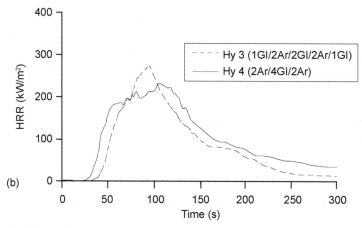

15.7 Heat release rate versus time curves for different hybrid composites at 50 kW/m² heat flux.

only at very high percent loadings (typically >50 wt%) which might have an adverse effect on the mechanical properties of the polymer matrix. Halogen-type fire retardants which yield flame inhibitors are most widely used. However, they may produce carcinogenic gases such as halogen-substituted dioxans and furans upon incineration and this has prompted development of more ecologically friendly fire retardants. The leading candidates for this purpose are phosphorus-, silicon- and nitrogen-containing fire retardants.

Additives like zinc borate, antimony trioxide, alumina trihydrate (ATH) are widely used with halogenated polyester, vinyl ester or epoxy resins[50,73–76] In naval applications brominated vinyl ester resin is commonly used, which during combustion generates smoke, carbon monoxide (CO) and toxic gases such as hydrogen bromide. Alternative resins with and without non-halogenated fire retardants are currently under development. Sorathia[77] has reported the use of siloxane powder, silicate nanocomposite and alumina trihydrate in non-brominated resin as a potential replacement of brominated resin composites. Alumina trihydrate (ATH) at 15 phr gave a 20 and 25% decrease in peak heat release rates, 24 and 13% decrease in average heat release rates, and 27 and 24% decrease in average mass loss rates at radiant heat fluxes of 50 and 75 kW/m^2, respectively. The results obtained indicated that the non-brominated epoxy vinyl ester possessed superior mechanical properties with lower smoke and CO generation than the brominated vinyl ester.

The presence of phosphorus and nitrogen offer enhanced flame retardancy at lower loadings and are becoming increasingly popular. Kim and co-workers[78] employed several kinds of encapsulated red phosphorus as flame retardants to epoxy moulding compounds (ortho-cresol novolac epoxy cured with phenol novolac resin). The encapsulating materials for red phosphorus included resol resin, melamine and TiO_2. Incorporation of these fire retardants significantly improves interfacial adhesion while effectively serving their primary purpose.

Hussain and co-workers[79] have used 9,10-dihydro-9-oxa-10-phosphaphenanthrene10-oxide (DOPO) (P-content = 14%), synthesised from a mixture of 3,5-diethyltoluene-2,4-diamine and 3,5-diethyltoluene-2,6-diamine (Ethacure-100) to flame retard an epoxy resin, di-glycidyl ether of bisphenol A (DGEBA). They prepared and characterised formulations having 1 and 3% phosphorus content using thermogravimetric analysis (TGA), dynamic mechanical thermal analysis (DMTA) and cone calorimetry. From the TGA analyses char yields were observed to increase with phosphorus content.

Morrey[40] in her PhD thesis has investigated the effect of a number of commercially available fire retardant additives (ATH, dimelamine phosphate, melamine pyrophosphate, zinc borate + boric acid, calcium sulphate, etc.) on flammability of carbon/epoxy and glass/phenolic composites. Most of the fire retardants at 20% resin weight fraction resulted in some reduction in PHRR and AvHRR, but had little effect on THR, indicating reduced potential fire propagation, but not necessarily reduced overall flammability of the resin (see Table 15.3). Inclusions of intumescents and blowing agents through the thickness of the composites tended to initiate delamination and early release of volatiles. There was a significant reduction in flexural strength and flexural modulus of the composite with addition of these additives. Additives with larger particle sizes were noted to produce greater reductions in mechanical performance. From this study, four fire retardant systems were created which were blends of different fire retardants. Selection of additives in each blend was related to their fire retardant mechanisms as shown below:

- System 1; a blend of additives that endothermically release water over a range of temperatures,

Table 15.3 Effect of additives at 20% loading on flammability and flexural mechanical performance of carbon/epoxy composite laminates[40]

Material	ªCone calorimetry		Mechanical properties	
	TTI (s)	PHRR (kWm^{-2})	Flexural strength (% gain/loss)	Flexural modulus (% gain/loss)
Batch 1				
CF/8552 – Control	32	484	944	55
CF/8552 + Ceepree	33	359	978(4)	53(−4)
CF/8552 + zinc borate	36	298	961(2)	55(0)
CF/8552 + Ultracarb	34	334	986(4)	60(9)
CF/8552 + Al(OH)$_3$	38	350	734(−22)	53(−4)
CF/8552 + ZnSn(OH)$_6$	30	226	782(−17)	44(−20)
CF/8552 + melamine	20	388	760(−19)	54(−2)
CF/8552 + Amgard NH	24	249	737(−22)	46(−16)
CF/8552 + boric oxide	22	244	698(−26)	47(−15)
Batch 2				
CF/8552 – Control	32	484	722	56
CF/8552 + System 1	29	274	728(1)	52(−7)
CF/8552 + System 2	24	221	728(1)	50(−11)
CF/8552 + System 3	22	248	801(11)	66(18)
CF/8552 + System 4	20	227	647(−10)	49(−16)

ªCone calorimetry irradiance = 75 kW/m^2
System 1 = a blend of additives that endothermically release water over a range of temperatures including alumina trihydrate (ATH), Al(OH)$_3$, zinc hydroxystannate, ZnSn(OH)$_6$ and 60% Ultracarb (a mixture of Mg$_4$(CO$_3$)$_3$(OH)$_2$.3H$_2$O and Mg$_3$Ca(CO$_3$)$_4$),
System 2 = potential char-forming intumescent systems containing a blend of low and high temperature blowing agents such as melamine monophosphate,
System 3 = a blend of a blowing agent that decomposes around the same temperature as the epoxy and an intumescent to promote char formation such as that of zinc borate and melamine monophosphate and,
System 4 = a blend of intumescent additives only to optimise the formation of a continuous intumescent layer such as that of melamine monophosphate, glass frits/carbonates and boric oxide.

- System 2; potential char-forming intumescent systems containing a blend of low and high temperature blowing agents,
- System 3; a blend of a blowing agent that decomposes around the same temperature as the epoxy and an intumescent to promote char formation and,
- System 4; a blend of intumescent additives only to optimise the formation of a continuous intumescent layer.

The cone calorimetric and mechanical (flexural) test results are reported here in Table 15.3. At 75 kW/m^2 irradiance, no improvements were observed in TTI, in fact the value was reduced for most of the samples. While enhanced reductions in PHRR were observed with increase in the additive fraction, the improvements from 10 to 20 wt% were larger than improvements observed from

20 to 40 wt% additives. All systems with the exception of System 4 retained or showed an improvement in the flexural strength at 20 wt% additive. However, diminished flexural modulii were observed for all systems at 20 wt% loadings.

Recent work in our laboratories has focused on the use of phosphorus and nitrogen- and organophosphorus-based flame retardants as well as halogenated ones to achieve flame retardance in unsaturated polyester[80] and epoxy resin systems.[57,58] Igwe and co-workers[80] studied the effect of additives, ammonium polyphosphate, alumina trihydrate, expandable graphite, talc and spiolite clay on physical, thermal and mechanical properties of glass-reinforced polyester composites. The water durability properties were also observed by immersing laminates in water for 21 days prior to testing. Polyester samples containing flame retardants absorbed more water than the control sample. The flammability behaviour of samples before and after they were immersed in water was assessed by cone calorimetry at $50 \, kW/m^2$ incident heat flux. Mixed, but insignificant changes in the time-to-ignition (TTI) and the peak heat release (PHRR) values were observed. Notable increments in the amount of smoke produced were observed for samples subjected to water treatment when compared to dry samples. However, the changes observed were similar for the control sample and the flame-retarded samples suggesting that there is no adverse effect on the physical properties of the polyester resin following the addition of flame retardants.

For epoxy cast resin laminates, we have studied the effect of phosphorus and nitrogen-containing, organophosphorus and halogenated flame retardants at different loading fractions. The limiting oxygen index and UL 94 results of different epoxy formulations are given in Table 15.4. LOI results for flame-retarded samples are significantly increased relative to the control sample, suggesting that the burning process is slowed down. Increasing the additive level from 4 to 8 wt% did not affect the LOI values of samples containing phosphorus and nitrogen, but notable increments were realised with organophosphorus and halogenated flame retardant additives. Addition of flame retardants at 4 and 8 wt% improved the UL 94 ratings from V-2 for the control to V-1 and V-0 respectively for most of the flame retarded formulations. Cone calorimetry ($50 \, kW/m^2$ external heat flux) behaviour, for which results are not shown here in, is typified by significant reductions in PHRR values. This is reflective of the LOI and UL 94 results.

Intumescent/FR fibre interactive fire retardant systems

In our earlier review,[5] we reported work on the development of novel glass/polyester and glass/epoxy composite materials incorporating intumescent and cellulosic fibres as fire retardants. These components are added either as pulverised additives to the resin or as an additional textile fabric layer to the composite structure. The composite structures thus produced show superior fire

Table 15.4 The limiting oxygen index values and UL 94 ratings for flame-retarded epoxy resin[57]

Sample	Flame retardants	LOI, vol %		UL 94 ratings	
		4% FR	8% FR	4% FR	8% FR
Epoxy resin	–	24.0[a]	24.0[a]	V2[a]	V2[a]
P and N-based FRs	APP	28.3	29.0	V1	V0
	MP	29.5	29.9	V1	V0
	MPP	28.6	28.5	V1	V0
	NH1197	30.0	29.8	V1	V1
	NH1511	29.8	30.0	V1	V0
Halogenated FRs	FR372	29.1	31.9	V1	V0
	FR245	28.3	31.0	V1	V0
Organophosphorous-based FRs	RDP	31.1	34.5	V1	V1
	BAPP	29.5	34.1	V0	V0
	TPP	28.1	31.8	V1	V0
	TTP	27.8	29.1	V1	V1

[a] Control sample with no additives
APP = ammonium polyphosphate; MP = melamine phosphate; MPP = melamine pyrophosphate; NH 1197 = phosphorylated pentaerythritol; NH 1511 = phosphorylated pentaerythritol with melamine ; FR 372 = tris (tribromo neopentyl) phosphate; FR 245 = tris(tribromo phenyl) cyanurate; RDP = resorcinol diphenyl phosphate; BAPP = bisphenol A diphenyl phosphate; TPP = triphenyl phosphate; TTP = tritolyl phosphate

retardant properties and retention of mechanical properties on exposure to heat/fire compared to control samples. Different thermal analytical and flammability studies have led us to believe that on heating, all components degrade by physically and chemically compatible mechanisms, resulting in interaction and enhanced char formation and hence, the laminates showed significantly superior fire retardant properties. For detailed information on this work, the reader is referred to our other publications.[8,41,81] In this work, additive levels of intumescent and Visil (an inherently FR cellulosic fibre, Sateri Fibres, Finland) have been kept at 5 wt% each. However, while the quantities of FR fibre and intumescent, and their relative ratios in the resin are important in maximising interaction, excessive amounts can change the mechanical properties detrimentally. Hence, to observe the effect of different levels and relative ratios of the two additives and to optimise the ratios of the matrix constituents to produce composite laminates with maximised fire retardant properties without compromising their mechanical performance, a number of samples were prepared and tested for their fire performance using a cone calorimeter.[82] Their mechanical performance was evaluated from tensile and flexural behaviour. The effect of increasing intumescent (Int) and Visil (Vis) on time-to-ignition and PHRR and tensile modulii are shown here in Fig. 15.8(a–c). TTI increased with increasing contents of Int and Vis and PHRR decreased with increasing content,

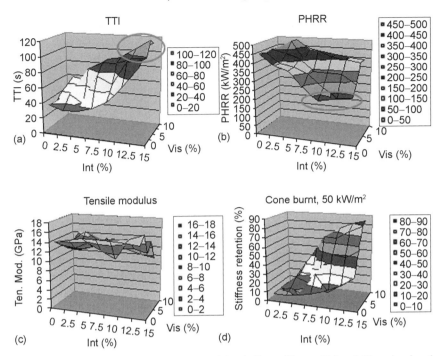

15.8 Surface plots showing combined effect of Int and Vis additives in glass/epoxy composites on fire: (a) TTI, (b) PHRR; and mechanical performance, (c) tensile modulus and (d) stiffness retention after cone calorimetric tests at 50 kW/m² heat flux.[82] (Reprinted from ref. 82 with permission from Elsevier.)

although at the highest Vis content, the value increased. Best results were shown with samples containing 15 wt% Int, and 2.5 and 5 wt% Vis. Flexural results were not adversely affected by high levels of Vis, but tensile modulii reduced slightly for samples containing higher than 7.5 wt% Vis.

15.5.3 Resin modification – new developments

As discussed above fire retardants can be added to the resin to reduce the flammability of the resin. However, many cured resins are already rather brittle in nature due to their high cross-linking density, and further addition of fire retardants often induces degradation of the overall physical and mechanical properties of the resultant composite.[83] An alternative approach is to incorporate fire retardant elements or functional groups, such as phosphorus, halogen, boron and phenol groups into the backbone of the resin.

In unsaturated polyester the use of halogenated resin or replacement of curing agent from styrene to bromostyrene is quite common. Presence of phosphorus in the backbone of epoxy resin can enhance its LOI from 22 to 28 vol%.[84] The halogen elements in the epoxy backbone such as chlorine in diglycidyl ether of

Bisphenol C (DGEBC), fluorine in diglycidyl ether of Bisphenol F (DGEBF), bromine in tetrabromobisphenol A (TBBA), also enhance the thermal stability of the epoxy resins. For example, the presence of chlorine in DGEBC enhances LOI up to 31 compared to 22 vol% in DGEBA.[85] Another approach to form inherently fire retarded epoxy resin has been made by reacting diphenyl silanediol with DGEBA, which results in a silicon-containing epoxy resin.[84] The silicon-containing epoxy exhibits higher char formation and an LOI of 35 vol%. Commercial DGEBA can be copolymerised with cresol novolac phenolic resin to achieve high thermal stability and fire retardancy.[84]

The proper choice of curing agents or hardeners for the resin can also enhance thermal stability and fire resistance of the resin. Phenol-formaldehyde[86] and aryl phosphinate anhydride[87] are examples of curing agents that can improve the fire resistance of epoxy resins.

Braun and co-workers[88] have used phosphorus-containing hardeners to produce fire retardant composites. They systematically and comparatively evaluated the pyrolysis of flame-retarded epoxy resins containing phosphine oxide, phosphinate, phosphinate and phosphate (with phosphorus contents of around 2.6 wt%) together with the fire behaviour of their carbon fibre-reinforced composites. With increasing oxidation state of the phosphorus, the amount of thermally stable residue increased while the release of phosphorus-containing volatiles decreased. The flammability of the composites was investigated using LOI and UL 94 tests and the fire behaviour studied with cone calorimetry at different radiant fluxes. The processing and the mechanical performance (delamination resistance, flexural properties and inter-laminar bonding strength) of the fibre-reinforced composites containing phosphorus were maintained at high levels and, in some cases, even improved. Here, the potential for optimising composite flame retardancy while maintaining or even improving the mechanical properties is highlighted.

The inclusion of the organophosphorus functionality within the polymeric resin structure can enhance its fire retardancy. Toldy and co-workers[89] incorporated aromatic organophosphorus compounds into the epoxy resin and also studied the effect of combining them with nanoparticles. By using a fully phosphorylated calixresorcinarene derivative, they significantly increased the limiting oxygen index (LOI) from 21 to 28 vol% and achieved a V-0 UL 94 rating. Espinosa and co-workers[90] modified novolac resins with benzoxazine rings and then cured them with isobutyl bis(glycidylpropylether) phosphine oxide (IHPOGly) as a cross-linking agent and could achieve V-0 rating with the UL 94 test. Previously the same authors had studied the synthesis and polymerisation of a novel glycidyl phosphinate, 10-(9,10-dihydro-9-oxa-10-phosphaphenanthrene-10-oxide)-2,3-epoxypropyl (DOPO-Gly).[91] Both of these materials were found to have high glass transition temperatures and retarded thermal degradation rates with excellent fire retardancy properties.

In addition to the above examples of recent resin modification work, there is considerable literature available in this field and to cover all the references is beyond the scope of this review. However, the reader is referred to a recent short review on phosphorus-containing epoxy monomers and resins with improved fire resistance properties by Levchik et al.[92]

15.5.4 Co-curing of different resins

In order to improve flame retardance of a resin, e.g. polyester, it can be blended with other resins having good flame retardance such as phenolic and epoxy resins. Chiu et al.[93] have blended an unsaturated polyester resin with a resole type of phenolic resin and studied thermal degradation behaviour and combustion properties of the resultant blend. Although unsaturated polyester resin is not quite compatible with phenol resin, an integral interpenetrated structure was formed by a dispersive mechanical stirring and mixing process. Thermogravimetric analysis, oxygen index, smoke density measurements and cone calorimetric studies showed not only improvements in thermal stability but also reductions in smoke and toxic gas evolution and heat release properties. The LOI values and thermal decomposition temperatures also increased with increase in phenolic resin content in the blend.

Cherian and co-workers[94] synthesised hybrid polymer networks based on unsaturated polyester resin and epoxy resins by reactive blending. The co-cured resin showed substantial improvement in toughness and impact resistance. Considerable enhancements of tensile strength and toughness were noticed at very low loadings of epoxidised phenolic novolac. Thermogravimetric analysis, dynamic mechanical analysis and differential scanning calorimetry showed substantial improvement in thermal stability of these blends.

15.5.5 Nanoparticle inclusion in resin and fibre components

The need to develop fire retarded resins with improved thermal stability releasing small quantities of smoke, heat and toxic gases has seen a rapid development of polymer/layered nanocomposites.[61,95–97] Nanodimensional organically modified clays can be uniformly dispersed within the polymer matrix and have frequently been shown to exhibit remarkable improvements in mechanical and material properties while achieving their primary purpose of reducing the flammability. Even though only 5–10 wt% nanoclay content is sufficient to reduce the peak heat release value by up to 70% for most resin systems, in most cases polymer/layered nanocomposites do not pass commercial combustion test requirements defined by LOI and UL 94, for example. However, their combination with conventional fire retardants is attracting a lot of research attention.[10] General observations lead to the conclusion that owing to their low combustion rates, nanocomposites can reduce the amount of classical fire retardants required to satisfy the industrial standards.

In our own laboratory, we have studied the effect of organically modified nanoclay with and without conventional flame retardants on the flammability of unsaturated polyester.[10,98] Montmorillonite clay modified with a series of organic modifiers was used. Some clays were fully exfoliated, some indicated combined nanostructures with ordered intercalation and partial exfoliation and some indicated only microcomposite structures, depending upon the type of organic modifier used. Selected cone calorimetric and flexural results for glass/polyester composites are shown in Table 15.5. Flexural moduli of the glass-reinforced samples were improved with the presence of nanoclays in the resin formulations, whereas inclusion of flame retardants, such as ammonium polyphosphate (APP), in the resin reduced the values. Introduction of clay in Resin/Clay/FR samples showed a small but not significant improvement compared to Resin/FR samples in their flexural performances, which is expected due to the high levels of flame retardant additives present. Cone calorimetric results for glass-reinforced composites (see Table 15.5) in general show slight increases in TTI values of samples with and without flame retardants. PHRR and smoke production are reduced with respect to the control sample. Despite higher TTI values and longer flame-out (FO) times, all the samples with Resin/Clay/FR formulation show reduced THR values suggesting reduced flammability. Composite laminate Resin/Clay B/APP shows the highest reduction (43%) in

Table 15.5 Flammability and flexural properties of glass-reinforced polyester-nanocomposite samples containing 5 wt% nanoclays with and without 20 wt% flame retardants

Sample	Cone calorimetry			Flexural modulus (GPa)
	TTI (s)	PHRR (kW/m^2)	THR (MJ/m^2)	
Res	36	401	31.0	14.9
Res/Clay 10A	41	370	35.3	16.3
Res/Clay B	39	358	33.8	15.6
Res/APP	37	245	28.4	8.8
Res/ATH	40	260	33.8	8.3
Res/NH	42	340	30.1	9.9
Res/NW	39	239	22.5	–
Res/Clay 10A/APP	39	287	26.5	9.5
Res/Clay B/APP	38	229	31.6	8.6
Res/Clay 10A/ATH	40	288	30.2	12.3
Res/Clay 10A/NH	40	291	30.8	7.9
Res/Clay 10A/NW	42	224	26.2	–

Note: Clay 10A is a commercial clay (Cloisite 10A, Southern Clay), MMT modified with dimethyl benzyl hydrogenatedtallow quaternary ammonium chloride
Clay B is MMT modified with vinyl triphenyl phosphonium bromide
APP = Ammonoium polyphosphate; NH = Melamine phosphate; NW = Dipentaerythritol/melamine phosphate ; ATH = Alumina trihydrate

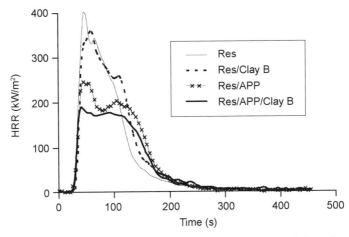

15.9 Heat release rate versus time curves for glass fibre-reinforced polyester resin (Res) and flame retarded (20 wt% APP) polyester resin with and without nanoclay B (montmorillonite modified with vinyl triphenyl phosphonium bromide) composites at 50 kW/m² heat flux.

PHRR value as compared to the control sample and this is probably due to presence of the phosphonium group in the organic modifier of functionalised clay. HRR versus time curves of resin, Resin/Clay (i) with and without APP are graphically shown in Fig. 15.9, which shows that clay in presence of APP is quite effective in reducing burning of the resin.

In our laboratories, Katsoulis and co-workers,[99] have studied the effects of different nanoclays and double-walled carbon nanotubes (DWNTs) on the fire performance and the thermal stability of an epoxy resin, tetraglycidyl – 4,4' diaminodiphenylmethane, (TGDDM). Limiting oxygen index (LOI), UL 94 and cone calorimetry results are presented in Table 15.6. Addition of nanoclays (Cloisite 30B, Nanomer I.30E, and a vinyl triphenyl phosphonium (VTP) intercalated montmorillonite clay) at 5 wt% loading resulted in increased LOI values when compared to the control. Only the VTP-containing nanoclay achieved a UL 94 rating, namely V-1, whereas all other samples were completely burnt. The peak heat release rate is reduced in some cases by as much as 25%, while time-to-ignition values remained more or less the same following the addition of nanoclays. The DWNT-containing epoxy formulations showed similar behaviour to that observed for nanoclay-containing formulations.

Nanoparticles in inorganic oxide form have also been used to improve the flame retardancy of polymeric resins.[100–102] In this class of nano-dimensional additives carbon nanotubes (CNT) are also included, which have been shown to improve both the flame retardancy and mechanical properties of polymer resin systems.[62–65] The nano-dimensionality of CNT is implicated in the enhanced mechanical property due to the improved interfacial interactions (bonding) with the resin matrix. Recently, nanofibres have also been shown to improve the

Table 15.6 LOI, UL 94 and cone calorimetric results of epoxy resin/clay nanocomposites and epoxy/flame retardant formulations

Sample	LOI vol %	UL 94	[a]Cone results		
			TTI (s)	PHRR (kW/m^2) (% red)[b]	THR (MJ/m^2) (% red)[b]
Epoxy	27.8	–[d]	24	823	56
Epoxy/30B	30.8	–[d]	25	867 (−5)	52(7)
Epoxy/I.30E	30.0	–[d]	24	712 (13)	56(0)
Epoxy/VTP	31.7	V-1	28	616 (25)	49(13)
Epoxy/DWNT[c]	29.6	V-1	21	727 (12)	53(5)

[a] Cone calorimetric results at an irradiance of 50 kW/m^2
[b] % reduction in PHRR and THR compared to the control sample
[c] Loading percent of DWNTs = 0.5%
[d] completely burnt
TTI, time-to-ignition; PHRR, peak heat release rate; (% red.), reduction in PHRR; THR, total heat release; The coefficient of variations in cone calorimetry data are less than 10% for all parameters.

thermal stability and flammability properties of polymer resin systems without compromising that mechanical strength and stiffness.[65] The authors, however, note that there are health and environmental hazards associated with nanodimensional materials and also that their properties are too complex to be fully explained in this article and therefore refer the reader to a review on nanostructured materials by Kuchibhatla *et al.*[103]

15.6 Mechanical property degradation during and after fire

Apart from the flammability performance issues discussed in Section 15.2, another factor limiting the use of polymer matrix composites (PMCs) as load-bearing structures is the retention of their original strength and stiffness during and after a fire. The requirements are application specific; for example, according to IMO (International Maritime Organisation regulations)/HSC (Code of Safety for High Speed Craft) all composite materials used in load-bearing structures on naval ships must retain their mechanical properties during a fire for 30–60 minutes.[104]

The stiffness of a composite is reduced substantially when it is heated above the glass transition temperature of the resin matrix. For glass/vinyl ester composites the strength reduction can be up to 50% at temperatures as low as 121 °C,[105] whereas after exposure to fire, the reduction can be up to 90%.[69,106] Thin laminates of glass/epoxy composites have been reported to have no post-fire strength.[69] Glass/resole phenolic composites, despite having outstanding fire retardant properties, have low retentions of mechanical properties as reported by

Mouritz and Mathys.[107] They exposed glass/phenolic laminates to various heat fluxes (25, 50, 75 and 100 kW/m^2) for times up to 1800 s using cone calorimetry and studied the post fire tensile and flexural properties. Even after exposure to low heat fluxes (<30 kW/m^2) for a short period of time and with no apparent signs of charring, the mechanical properties were reduced by 30%. On exposure to high heat fluxes, the composite charred, but by the time it ignited, it had lost up to 50% of its original stiffness and strength. This is due to chemical degradation of the phenolic resin matrix. The char formed is usually very porous and brittle and provides minimal structural support to the fibres. Hence, the post-fire tension and flexural properties of phenolic composites are similar to other composites despite superior fire resistance.[108]

Kourtides[109] has studied the effect of temperature on the mechanical properties (flexural, tensile, compressive, and short-beam shear-strength) of graphite fabric-reinforced laminates prepared with four different matrix resins: epoxy, phenolic-Xylok®, bismaleimide A, and bismaleimide B. The samples were heated at selected temperatures for 30 minutes prior to testing. The residual flexural modulus and tensile strength at selected temperatures for all the samples are presented in Table 15.7. At room temperature the highest mechanical properties were obtained with epoxy followed by bismaleimide B. At elevated temperatures, i.e. >150 °C, bismaleimide A shows the greatest loss in flexural and tensile strength and this has been attributed to the loss of residual solvent from the composite which would otherwise act as a plasticiser. While bismaleimide B and phenolic-Xylok® have lower mechanical properties than epoxy composites at ambient temperatures, they are primarily designed to perform better at higher temperatures. Thus bismaleimide B and phenolic-Xylok® composites retained their mechanical properties at elevated temperatures, >150 °C, while a significant mechanical property degradation was observed for epoxy based composites. Epoxy composites demonstrated the worst flame resistance of all the composites giving lowest neat resin char yield, 38 wt%, as measured from thermogravimetric analysis at 900 °C. The low char yield from the epoxy resin when compared to other resin system suggests that there is not enough bonding matrix to support reinforcement fibres after high temperature exposure of the composites.

Table 15.7 Effect of temperature on the residual flexural and tensile strength of graphite fabric-reinforced polymer composites[109]

Sample	Flexural strength (MN/m^2)			Tensile strength (MN/m^2)		
	23 °C	100 °C	200 °C	23 °C	75 °C	150 °C
Epoxy	933	676	333	694	630	491
Phenolic-Xylok®	524	505	448	565	569	569
Bismaleimide A	829	371	138	519	519	343
Bismaleimide B	619	590	562	–	–	–

Ulven and Vaidya[110] have studied the effect of fire on the low velocity impact (LVI) response of glass/vinyl ester laminates and balsa wood-core, sandwich composites with glass/vinyl ester face skins. The LVI response parameters, peak force and contact stiffness decreased by 20–30% for glass/vinyl ester laminate and by 65–75% for a glass/VE balsa core sandwich laminate, both subjected to a 100 s period of fire exposure. Greater reductions in post-fire properties in sandwich structures arise because on exposure to heat/fire the phenolic skin and natural polymer in balsa wood core char, the residual char is very porous and brittle with very low structural integrity. A lower density balsa wood core was observed to insulate the face sheet more than that by the one with higher density. Insulation provided by the core resulted in accumulation of heat in the facesheet, causing greater thermal damage and hence, less retention of contact stiffness. Mouritz and Gardiner[111] have studied the compression properties of fire-damaged sandwich composites of glass/vinylester face sheets with poly(vinyl chloride) (PVC) and phenolic foam cores. Although the laminate with phenolic foam core had superior fire resistance properties, post-fire residual properties of both laminates degraded substantially before they ignited and started burning.

Recently we have examined and reported the effect of fibre type on the stiffness retention of composites exposed to heat for a short period of time.[43,44] The flexural stiffness retention results of all samples reported in Table 15.2 and discussed in Sections 15.3.2 and 15.4.2 after exposure to 50 kW/m^2 heat flux for 40 s are shown in Fig. 15.10. Of all the samples the glass laminate has the maximum stiffness retention (~55%). This behaviour may be due to the delayed

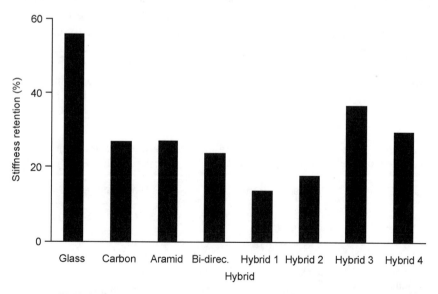

15.10 Stiffness retention for partly burnt samples mentioned in Table 15.2.[44]

time to ignition (34 s) relative to carbon fibre-reinforced laminates (25 s) which means the latter will incur more damage during the exposure time. For hybrid samples, stiffness retention is higher for the samples containing aramid fibre, which may be due to slow burning of the latter. The values are, however, higher than for an all-aramid sample (Sample 3). The rate at which the modulus deteriorates also is function of the fibre type and their respective position within the composites. The sample with aramid fibres (Hybrid 4) on the outer surface possibly deteriorates more quickly than the system with glass fibres on the outer surface (Hybrid 3).

15.7 Post-fire mechanical performance of fire retardant composites

As discussed above, the composites under heat/fire lose their mechanical stiffness significantly. However, if the surface of the laminate is protected from the heat, or fire retardant additives present in resins slow down or prevent the commencement and acceleration of combustion, the retention of stiffness and strength could be higher compared to unmodified laminates. This section reviews the mechanical performance studies of composites fire retarded by different methods.

Sorathia and coworkers[71] have studied the effect of different types of thermal barriers on flammability of glass/vinyl ester, graphite/epoxy, graphite/bismaleimide and graphite/phenolic composite laminates. Thermal barrier treatments used include ceramic fabric, ceramic and intumescent coatings, hybrids of ceramic and intumescent coating, silicone foam and phenolic skins. All the composite laminates were tested by different standard methods for flammability including cone calorimetry at 25, 75 and 100 kW/m^2, and residual flexural strength after exposing to a radiant heat source at 25 kW/m^2 for 20 minutes in a flaming mode. Selected cone and residual flexural test results for glass/vinyl ester and graphite/epoxy are reproduced here in Table 15.8. The graphite/epoxy laminates with no thermal barrier completely delaminated after the fire exposure and hence, showed no residual strength. All laminates with a water-based intumescent coating by itself or in combination with ceramic coating showed higher retention of residual strength compared to the other barrier systems used. This observation is consistent with the flammability test results of these laminates as seen from Table 15.8. Laminates with these coatings have higher TTI and lower PHRR values compared to all other laminates. This shows that the retention of mechanical properties of the laminate depends upon the thermal insulative effectiveness of the surface barrier treatment.

In our own work we have studied the effect of heat and fire on intumescent/FR fibre interactive fire retardant systems discussed in Section 15.4.3. A number of samples containing different amounts of Visil (Vis) and intumscent (Int) (see Section 15.4.3 for sample details) were exposed to different heat fluxes (two

Table 15.8 Effect of different fire barriers on flammability and residual flexural strength of glass/vinyl ester and graphite/epoxy composites[71]

Fire barrier	Glass/vinyl ester			Graphite/epoxy		
	TTI[a] (s)	PHRR[a] (kW/m^2)	Residual strength[b] (%)	TTI[a] (s)	PHRR[a] (kW/m^2)	Residual strength[b] (%)
Control	22	498	14	53	197	1
Ceramic fabric	55	213	26	45	156	1
Ceramic coating 1	88	344	28	105	179	13
Intumescent coating (water-based)	450	204	45	628	76	77
Hybridised ceramic and intumescent coating	445	139	40	264	97	14
Silicone foam	34	435	20	32	270	8
Intumescent coating (solvent-based)	80	288	22	90	171	20
Ceramic coating 2	170	289	20	140	119	3
Phenolic skin	88	146	14	–	–	0

[a] = at 75 kW/m^2.
[b] = Residual flexural strength after exposure at 25 kW/m^2 for 20 min.

replicates per condition) for a short period of time so that char formation is restricted to a few layers of composites only.

The conditions used were:

- 25 kW/m^2 – exposed for 80 s
- 35 kW/m^2 – exposed for 60 s
- 50 kW/m^2 – exposed for 45 s or if the sample ignited before this period, the sample was removed and flames were extinguished.

Short exposure times were selected to restrict the damage to the first few layers only. The damaged area increases with increasing heat flux, which once the sample ignited, became difficult to control. In order to determine the residual flexural stiffness of completely burnt samples, specimens were exposed to the cone heater for 300 s at 50 kW/m^2. Results indicated that the Visil and intumescent levels had minimal influence on the stiffness retention of the heat-damaged samples. This was because the surface temperature achieved was possibly not high enough for the Visil and intumescent chemicals to interact. In the case of the cone-burnt samples, the effect of additive components on stiffness retention is significant as shown in Fig. 15.8(d), the stiffness retention increased with increasing additive contents. This is because both the intumescent and Visil additives interacted completely, leading to char, which is less prone to oxidation. The best results were shown by samples with high contents of both intumescent and Visil.

15.8 Future trends

The demand for new materials in the aerospace, automotive and marine industries with superior specific strength and stiffness will sustain the accelerating research and development in the area of polymer matrix composites. Within the most common composite markets where the cheaper and more flammable polyester resins are used, smoke generation is high. Consequently smoke suppression is a significant fire performance factor. It is here that the traditional antimony-bromine formulations are weak and considerable interest lies in finding alternatives with comparable flame retarding properties but with enhanced smoke reduction. From the above discussion it can be concluded that choices of resin and fibre are crucial in determining the flammability properties of the whole structure. If flame retardant chemicals, which are compatible with both fibres and resin matrix, are selected, resulting effects can be synergistic. The role of nanoclays and nanocomposite structures within the macrocomposite is currently being addressed by a number of research teams including our own. The use of nanotechnology in development of novel surface coatings, which act as effective thermal barriers, is the way forward. Development of novel nano-sized particulate fire retardants will probably present a real breakthrough in this area. The next few years promise much excitement in the discovery of novel fire resistant systems that will have superior performance to the current systems and will be based on both present and developing fire science understanding.

While development of new materials is rapid, their implementation requires rigorous testing such as typical full-scale fire tests. These types of tests are expensive and in order to reduce the financial burden on small businesses, composite system fire screening protocols using small-scale samples to predict full-scale fire performance are a necessity. Cone calorimetry and lateral ignition flame spread tests are some of the small-scale tools that can be used to evaluate the fire performance of new technologies.

On the other hand, concerted efforts are underway to develop thermomechanical models for predicting time-to-failure of fibre-reinforced polymer laminates loaded in both tension and compression and exposed to a radiant heat.[112–116] Use of mathematical models allows the prediction of temperature profiles through-the-thickness of composites, exposed to fire. Using temperature-dependent properties of the material, it is possible then to calculate the reduction in stiffness and strength using the laminate theory. This is the most efficient and cost effective method to access fire performance of new composites without performing extensive experimental work. This, too, is currently being addressed by a number of research teams including our own. Computer-aided, mathematical prediction of mechanical property degradation with temperature is set to become the most popular screening tool for fire performance of new technologies as many new composite materials are inevitably going to be produced.

15.9 References

1. Kandola B K and Horrocks A R, 'Composites', Horrocks A R and Price D, editors. *Fire Retardant Materials*. Cambridge: Woodhead Publishing Ltd; 2001, Chapter 6, pp. 182–203.
2. Kandola B K and Horrocks A R, 'Flame Retardant Composites – A Review, The Potential for Use of Intumescents?', Camino G, Bras M Le, Bourbigot S and Delobel R, editors, *Fire Retardancy of Polymers: The Use of Intumescence*. The Royal Society of Chemistry, Cambridge, 1998.
3. Bisby L A and Kodur V K R, 'Evaluating the fire endurance of concrete slabs reinforced with FRP bars: Considerations for a holistic approach', *Composites Part B: Eng.*, 2007, **38**, 547–558.
4. http://www.cmt-ltd.com. Metal matrix composites.
5. Horrocks A R and Kandola B K, 'Flammability and Fire Resistance of Composites' in Long AC (Ed), *'Design and Manufacture of Textile Composites'*, Cambridge, The Textile Institute, Woodhead Pub. Ltd., 2005.
6. Mouritz, A P, *'Fire Safety of Advanced Composites for Aircraft'*, Australian Transport Safety Bureau Research and Analysis Report, 2006.
7. Grenier A T. (a) 'Fire Characteristics of Cored Composite Materials for Marine Use', MSc Thesis, Worcester Polytechnic Institute, US, 1996.; (b) Grenier A T, Dembsey N A, Barnett J R, 'Fire Characteristics of Cored Composite Materials for Marine Use – Ignition', *Fire Safety J.*, 1998, **30**, 137–159.
8. Kandola B K, Horrocks A R, Myler P and Blair D, 'Thermal Characterisation of Thermoset Matrix Resins', In : *Fire and Polymers*, ed. Nelson, G.L. and Wilkie, C.A. *ACS Symp. Ser.*, 2001, 344.
9. Baudry A, Dufay J, Regneir N and Mortaigne B, 'Thermal degradation and fire behaviour of unsaturated polyester with chain ends modified by dicyclopentadiene', *Poly.Degrad. Stab.*, 1998, **61**, 441–452.
10. Nazare S, Kandola B K and Horrocks A R, 'Cone Calorimetric Study of Flame-Retardant Unsaturated Polyester Resin Incorporating Functionalised Nanoclays', *Poly. Adv. Tech.*, 2006, **17**, 294–303.
11. Baley C, Perrot Y, Davies P, Bourmaud A and Grohens Y, 'Mechanical properties of composites based on low styrene emission polyester resins for marine applications', *Appl. Comp. Mater.*, 2006, **13**, 1–22.
12. Mathur V K, 'Composite materials from local resources', *Construction Building Mater.*, 2006, **20**, 470–477.
13. Kootsookos A and Mouritz A P, 'Seawater durability of glass-and carbon-polymer Composites', *Comp. Sci. Tech.*, 2004, **64**, 1503–1511.
14. Cain J J, Post N L, Lesko J J, Case S W, Lin Y-N, Roffle J S and Hess, P E, 'Post-curing effects on marine VARTM FRP composite material properties for test and implementation', *J. Eng. Mater. Tech.*, 2006, **128**, 34–40.
15. Lua J, Key C T, Hess P E, Jones B J, Lopez-Anido R and Dagher H J, 'A virtual testing methodology for marine composites to develop design allowables', *Composites 2004 Convention and Trade Show American Composites Manufacturers Association*. October 6–8, 2004, Tampa, Florida, USA.
16. Marsh G, 'Composites boost patrol craft performance', *Reinforced Plastics*, 2006, **50**, 18–22.
17. Marsh G, 'GKN aerospace extends composites boundaries', *Reinforced Plastics*, 2006, **50**, 24–26.
18. Harman A B and Wang C H, 'Improved design methods for scarf repairs to highly

strained composite aircraft structure', *Compos. Struct.*, 2006, **75**, 132–144.
19. Bezazi A, Pierce S G, Worden K and Harkati E H, 'Fatigue life prediction of sandwich composite materials under flexural tests using a Bayesian trained artificial neural network', *Inter. J. Fatigue*, 2007, **29**, 738–747.
20. Marsh G and Jacob A, 'Trends in marine composites', *Reinforced Plastics*, 2007, **51**, 22–27.
21. Hyo Jin Kim H-J and Seo D-W, 'Effect of water absorption fatigue on mechanical properties of sisal textile-reinforced composites', *Inter. J. Fatigue*, 2006, **28**, 1307–1314.
22. Kumara S A, Balakrishnan T, Alagar M and Denchev Z, 'Development and characterization of silicone/phosphorus modified epoxy materials and their application as anticorrosion and antifouling coatings', *Prog. Organic Coatings*, 2006, **55**, 207–217.
23. Bahramian A R, Kokabi M, Famili M H N and Beheshty M H, 'Ablation and thermal degradation behaviour of a composite based on resol type phenolic resin: Process modeling and experimental' *Polymer*, 2006, **47**, 3661–3673.
24. Wang M, Weia L and Zhao T, 'A novel condensation–addition-type phenolic resin (MPN): Synthesis, characterization and evaluation as matrix of composites', *Polymer*, 2005, **46**, 9202–9210.
25. Marsh G, 'Composites fight for share of military applications', *Reinforced Plastics*, 2005, **49**, 18–22.
26. Tsai S W and Hahn H T, *Introduction to Composite Materials*, Technomic Pub. Co. Inc., 1980, Chapter 5. p 167.
27. Chasseaud P T, 'Adhesives, syntactics and laminating resins for aerospace repair and maintenance applications from Ciba Specialty Chemicals', *Inter. J. Adhesion Adhesives*, 1999, **19**, 217–229.
28. FAA fire testing; http://www.faafiretesting.com.
29. Ingason H, 'Model scale railcar fire tests', *Fire Safety J.*, 2007, **42**, 271–282.
30. Kang K, 'Validation of FDS for the prediction of medium-scale pool fires. *Fire Safety J.*, 2007, **42**, 127–138.
31. Federal railroad administration; http://www.fra.dot.gov.
32. BS 5852: 2006 Methods of test for assessment of the ignitability of upholstered seating by smouldering and flaming ignition sources; http://www.bsi-global.com/en/.
33. ISO 5659-2: Plastics – Smoke generation – Part 2: Determination of optical density by a single-chamber test; *http://www.iso.org/iso/home.htm*.
34. Code of Federal Regulations, Title 46, Subchapter T; *http://www.access.gpo.gov*.
35. Improved Fire- and Smoke-Resistant Materials for Commercial Aircraft Interiors: A Proceedings. 1995, *National Materials Advisory Board (NMAB)*.
36. SOLAS, Consolidated Edition, 2004, consolidated text of the International Convention for the Safety of Life at Sea, 1974, and its Protocol of 1978: articles, annexes and certificates.
37. Fire Test Procedure (FTP) Code, International Code for Application of Fire Test Procedures, Resolution MSC.61 (67), International Maritime Organization, London, 1998.
38. Kourtides D A, Gilwee Jr W J and Parker J A, (a) 'Thermochemical characterisation of some thermally stable thermoplastic and thermoset polymers', *Polym. Eng. Sci.*, 1979, **19**, 24–29; (b) 'Thermal response of composite panels', *Polym. Eng. Sci.*, 1979, **19**, 226–231.
39. Gilwee W J, Parker J A and Kourtides D A, 'Oxygen index tests of thermosetting resins', *J. Fire Flamm.*, 1980, **11**, 22–31.

40. Morrey E L, PhD Thesis, Imperial College of Science, Technology and Medicine, London, 2001.
41. Kandola B K, Horrocks A R, Myler P and Blair D, 'The effect of intumescents on the burning behaviour of polyester-resin-containing composite', *Composites: Part A*, 2002, **33**, 805–817.
42. Kandola B K, Horrocks A R and Rashid M R, 'Effect of reinforcing element on burning behaviour of fibre-reinforced epoxy composites', In *Recent advances of flame retardancy of polymeric materials*, Ed. Lewin M; Proceedings of the 17th Conference, BCC, Stamford, Connecticut, USA, 2006.
43. Kandola B K, Myler P, Horrocks A R, Herbert K and Rashid M R, 'Effect of fibre type on fire and mechanical behaviour of hybrid composite laminates', *SAMPE Fall Technical Conference*, November 6–9, Dallas, Texas, USA. 2006.
44. Kandola B K, Myler P, Kandare E, Herbert K and Rashid M R, 'Fire and mechanical behaviour of hybrid composite laminates', *8th International Conference on Textile Composites (TEXCOMP-8)*, Nottingham, UK, 16–18 Oct 2006.
45. Brown J R and Mathys Z, 'Reinforcement and matrix effects on the combustion properties of glass reinforced polymer composites', *Composites Part A*, 1997, **28A**, 675–681.
46. Hshieh F Y and Beeson H D, 'Flammability testing of flame-retarded epoxy composites and phenolic composites', Proc. Int. conf, *Fire Safety*, 1996, **21**, 189–205.
47. Brown J R, Fawell P D and Mathys Z, 'Fire-hazard assessment of extended-chain polyethylene and aramid composites by cone calorimetry', *Fire Mater.*, 1994, **18**, 167–172.
48. Gibson A G and Hume J, 'Fire performance of composite panels for large marine structures', *J. Plastics Rubber and Composites*, 1995, **23**, 175–183.
49. Mouritz A P and Gibson A G, *Fire properties of polymer composite materials*, Springer, Dordrecht Netherlands, 2006.
50. Scudamore M J, 'Fire performance studies on glass-reinforced plastic laminates', *Fire Mater.*, 1994, **18**, 313–325.
51. Xiao J, Chena J-M, Zhoua H-D and Zhang Q, 'Study of several organic resin coatings as anti-ablation coatings for supersonic craft control actuator', *Mater. Sci. Eng. A*, 2007, **452–453**, 23–30.
52. Kahramana M V, Kayaman-Apohana N, Arsub N and Güngöra A, 'Flame retardance of epoxy acrylate resin modified with phosphorus containing compounds', *Prog. Organic Coating*, 2004, **51**, 213–219.
53. Wang X and Zhang Q, 'Synthesis, characterization, and cure properties of phosphorus-containing epoxy resins for flame retardance', *Eur. Poly. J.*, 2004, **40**, 385–395.
54. Jain P, Choudhary V and Varma I K, 'Phosphorylated epoxy resin: effect of phosphorus content on the properties of laminates', *J. Fire Sci.*, 2003, **21**, 5–16.
55. Jain P, Choudhary V and Varma I K, 'Effect of structure on thermal behaviour of epoxy resins', *Eur. Poly. J.*, 2003, **39**, 181–187.
56. Jeng R-J, Shau S-M, Lin J-J, Su W C and Chiu Y-S, 'Flame retardant epoxy polymers based on all phosphorus-containing components', *Eur. Poly. J.*, 2002, **38**, 683–693.
57. Biswas, B, PhD Thesis, 'Fire retardation of carbon fibre-reinforced epoxy composites using reactive flame retardants', The University of Bolton, UK. 2007.
58. Biswas B, Kandola B K, Horrocks A R and Price D, 'A quantitative study of carbon monoxide and carbon dioxide evolution during thermal degradation of flame retarded epoxy resins', *Poly. Degrad. Stab.*, 2007, **92**, 765–776.

59. Liu W, Varley R J and Simon G P, 'Understanding the decomposition and fire performance processes in phosphorus and nanomodified high performance epoxy resins and composites', *Polymer*, 2007, **48**, 2345–2354.
60. Perez R M., Sandler J K W, Altstädt V, Hoffmann T, Pospiech D, Ciesielski M, Döring M, Braun U, Balabanovich A I and Schartel B, 'Novel phosphorus-modified polysulfone as a combined flame retardant and toughness modifier for epoxy resins', *Polymer*, 2007, **48**, 778–790.
61. Chowdhury F H, Hosur M V and Jeelani S, 'Studies on the flexural and thermomechanical properties of woven carbon/nanoclay-epoxy laminates', *Mater. Sci. Eng. A*, 2006, **421**, 298–306.
62. Yokozeki T, Iwahori Y and Ishiwata S, 'Matrix cracking behaviors in carbon fiber/epoxy laminates filled with cup-stacked carbon nanotubes (CSCNTs)', *Composites: Part A*, 2007, **38**, 917–924.
63. Zhou Y, Pervin F, Lewis L and Jeelani S, (a) 'Fabrication and characterization of carbon/epoxy composites mixed with multi-walled carbon nanotubes', *Mater. Sci. Eng. A*, 2006, doi:10.1016/ j.msea.2006.11.066; (b) 'Experimental study on the thermal and mechanical properties of multi-walled carbon nanotube-reinforced epoxy', *Mater. Sci. Eng. A.*, 2007, **452–453**, 657–664.
64. Park K-Y, Lee S-E, Kim C-G and Han J-H, 'Application of MWNT-added glass fabric/epoxy composites to electromagnetic wave shielding enclosures', *Compos. Struct.*, 2007, **81**, 401–406.
65. Meguid S A and Sun Y, 'On the tensile and shear strength of nano-reinforced composite interfaces', *Mater. Design*, 2004, **25**, 289–296.
66. Wang X, Han E and Ke W, 'An investigation into fire protection and water resistance of intumescent nano-coatings', *Surf. Coatings Tech.*, 2006, **201**, 1528–1535.
67. Duquesne S, Magnet S, Jama C and Delobel R, 'Intumescent paints: fire protective coatings for metallic substrates', *Surf. Coatings Tech.*, 2004, **180–181**, 302–307.
68. Fire-resistant barriers for composite materials. United States Patent 5236773; 5284697.
69. Tewarson A and Macaione D P, 'Polymers and composites – an examination of fire spread and generation of heat and fire products', *J. Fire Sci.*, 1993, **11**, 421–441.
70. Sorathia U, Gracik T, Ness J, Durkin A, Williams F, Hunstad M and Berry F, 'Evaluation of intumescent coatings for shipboard fire protection' in *Recent Advances in Flame Retardancy of Polymeric Materials*, Vol XIII, ed. Lewin M; Proc. of the 2002 Conf, BCC, Stamford, Conn., 2002.
71. Sorathia U, Rollhauser C M and Hughes W A, 'Improved fire safety of composites for naval applications', *Fire. Mater.*, 1992, **16**, 119–125.
72. Chukwudolue C, 'Mechanical property degradation of glass reinforced unsaturated polyester resin with and without fire-retardant surface coating', MSc Thesis, University of Bolton, 2008.
73. Stevart J L, Griffin O H, Gurdal Z and Warner G A, 'Flammability and toxicity of composite materials for marine vehicles', *Naval Engn. J.*, 1990, **102 (5)**, 45–54.
74. Morchat R M and Hiltz J A, 'Fire-safe composites for marine applications', Proc 24th Inter. Conf *SAMPE Tech.*, 1992, **24**, T153–T164.
75. Morchat R M, *'The Effects of Alumina Trihydrate on the Flammability Characteristics of Polyester, Vinylester and Epoxy Glass Reinforced Plastics'* Techn. Rep. Cit. Govt. Rep. Announce Index (U.S.), 1992, **92 (13)**, AB NO 235, 299.
76. Nir Z, Gilwee W J, Kourtides D A and Parker J A, 'Rubber-toughened

polyfunctional epoxies: brominated vs nonbrominated formulated for graphite composites', *SAMPE Q.*, 1983, **14 (3)**, 34–38.
77. Sorathia U, Ness J and Blum M, 'Fire safety of composites in the US Navy', *Composites: Part A*, 1999, **30**, 707–713.
78. Kim J, Yooa S, Baea J-Y, Yunb H-C, Hwangb H and Kongb B-S, 'Thermal stabilities and mechanical properties of epoxy molding compounds (EMC) containing encapsulated red phosphorous', *Poly. Degrad. Stab.*, 2003, **81**, 207–213.
79. Hussain M, Varley R J, Mathus M, Burchill P and Simon G P, 'Development and characterization of a fire retardant epoxy resin using an organo-phosphorus compound', *J. Mater. Sci. Lett.*, 2003, **22**, 455.
80. Igwe U, 'Effect of additives on physical, thermal and mechanical properties of glass-reinforced polyester composites', MSc thesis, University of Bolton, UK, 2006.
81. Kandola B K, Horrocks A R, Myler P and Blair D, 'Mechanical Performance of Heat/Fire Damaged Novel Flame Retardant Glass-Reinforced Epoxy Composites', *Composites Part A*, 2003, **34**, 863.
82. Kandola B K, Myler P, Horrocks A R and El-Hadidi M, 'Empirical and numerical approach for optimisation of fire and mechanical performance in fire-retardant glass-reinforced epoxy composites', *Fire Safety J.*, 2008, **43(1)**, 11–23.
83. Troitzsch J, International plastics flammability handbook, München: Hanser; 1990.
84. Wang C S and Lin C H, 'Synthesis and properties of phosphorus-containing epoxy resins by novel method', *J. Polym. Sci: Polym. Chem. Ed*, 1999, **37**, 3903–3909.
85. Lyon R E, Castelli L M and Walters R, 'A Fire-Resistant Epoxy, DOT/FAA/AR-01/53, Final Report', Federal Aviation Administration (FAA), Washington, September, 2001.
86. Shieh J Y and Wang C S, 'Synthesis of novel flame retardant epoxy hardeners and properties of cured products', *Polymer*, 2001, **42**, 7617–7625.
87. Cho C S, Fu S C, Chen L W and Wu T R, 'Aryl phosphinate anhydride curing for flame retardant epoxy networks', *Polym. Inter*, 1998, **47**, 203–209.
88. Braun U, Balabanovich A I, Schartel B, Knoll U, Artner J, Ciesielski M, Döring M, Perez R, Sandler J K W, Altsta V, Hoffmann T and Pospiech D, 'Influence of the oxidation state of phosphorus on the decomposition and fire behaviour of flame-retarded epoxy resin composites', *Polymer*, 2006, **47**, 8495–8508.
89. Toldy A, Tóth N, Anna P and Marosi G, 'Synthesis of phosphorus-based flame retardant systems and their use in an epoxy resin', *Poly. Degrad. Stab.*, 2006, **91**, 585–592.
90. Espinosa M A, Galià M and Cádiz V, 'Novel phosphorilated flame retardant thermosets: epoxy–benzoxazine–novolac systems', *Polymer*, 2004, **45**, 6103–6109.
91. Alcón M J, Espinosa M A, Galià M and Cádiz V, 'Synthesis, characterization and polymerization of a novel glycidyl phosphinate', *Macromol. Rapid Commun.*, 2001, **22**, 1265–1271.
92. Levchik S, Piotrowskia A, Weilb E and Yaob Q, 'New developments in flame retardancy of epoxy resins', *Poly. Degrad. Stab*, 2005, **88**, 57–62.
93. Chiu H T, Chiu S H, Jeng R E and Chung J S, 'A study of the combustion and fire-retardance behaviour of unsaturated polyester/phenolic resin blends', *Polym. Degrad. Stab.*, 2000, **70**, 505–514.
94. Cherian A B, Varghese L A and Thachil E T, 'Epoxy-modified unsaturated polyester hybrid networks', *Eur. Polym. J.*, 2007, **43**, 1460–1469.
95. Camino G, Tartaglione G, Frache A, Manferti C and Costa G, 'Thermal and combustion behaviour of layered silicate-epoxy nanocomposites', *Poly. Degrad. Stab.*, 2005, **90**, 354–362.

96. Kandare E, Chigwada G, Wang D, Wilkie C A and Hossenlopp J M, 'Nanostructured layered copper hydroxy dodecyl sulfate: A potential fire retardant for poly(vinyl ester) (PVE)', *Poly. Degrad. Stab*, 2006, **8**, 1781–1790.
97. Kandare E, Chigwada G, Wang D, Wilkie C A and Hossenlopp J M, 'Probing synergism, antagonism, and additive effects in poly(vinyl ester) (PVE)' composites with fire retardants', *Poly. Degrad. Stab.*, 2006, **91**, 1209–1218.
98. Kandola B K, Nazare S and Horrocks A R, 'Thermal degradation behaviour of flame-retardant unsaturated polyester resins incorporating functionalised nanoclays. In: *Fire retardancy of polymers: The use of mineral fillers in micro- and nanocomposites*, Ed. M Le Bras *et al*, Royal Chemical Society, Cambridge, 2005, pp. 147–160.
99. Katsoulis C, Kandare E, Kandola B K and Myler P, 'Enhancement of fire performance of fibre-reinforced epoxy composites using layered-silicate clays and fire retardants', paper in preparation.
100. Lin C H, Feng C C and Hwang T Y, 'Preparation, thermal properties, morphology, and microstructure of phosphorus-containing epoxy/SiO_2 and polyimide/SiO_2 nanocomposites', *Eur. Poly. J.*, 2007, **43**, 725–742.
101. Laachachi A, Leroy E, Cochez M, Ferriol M and Lopez Cuesta J M, 'Use of oxide nanoparticles and organoclays to improve thermal stability and fire retardancy of poly(methyl methacrylate)', *Poly. Degrad. Stab.*, 2005, **89**, 344–352.
102. Chen Y, Wang Q, Yan W and Tang T, 'Preparation of flame retardant polyamide 6 composite with melamine cyanurate nanoparticles in situ formed in extrusion process', *Poly. Degrad. Stab.*, 2006, **91**, 2632–2643.
103. Kuchibhatla S V N T, Karakoti A S, Bera D and Seal S, 'One dimensional nanostructured materials', *Prog. Mater. Sci.*, 2007, **52**, 699–913.
104. Guitierrez J, Parneix P, Bollero A, Høyning B, McGeorge D, Gibson G and Wright P, 'Fire Performance Of Naval Composite Structures', Proceedings of Fire and Materials Conference, San Francisco, USA, 2005.
105. Sorathia U, Lyon R, Gann R and Gritzo L, 'Materials and fire threat', *SAMPE J.*, 1996, **32(3)**, 8–15.
106. Asaro R J and Dao M, 'Fire degradation of fiber composites', *Marine Tech.*, 1997, **34**, 197–210.
107. Mouritz A P and Mathys Z, 'Mechanical properties of fire-damaged glass-reinforced phenolic composites', *Fire. Mater.*, 2000, **24**, 67–75.
108. Mouritz A P and Mathys Z, 'Post-fire mechanical properties of marine polymer composites', *Compos. Struct.*, 1999, **47**, 643–653.
109. Kourtides D A, 'Processing and flammability parameters of bismaleimide and some other thermally stable resin matrices for composites', *Polym. Comp.*, 1984, **5**, 143–150.
110. Ulven C A and Vaidya U K, 'Impact response of fire damaged polymer-based composite materials', *Composites: Part B*, 2008, **39**, 92–107.
111. Mouritz A P and Gardiner C P, 'Compression properties of fire-damaged polymer sandwich composites', *Composites: Part A*, 2002, **33**, 609–620.
112. Kandare E, Kandola B K, Staggs J, Myler P and Horrocks R A, 'Mathematical prediction of thermal degradation behaviour of flame-retarded epoxy resin formulations', *Interflam2007, 11th Inter. Confer. Fire. Sci. Eng.*, September, 2007, University of London, UK.
113. Feih S, Mathys Z, Gibson A G and Mouritz A P, 'Modelling the tension and compression strengths of polymer laminates in fire', *Compos. Sci. Tech.*, 2007, **67**, 551–564.

114. Gibson A G, Wu Y-S and Evans J T, 'The Integrity of Polymer Composites during and after fire', *J. Compos. Mater.*, 2004, **38**, 1283–1307.
115. Mouritz A P, 'Simple models for determining the mechanical properties of burnt FRP composites', *Mater. Eng.*, 2003, **A359**, 237–246.
116. Gibson A G, Wu Y-S, Evans J T and Mouritz A P, 'Laminate theory analysis of composites under load in fire', *J. Compos. Mater.*, 2006, **40**, 639–658.

16
Improving the fire safety of road vehicles

M M HIRSCHLER, GBH International, USA

Abstract: The number of fires in vehicles is in the same range as that in buildings. Moreover, well over 70% of vehicle fire losses occur in road vehicles, and over 90% of those happen in private cars, often from minor ignition sources (i.e., not involving the fuel tank). Regulatory requirements in most developed countries are based on an old flame spread test, developed as FMVSS 302, shown not to be suitable for the fire hazard of vehicles in the 21st century. The US NFPA Technical Committee on Hazard and Risk of Contents and Furnishings is working on a guide to investigate means for decreasing fire hazards associated with road vehicles. Key problems identified include the need for including reaction-to-fire requirements for materials, particularly in the passenger and engine compartments. The importance of material heat release and of the overall fire safety of the entire vehicle was also identified.

Key words: automobiles, fire safety, heat release, ignitability, motor vehicles.

16.1   Introduction

The number of fires in vehicles in developed countries (detailed data are available from the US, the UK and Canada) is in the same range as that in buildings. Moreover, generally well over 70% of vehicle fire losses occur in road vehicles, and over 90% of those happen in private cars. Many examples exist of multiple fire fatalities resulting from minor ignition sources (i.e., not involving the fuel tank) in private cars. Fire statistics from the US National Fire Protection Association (NFPA) show very high annual average US vehicular losses (1980–1998): 433 000 fires, 679 civilian fire fatalities, 2990 civilian fire injuries and $959.0 millions fire losses.[1]* In structures, the numbers, from 1980 to 1998,[1–2] are 682 200 fires, 4,440 civilian fire fatalities, 23 014 civilian fire injuries, and $6438.3 millions fire losses. The data indicates that the number of vehicular fires is almost two-thirds of the number of structure fires (Table 16.1) and that over 1 in 12 fire fatalities occur in passenger road vehicles. The heat release rate obtained from a burning automobile has been shown to be in the range of 1.5 to 8.0 MW, roughly the same order as heat released from a fully

---

* The US fire statistics collected by NFPA have a fundamental discontinuity between before and after 1998 due to the changes introduced into the NFIRS (US National Fire Incident Reporting System) in 1999 when they went to version 5.0 of their software. Most of the fire statistics reported in this work will address the information before 1999.

Table 16.1 Fires in US 1980–1998 – yearly average

	Fires	Civilian fire fatalities	Civilian fire injuries	Property damage (millions $)
Total structure fires	682 200	4 440	23 014	6 438.3
Overall vehicle fires	433 000	679	2 990	959.0
Passenger road vehicle fires	332 640	437	2 048	551.7
Passenger road vehicle fires in the period 1994–98	295 170	330	1 403	692.6
Automobile fires (1994–98)	280 550	302	1 161	609.8

involved room in a home. In fact, a study by Janssens showed that 34 full-scale calorimeter heat release tests have been conducted on road vehicles between 1994 and 2006 and have yielded data for fire safety engineering calculations.[3]

In 2003 there were some 230 million registered vehicles in the US, according to the statistics collected by the regulatory authority: Federal National Highway Traffic Safety Administration (NHTSA). Over 135 million of those vehicles were passenger cars and over 87 million were light trucks, including pickup trucks, sports utility vehicles (SUVs) and minivans, according to the US Bureau of Transportation Statistics (BTS). Such cars have much more plastic and much less steel than they used to have in earlier times. The increased use of plastics has made cars much more fuel-efficient, by lowering vehicle weight, and allowed more flexibility in design and aesthetics, but at the expense of increasing the potential for fire losses. Significant recent work has highlighted some critical fire safety deficiencies of the materials used in this area.[4–6] In particular, such work showed problems in terms of both the reaction-to-fire properties of automobile materials used in both the passenger and engine compartments in private cars. The work showed that many of these materials are highly combustible and have poor fire performance and that the fire resistance properties (in terms of a standard time-temperature curve) of the so-called barrier materials used is virtually non-existent. For example, real-scale tests on actual cars showed that small fires that start in the engine compartment, even if they don't involve the vehicle's liquid fuel, would penetrate into the passenger compartment through the 'engine cover' in less than 2 minutes and that the 'engine cover' would actually burn at low heat fluxes, and not offer any type of real barrier.[5] This 'engine cover' used to be called a 'firewall' or 'bulkhead' and is now often designated by automobile manufacturers as the 'passenger compartment engine access cover'. Figure 16.1 shows the temperature reached in the front and rear seats, headliner, duct, vent and carpet following ignition with a small ignition source in the engine compartment: temperatures exceeding 700 °C were reached quickly in all cases.[5]

It has also been shown that the materials used for car headliners, foam fillings

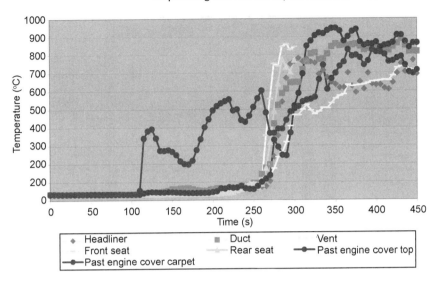

16.1 Real scale test car fire temperatures.

and, especially, ducts, are of particularly poor fire performance, with high heat release rates.[4]

16.2 Regulatory requirements

The basic approach to fire safety requirements for private passenger cars and trucks in the US (and actually throughout the developed world) has been the prevention of ignition of interior passenger materials by the effect of cigarette ignitions. This is based on the fact that people often discarded lit cigarettes in cars when the regulations were initially developed in the late 1960s. At the time this approach was developed, it was considered that cars and trucks can stop very rapidly and that they would be evacuated almost immediately. Thus, the US Federal National Highway Traffic Safety Administration (NHTSA) developed a small-scale regulatory fire test (FMVSS 302[7]), with a very mild ignition source for such materials. This is the same test that is still the only regulatory test, both in the US and internationally, for the fire safety of car materials. The FMVSS 302 test can virtually be considered an international standard, as it has been adopted by many countries, in different ways and under a number of designations, including: ISO 3795, BS AU 169 (UK), ST 18-502 (France), DIN 75200 (Germany), JIS D 1201 (Japan), SAE J369 (automotive industry) and ASTM D 5132.

Cars of the 1960s contained some 10 kg of plastics and large amounts of noncombustible materials. Since then, however, there has been a considerable growth in the use of combustible materials. For example, by the year 2000 cars contained well over 90 kg of plastic materials. This over 10-fold increase in

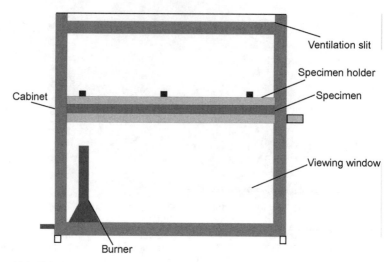

16.2 FMVSS 302 test equipment schematic.

combustible materials has the potential to increase fire hazard, and yet the regulatory approach to fire safety has not changed.

The FMVSS 302 test contains a burning rate requirement for the flammability of interior materials in motor vehicles, which was proposed in 1969 and became effective in September 1972. The test method (see Fig. 16.2) exposes a horizontal material sample to a Bunsen burner flame at one end and determines the horizontal rate of spread away from the flame, with the material laid on a grid. In order to be acceptable, the flame spread rate is not allowed to exceed 102 mm/min,[7] a criterion that is almost always met by thermoplastic materials, particularly if they melt and drip (even with flames) through the grid, without spreading flame to the end of the material.

It is interesting to note that the NHTSA web site contains extensive information on research conducted on fire safety of private cars as a result of the agreement between the US Federal government and General Motors.[8] NHTSA has recently recognized that there is a serious fire safety problem and has undertaken several research programs, in conjunction with Southwest Research Institute,[9–10] to investigate the fire performance of automotive materials and the potential for alternative tests for material regulation, probably by using the cone calorimeter (ASTM E 1354[11]), a source of heat release information. Information on comparisons between the cone calorimeter peak rate of heat release of a whole series of car interior materials and random plastics is shown in Fig. 16.3.[4] These data highlight the poor fire performance of car materials, compared with that of random plastics. A comparison of the heat release of a polypropylene car duct material and a fire retarded polypropylene alternative material also showed the improvements achievable with better materials.[5]

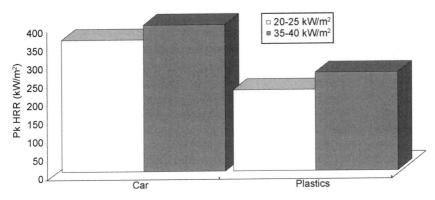

16.3 Pk HRR: cone calorimeter medians for car components vs. random plastics.

16.3 United States fire loss statistics

The major factors contributing to ignition in US road vehicle fires between 1999 and 2003, according to NFPA statistics, are shown in Table 16.2.[12] The largest factor associated with fires is mechanical failure or malfunction, while the largest factor associated with fire fatalities is operational deficiency (mainly collisions).

In the US, during the period between 1994 and 1998, there were approximately 300 000 passenger road vehicle fires (according to NFPA statistics). Table 16.3 shows data on fires and fire fatalities associated with fires starting in

Table 16.2 Major factors contributing to ignition in US road vehicle fires between 1999 and 2003

Factor	% of fires	% of fire fatalities
Mechanical failure or malfunction (including leaks or breaks, backfire and equipment worn out)	49	8
Electrical failure or malfunction (including short circuit, ground faults and arcs)	23	1
Misuse of material or product (including spill of flammable or combustible liquid, heat source too close to combustibles, abandoned materials and misuse)	11	14
Operational deficiency (including collision and equipment not being operated properly)	11	58
Fire spread or control	6	6
Design, manufacturing or installation deficiency	1	0
Other	6	13

Table 16.3 US passenger road vehicle fires, by area of fire origin between 1994 and 1998

Area of fire origin	Fires		Civilian fire deaths	
Engine, running gear or wheel area of vehicle	201,300	68.20%	134	40.50%
Passenger area	52,050	17.60%	65	19.60%
Exterior surface of vehicle	8,230	2.80%	8	2.30%
Unclassified vehicle area	5,630	1.90%	18	5.60%
Trunk or load carrying area of vehicle	5,040	1.70%	10	3.10%
Operating or control area of vehicle	4,010	1.40%	3	0.90%
Fuel tank or fuel line of vehicle	3,940	1.30%	64	19.50%
On or near highway or street	3,360	1.10%	10	3.20%
Unclassified area of origin	2,840	1.00%	2	0.70%
Other known area	8,780	3.00%	15	4.70%
Total	295,170	100.00%	330	100.00%

various areas. All the important ignition factors of those road vehicle fires are shown in Table 16.4. They include four large ignition factors, and some minor ones, such as other electrical faults, hot surfaces and overheating. The major ignition factors are: (1) part failure (19%), (2) short circuit or ground fault (>18%), (3) incendiary or suspicious (arson) (almost 17%) and (4) collisions or overturns (<2%). The key point is that the data in the tables makes it clear that there are many fires and fire fatalities that are not associated with fuel in the fuel tanks.

16.4 State of the art of regulation and guidelines

The Fire Protection Research Foundation of NFPA (FPRF) issued a white paper on 'Fire and Transportation Vehicles – State of the Art of Regulatory Requirements and Guidelines' in 2004.[13] In that white paper the Research Advisory Council addresses highway vehicles and states: 'Highway vehicles are highly regulated, perhaps more than any other consumer product, with the exception of fire safety.' The section on conclusions and recommendations of the FPRF white paper contains the following recommendations for highway vehicles.

- Highway vehicles have minimal fire safety regulation for interior (and none for exterior) combustible materials. The regulatory test, from NHTSA, was intended to provide protection from cigarette ignition and has not changed for over 30 years.
- This type of regulation, although originating in the US, has been adopted by all major countries worldwide, including the European Union, Japan and Canada.
- Highway vehicles must meet regulation concerning the integrity of fuel

Table 16.4 Ignition factors for highway vehicle fires in the USA between 1994 and 1998

Ignition factor	Fires		Civilian deaths		Civilian injuries		Property damage	
	#	%	#	%	#	%	$ (millions)	%
Part failure, leak or break	63,260	19.0	21	4.9	278	15.9	139.8	16.0
Short circuit or ground fault	60,690	18.3	4	1.0	196	11.2	131.8	15.1
Incendiary or suspicious	54,900	16.5	38	8.8	93	5.3	220.9	25.2
Backfire	34,690	10.4	4	1.0	171	9.8	52.1	6.0
Unclassified or unknown-type mechanical failure or malfunction	29,310	8.8	10	2.3	96	5.5	77.0	8.8
Electrical failure other than short circuit or ground fault	21,260	6.4	2	0.6	46	2.6	59.8	6.8
Lack of maintenance	7,670	2.3	1	0.3	22	1.3	8.6	1.0
Fuel spilled or unintentionally released	7,360	2.2	23	5.4	132	7.6	21.7	2.5
Property too close	7,220	2.2	10	2.2	18	1.0	22.4	2.6
Unclassified ignition factor	6,900	2.1	16	3.6	38	2.2	21.2	2.4
Collision, overturn or knock down	6,160	1.9	263	60.6	276	15.8	51.8	5.9
Combustible too close to heat	6,060	1.8	6	1.3	52	3.0	10.7	1.2
Abandoned material	5,190	1.6	6	1.4	53	3.0	9.9	1.1
Unclassified or unknown-type operational deficiency	3,450	1.0	3	0.7	26	1.5	9.8	1.1
Other known ignition factor	18,170	5.5	25	5.8	250	14.3	38.0	4.3
Total	332,280	100	434	100	1,747	100	875.6	100

systems, especially the fuel tank. The regulation for school buses has been made slightly more severe than that for other highway vehicles.
- Fire safety of buses fully mirrors that of underground rail in terms of the voluntary Federal Transit Administration (FTA) guidelines.
- There has been an increase in the amount of combustible materials used for highway vehicles, without a corresponding change in the regulations.
- There are no mandatory fire safety requirements for the electrical system (including the wiring) or the drive train, but manufacturers are tending to use their own internal specifications.
- The developed world is moving towards high energy (42 volts) ignition systems, but the electrical system is still unregulated, and there are no rules governing the added fire hazard from such systems.
- There are no egress regulations for passenger cars and vans, including sports utility vehicles, as it is expected that egress should not take very long, because of the proximity of the driver and/or passenger to the doors. Furthermore, such vehicles tend to be able to come to a complete stop within a short time, theoretically allowing occupants immediate egress. In reality, however, following a collision, it is often found that occupants are injured and require assistance to evacuate.
- Evacuation from passenger cars and vans would be aided by structural compartmentalization that gives added time before fires penetrate from the engine compartment or the storage compartment into the passenger compartment.
- The fire loss record of public transport vehicles appears to be reasonably good; however, the fire losses in private cars and vans represents an overwhelming fraction of all fire losses in transportation vehicles: 74% of all transportation vehicle fires and 95% of all highway vehicle fires.
- In spite of this series of shortcomings, there does not appear to be movement by the authority having jurisdiction to increase fire safety of highway vehicles.

The FPRF Research Advisory Council on Transportation Vehicles also decided that highway vehicles, especially automobiles, represent the most important fire safety issue in the area of transportation, and they are preparing to develop a second white paper with recommendations for action in that area.

16.5 Activities of the US NFPA Hazard and Risk of Contents and Furnishings Technical Committee

One of the National Fire Protection Association (NFPA) technical committees is entitled 'Hazard and Risk of Contents and Furnishings' and is known by the acronym HAR. The technical committee is responsible for providing guidance

Improving the fire safety of road vehicles 451

to other NFPA technical committees on areas associated with fire hazard and fire risk of contents and furnishings. Its scope reads as follows:

> This Committee shall have primary responsibility for documents on fire hazard calculation procedures for use by other Committees in writing provisions to control the fire hazards of contents and furnishings. This Committee shall also provide guidance and recommendations to Committees in assessing the fire hazard of contents and furnishings. It shall establish classification and rating systems, request the development and standardization of appropriate fire tests, and identify and encourage necessary research as it relates to the fire hazards of contents and furnishings. It shall act in a liaison capacity between NFPA and the committees of other organizations with respect to the hazard of contents and furnishings.

In 1996 the committee developed NFPA 555, 'Guide on Methods for Evaluating Potential for Room Flashover', which has most recently completed a 2008 revision.[14] Separately, the committee is also working on the development of a document expected to be entitled NFPA 557, Standard Fire Loads for Engineering Design of Structural Fire Resistance in Buildings.

In 2002, the committee requested, and obtained, permission from NFPA Standards Council, to develop a document for vehicles (especially highway vehicles) intended to be parallel to NFPA 556, that would focus on ways to lower fire hazard for occupants of such vehicles. A key element in the rationale for developing such a guide is the fact that NFPA issues three documents that reference FMVSS 302, namely: NFPA 1192, Standard on Recreational Vehicles, NFPA 1901, Standard for Automotive Fire Apparatus, and NFPA 1906, Standard for Wildland Fire Apparatus. The document, which has been assigned the designation NFPA 556, was originally expected to be issued in the Annual 2006 cycle. However, NFPA members felt that more work was needed and returned the document to committee. The document is now expected to be assigned to the Annual 2010 cycle, after the committee has addressed input from the automotive industry and other public sources. The latest draft of NFPA 556 is entitled 'Guide on Methods for Evaluating Fire Hazard to Occupants of Passenger Road Vehicles'. The draft 2007 document, which has not yet been approved by the committee, consists of 12 chapters, as follows:

- Chapter 1: Administration: Scope, Purpose, and Applicability
- Chapter 2: Referenced Publications
- Chapter 3: Definitions
- Chapter 4: Types of Vehicles
- Chapter 5: General Description of Passenger Road Vehicle Fires and Background Information
- Chapter 6: Approach to Evaluating Vehicle Fire Hazard
- Chapter 7: Objectives and Design Criteria
- Chapter 8: Selecting Candidate Designs

- Chapter 9: Typical Fire Scenarios to be Investigated
- Chapter 10: Evaluation Methods and Tools
- Chapter 11: Individual Fire Scenarios
- Chapter 12: Further Guidance.
- Annexes: Explanatory Material, including a discussion of the use of fire retardants and of data from individual fire scenarios, and informational references.

16.6 2007 Draft Guide NFPA 556

16.6.1 General aspects

This work describes the 2007 draft of the NFPA 556 Guide and is based on the committee recommendations (which are tentative and were not approved) and on the opinions of the author of this work.

The draft NFPA 556 Guide suggests ways to assess fire safety in cars and to improve the probability that fires in road vehicles, especially cars, become less severe or otherwise have less serious effects on the vehicle occupants. The draft NFPA 556 document is one of the more important systematic approaches to understanding ways to improve fire safety in passenger road vehicles and will now be discussed in further detail. The document identifies five broad fire scenarios for ways in which fire effects can start and eventually reach the passenger compartment.

- Fires starting inside the passenger compartment. They can start at the instrument panel, seat, headliner, floor (carpet) or one of the doors to the passenger compartment.
- Fires starting in the engine compartment. They can penetrate into the passenger compartment through the passenger compartment engine access cover (formerly known as firewall), the ductwork or the windshield.
- Fires starting in the trunk or load carrying area.
- Pool fires resulting from fuel tank failure and burning under the vehicle.
- Fires resulting from other external heat sources.

Chapters 1–4 set the scene and include all the NFPA boiler plate information. Probably the most important information is the purpose, which, at least tentatively, reads as follows: 'The purpose of this document is to provide guidance for persons investigating methods to decrease fire hazard or fire risk in passenger road vehicles, by providing additional time for occupants of the passenger road vehicle to be able to exit or be rescued in case of the occurrence of a fire involving the vehicle.' The applicability of the guide is expected to be limited, but it addresses the key vehicles and portions thereof. It is intended to apply to passenger road vehicles used to transport people who are either drivers or passengers. If approved, this guide would apply to all portions of a passenger road vehicle that have the potential to affect the fire safety of drivers or passengers. It is not intended that the provisions of the guide be applied to compartments in vehicles such as ships, trains, airplanes, or off-road vehicles, irrespective of whether they are or are not intended for use by human passengers or drivers. Chapter 4 describes which vehicles are and are not covered in this document on road vehicles, and it would apply to passenger road vehicles, namely vehicles that travel on public roads or highways, including automobiles (among them pickups, minivans and sports utility vehicles), buses (among them school buses), fire department vehicles, trackless trolleys, and motor homes or recreational vehicles.

Chapters 5 through 10 provide the background and methodology for Chapter 11, intended as guidance for detailed investigation of the various fire scenarios discussed above and Chapter 12 which gives overall guidance.

16.6.2 Background information for automobile fire hazard

Chapter 5 incorporates a variety of fire and passenger road vehicle statistics (from NFPA,[12] similar to those discussed in Section 16.3, and from the Alliance of Automobile Manufacturers[15]) and descriptions of the types of plastic materials usually found in road vehicles (based on a survey by Tewarson[16]). It also discusses the current lone prescriptive fire test required, namely FMVSS 302, and some performance-based considerations and general tenability criteria. As discussed earlier, the weight of polymeric materials used in both the engine compartment and the passenger compartment has increased from *ca.* 10 kg/car (22 lbs/car) in 1960 for US automobiles[17] to more than 91 kg/car (200 lbs/car) in 1996.[16] The substitution of metal components by plastic ones has, of course, dramatically increased the vehicle fuel load.

This is one of the most extensive chapters as it contains all the background information, including descriptions of the materials presently in use in cars and also the tenability criteria that are most relevant to fire hazard assessments, based on the work in HAZARD I,[18] as shown in Table 16.5.

16.6.3 Methodology to decrease automobile fire hazard

Chapter 6 discusses the approach to evaluating vehicle fire hazard, while taking into account the fact that optimizing materials for fire performance may result in the degradation of some other fire properties. It is therefore critical to ensure that all the properties that affect overall vehicle safety, fuel economy, emissions, manufacturability, utility, and durability, are considered when selecting materials. The chapter is intended to have a table of key material and vehicle properties. This chapter will also address the basic fire-performance-based approach to fire safety. The performance-based approach employs a systematic analysis, which will be depicted in a flow diagram, and will follow the

Table 16.5 Tenability criteria

Hazard	Incapacitation criterion	Lethality criterion
Smoke toxicity Ct (g min/m^3)	450	900
Smoke toxicity FED	0.5	1
CO concentration (ppm min)	45,000	90,000
Convected heat/temperature (EC)	65	100
Radiated heat/heat flux (kW min/m^2)	1.0	2.5

Notes:
(1) Smoke toxicity Ct: concentration-time product of toxic gases. If exposure is 30 minutes, smoke toxicity criteria will be 15 g/m^3 for incapacitation and 30 g/m^3 for lethality.
(2) Smoke toxicity FED: fractional effective dose of toxic insult required to cause lethality (if FED = 1).

guidelines from the SFPE Guide for Performance Based Design and Analysis,[19] for new and existing designs, and/or the type of framework provided by ASTM E 1546,[20] in a generic way, or by ASTM E 2061[21] for the application of this framework to a rail transportation vehicle. The process begins with the establishment of fire performance design criteria that establish the limiting hazard levels for the desired fire safety (as discussed in Chapter 7). Next the candidate design for the vehicle components or systems to be evaluated is established (Chapter 8). The relevant fire scenario for the specific analysis is selected including the scenario elements in the flow diagram, with the key generic scenarios described in Chapter 9. It is essential to realize that performance-based evaluation requires information regarding the expected fire conditions, which can be developed through tests or predictions/fire modeling. Tests can be conducted on materials, composites, fuel packages, sub-systems or full-scale vehicles. Calculation methods and simulations can be based on fire properties such as heat release rate, ignitability or combustion product yield. In order to assess fire hazard, the test results, or the fire conditions predicted by the calculations, can then be compared to the design criteria to see whether the desired criteria have been met. If the design criteria have not been met, there is a need to reconsider the approach. For example, the objectives and design criteria or fire selected fire scenario can be reassessed, additional fire test data can be collected or the candidate design can be modified and the fire performance-based process will then need to be repeated. Figure 16.4 shows the concepts in the performance-based approach used in the draft guide.

Chapter 7 addresses specific design and performance objectives and criteria. This section incorporates a combination of the overall tenability criteria (from Chapter 5), the specific fire scenarios (from Chapter 9) and the considerations (from Chapter 6) that are most relevant to fire safety of passenger road vehicles.

Improving the fire safety of road vehicles 455

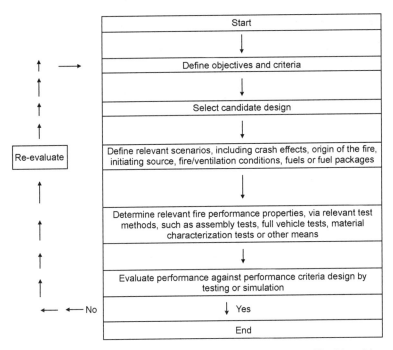

16.4 Flow chart for the performance-based approach to be used in this guide.

Chapter 8 addresses how to select candidate designs for fire-safe vehicles. The candidate designs to be evaluated by the performance-based method can be a single material, such as a candidate headliner or instrument panel facing (for example, by comparing it to one in use), or a complete package, such as an upholstered seating system. These design elements can be tested for fire properties as input to fire hazard calculations or simulations. Evaluation of the expected conditions with regard to a complete vehicle can be studied by full-scale testing of complete vehicle or by simulation techniques using data from small- and intermediate-scale testing of materials and fuel packages.

Chapter 10 is primarily intended to show the most relevant fire tests, calculation methods and guidance documents that exist for use with passenger vehicle materials (Table 16.6). In terms of fire properties, clearly ignitability and heat release are going to be the key properties needed to develop a proper fire hazard assessment or fire-performance-based design. Heat release can be measured in various scales (full scale, fuel package scale and individual material scale), with the latter primarily based on the cone calorimeter, ASTM E 1354[11]). Other recommended tests principally address individual materials. In the case of electrical and optical fiber cables contained in the engine compartment, it is likely that a vertical cable tray test would be recommended as the key performance property (UL 1685[22]). Cargo compartment lining materials and other fabric or foam materials likely to be subject to smoldering will probably be

Table 16.6 Recommended test methods and guidance for use in automobile fire safety

Vehicle component	Evaluation tool	Comments
Bulk of materials	ASTM E 1354/ISO 5660	Cone calorimeter
Bulk of materials	ASTM E 1321/ISO 5658-2	LIFT apparatus
Interior materials	FMVSS 302	Regulatory test (not relevant to fire hazard)
Seat materials	ASTM E 1474	Cone calorimeter for upholstery composites
Carpets/floor coverings	ASTM D 2859	Pill test
	ASTM E 648 (NFPA 253)	Critical radiant flux
Wire and cable	UL 1685	Vertical cable tray fire test
	ASTM D 6113	Cone calorimeter for electrical materials
	UL VW-1	Vertical wire test
Fire resistance – fuel spill from underneath	ASTM E 119 (NFPA 251 or UL 263)	Fire resistance time-temperature curve tests
	ASTM E 1529 (UL 1709)	Fire resistance for hydrocarbon fires
Fire stops in the undercarriage	ASTM E 814 (UL 1479)	Firestop test
Foams and fabrics (smoldering)	ASTM E 1353 (NFPA 260)	Cigarette ignition – component test
Foams and fabrics (smoldering)	ASTM E 1352 (NFAP 261)	Cigarette ignition – composite test
Windshields	NFPA 257 (UL 9)	Fire resistance of glazing (excluding hose stream test)
Individual fuel packages	NFPA 289	Engine compartment and passenger compartment furniture calorimeter
Flat materials	ISO TS 17431	Title: Fire Tests Reduced Scale Model Box Test
Transmission through the bulkhead (dash panel) and the windshield	ASTM E 1354/ISO 5660	Cone calorimeter
	ASTM E 1623	Intermediate scale calorimeter (ICAL)
	EN13823 (SBI)	Single burning item (SBI)
Plastic fuel tanks	ECE R34.01 Annex 5	Full fire test of tanks
Batteries	SAE J2464	
Guidance	ASTM E 603	Guidance for large-scale tests
Guidance	ASTM E 2061	Guidance for fire hazard assessment in transportation vehicles
Guidance	ASTM E 2067	Guidance for conducting large-scale heat release tests
Guidance	ASTM E 2280	Guidance for fire hazard assessment in a compartment
Guidance	ASTM E 1546	Guidance for fire hazard assessment
Other evaluation methods	As appropriate	

required to exhibit smoldering ignition resistance testing, either based on the component ASTM E 1353 test[23] (also known as the Upholstered Furniture Action Council (UFAC) test or on the composite test, ASTM E 1352[24]). The guide will probably recommend testing floor covering materials to assess flaming ignition by the types of tests that carpets usually need to pass, namely the methenamine pill test, ASTM D 2859,[25] a US regulatory test for carpets and rugs, as well as critical radiant heat flux for flame spread by the flooring radiant panel test, ASTM E 648.[26] The guide is also likely to recommend that vehicle fuel tanks meet the fire test requirements of fire exposure testing conducted per European Standard ECE R34.01, Annex 5, for plastic fuel tanks,[27] which requires fuel tanks to withstand a two-minute fire exposure without any liquid fuel leakage. Fire resistance testing is also recommended for preventing the penetration of fires through the engine cover, the windshield or the car undercarriage. This can take the form of testing with a standard fire resistance test such as ASTM E 119,[28] a hydrocarbon fire test such as ASTM E 1529,[29] or a fire stop test such as ASTM E 814[30] or a combination of these.

16.6.4 Approaches to key fire scenarios

General

The key issue for passenger fire safety is the fire in the passenger compartment. Thus, fires can start there or penetrate into that compartment. Chapter 9 considers the type of fire scenarios that most frequently occur with road vehicles. In general, whether moving or stationary, passenger car fires can originate (1) inside the passenger compartment, (2) in the engine compartment, (3) in the trunk or load carrying area and (4) in the vicinity of the vehicle (such as in a pool fire). Fires that originate in the engine compartment are very likely to spread to the passenger compartment through the bulkhead (engine cover) between the engine compartment and passenger compartment, as well as through the windshield or through the ductwork. Fire propagation from the engine compartment into the passenger compartment depends on the size and numbers of openings in the bulkhead (e.g., brake pedal, wire harness, heater core, heating-ventilation air conditioning system or HVAC, etc.). The materials used to seal the openings in the bulkhead are typically combustible and are usually of no consequence when exposed to a large fire. Directly on the other side of the bulkhead are polyolefin or ABS HVAC ducts, which transverse the length of the dashboard and provide direct openings to the passenger compartment. A summary of 13 collision-related fires showed that fires originating in the engine compartment usually reached the passenger compartment in less than eight minutes (and in some cases in times as low as 2–4 minutes).[31]

Chapter 11 discusses in detail all of the key fire scenarios: fires starting in the passenger compartment, engine compartment, cargo compartment and elsewhere. In every compartment, extensive informational data is presented on fire performance of some materials that are typically involved.

Fires starting in the passenger compartment

Clearly, the fires starting in the passenger compartment are those most immediately dangerous to the passengers, as time available for escape is minimal. This is of special consequence, of course, when the passengers, or driver, are injured or in some other way incapacitated or suffering from decreased escape capability. The fires studied in this chapter are fires starting in the instrument panel, seat, floor, headliner, and compartment door. Once the fire has started, the spread of fire inside the passenger compartment is directly related to the quantity, composition, orientation, configuration and fire properties of the materials in the passenger compartment. Potential ignition sources include: smoldering of cigarettes or other smoking materials, electrical short circuits or electrical malfunctions, ignition of electrical dashboard components, of aftermarket consumer electronics and their power connections, and of heating elements in seats. Combustible materials brought into the passenger compartment present additional potential sources of ignition and fuel. For example, a collision might cause the release of a liquid fuel brought into the passenger compartment which could then be ignited and spread flame to components in the passenger compartment.

The passenger compartment is the only area in the vehicle where regulatory fire safety requirements are available, namely the need for all materials within 13 mm of the passenger compartment to meet a flame spread rate of 10 mm/min when tested to FMVSS 302.

Fires starting in the engine compartment

Fires starting in the engine compartment can spread into the passenger compartment. Many more fires start in the engine compartment than in the passenger compartment (68% vs. 18%). As discussed earlier, the spread can occur through the passenger compartment engine access cover, ductwork or the windshield. Collision damage can provide alternative paths for fire penetration into the passenger compartment. These fires are typically started in one of three ways: via an electrical fault following a collision, via an electrical fault independent of collision and via non-electrical ignition sources. One type of product most widely used in the engine compartment is electrical wires and cables; fire safety requirements for such products are widely available in many environments, including the National Electrical Code (NFPA 70, for building applications) and codes and standards associated with other transportation vehicles (as discussed in the FPRF white paper[13]). Some manufacturers use voluntary fire test requirements for electrical wires and cables, but such requirements are not uniformly applied, if at all.

The most interesting scenario is the one where the fire penetrates through the passenger compartment engine access cover. Traditional passenger road vehicles used to have separations, called firewalls, between the engine compartment and the passenger compartment. Such separations used to be of steel construction, with openings for ducting or cabling, which were, in turn, sealed off with the intention of preventing passage of fire (flames) or smoke between compartments. In some vehicles, such as in vans, where engines extend in part into the passenger compartments, the passenger compartment is separated by an engine cover that is combustible. This might allow fire to penetrate into the passenger compartment. Thus, in the event of a vehicle fire, conditions inside the passenger compartment can become untenable quite rapidly, particularly if passenger mobility is impaired by injury. Engine covers offer two types of ways in which a fire can penetrate from the engine compartment into the passenger compartment: (1) through openings in the engine cover and (2) through destruction of the engine cover. If the engine cover were constructed of non-combustible materials or if it offered a fire resistance rating (for example, a fire resistance rating equivalent to a 15-minute fire barrier), then it is likely that penetration through the engine cover would only occur: (a) if the engine cover has been damaged (perhaps as a result of a collision) or (b) by penetration through openings used for cables or ducts. This would typically have been the case with engine covers made of steel construction, which formed real firewalls. In one van involved in a fire, the engine cover was analyzed and cone calorimeter fire tests were conducted on it. The engine cover was composed of two materials: (a) a fibrous insulation material sandwiched between two layers of aluminum foil, and with 4.2% of polyester binder, and (b) a molded plastic material, containing 42% plastic, composed of a styrene-butadiene rubber (70%) and poly(vinyl acetate) (27%).[5] Figure 16.5 shows cone calorimeter results, indicating that the material was easily combustible.

In the case of the fire penetrating through the duct, these ducts might provide a path for spread of fire and combustion products into the passenger compartment. Three possible means exist for fire spread into the passenger compartment. If the duct is non-combustible, fire can spread through the duct's design openings. Fire can also extend through ducts by burning through the material if the duct is combustible. Alternately, it might pass through any duct openings caused by collisions. Once in the ducts, fire may extend to the underside of the dashboard assembly or penetrate through the dashboard via the HVAC discharge vents, located at the base of the windshield or at the front of the dashboard. When the fire extends under the dashboard, the component materials might ignite and the subsequent fire spread has the same effect as a fire initiated under the dashboard in the passenger compartment. If the path of fire extension is through the vents at the base of the windshield, the next materials ignited could be the visors or the headliner.

If the fire penetrates through the windshield, there are three basic modes of

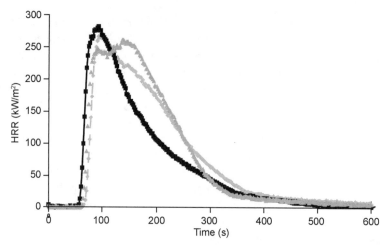

16.5 Cone calorimeter heat release rate of van engine cover molding @ 40 kW/m^2.

flame spread from the engine compartment to the passenger compartment:

1. radiation and convection from the engine compartment fire to the windshield, followed by subsequent re-radiation from the windshield to the dashboard inside the passenger compartment;
2. similar to the previous mode, except that upon radiative and convective heating of the windshield from the engine compartment fire, thermal stress fractures develop in the windshield and the inner polymer layer is vaporized, and
3. for post-collision fires, when the windshield is cracked or broken, but not penetrated, the polymer layer vaporizes and the windshield is weakened until it collapses and fragments drop onto the dashboard.

Fires starting in the storage compartment

Fires starting in the storage compartment can also penetrate into the passenger compartment, but they rarely occur. Automobiles generally have storage compartments, primarily in the rear of the vehicle, but in certain vehicles such as hatchbacks, minivans and sports utility vehicles, the compartments are not physically separated from the passenger compartment. Only 1.7% of all fires start there.[1] Non-collision ignition sources include careless use of cigarettes or other smoking materials and other heated equipment, as well as dangerous cargo. Fire spread is most likely to occur through the upholstered boundary, in passenger vehicles, or the rear window, in trucks. In the event that the trunk or load carrying area is maintained in a configuration separate from the passenger compartment, fire spread may occur through the flammable upholstery, stereo system components, or other wiring or air conditioning components. Fire spread and generation of products of combustion into the passenger compartment will be noticed by vehicle occupants sooner in the event that the rear seats are lowered in which case the trunk or load carrying area essentially becomes an extension of the passenger compartment.

Pool fires burning under the vehicle

Pool fires resulting from fuel tank failure and burning under the vehicle may result from collisions associated with passenger vehicles. These collisions can cause automotive fluids to be spilled within the engine compartment, near fuel system components, or on the ground underneath the vehicle. Ignition of these fluids can occur by any of the ignition processes outlined in the previous chapters. It should be noted that the fuel for the pool fire can originate from any vehicles involved in the collision. Either vehicle's headlights or running lights could act as an ignition source. The hazard posed by the pool fire depends, in large part, on the volume of the fluid spill. The hazard to occupants due to large pool fires, involving a substantial portion of the vehicle, are primarily due to the external fuel load and associated fire; the fire performance of the vehicle itself is of secondary importance in these cases. The heat release depends upon the surface area of the spill. The area of the spill is highly dependent upon the slope and other characteristics of the surface onto which the fuel is spilled. If the surface area of the spill is known, the heat release rate and radiant flux of the fuel can be determined. The surface area of the spill can be calculated based on an estimate of the fuel volume spilled and spill depth. A fuel tank fire test is available in the European community,[27] but is only occasionally used in the US, and then on a voluntary basis.

Other fire scenarios

Finally, fires can also result from other external heat sources, which are typically caused by vehicle malfunctions or by human causes. External heat sources (other than abandoned material) account for a small but significant fraction of fires. However, such fires account for a consistent fraction of all fire losses (around 4%), which is very different from most other sources. The main strategies for minimizing the effects of external heat sources involve hardening of the vehicle exteriors to minimize exterior ignitions. However, even if vehicle exteriors were completely resistant to ignition, external heat sources could still cause fire penetration into the passenger compartment by way of openings (such as open windows). Moreover, external heat sources can also cause ignitions in the engine and storage compartment. This would then lead to the type of fires addressed earlier.

16.6.5 Draft recommendations

Chapter 11 is likely to contain recommended types of mitigation strategies. In general those strategies tend to follow the concepts of: decreasing ignition propensity of materials, decreasing heat release of materials and designing improvements in vehicle engineering, while ensuring that the changes do not affect the vehicle safety or performance in terms of issues other than fire. In terms of fire properties, clearly it is principally important to reduce ignitability, heat release and the potential for flame spread, which is why the test most widely considered for assessing material and product fire properties would be the cone calorimeter, ASTM E 1354.[11]

The guidance chapter in this document is intended to discuss a summary of the guidance offered as ways to minimize fire hazard for each individual scenario. There have been a number of examples in the literature of full-scale fire tests conducted to assess the fire performance of road vehicles, some of which are referenced in the document. Such tests have often analyzed one or more of the scenarios outlined in the guide as most likely to cause harm to road vehicle occupants. Quantitative full-scale tests are most useful if they assess heat release properties. Therefore, in addition to the specific tests mentioned earlier for specific components, the committee tended to recommend tests based on heat release, at various levels: full-scale, fuel package scale and individual material scale.[11] It is important to mention a screening test, such as the mass loss cone (ASTM E 2102[32]), which can correlate with cone calorimeter results, but which has not been recommended by the committee for use as it does not provide data that are quantitative enough for fire hazard assessment.

The guidance chapter is intended to stress the performance-based approach and contrast it with the continued use of FMVSS 302 as the sole fire safety tool. The exclusive use of FMVSS 302 is unlikely to be consistent with significant decreases in fire losses associated with road vehicles. FMVSS 302 was initially intended to solve the problem of smoldering ignition caused by cigarettes and it has been effective in doing so. However, with the growing use of combustible materials in road vehicles, such a mild flaming ignition test is clearly insufficiently severe to ensure that materials meeting it would allow enough time for escape to passengers and drivers in the case of a fire. Fire hazard will most likely decrease either if materials or products with better fire properties are used or if the vehicle is redesigned to minimize the speed of fire development, particularly into the passenger compartment. The major fire properties that should be controlled in the materials and products are: ignitability, heat release and smoke obscuration, of which heat release is the most critical one and the easiest to scale up and predict. Quantitative full-scale tests which assess heat release properties, particularly when based on guidance from ASTM E 603[33] and ASTM E 2067,[34] are particularly helpful in assessing the potential fire safety of road vehicles. However, whenever such full-scale tests are performed, it is most advisable to maximize the measurements made, so as to also get information on potential drawbacks of alternative designs, with respect to properties other than heat release.

It must be recognized that, although the conduction of full-scale tests is the most accurate way of understanding both where deficiencies in fire safety are present in a road vehicle and also to develop mitigation strategies, their high cost means that other strategies need to be used also. For example, testing of sections, such as individual road vehicle compartments or fuel packages in a furniture calorimeter (i.e., under a hood, above a load cell), will be a way of understanding the interactions between the materials and products contained in the various sections of the road vehicle. In terms of screening tests, the mass loss cone fire test, ASTM E 2102,[32] is useful since: (a) it provides ignitability data under the same fire exposure conditions as in the cone calorimeter, (b) the mass loss data from the test probably parallels the heat release data from the cone calorimeter and (c) the instrument is available at significantly lower cost than the cone calorimeter.

Testing the fire properties of materials or products searching for an individual fire property should always be accompanied by an overall analysis that indicates that a resulting improvement, or apparent improvement, in the fire property assessed will be accompanied by an actual improvement in fire safety in the road vehicle. The scenario sections all indicate that there are some engineering design approaches that can be used to mitigate the effects of fire on road vehicle occupants. Such engineering solutions should be based on an overall fire performance evaluation. Battipaglia *et al.*[10] used the ASTM E 1546 framework for assessing the fire hazard of automotive materials in the engine compartment of a passenger road vehicle following a collision. This type of engineering analysis is particularly important when considering the use of fire test methods that are either unsuitable to generate fire test results in engineering units or that have been shown not to be adequately predictive of real-scale fire performance.

16.7 Alternate approach

An alternate approach to fire safety in cars was recommended by a multiple author coalition.[35] These authors analyzed the auto industry reports, the scientific literature, and statistical data, and concluded that measures need to be taken to improve survivability in automobile fires. The Federal Motor Vehicle Safety Standard 302, which was introduced almost 40 years ago to regulate the flammability of interior materials, is now no longer adequate. However, improvements in the crashworthiness of automobiles and their fuel tanks and the increased use of combustible materials have also changed the motor vehicle fire scenario significantly. In particular, the primary threat has changed from ignition of a small quantity of combustible interior materials by a lit cigarette in 1960 to ignition of a large quantity of combustible interior and exterior materials by an impact-induced fire at present. The authors therefore suggest that FMVSS 302 is no longer relevant to automobile fire safety and recommend improved standards based on objective criteria for fire safety, including especially 'fireworthiness' at the system/vehicle level as is routinely done for crashworthiness.

They also point out the significant advantages to be obtained simply by reducing heat release. Thus, they calculate that, for a frontal impact with 10–15 minutes of fire growth in the engine compartment and two minutes of fire growth in the interior compartment before untenable conditions, a reduction in heat release of under hood and interior materials by a factor of two would result in a 10-minute increase in time to flashover. A similar calculation for rear or side impact or rollover, in which fire penetrates the interior within two minutes and flashover occurs two minutes thereafter, a decrease in the heat release rate of under hood and interior plastic materials by a factor of about four would provide 10 minutes of additional escape time.

Thus, these authors recommend the use of heat release rate reductions as the key tool to improve fire safety.

16.8 Conclusions

Draft guide NFPA 556 clearly highlights the serious problems associated with the use of FMVSS 302 as the sole test for assessing the fire safety of materials in road vehicles. Moreover, the guide is intended to show that the materials and designs used in passenger road vehicles should be assessed for their fire performance characteristics.

The sector of the fire safety community interested in vehicle fires is particularly concerned with the high fire losses in passenger road vehicle fires (especially fatalities) which are not associated with fuel tank fires. Draft guide NFPA 556, although not complete, already contains sufficient elements that it may, it is hoped, inspire renewed interest in vehicular fire safety by all relevant parties, after a long period when there was very little public activity.

16.9 References

1. Ahrens, M., *U.S. Vehicle Fire Trends and Patterns*, NFPA, Quincy, MA, August 2001.
2. Ahrens, M., *The U.S. Fire Problem Overview Report: Leading Causes and Other Patterns and Trends*, NFPA, Quincy, MA, June 2001.
3. Janssens, M.L., 'Database on full-scale calorimeter tests on motor vehicles', presentation to NFPA Technical Committee on Fire Hazard and Fire Risk of Furnishings and Contents on October 23, 2006, Detroit, MI.
4. Grayson, S.J. and Hirschler, M.M., 'Fire Performance of Plastics in Car Interiors', *Flame Retardants 2002*, February 5–6, 2002, London, Interscience Communications, London, UK, 2002.
5. Hirschler, M.M., Hoffmann, D.J., Hoffmann, J.M. and Kroll, E.C., 'Fire Hazard

Associated with Passenger Cars and Vans', Fire and Materials Conf., San Francisco, CA, Jan. 27–28, 2003, Interscience Communications, London, UK, pp. 307–319.
6. Janssens, M.L. and Huczek, J.P, 'Comparison of Fire Properties of Automotive Materials', in Business Communications Company Fourteenth Ann. Conference on Recent Advances in Flame Retardancy of Polymeric Materials, June 2–4, 2003, Stamford, CT, Ed. M. Lewin, Norwalk, CT, 2003.
7. FMVSS 302, 'Motor Vehicle Safety Standard No. 302, Flammability of Materials – Passenger Cars, Multipurpose Passenger Vehicles, Trucks and Buses', National Highway Traffic Safety Administration, Washington, DC. [Code of Federal Regulations § 571.302, originally Federal Register 34, No. 229, pp. 20434–20436 (December 31, 1969)].
8. National Highway Traffic Safety Administration (NHTSA), 'Car and Van Fire Research Work Conducted under the General Motors Corporation – Settlement Agreement – Section B. Fire Safety Research' at the web site: http://dms.dot.gov/, Docket # 3588.
9. Battipaglia, K., Huczek, J., Janssens, M., Miller, M. and Willson, K., 'Fire Properties of Exterior Automotive Materials', in *Proc. Flame Retardants 2004*, 11th Int. Conference, January 27–28, 2004, London, UK, Interscience Communications, UK, pp. 253–262.
10. Battipaglia, K., Huczek, J., Janssens, M. and Miller, M., 'Development of a Method to Assess the Fire Hazard of Automotive Materials', in *Proc. Interflam '04*, 10th Int. Fire Safety Conference, July 5–7, 2004, Edinburgh, UK, Interscience Communications, UK, pp. 1587–1596.
11. ASTM E 1354, *Standard Test Method for Heat and Visible Smoke Release Rates for Materials and Products Using an Oxygen Consumption Calorimeter (cone calorimeter)*, Vol. 04.07, Amer. Soc. Testing & Mater., West Conshohocken, PA.
12. Ahrens, M., *U.S. Vehicle Fire Trends and Patterns*, NFPA, Quincy, MA, October 2006.
13. NFPA Fire Protection Research Foundation, 'White Paper on Fire and Transportation Vehicles – State Of The Art Of Regulatory Requirements And Guidelines', Transportation Vehicles Research Advisory Council, Quincy, MA, 2004.
14. NFPA 555, *Guide on Methods for Evaluating Potential for Room Flashover*, National Fire Protection Association, Quincy, MA.
15. *Motor Vehicle Facts and Figures 2006*, Alliance of the Automobile Manufacturers, Ward's, Southfield, MI, 2006.
16. Tewarson, A., *A Study of the Flammability of Plastics in Vehicle Components and Parts*, Factory Mutual Research Corp., Norwood, MA, October 1997.
17. Fire Safety Aspects of Polymeric Materials, Volume 8, Land Transportation Vehicles, Report of the Committee on Fire Safety Aspects of Polymeric Materials, National Materials Advisory Board, Publication NMAB 318-8, National Academy of Sciences, Washington, DC, 1979.
18. Peacock, R.D., Jones, W.W., Bukowski, R.W., and Forney, C.L., 'Technical Reference Guide for the HAZARD I Fire Hazard Assessment Method, Version 1.1,' *NIST Handbook 146*, Vol. II, Natl. Inst. Stand. Technol., Gaithersburg, MD, 1991 and Peacock, R.D., Jones, W.W., Forney, G.P., Portier, R.W., Reneke, P.A., Bukowski, R.W., and Klote, J.H., *An Update Guide for HAZARD I, Version 1.2*, NISTIR 5410, Natl. Inst. Stand. Technol., Gaithersburg, MD, 1994.
19. SFPE *Engineering Guide to Performance-Based Fire Protection: Analysis and Design of Buildings*, Society for Fire Protection Engineers, Bethesda, MD, May 2005.

20. ASTM E 1546, *Standard Guide for Development of Fire-Hazard Assessment Standards*, Volume 04.07, ASTM International, West Conshohocken, PA.
21. ASTM E 2061, *Standard Guide Fire Hazard Assessment of Rail Transportation Vehicles*, Volume 04.07, ASTM International, West Conshohocken, PA.
22. UL 1685, *Standard for Safety for Vertical-Tray Fire-Propagation and Smoke-Release Test for Electrical and Optical-Fiber Cables*, Underwriters Laboratories, Inc., 333 Pfingsten Road, Northbrook, IL
23. ASTM E 1353, *Standard Test Method Standard Test Methods for Cigarette Ignition Resistance of Components of Upholstered Furniture*, Volume 04.07, ASTM International, West Conshohocken, PA.
24. ASTM E 1352, *Standard Test Method for Cigarette Ignition Resistance of Mock-Up Upholstered Furniture Assemblies*, Volume 04.07, ASTM International, West Conshohocken, PA.
25. ASTM D 2859, *Standard Test Method for Ignition Characteristics of Finished Textile Floor Covering Materials*, Volume 04.07, ASTM International, West Conshohocken, PA.
26. ASTM E 648, *Standard Test Method for Critical Radiant Flux of Floor-Covering Systems Using a Radiant Heat Energy Source*, Volume 04.07, ASTM International, West Conshohocken, PA.
27. European Standard ECE R34 Annex 5, *Fire Risks – European Economic Community Regulation - Fire safety of plastic fuel tanks for automobiles* (ECE R34 Annex 5, RREG 70/221/EWG, 2000/8/EG).
28. ASTM E 119, *Standard Test Methods for Fire Tests of Building Construction and Materials*, Volume 04.07, ASTM International, West Conshohocken, PA.
29. ASTM E 1529, *Standard Test Methods for Determining Effects of Large Hydrocarbon Pool Fires on Structural Members and Assemblies*, Volume 04.07, ASTM International, West Conshohocken, PA.
30. ASTM E 814, *Standard Test Method for Fire Tests of Through-Penetration Fire Stops*, Volume 04.07, ASTM International, West Conshohocken, PA.
31. Scheibe, R., Angelos, T., *Motor Vehicle Collision-Fire Analysis Methods and Results*, Washington State Transportation Center, November 17, 1998.
32. ASTM E 2102, *Standard Test Method for Measurement of Mass Loss and Ignitability for Screening Purposes Using a Conical Radiant Heater*, Volume 04.07, ASTM International, West Conshohocken, PA.
33. ASTM E 603, *Standard Guide for Room Fire Experiments*, Volume 04.07, ASTM International, West Conshohocken, PA.
34. ASTM E 2067, *Standard Practice for Full-Scale Oxygen Consumption Calorimetry Fire Tests*, Volume 04.07, ASTM International, West Conshohocken, PA.
35. Digges, K.H., Gann, R.G., Grayson, S.J., Hirschler, M.M., Lyon, R.E., Purser, D.A., Quintiere, J.G., Stephenson, R.R. and Tewarson, A., 'Improving Survivability in Motor Vehicle Fires', in *Interflam 2007* Conference, Royal Holloway College, University of London, UK, Sep. 3–5, 2007, Interscience Communications, London.

17
Firefighters' protective clothing

H MÄKINEN, Finnish Institute of Occupational Health, Finland

Abstract: This chapter discusses needs for heat protection and types and design of clothing in different tasks firefighters are performing. Requirements for heat protection and methods of testing according to ISO, CEN and NFPA are reviewed. Some future trends concerning firefighters' work, new materials and applications of wearable electronic technology in protective clothing are also discussed.

Key words: firefighters' protective clothing, requirements of heat protection, testing of heat protection, ISO, EN and NFPA standards.

17.1 Introduction

A firefighter is a worker whose main job is to respond to emergencies in many different locations with a view to saving life, performing rescue, and minimizing damage to property. Preparation for responding and prevention are also important aspects of this work.[1] The role of protective clothing and other personal protective equipment (PPE) is fundamental for firefighters' safety when they are balancing the need for life and property with protection in dynamic and increasingly diverse alarm situations. Firefighters responding to a fire call may now find that they are also exposed to chemical and biological toxins.[2] In addition to the need for different levels of heat and fire protection, this means more complex requirements for protective clothing and other personal protective equipment. In basic firefighting operations, the clothing should protect the user against possible flame impingement, high air temperatures, radiant heat and accidental contact with chemicals, while also providing water resistance or repellency and some level of mechanical protection. At the same time it should allow the user to carry out his/her duties without undue stress being caused by clothing and other personal protective equipment.[3] The traditional protective clothing for structural firefighting has been designed to provide protection from fire and heat, but not necessarily from all the new challenges originating in today's rescue situations.[2]

This chapter focuses on the needs for heat and fire protection of protective clothing used for different tasks undertaken by firefighters concerning materials used for protection, levels of performance defined in standards as well as test methods used to show compliance. Future trends and needs for development are discussed, and the chapter will also provide some examples of current developments in the field.

17.2 Different tasks and requirements for fire and heat protection

Generally, the tasks of firefighters are categorized into the following groups based on types of hazards and level of heat and fire exposure: specialized firefighting, structural firefighting, wildland firefighting, hazardous materials response, and rescue work. Types of rescue are further divided into four main types consisting of rope rescue, water rescue, rescue from vehicles, and plant and special rescue. Naturally, statements on fire protection are not applicable to water rescue.

Potentially the most dangerous element in firefighters' work is exposure to temperature extremes, the primary thermal exposures being open flame contact, thermal radiation from flames, smoke, hot gas convection, conduction from high temperature surfaces, falling materials, debris and hot objects.[4] However, the worst situations are major fires with very high exposure, which are rare. For example, in Hamburg, a city with about 1.8 million inhabitants, major structural and industrial fires occur with a frequency of one to two per year for each firefighter.[5] Therefore, the tendency to design the level of protection of personal protective equipment for actual fires and against the worst possible scenarios has led to the over-protection of firefighters for the main part of their working time.

17.2.1 Types of heat transfer

In high temperature exposure situations, burn injuries can occur when radiative, convective, or conductive heat, or a combination of these forms, is transferred through clothing layers, but also at gaps in personal protective equipment that may be located at wrist, neck or face, if the compatibility between different types of personal protective equipment is not good or the wearer has not dressed correctly.

The level of thermal radiation depends on the temperature difference and the distance between the radiating source and the targeted surface, and the reflectivity of the surface. Firefighters can be exposed to the highest level of thermal radiation during specialized firefighting, such as proximity, approach, and entry firefighting. Heat transfer by convection occurs through the movement of hot air. Convection also affects the transfer of heat within layers of clothing and between clothing and the body. Heat transfer by conduction takes place when the surface of clothing is in contact with a hot surface and may be significant when clothing layers are pressed against the body, for example. Compression that occurs when kneeling permits greater transfer of heat between clothing. Exposure to molten substances and hot solids forms another type of heat transfer by conduction. Prolonged fires may melt metals, which can either drip or pool and thus expose the firefighters to this hazard. Also, increased use of synthetic plastic materials in building construction is a source of similar risks.[6] In exposure situations, protective clothing is also exposed to moisture originating from human perspiration, hose spray and possibly the weather conditions. In practice, the protection system is very complex: the fabric moves, the layers between are compressed altering their spacing and ease of garment movement, and there is dynamic movement of moisture in the protective clothing while it is being used and as its temperature is elevated in fire environments.[7]

17.2.2 Heat exposure in different types of firefighting operations

High heat exposures are a principal hazard in most structural and related fires. There are different ways to classify heat exposure, but generally firefighters are at risk from all forms of heat transfer during a fire response with the intensity varying according to their role at the fire. Its characteristics[6,8] divide the thermal conditions into three categories namely, routine, hazardous and emergency, each defined by a range of air temperature and a range of radiant flux.

- *Routine conditions*: Air temperature up to 50–60 °C, radiant heat flux up to 1.4–1.6 kW/m^2.
- *Hazardous conditions*: Air temperature from 50 °C to 300 °C, radiant heat flux from 1.4 kW/m^2 to 8 kW/m^2.
- *Emergency conditions*: Air temperature from 300 °C to 1000 °C, radiant heat flux from 8 kW/m^2 to 200 kW/m^2.

Similar categories are given by Abbot, Coletta and Foster and Roberts, referred to by Barker et al.[9] The tolerance time in routine environments is 5–60 minutes. In these conditions, firefighters are operating with hoses or otherwise fighting fires from a distance. Under hazardous conditions the tolerance time is 1–10 minutes. In such conditions at a lower level of heat exposure, firefighters are ventilating a fire without water, whilst at a higher level they are entering a burning building. Emergency conditions may be encountered during flashover in a building fire. The tolerance time is 5–20 seconds. The protective clothing should provide the firefighter with enough time to escape. It has been estimated that it takes 3–10 seconds to escape from aircraft or vehicle crash fuel spills with heat flux intensities peaking between 167 and 226 kW/m^2.[10]

Specialized firefighting includes exposure to high heat from radiant sources, such as aircraft fires, bulk flammable gas fires, and bulk flammable liquid fires. In large fuel fires, heat flux levels can be between 90 and 230 kW/m^2.[11] Also levels of convective and conductive heat may be high and so specialized thermal protection is necessary.

Structural firefighting entails activities of rescue, fire suppression, and property conservation in buildings and enclosed structures. It involves a relatively complex set of different hazards, such as extreme heat and the potential for flame contact, exposure to steam or scalding water, contact with hot surfaces, solids, and molten metal, severe physical hazards, the possibility for disorientation and entrapment within the structure due to poor visibility, high heat and stressful conditions and heat stress from high ambient heat and the encapsulating effect of protective clothing. Also several other hazards can be involved, such as chemicals, body fluids, fall from heights and contact with live electric power lines, depending on the nature of the fire.[6] In fire situations Lawson[12] has measured heat fluxes as high as 30 kW/m^2 with a temperature of 175 °C, and Rossi[13] measured radiant heat fluxes typically between 5 and 10 kW/m^2, with temperatures of 100–190 °C at 1 m above ground level. The main function of protective clothing is heat protection during its operational time, but some level of chemical and mechanical protection is also needed as well as thermal comfort.

Wildland firefighting involves activities of fire suppression and property conservation in woodlands, forests, grasslands, brush, prairies, and other such vegetation, or combination of these in a fire situation, but not within buildings or structures. Wildland firefighters generally work for long stretches (8–16 hours/day) exposed to a general radiant heat flux between 1 kW/m^2 and 8 kW/m^2, although during increased fire activities of shorter duration they may undergo exposure to a radiant heat flux of 8 kW/m^2 to 20 kW/m^2, and during extreme conditions to a radiant heat flux of 20 kW/m^2 to 100 kW/m^2. Temperatures can range from an ambient air temperature 25–49 °C up to 1200 °C.[6] Protective clothing for wildland firefighting should simultaneously permit evaporation of sweat, shield from radiant heat, and minimize risk of burn injury.

Rescue work in traffic accidents now forms the main part of work for many fire brigades. For example, in 2006 10% of all alarms in Finland were traffic accidents,[14] and in Sweden 25%.[15] In the case of traffic accidents, the environmental temperature depends on weather conditions, but there is often risk of fire and explosions from fuel tanks. Urban search and rescue is a relatively new title for a field of special heavy and technical rescue tasks. All the main hazard types of physical, thermal, flame and heat, chemical, and biological, are included in this task type. The degree of the hazard varies with the task and type of activity to be performed.[6]

In hazardous materials responses, the material and design of the protective clothing must resist the most common liquid and gaseous chemicals. In addition, protection against flames and cold may be needed, depending on the type of chemical and other hazards.

Protective clothing for first responders was introduced especially in the United States last year as a part of firefighters' protective clothing. First responders must be ready to face exposure to a variety of specific agents that have now become part of emergency activities. These threats are commonly labelled CBRN for chemical, biological, radiological and nuclear exposures and involve chemical warfare agents, toxic industrial chemicals, and biological agents. The development of a new generation of structural firefighting personal protective equipment offering CBRN protection is currently underway in the United States.[2] Also in Europe CEN BT/WG161, 'Protection and Security of the Citizen' (http://www.cen.eu/cenorm/businessdomains/businessdomains/security+and+defence/security/btwg161.asp) has started work concerning the requirements of a Security of Citizen and Emergency Service that will also address the role of firefighters and special requirements for protection.

17.3 Types and design of clothing required for protection

General design requirements for all types of protective clothing are that there are no restrictions of movement and that the garments are compatible with all other personal protective equipment that may be necessary. Closure systems, labels, accessories and retro-reflective materials must be heat and flame resistant, and they must not impair the clothing's performance. Closure systems, pockets, ends of the sleeves and trouser legs should be designed to protect the wrist and the lower leg and to prevent the entry of burning debris. Hardware penetrating the outer material must not be exposed on the innermost surface of the component assembly.

According to EN 1486, protective clothing for specialized firefighting must provide protection against radiant heat and flame impingement to the full body, including the head, hands, and feet, and shall therefore consist of garment(s), hood, integrated or not, gloves and overboots. This type of protective clothing is designed to be used with respiratory protection and the designs may vary as to whether the breathing apparatus is to be worn inside or outside of the protective clothing.

The levels of protection in protective clothing for structural firefighting in accordance with EN 469 may contain multilayer materials, material combinations, component assemblies or a series of separate garments. A two-piece suit must contain an adequate overlap between the jacket and trouser during different movements. Furthermore, an anti-wicking barrier can be used as part of an interlining at the edge part of a moisture barrier, to avoid the transfer of liquid from outside the garment to inside the garment, for instance at the end of the sleeves, the trouser legs or bottom of a jacket. NFPA 1971 has more specific requirements for the design of, for example, suits for a trim configuration.

According to EN 15614, protective clothing for wildland firefighting may be a coverall, a two-piece suit including an interface area, or a number of inner and/or outer garments designed to be worn together. The collar must be able to remain in the vertical position when it is set upright. All protective clothing encircling the neck must have a closure system at the level of the line of the collar. A two-piece suit must be contain an interface overlap area of at least 15 cm between the jacket and the trousers. After being opened, the protective flaps of external pockets shall be required to overlap the pocket by a minimum of 20 mm.

No design requirements exist for protective clothing for rescue tasks at the moment. Considering the requirements of this task, some level of mechanical protection is needed especially at the knees and arms. For example in Tokyo, Japan, where urban rescue is an important rescue activity, the firefighters use two-piece protective clothing assemblies of aramid fibre with additional mechanical protection on shoulders, knees, elbows and the front of trousers.

17.4 Materials used in firefighters' clothing

The outer layer of the protective garment for specialized firefighting reflects radiant heat, utilizing aluminized outer surfaces. Protective clothing ensembles for structural firefighting typically consist of a flame-resistant outer shell material and inner liner generally composed of a moisture barrier, a thermal barrier and lining material. The complex system is intended to offer a balance of protection, comfort and value. Consequently, the materials in different layers are designed and selected to form a modular system,[16] where air gaps between material and garment layers play a critical role in providing thermal insulation. Holmer et al.[17] found that small variations in the thermal properties of protective clothing had little or no effect on heat exchange and did not affect the resulting thermal strain. The clothing for wildland firefighting and rescue work is typically a one- or two-layer garment with reinforcements at certain areas.

17.4.1 Outer shell

The outer shell is the first line of defence against the hazards faced by the firefighter. It should provide basic protection against exposure to flame and heat and other environmental hazards and have sufficient mechanical resistance to cuts, snags, tears and abrasion.[10] These high level requirements are normally met only by strong inherently flame retardant fibres, such as aromatic polyamides (meta or para-linked aramids), polybenzimidazole (PBI), polybenzoxazole (Zylon or PBO), and polyamide-imide fibre (Kermel). Similar types of fibres are offered by different manufacturers for meta-aramid, examples are Nomex (DuPont), Conex (Teijin), Fenilon (Russian), and Apyeil (Unitika) and para-aramid fibre, examples are Kevlar (DuPont), Twaron (Akzo Nobel). Other fibres have unique manufacturers such as Technora (Teijin), Kermel (Rhodia Performance Fibres), PBI (Celanese), and PBO (Toyobo). The advantage of PBI is that it absorbs more moisture than cotton, and has a wearer comfort rating equivalent to that of 100% cotton.[18] The recently developed PBO has outstanding tensile properties but its price is about double that of Kevlar.[19] The high price of the fibre limits its use in blends; otherwise it might be used similarly with para-aramid fibres in blends with meta-aramids to increase durability, such as Nomex III (blend of Nomex and Kevlar (95/5%)), X-fire (blend of Teijin Conex and Technora), Millenia^{TM}XT (60% para-aramid/40%

PBO) and PBI Gold (blend of PBI and para-aramid). A small amount (e.g. 2%) of antistatic fibre is often added to the blends. To increase moisture absorption, the blends for firefighters' outer shell fabrics may contain flame-retardant viscose and flame-retarded wool. Typical constructions of the outer fabrics are twill or ripstop woven fabrics with a mass of 195–270 g/m^2.

Aluminized fabrics are normally used as outer shell fabrics for specialized firefighting. In some countries, such as Japan, they are used also in protective clothing for structural firefighting.[20] They cause higher physiological stress than materials of conventional materials, however. The basic material in aluminized fabrics can be flame retardant cotton, flame retardant wool, fibreglass or aramid in woven or knitted constructions. There are also different ways of introducing the aluminium layer. For instance, the aluminized layer as a thin foil can be glued using different types of adhesive, e.g. polyurethane, neoprene or silicon polymer or as an aluminized film, e.g. transfer film, one-sided vaporized PET-film, both-sided vaporized PET-film, aluminium metalfilm sheet. These possible modifications have the effect that the properties of the clothing may vary significantly with the type of product used.[21]

In addition to the above fibres and blends, in garments for wildland firefighting, some lower performance materials such as fabrics with flame-retardant finishes, such as Proban and Pyrovatex® for cotton, are used either alone or in blends. They are required to retain the flame-retardant properties after 50 launderings (ISO 15384).

17.4.2 Thermal liner

Air trapped between material layers forms the main thermal protection, because fibres conduct heat 10–20 times faster than still air. This is the determining principle in the construction of a thermal liner which should prevent the transfer of heat from the environment to the body by slowing down the passage of heat from the outside to the inside of the garment. The products used for firefighters' clothing can consist of a spunlaced, nonwoven felt or batting quilted or laminated to a woven lining fabric. It can also be a knitted fabric between the outer shell and the lining to give the highest insulation against heat, but at the same time allowing the escape of moisture due to perspiration. The thermal liner is normally made of inherently flame-retardant fabrics or their blends as explained above regarding outer shell materials. A similar fibre content of the thermal liner and outer shell fabric make the laundering of the garment easier.

Non-textile insulation materials are also used to construct thermal liners. In the Airlock® product from the W L Gore company, air cushions replace the traditional textile insulation. The method is a combination of a moisture barrier and thermal protection. 'Spacers' made of foamed silicone on the GORE-TEX moisture barrier create the insulating air buffer in the material.[22]

17.4.3 Moisture barrier

A moisture barrier is designed to reduce the amount of water from the environment to penetrate to the inside of the garment. It also provides protection against many common liquids such as common chemicals and bloodborne pathogens. It is obligatory in some countries, whereas in other countries firefighters prefer suits without moisture barriers because of their need to maximize thermal comfort. The moisture barrier in firefighters' clothing can be:

1. laminated or coated to the inside of the outer shell fabric: this type can decrease the durability of the whole garment because the moisture barrier may be much more easily punctured or torn thus allowing water penetration;[3]
2. a lightweight knitted material or web, and the structure is inserted loosely between the outer fabric and the liner; or
3. on the outside of the thermal liner.

The moisture barrier can be a microporous or hydrophilic membrane or coating.[23] GORE-TEX®, CROSSTECH® and TETRATEX® are textile laminates incorporating microporous polytetrafluoroethylene. PORELLE®, PROLINE® and VAPRO® are microporous polyurethane laminates with textiles. BREATHE-TEX PLUS®, STEDAIR 2000® are hydrophilic polyurethane laminates or coated fabrics, and SYMPATEX® is a hydrophilic polyester laminate. Microporous and hydrophilic coatings are normally polyurethane products. ACTION® is an example of a polyurethane coating. Neoprene (NEOGUARD®) and polyvinyl chloride (PVC) are non-breathable moisture barrier products.

17.4.4 Accessory materials

Materials to improve visibility are important accessory materials. Fluorescent materials are used to enhance day-time visibility and retro-reflective materials, night-time visibility. In future, the need for materials to improve the visibility in firefighters' protective clothing will increase together with new developments of the special garments for rescue tasks. Even though there are hundreds of types of fluorescent materials on the market, only a few materials meet the requirements of heat resistance set for firefighters' protective clothing. The colours of these materials are very sensitive when exposed to the smoky environments of firefighting. Their usage time can thus be short. Retro-reflective material and fluorescent colours are often combined in the trim near the hands, head and feet of garments where motion increases the visibility. The manufacturers of reflective materials (e.g 3M Scotchlite^{TM}, Reflexite®, Unitika®) have heat-resistant products for firefighting applications. The basic requirement of other accessory materials, such as zippers and braces (suspenders), is resistance to heat.

Antiwicking barrier materials consist of, for instance, PU-coated FR fabrics. Reinforcements for mechanical protection are highly durable fire-resistant materials such as silicon coated para-aramide, and are used especially on the knee areas.

17.4.5 Underwear

The basis of the overall protective clothing systems is underwear that keeps the skin as dry as possible and should be easy to wear in all the required combinations, thus constituting an important element of the protection. It has great impact on the thermal and moisture transfer.[24] Today no actual regulations exist for firefighters' underwear, but its protective function should not be overlooked. The common recommendation is that the layer next to the skin should not made of pure thermoplastic and fusible fibres. Therefore, most underwear materials with good moisture transport properties and used especially in sport activities are not suitable for firefighting. Flame retardant underwear may be made from 100% P84 or blended with lower cost fibres like flame retardant viscose. For example, a 50/50 P84/Viscose FR (Lenzing) blend is used for knitted underwear with high moisture absorbency.[19] In a study by Ilmarinen et al.[25] concerning physiological parameters of firefighters' clothing, 66% PES 34% WOOL (Sportwool®) underwear was rated excellent when wet compared to the traditional cotton underwear. Polyester part binds on the wool fibres when exposed to fire.

17.5 Measuring flame and thermal performance

The properties and suitability of protective clothing for different purposes can be measured at the following levels:

- material measurements in the laboratory;
- biophysical measurements using manikins in the laboratory;
- measurements in the laboratory using human subjects;
- measurements in the field using human subjects;
- wearer trials in the field by intended end users, including questionnaires.

Most requirements mentioned in the standards are based on material tests. Material tests are more accurate and reproducible. Human-sized dummies or manikins equipped with sensors for measuring heat or cold protection represent biophysical measurements. Today, there are also plans to introduce practical performance tests to be carried out in a laboratory using test persons for measuring of standards for protective clothing. The purpose of the test is to quantify the physiological impact and hindrance of complete garments for firefighting under standardized working conditions by assessing wearability and physiological load. For safety and ethical reasons, exposure to heat is limited.

Trials in the field, with physiological measurements and questionnaires, are normally performed during research programmes or the designing of new types of products; these can also be carried out during the selection process of protective clothing to be adopted at a fire brigade.

International and national standards specify performance requirements for firefighters' protective clothing for structural firefighting (ISO 11613:1999, EN 469:2005, NFPA 1971 (2000), wildland firefighting ISO/FDIS 15384:2005, NFPA 1977:1998, EN 15614:2007), specialized firefighting (ISO 15538:2001, EN 1496:2007, NFPA 1976), shipboard firefighting (ISO/DIS 22488) and fire hoods (EN 13911:2004). ISO 11613 contains two separate approaches: Approach A refers to European standards and approach B to NFPA standards. In Japan, different cities have had their own specifications. Today they are attempting to create a united standard based on Approch A of ISO 16113: 1999.[26] These standards define the minimum performance levels for fire and heat protection and test methods to evaluate the conformity against different types of hazards. Most of the requirements are based on material tests. Also, ISO 11613 is under revision in ISO TC 94 SC 14. The new version will contain all personal protective equipment for firefighters excluding respiratory protection.

European standards (EN) are prepared under a mandate given to CEN by the European Commission and the European Free Trade Association, and support the essential requirements of EU Directive 89/686/EEC.[27] Materials used in the manufacture of a garment that may come in contact with the wearer's skin should not be likely to cause irritation or to have any other adverse effect to health, and they should not cause discomfort to the wearer when worn. The basic requirements for fire and heat protection are that the constituent materials and other components suitable for protection must provide appropriate protection and be sufficiently incombustible to preclude any risk of spontaneous ignition under the foreseeable conditions of use. The complete PPE ready for use must not cause pain or health impairment threshold under the foreseeable conditions of use.

Before testing, the materials are pretreated and conditioned in the standardized temperature of (20 ± 2) °C, NFPA (21 ± 3) °C, and a relative humidity of $(65 \pm 5)\%$ for at least 24 hours. In accordance with the EN 469, the test materials shall be washed and dried or dry-cleaned according to the instructions of the care labelling and the manufacturer's instruction. If the information is missing, five cycles of washing are used. According to NFPA, the specimens shall be subjected to 10 cycles of washing and drying within a procedure simulating home laundering.

17.5.1 Measuring flammability

The flame spread test is the basic test for all kind of heat protective clothing. The EN standard for firefighters' protective clothing refers to ISO 15025 procedure

A. According to this method, a flame of 25 mm in height is exposed to the surface of the sample at a distance of 17 mm for 10 seconds. Afterflame and afterglow time are measured and shall be ≤ 2 s. No flaming to top or side edge, molten debris, or hole formation is allowed except for a layer used for specific protection such as liquid protection. The flame is applied to both sides of the component assembly, including wristlets and seams as well as other additional material. NFPA 1977 requires bottom edge exposure. No melting or dripping is allowed, afterflame time must be ≤ 2 s and char length ≤ 100 mm. All material layers used in the garment are individually tested. The final draft of European standard for wildland firefighting (prEN 15614) requires surface ignition for the basic material, but for hemmed specimens of single layer garments, edge ignition is also required.

According to EN 469, the component assembly of the outer garment is tested by applying the test flame to the surface of the outer garment and to the surface of the innermost lining. Also seams and wristlet material are tested.

17.5.2 Measuring protection against radiant heat

According to European standards, protection against convective and radiant heat is measured separately with different test methods. NFPA 1977 uses a combined method, the Thermal Protective Performance (TPP) testing method.

EN ISO 6942 specifies two complementary methods (method A and method B) for determining the behaviour of materials for heat protective clothing when exposed to radiant heat. Method A serves for visual assessment of any changes in the material after action of heat radiation of $10\ kW/m^2$ for three minutes. Any changes (e.g., discoloration, deposits, smouldering, charring, rupture, melting, shrinkage, sublimation) are noted separately for each layer of the sample. Method A is also used as a pre-treatment of the material samples for subsequent tensile strength testing. The minimum value for tensile strength after this pre-treatment is 450 N. With method B, the protective performance in Radiative Heat Transfer Index (RHTI) of the materials is determined. The radiation is produced by six silicon carbide heating rods, and a copper disc/copper-constantan thermocouple is used to measure the temperature behind the test samples.

Heat flux densities of $40\ kW/m^2$ or $20\ kW/m^2$ are used to measure performance against radiant heat. Table 17.1 summarizes the requirements given for protection against radiant heat in different standards. The level of performance is marked in the label of EN 469 suit as shown in Fig. 17.1.

In NPFA 1976, the radiant heat test is based on ASTM F 1939. The heat flux delivered by five quartz lamps is controlled through a variable transformer. A copper calorimeter-based sensor similar to the one used in TPP testing is employed to measure heat transfer through the specimen. The heat flux is $8.4\ kW/m^2$.

Table 17.1 Summary of the requirements given for protection against radiant heat in different standards

	$RHTI_{24}$	$RHTI_{24} - RHTI_{12}$
Specialized EN 1486, 40 kW/m²	≥ 120 s	
Specialized ISO 15538, 40 kW/m²	level 1 ≥ 60 s	
	level 2 ≥ 120 s	
Structural EN 469, 40 kW/m²	level 1 ≥ 10 s	level 1 ≥ 3 s
	level 2 ≥ 18 s	level 2 ≥ 4 s
Wildland EN 15614, 20 kW/m²	≥11 s	≥ 4 s
Wildland 15384, 20 kW/m²	≥ 11 s	≥ 4 s
ISO/DIS 22488, 40 kW/m²	Jacket ≥ 22 s	Jacket ≥ 6 s
	Trousers ≥ 14 s	Trousers ≥ 4 s
EN 13911:2004, 20 kW/m²	≥ 11 s	1 ≥ 3 s

Identification of the manufacturer

CE 1234

Name of the product
EN 469:2005

Xf2, Xr2, Y2, Z2

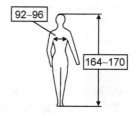

(if re-impregnation of the outer material is required, the number of washes before re-impregnation shall be clearly stated on the marking)

17.1 An example of marking on a suit in accordance with EN 469.

17.5.3 Measuring protection against convective heat

Heat transmission on exposure to flame in EN and ISO standards is determined using ISO 9151:1995. In this method, a horizontally oriented test specimen is partially restrained from moving and subjected to an incident heat flux of

Table 17.2 Summary of the requirement given for protection against convective heat in different standards

	HTI_{24}	$HTI_{24} - HTI_{12}$
Specialized EN 1486	≥ 21 s	NA
Specialized ISO 15538	level 1 ≥ 13 s	
	level 2 ≥ 21 s	
Structural EN 469	level 1 ≥ 9 s	level 1 ≥ 3 s
	level 2 ≥ 13 s	level 2 ≥ 4 s
Wildland EN 15614	NA	NA
Wildland 15384	≥ 11 s	≥ 4 s
ISO/DIS 22488	Jacket ≥ 13 s	Jacket ≥ 3 s
	Trousers ≥ 10 s	Trousers ≥ 3 s
EN 13911:2004	≥ 8 s	≥ 3 s

80 kW/m² from the flame of a gas burner placed beneath it. The heat passing through the specimen is measured by means of a small copper calorimeter on top of and contact with the specimen.

The time, in seconds, it takes for the temperature in the calorimeter to rise 24 ± 0.2 °C is recorded. The mean result for three test specimens is calculated as the Heat Transfer Index (HTI). For firefighters' clothing classification, the time in seconds for a 12 °C temperature rise is also reported according to EN and ISO requirements. Table 17.2 summarizes the requirements given in different standards.

NFPA defines a Thermal Protective Performance (TPP) parameter to establish the thermal insulation qualities of the material combination (ISO 17492). The test uses a bank of quartz radiant tubes and two Meker burners as a heat source. These two modes of heating are balanced to provide a 50/50 radiant and convective heat source. A copper disk calorimeter is placed against the back surface of the test specimen and the outer shell material is directed toward the heat source.[28] At least a TPP rating is required when the outer surface of the combination is exposed to a heat flux of 84 kW/m² (50% radiative and 50% convective heat) for a minimum of 17.5 s. At this level of protection, a firefighter would have about 10 seconds to escape from flashover conditions before getting second-degree burns.[29] A lower requirement of a minimum TPP of 20 has been later established for interface items such as protective hoods, garment sleeve wristlets, and glove wristlets.

17.5.4 Measuring protection against contact heat

NFPA sets a requirement for conductive compressive heat resistance of the reinforced knee and shoulder regions. In the test, a hot plate temperature of 280 °C and a contact pressure of 55.1 kPa for knees and 13.8 kPa for shoulders are used. The time required for a 24 °C temperature rise is measured and the requirement is $t_{24} \geq 13.5$ s. Moisture is applied to the thermal barrier by placing the sample between two pieces of blotting paper that have been wetted under controlled conditions.

In EN standards, a requirement for contact heat protection is specified only for specialized firefighting clothing (EN 1486). It uses the EN 702 (ISO 12127) test method in which a heating cylinder is heated up to the contact temperature and the specimen is placed on the calorimeter. The heating cylinder is lowered on to the specimen supported by the calorimeter or, alternatively, the calorimeter with the specimen is lifted up to the heating cylinder. In each case the operation shall be carried out at a constant speed. By monitoring the temperature of the calorimeter, the threshold time is determined and EN 1486 requires a contact temperature of 300 °C and a threshold time of at least of 15 seconds. ISO 15538 specifies two levels; level $1 \geq 10$ and level $2 \geq 15$ s.

17.5.5 Measuring thermal resistance of materials

ISO 17493 tests the heat resistance and thermal shrinkage for each material used in a garment, including the wristlet. Material specimens (375×375 mm) are suspended in a hot air circulating oven for five minutes at the specified test temperature of either 185 ± 5 °C or 260 ± 5 °C. Any ignition, melting, dripping, separation or shrinkage of the specimen is recorded. The exposure temperature varies depending on the standard. The main requirement is that materials to be used next to the skin must be exposed to a temperature of 260 °C. No melting, dripping, separation, or ignition is allowed. Table 17.3 summarizes the requirements for thermal resistance by various different standards.

17.5.6 Thread heat resistance

Some standards (EN15614:2007) specify heat resistance for sewing thread used in seams. One principle test method is EN ISO 3146 and the thread must not

Table 17.3 Summary of the requirement given for thermal resistance of materials in different standards

	Test temperature °C	Requirement
Specialized EN 1486 and ISO 15538	255 ± 10	Shrinkage $\leq 5\%$
Structural EN 469	185 ± 5	Shrinkage $\leq 5\%$
NFPA 1971	260 ± 5	Shrinkage $\leq 10\%$
Wildland EN 15614		
Wildland ISO 15384 ISO/DIS 22488	260 ± 5	Shrinkage $\leq 10\%$
EN 13911:2004	260 ± 5	Shrinkage $\leq 10\%$

melt at a temperature less than 260 ± 5 °C. No melting is also required according to NFPA standards.

17.5.7 Measuring flame and thermal performance of the complete clothing system

Measuring flame and thermal performance of the complete outfit is specified as an optional test method in EN 469:2005 and is described in ISO 13506 2008. The aim of this method is to evaluate the interaction of material behaviour and garment design. It uses a stationary full-sized male manikin, 1830 ± 40 mm in height. In a laboratory simulation, the dressed manikin is exposed to a simulated flash fire with controlled heat flux, duration and flame distribution. The average incident heat flux density to the exterior of the garment is 84 kW/m^2 for an exposure time of at least 8 s. Measuring data is gathered for single layer garments for a period of 60 s, and for 120 s for multilayer assemblies. A minimum of 100 heat sensors are fitted over the manikin surface to measure the energy transferred to the whole area during the data gathering period. The sensors must have the capacity to measure the incident heat flux density over a range from zero to 200 kW/m. The hands and feet of the manikin are without sensors. The nature and extent of damage that would occur to the human skin from the exposure are calculated, including the time it would take for the person to feel pain or develop first-, second- or third-degree burn injuries.

ISO 13506 does not include a female manikin. This is despite the fact that the number of females in fire risk occupations is increasing. In order to create a female manikin, anthropometric data on the UK female population has been collected and BTTG in UK has designed, built and commissioned a heat sensing manikin test facility which is able to operate comparably in male (Ralph) and female (Sophie) mode. The first generation of 'Sophies' and revised 'Ralphs' will operate in a mode equipped with sensors in each hand.[30]

The manikin measurements are useful especially for three types of evaluation:

1. comparison of garments or assemblies of materials;
2. comparison of garment or assembly designs; and
3. evaluation of any garment or assembly prototype for a particular application or to a specification.

17.5.8 Measurements with test subjects in laboratory

In addition to material tests, the clothing parameters can also be determined in scenarios simulating relevant conditions. TNO in the Netherlands has proposed a subjective test aiming to define the protective capacity of the protective clothing when exposed to radiant heat loads of 7 kW/m^2 produced by propane burners. The tests using human subjects serve more as 'evaluation tools'. They can show typical weak spots in the clothing design.[31] CEN TC 162 WG 2 is working to define a physiological test for firefighters' protective clothing as a part of the EN 469 revision.

17.5.9 Consideration on other needs for testing

The effect of moisture

Despite improvements in the testing standards used for firefighters' protective clothing, there are still some areas of such clothing performance that are not well understood, and significant limitations exist in the standards for evaluating all the different hazards to which firefighters are exposed.[32] One of these limitations is the effect of humidity between the clothing layers and absorbed in the component fibres and its influence on the speed of heat transfer with different mechanisms. So far, few standards account for the effect of moisture. Numerous studies have been carried out to understand the effect of moisture in the garment systems, and related literature mentions several factors that affect the performance of clothing. These factors include the amount of water, its location in the material system, the type of heat transfer, the intensity and duration of the exposure and the type of materials and their condition during testing.[33–37] Unfortunately, the results are often contradictory.[24] However, the presence of moisture tends to reduce differences in performance among fabric systems, in many cases reducing heat transfer. But in some cases and especially in the internal layers, moisture may actually increase heat transfer and thus decrease the protection.[38] Barker[39] concluded that the addition of moisture negatively impacts the predicted burn protection to the greatest degree when the moisture is added at a comparatively low level of approximately 15% of turnout composite weight. As the moisture level increases beyond this critical level, predicted second-degree burn time increases, approaching values measured for dry composites. Test method development is still needed.

The influence of steam has not been discussed widely. Steam flowing from the outside towards the body, or evaporating from the textile layers can lead to steam burns. These may be more severe than dry burns, as the hot moisture can be partly absorbed by the skin and transferred to the deeper skin layers. In general impermeable materials offered better protection against hot steam than semi-permeable ones.[24] The transfer of steam was reported to depend on the water vapour permeability of the samples, but also on their thickness. Increasing the thickness of the samples with a spacer increased the protection capacity of the impermeable samples more than that of semi-permeable materials. Materials with good water vapour absorbency tended to offer better protection against hot steam.[24] Higher amounts of moisture in underwear might also contribute to an increased risk of steam burns and that accumulated in cotton underwear strongly depended on the characteristics of the neighbouring layer of clothing.[40]

Stored heat energy

Structural firefighters are more likely to receive second-degree burns while working in prolonged low thermal exposure conditions than in flashover conditions. Such exposure typically lasts for several minutes, and usually does not degrade the outer-shell fabric of protective clothing. One potential source for causing such burns is the thermal energy stored in the firefighter's protective clothing as a result of prolonged exposure in a structural fire within a room that has not reached a flashover condition.[9] Eni[41] has developed a novel laboratory apparatus and test protocols for measuring the contribution of both transmitted and discharged stored energy on thermal protective performance. It uses a TPP tester with preprogrammed procedure with an exposure of 21 kW/m^2 and defines the minimum value of exposure time or MET value. The Store Energy Index = (RPP burn time − MET)/MET. In the absence of moisture, protective clothing with thicker liners typically has higher stored energy while its presence decreased the stored energy for most systems, but increased the amount of thermal energy transmitted.

17.6 Future trends

17.6.1 Firefighters' work in future

New technology for extinguishing fires and for searching for people in fires will be available in the near future and so the need to expose firefighters to heat may decrease. Yet, high performance protective clothing systems will continue to be needed to address diverse exposures. In order to prevent daily overprotection and hence overstrain in firefighters and to promote work ability and comfort, the use of a clothing system with layered protective clothing comprising easily attachable intermediate and external protective layers designed to address the required level of the hazards in question may be the best functional solution. For this purpose, new advanced high-performance materials and new technologies offer many novel possibilities for developing more functional protective clothing.

17.6.2 New materials

New fibres with integrated nanoparticles[42] or with bicomponent fibres[43] to reduce their flammability or improve their mechanical stability and a new generation of plasma treatments of fabrics and fibres[44] to tailor their functionality offer potential methods for producing passive materials in the future. New materials with adaptive functions, such as phase transition or change of shape, are currently being developed further to better fit the particular needs of the work environment and are likely to constitute active materials of the future.[45]

Hydroweave™ by Aquatex Industries[46] is one example of today's active materials. Hydroweave™ employs a multi-layer construction consisting of an outer woven fabric shell, a fibrous batt containing a water-absorbent polymer, and a conductive, microporous film on a light fabric substrate. The water-absorbent polymer with finite water absorption capacity is distributed evenly throughout the material batt. It incorporates functions for cooling and insulation from heat and also meets the performance criteria established by NFPA 1971 for TPP (in an activated condition, TPPC 100) and radiant heat resistance test at $20\,kW/m^2$. Despite the potential benefits of both heat-stress reduction and improved thermal insulation, some concerns have been raised about the potential for steam generation or scalding temperatures due to water held in the Hydroweave material. It is recommended that supervisors and other persons responsible for employee safety should take into account the specific hazards and circumstances of their own situation in applying these results.[47]

17.6.3 Garments with wearable electronic technology

Many developments include smart electronics such as sensors, electronic infrastructure, communication interface and energy supply placed into the clothing system in order to increase the functionality and the protective effect. New technologies for the durable incorporation of sensors for vital function, position/ localization and environment monitoring in textiles, applied as thin films on single fibres or as labels embedded into the fabric structure and with a focus on re-usability, are being developed.[48,49]

Examples of the development for firefighters' protective clothing are from the Netherlands and Finland. In the Dutch design, the suit is fitted with sensors that measure the wearer's core temperature, skin temperature, and heart rate. The fire commander outside the immediate hazard zone receives these data via a wireless connection. When the temperature starts to become critical, the firefighter can be instructed to turn back.[50]

The so-called Finnish Clothing Area Network (CLAN)[51] has enabled the development of smart clothing for firefighters in which the smart fire suit measures information from the firefighter and also from the ambient environment. Temperature measurements from the ambient air, inside the fire suit, glove and boot sole are possible and the firefighter's heart rate and the amount of the remaining air in the compressed air bottle may also be measured. The data and power transfer inside the fire suit as well as between the suit and separate pieces of clothing (e.g., glove and boot) are carried out via the CLAN network developed for sensor communication. Data from the CLAN network is transferred to the main module inside the fire suit, called the ClanBox. From the ClanBox the data is sent in real time to a portable laptop computer located in a leading fire engine. The laptop runs the ClanWare software, which processes and visualizes the sensor data. Thus the rescue operation supervisor can control

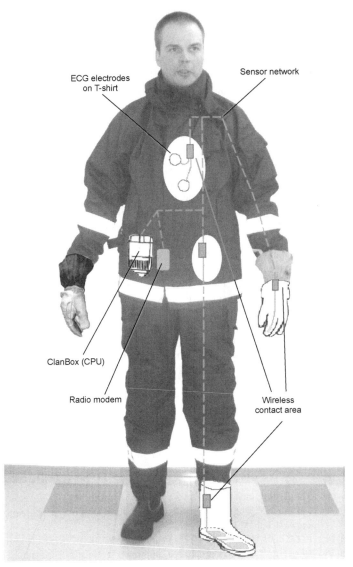

17.2 A working prototype of the smart fire suit.

the rescue operation and if necessary, call the firefighter away from the ongoing situation. A working prototype of the smart fire suit (Fig. 17.2) has been tested in practice in live-fire training exercises, and the results were promising. The demanding environment was given particular emphasis when designing the electronics. The attitude among firemen towards the smart fire suit has also been very positive: they perceived new means to increase occupational safety as highly important.

17.6.4 Need for test method development

Finally, parallel to this development work, the applicability of the current standards to new products should be investigated, since entirely new types of requirements and testing methods are perhaps needed to be able to investigate these proposed new multi-functional materials.

17.7 References

1. International Labour Organization, International Hazard Datasheets on Occupation, 'Fire-fighter', Occupational Hazard Datasheets. http://www.ilo.org/public/english/protection/safework/cis/products/hdo/htm/firefightr.htm
2. Stull J O, Haskell W E and Shepherd A M (2006), 'Approaches for incorporating CBR requirements as a part of protective ensemble standards for emergency responders' in *Protective clothing – towards balanced protection. Proceedings of the 3rd European Conference on Protective Clothing (ECPC) and NOKOBETEF 8*, May 10–12; Gnydia. [CD-ROM] Gdynia.
3. The JOIFF Standard (2006), *Guidance handbook on Personal Protective Equipment to protect against heat and flame*, The Organisation for Emergency Services Management, www.fulcrum-consult.com.
4. Lawson, J R (1996), *Firefighters' protective clothing and thermal environments of structural firefighting*, NISTIR 5804, National Institute of Standards and Technology, Gaithersburg, MD, August 1996.
5. Wagner N L (2006), 'Mortality and life expectancy of professional firefighters in Hamburg, Germany: a cohort study 1950–2000', *Environmental health: A Global Access Science Source*, 5, 27. http://www.ehjournal.net/contents/5/1/27.
6. Project Heroes (2003), *Homeland Emergency Response Operational and Equipment Systems. Task 1: A review modern fire service hazard and protection needs*, Occupational Health and Safety Division, International Association of Firefighters (IAFF), 1750 New York Avenue, N.W. Washington, DC 20006, 13 October 2003.
7. DeMars K A, Henderson W P and Liu M (2002), 'Thermal measurements for firefighters' protective clothing', in Gritzo L A, Alvares N J, *Thermal Measurements: The Foundation of Fire Standards*, ASTMSTP 1427, American Society for Testing and Materials, West Conshohocken, PA.
8. Hoschke B N (1981), 'Standards and specifications for firefighter's clothing', *Fire Safety Journal*, 4, 125–137.
9. Barker R I, Guert C, Behnke W P and Bender M (2000), 'Measuring the thermal energy stored in firefighter protective clothing', in Nelson C N and Henry N W, *Performance of protective clothing: Issues and Priorities for the 21st Century*, Seventh Volume, ASTM STP 1386, American Society for Testing and Material, West Conshohocken, PA.
10. Holmes D A (2000a), 'Textiles for survival', in Horrocks A R and Anand S C, *Handbook of technical textiles*, The Textile Institute, Woodhead Publishing Limited, 461–489.
11. Morse H L, Thompson J G, Clark K J, Green K A and Moyer C B (1973), *Analysis of the thermal response of protective fabrics*, Technical report AFML-Tr-73-17. Air Force material laboratory, Air Force Systems Command, Wright-Patterson Air force base, Ohio 45433, January 1973.
12. Lawson J R (1997), 'Firefighters' Protective Clothing and Thermal Environments of

Structural Firefighting', in Stull J O and Schope A D, *Performance of protective clothing: Sixth volume, ASTM STP 1273*, American Society for Testing and Materials, 335–352.
13. Rossi R (2003), 'Firefighting and its influence on the body', *Ergonomics*, 46 (10), 1017–1033.
14. Mäkinen H and Kervinen H (2005), 'The present situation and future trend of Finnish firefighters' protective clothing' in *Proceedings of Fourth NRIFD Symposium – International Symposium on Protective Clothing for Fire Fighing Activities*, March 9–11, 211-217. National Research Institute of Fire and Disaster Mitaka, Tokyo, Japan.
15. Insatsstatistik 2006, 'Var fjärde insats vid trafikolycka' *Sirenen*. Räddningsverkets tidning Nr 2 mars 2007, 16–17.
16. Bader Y and Capt A (2005), 'European firefighter clothing: Trends and technical evolutions' in *Proceedings of Fourth NRIFD Symposium – International Symposium on Protective Clothing for Firefighting Activities*, March 9–11, National Research Institute of Fire and Disaster Mitaka, Tokyo, Japan, 125–129.
17. Holmer I, Kuklane K and Chuansi G (2006), 'Test of Firefighter's Turnout Gear in Hot and Humid Air Exposure' *International Journal of Occupational Safety and Ergonomics (JOSE)* 12, 3, 297–305.
18. Bajaj P (2000), 'Heat and flame protection', in Horrocks A R and Anand S C *Handbook of technical textiles*, The Textile Institute, Woodhead Publishing Limited, 223–263.
19. Horrocks A R (2006), 'Thermal (heat) and fire) protection' In: Scott RA editor. *Textiles for protection*. Boca Raton, FL, CRC Press, 398–440.
20. Tochihara Y, Fujita M, Chou C-M, Ogawa T (2005), 'Physiological strain of workers wearing protective clothing – requirements for research to reduce the discomfort of firefighters' in *Proceedings of Fourth NRIFD Symposium – International Symposium on Protective Clothing for Fire Fighing Activities*, National Research Institute of Fire and Disaster Mitaka, Tokyo, Japan, March 9–11, 1–7.
21. Assman S (2005), 'Performance and limitations of Aluminised PPE for promixity firefighting' in *Proceedings of Fourth NRIFD Symposium – International Symposium on Protective Clothing for Firefighting Activities*, March 9–11, National Research Institute of Fire and Disaster Mitaka, Tokyo, Japan, 109–115.
22. Hocke M, Strauss L, Nocker W (2000), 'Firefighter garment with non textile insulation', in Kuklane K and Holmer I, *Proceedings of NOKOBETEF 6 and 1st European Conference on Protective Clothing held in Stockholm*, Sweden, May 7–10, 2000, *Arbete och Hälsa* (8) 293–295.
23. Holmes D A (2000b), 'Waterproof breathable fabrics', in Horrocks A R and Anand S C *Handbook of technical textiles*, The Textile Institute, Woodhead Publishing Limited, 281–315.
24. Rossi R, Indelicato E and Bolli W (2004), 'Hot steam transfer through heat protective clothing layers', *International Journal of Occupational Safety and Ergonomics (JOSE)*, 10 (3), 239–245.
25. Ilmarinen R, Mäkinen H, Lindholm H, Punakallio A and Kervinen H (2006), 'Thermal strain in firefighters while wearing task-fitted protective clothing vs. EN 469 protective clothing during a prolonged job-related rescue drill' in *Proceedings of the 3rd European Conference on Protective Clothing (ECPC) and NOKOBETEF 8*, May 10–12, Gnydia. [CD-ROM]Gdynia.
26. Yanai E, Shinohara M and Hatamo T (2005), 'Japanise Standard for firefighter's protective clothing and their protective performance against flame using

Instrumented manikin' in *Proceedings of Fourth NRIFD Symposium – International Symposium on Protective Clothing for Fire Fighing Activities*, National Research Institute of Fire and Disaster Mitaka, Tokyo, Japan, March 9–11, 227–234.
27. Council Directive of 21 December 1989 on the approximation of the laws of the Member States relating to personal protective equipment (89/686/EEC), Official Journal of European Communities 30.12.89.
28. DeMars K A, Henderson W P and Lui M (2002), 'Thermal measurements for firefighters' protective clothing', in Gritzo L A and Alvares N J, *Thermal Measurements: The Foundation of Fire Standards, ASTMSTP 1427*, American Society for testing and Materials, West Conshohocken, PA.
29. Krasny J F, Rockett J A, and Huang D (1988), 'Protecting fire-fighters exposed in room fires: comparison of results of bench scale test for thermal protection and conditions during room flashover', *Fire Technology*, 24 (Feb), 5–19
30. Eaton P, Healey M and Sorensen N (2006), 'A new facility for testing the fire protective performance of ensembles of PPE' *in Proceedings of the 3rd European Conference on Protective Clothing (ECPC) and NOKOBETEF 8*, May 10–12; Gnydia. [CD-ROM] Gdynia.
31. Havenith G and Heus H (2004), 'A test battery related to ergonomics of protective clothing', *Applied Ergonomics*, 25, 3–20.
32. Stull J O (2005), 'Effect of moisture on firefighter protective clothing thermal insulation' in: *Proceedings of Fourth NRIFD Symposium – International Symposium on Protective Clothing for Fire Fighing Activities*, National Research Institute of Fire and Disaster Mitaka, Tokyo, Japan, March 9–11, 2005; 49–65.
33. Stull O J (1997), 'Comparison of conductive heat resistance and radiant heat resistance with thermal protective performance of firefighter protective clothing' in Stull J O and Schope A D, *Performance of protective clothing: Sixth volume, ASTM STP 1273*, American Society for Testing and Materials, 248–268.
34. Prasad K, Twilley W and Lawson J R (2002), 'Thermal Performance of Firefighters' Protective Clothing 1. Numerical Study of Transit Heat and Water Vapour transfer', NISTIR 688, 32 p. *http://www.fire.nist.gov/bfrlpubs/fire02/PDF/f02077.pdf*.
35. Lawson L K., Crown E M, Ackerman M Y and Dale J D (2004), Moisture Effects in Heat Transfer through Clothing Systems for Wildlands Firefighters, *International Journal of Occupational Safety and Ergonomics (JOSE)*, 10 (3).
36. Veghte J H (1984), 'Effect of moisture on the burn potential in firefighters' gloves', *Fire Technology*, 23 (4) 313–322.
37. Mäkinen H, Smolander J and Vuorinen H (1988), 'Simulation of the Effect of moisture Content in Underwear and on the Skin Surface on Steam burns of Fire-Fighters, in Mansdorf S Z, Sager R, Nielsen A P, *Performance of Protective Clothing: Second Symposium, ASTM STP 989*, American Society for Testing and Materials, Philadelphia, 415–421.
38. Crown E M, Lawson L K, Dale J D and Ackerman M Y (2005), Moisture effects in heat transfer clothing systems: implications for performance standards' in *Proceedings of Fourth NRIFD Symposium – International Symposium on Protective Clothing for Firefighting Activities*, National Research Institute of Fire and Disaster Mitaka, Tokyo, Japan, March 9–11, 35–47.
39. Barker R L, Guerth-Schacher C, Grimes R V and Hamouda H (2006), 'Effects of moisture on the thermal protective performance of firefighter protective clothing in low-level radiant heat exposures', *Textile Research Journal*, 76, 27 http://trj.sagepub.com/cgi/content/abstract/76/1/27.
40. Keiser C, Becker C and Rossi R (2006), 'Analysis of the distribution of sweat in

firefighters' protective clothing layers' in *Protective clothing - towards balanced protection. Proceedings of the 3rd European Conference on Protective Clothing (ECPC) and NOKOBETEF 8*, May 10–12, Gnydia. [CD-ROM] Gdynia.
41. Eni E U (2005), Developing test procedures for measuring stored thermal energy in firefighters protective clothing. Thesis submitted to the Graduate Faculty of North Carolina State University, Textile Engineering, Raleigh 2000. http://www.lib.ncsu.edu/theses/available/etd-07262005-165310/.
42. Holme Ian (2005), 'Nanotechnologies for Textiles, Clothing and Footwear'. *Textiles Magazine* 32 (1) 7–11.
43. Veerle Herrygers Veerle (2005), 'Use of bicomponent yarns & fibres: current possibilities for innovation. 3th International Avantex-Symposium, Frankfurt 6–8 June 2005.
44. Höcker Hartwig (2002), 'Plasma treatment of textile fibers', *Pure Appl. Chem.*, 74 (3) 423–427.
45. http://www.tut.fi/units/ms/teva/projects/intelligenttextiles/index3.htm.
46. www.hydroweave.com.
47. Stull J O (2000), *Special Radiant Testing of Hydroweave™ Material to Determine Potential for Steam Release*. Prepared for: Scott Bumbarger: AquaTex Industries 307 Wynn Drive NW Huntsville, AL 35805.
48. Stresse H, John L and Kaminorz Y (2005), 'Micro system technologies for smart textiles' in Avantex-symposium 6.-8.6.2005, Messe Frankfurt, Frankfurt am Main, [CD-ROM].
49. Linz T, Kallmayer C (2005), 'Integration of microeletrinics into textiles' in *Avantex-symposium 6.-8.6.2005*, Messe Frankfurt, Frankfurt am Main, [CD-ROM].
50. *TNO Magazine*, December 2006.
51. Kärki S, Honkala M, Cluitmans L, Cömert A, Hännikäinen J, Mattila J, Peltonen P, Piipponen K, Tieranta T, Väätänen A, Hyttinen J, Lekkala J, Sepponen R, Vanhala J and Mattila H (2005), 'Smart clothing for firefighters' in Talvenmaa P, International Scientific Conference '*Intelligent Ambience and Well-Being*', 19–20 September, Tampere, Finland, Tampere University of Technology, Sinitaival 6, FI-33720 TAMPERE, Finland.

17.8 Appendix: Standards

European Committee on Standardization (CEN)

EN 469 (2005) 'Protective clothing for firefighters – Performance requirements for protective clothing for fire-fighting', European Committee for Standardization, Rue de Stassart 36, B-1050 Bruxelles.

EN (702) 1994 'Protective clothing – Protection against heat and flame – test methods: Determination of the contact heat transmission through protective clothing or its materials', European Committee for Standardization, Rue de Stassart 36, B-1050 Bruxelles.

EN 1486 (2007), 'Protective clothing for firefighters. Test methods and requirements for reflective clothing for specialized firefighting', European Committee for Standardization, Rue de Stassart 36, B-1050 Bruxelles.

EN 15614 (2007), 'Protective clothing for firefighters – Laboratory test methods and performance requirements for wildland clothing.'

EN 13911 (2004), 'Protective clothing for firefighters – Requirements and test methods for fire hoods for firefighters' European Committee for Standardization, Rue de Stassart 36, B-1050 Bruxelles.

International Standards Organization (ISO)

ISO 3146 (2000), 'Plastics – Determination of melting behaviour (melting temperature or melting range) of semi-crystalline polymers by capillary tube and polarizing-microscope methods'. International Organization for Standardization.

ISO 6942 (2002), 'Clothing for protection against heat and fire – Evaluation of thermal behaviour of materials and material assemblies when exposed to a source of radiant heat', International Organization for Standardization.

ISO 9151 (1995), 'Protective clothing against heat and flame – Determination of heat transmission on exposure to flame', International Organization for Standardization.

ISO 11613 (1999), 'Protective clothing for firefighters – Laboratory test methods and performance requirements', International Organization for Standardization.

ISO 12127 (1996) 'Clothing for protection against heat and flame – Determination of contact heat transmission through protective clothing or constituent materials', International Organization for Standardization.

ISO 15025 (2000), 'Protective clothing – Protection against heat and flame – Method of test for limited flame spread', International Organization for Standardization.

ISO 17492 (2002), 'Clothing for protection against heat and flame – Determination of heat transmission on exposure to both flame and radiant heat', International Organization for Standardization.

ISO 13506 (2008), 'Protective clothing against heat and flame – Test method for complete garments – prediction of burn injury using an instrumented manikin', International Organization for Standardization.

ISO/FDIS 15384 (2005), 'Protective clothing for firefighters – Laboratory test methods and performance requirements for wildland clothing' International Organization for Standardization.

ISO 15538 (2001), 'Protective clothing for firefighters – Laboratory test methods and performance requirements for protective clothing with a reflective outer surface', International Organization for Standardization.

ISO 17493 (2000), 'Clothing and equipment for protection against heat – test method for convective heat resistance using a hot air circulating oven', International Organization for Standardization.

ISO/DIS 22488 (2003), 'Ship and marine technology – Shipboard fire-fighters' outfits (protective clothing, gloves, boots and helmet)', International Organization for Standardization.

National Fire Protection Association (NFPA)

NFPA 1971 (2000), 'Standard on Protective Ensemble for Structural Firefighting', 2000 edition, National Fire Protection Association, 1 Batterymarch Park, PO Box 9101, Quincy, MA 02269-9101.

NFPA 1976 (2000) 'Standard on Protective Ensemble for Proximity Firefighting', 2000 edition. National Fire Protection Association, 1 Batterymarch Park, PO Box 9101, Quincy, MA 02269-9101.

NFPA 1977 (1998), 'Protective clothing and equipment for wildland firefighting', 1998 edition. National Fire Protection Association, 1 Batterymarch Park, PO Box 9101, Quincy, MA 02269-9101.

18
Fire protection in military fabrics

S NAZARÉ, University of Bolton, UK

Abstract: Textiles and their composites are extensively used in protecting military personnel from fire hazards. This chapter first discusses different fire hazards and types of burn injuries in the military environment followed by existing fire resistant solutions for military clothing. Performance requirements and standard test methods for flame retardant textiles used for clothing and non-clothing applications are also discussed.

Key words: fire hazards, flash fire protection, military clothing, burn injuries, flame retardant textiles, flammability standards.

18.1 Introduction

The military is one of the most demanding consumers of textile materials with the most critical requirements. Since military forces are subjected to unpredictable environments at very short notice, the general requirements for military protective textiles are diverse and critical. While the most important requirement is to keep the armed forces personnel safe, textiles used in military applications are required to fulfil certain physical, environmental, camouflage and fire resistant characteristics. The only route to accomplish such vital requirements is through continued scientific research and development, which is mainly driven by the changing nature of wars and not least by the immediate requirements of the soldiers on the battlefield, for example. The British Army is undertaking a significant project on clothing developments for military personnel such as Personnel Equipment and Common Operational Clothing (PECOC). One of the main organisations dedicated to the research and development of protective textiles for soldiers is the US Army Natick Soldier Centre (NSC) in Natick, Massachusetts. In order to improve the performance, countries such as France, Germany, Netherlands, Sweden and Russia, too, are aiming to modernise clothing and equipment used by their armed forces.

Military clothing has undergone substantial changes over the past two centuries. Today's drab but functional military uniforms contrast sharply with the highly decorative although impractical garments worn during combat in the 18th and 19th centuries. Changes in modes of battle dress since man has engaged in war have been prompted by changes in the way that wars are fought. Since the Second World War improvements in fabric technology have led to dramatic changes in clothing worn by service personnel. The advancements in fabric technology are not competent to defy the challenges offered by the developments accomplished in weaponry. The armed services address this imbalance by developing uniforms which improve performance and comfort of service personnel, while helping to save lives during combat. Although military budgets are under pressure, fear of the public's response to war casualties has prompted heavy investment in clothing and equipment, which are capable of boosting the survival chances of individual servicemen and -women. Advances in materials have improved the performance of military clothing while providing high levels of protection.

Severe fire situations can particularly occur in combat vehicles on land, in air or sea and the deployment of armoured vehicles since the First World War has exposed crew personnel to the extreme hazards of fire and explosion caused by fuel, hydraulic fluid and munitions. Statistical studies on armoured vehicle crew casualties during Second World War reported that burn injuries accounted for about 25% of crew casualties.[1] Since then, large numbers of burn casualties were in Vietnam, during the Arab/Israeli Wars and in the Falkland Islands conflict, predominantly caused by bombs or by guided missiles. Technological advances have increased lethality and accuracy of these weapons, leading to high rates of blast injuries and burn injuries associated with fires. Armoured vehicles, ships and military aircraft share the particular danger of high explosive penetration into a confined space, with the potential for further explosion of fuel and explosives stored nearby. In addition to conventional munitions, devices causing burns span a range from small, homemade bombs to modified conventional incendiary weapons such as large charge-filled howitzer rounds to fuel tanks with detonators. Incidentally, modern warfare on land is often fought in an urban environment which is significantly different from that found on the conventional open battlefield. This often leads to large amount of collateral damage to civilian populations. In the present so-called war against terrorism in Iraq and Afghanistan, where the forces are operating predominantly on land, burn injuries are becoming an exceedingly common cause of fatalities amongst the soldiers as they are subjected to unpredictable attack by improvised and conventional incendiary explosive devices. To address this problem, research now is directed towards developing lightweight, low cost flame resistant uniforms for infantry soldiers in addition to flexible body armour.

Detailed accounts on protective clothing[2] and recent developments in military clothing,[3] in general, are available and of particular relevance are the new approaches in flame retardancy of textiles, addressed by Bourbigot in Chapter 2. However, the present chapter focuses on the particular incidence of fire hazards and the type of fire/flame protection required from textile materials in military environments. Recent developments in combat uniforms, particularly in terms of fire protection, have also been addressed.

18.2 Fire hazards

Flame and thermal radiative threats are hazards that soldiers are frequently exposed to on the battlefield, naval personnel experience when subjected to missile attacks in the confinements of the vessel, and air personnel experience when under attack in aircraft at high altitude or aboard helicopters closer to the ground altitude where a particularly hazardous environment is encountered in a more confined place such as a cockpit. Crown et al.[4] have recently reviewed fire hazards associated with military aircraft and helicopters, highlighting the difficulty of escape from the aircraft flying at high altitudes. In addition to missile attacks, aircraft fire hazards include in-flight, take-off, landing, refuelling and post-crash fires. Post-crash fires from ruptured fuel tanks are also common in aviation environments. Moreover, military aviation fires are fiercer than those encountered in civil aviation, primarily because fuels used have specifically formulated compositions to meet specific operational needs. Solid fuels used in missiles, for example, consist of pure hydrocarbons or mixtures of few hydrocarbons.[5] Catastrophic fires may also occur in an aviation environment as a consequence of unserviceable or untested equipment added during the urgency of military conflicts.

The submarine environment is a particularly hazardous problem within the navy as vessel safety relies on a closed cycle air purification system. Once fire breaks out, toxic fumes and smoke may overwhelm the air purification system resulting in fatalities due to inhalation of toxic gases and smoke. This hazard is aggregated by the clothing and textiles in bulk storage. Crew clothing ignition is also a major problem inside the submarine since the fire can spread very quickly in an enclosed compartment. For example, on 18 June 1984, a fire broke out in a Russian submarine, K-131 the cause of which was reported to be clothing ignition of one of the members of the crew while working on some electrical equipment. The fire spread to the adjacent compartment and caused the death of 13 members of the crew.[6]

For protection against flame and intense heat, it is often necessary to identify the nature of the hazard and the probability of its occurring and hence risk to life. However, identification and definition of such fire/flame hazards is complex. Even though fire hazards can be simply defined by ignition temperature of the fabric, the temperature cannot provide the rate of heat input experienced by the clothing system.[7] The NATO Allied Combat Clothing Publication[8] (ACCP) has therefore defined main fire hazards in terms of heat flux rates from different sources as:

1. Burning fuel $150\,kW/m^2$ 7–12 s reasonable escape time
2. Exploding ammunition $200\,kW/m^2$ 5 s before survivability is unlikely
3. Penetration by warheads 500–$560\,kW/m^2$ 0.3 s maximum duration of event
4. Nuclear thermal pulse 600–$1300\,kW/m^2$ less than 0.1 s escape time

The heat flux levels of such hazards vary and depend on the type of weapon. Threat analysis[9] has shown that the majority of battlefield fire hazards are in the vicinity of a fire-level with a heat flux ranging from 2 to 100 kW/m^2. The UK Home Office Fire experimental Unit studied[10] protective properties of fire fighters' ensemble in a worst-case scenario, namely an ambient temperature in excess of 235 °C and a flux greater than 10 kW/m^2. This condition was considered as life threatening and beyond the protected fire fighters' operability limit and in a military scenario, injury is certain and death highly probable if the exposure time is prolonged.

Fire hazards in a military environment are defined below:

- **Detonation:** This is a result of incendiary weapons colliding with explosive fuel, especially when stored in tanks. Confinement of the reaction leads to increase in pressure, rate of reaction and temperature resulting in a detonation.
- **Deflagration:** Failure to detonate fuel-air explosives may generate large amounts of heat by deflagration of the fuel. Deflagration is a surface phenomenon with the reaction proceeding towards the non-reacted material, usually petrol, along the surface at subsonic velocity. It is a rapid chemical combustion reaction in which the output of heat is sufficient to enable the reaction to proceed and be accelerated without input of heat from another source. In such a fire hazard, the casualty is exposed to considerable thermal energy as a combination of radiant and convective energy. The effect of a true deflagration under confinement is often an explosion.
- **Explosion:** The explosion of fuel-air explosives leads to formation of a fireball constituting hot products of combustion. A fireball burning with an orange flame can reach temperatures up to 1100 °C. Although the diameters of such fireballs are predictable, the heated air surrounding them may flow outwards, thus increasing the incidence of convection, heat causing burns beyond the fireball.
- **Flashover:** This may be defined as a point where all of the combustible material has ignited within the bounds of the fire zone. At this point, the fire usually reaches a steady state and maximum energy output is usually defined as a peak heat release rate. Flashovers have both convective and radiant components and the duration and intensity of thermal flux determines the extent of burn injury.

Military deployment also exposes personnel to accidental injury unrelated to combat. This is often associated with road traffic accidents, smoking and the use of unguarded flames for lighting, cooking and heating in hazardous military environments,[11] any of which could lead to thermally hazardous environments that vary in terms of the type, amount and duration of heat exposure. With regard to hazards associated with clothing textiles in particular, clothing while shielding the wearer from heat radiation, can ignite causing thorough burn injury, particularly through conduction and convection.[11] Fire hazards involving clothing are known to be severe, since typical textile flame temperatures range from 600 to 1000 °C[12] causing higher degree burn injuries which could be fatal or disable the wearer permanently. Because convection carries flames and burn products vertically upwards, the so called 'chimney effect', this increases the incidence of burn injury at the waistband, under the arms and around the neck and around the face and head.

18.3 Military burn injuries

Typically, burn injuries constitute 5–20% of combat casualties in conventional warfare. A statistical analysis[13] on the burns sustained in combat explosions in the ongoing war in Iraq and Afghanistan (April 2003–April 2005) has revealed that burn injuries are still a persistent threat in military environment. The percentage distribution of cause of burn injury among the casualties burned in explosions in Iraq and Afghanistan is shown in Fig. 18.1. It can be seen that detonation of improvised devices caused the majority of burn injuries, while those from conventional weaponry were less common.

Heat burns from radiation and convection are the principle injury outcome in military fire hazards. Burn injuries in the military environment are divided into four groups:

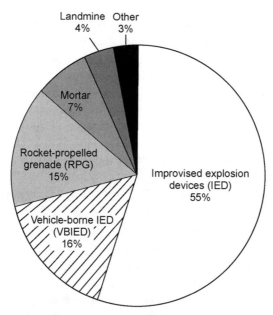

18.1 Percentage distribution of various causes contributing to burn injuries in Iraq and Afghanistan.[13] Note: IED is an improvised explosive device.

1. Combat-related burns directly due to weapons;
2. Combat-related burns from ignition of fuel, explosives, propellants and clothing;
3. Industrial-type accidental injuries due to working with vehicles, fuel, hot liquids and electrical equipment;
4. Off-duty accidental injuries.

Combat-related burn incidents are recognised and well documented. Accidental burns associated with handling of fuel, oil, munitions and other flammable materials during peacetime deployment and training do not often draw much media attention. Pruitt et al.[14] reported 13 047 military burn injuries out of which about half the burns were accidental (54%) and the rest (46%) were combat related. A recent study on combat and non-combat burns from ongoing US military operations[15] revealed that combat burns had a greater percentage ($8 \pm 16\%$) of 3rd degree burns compared to non-combat burns ($5 \pm 12\%$). However, the anatomical distribution of burns was similar between combat and non-combat casualties with hands and face being most frequently affected areas in both groups. The pattern of burn injury among casualties burned during explosions, as well as high risk surface area of body, are shown in Fig. 18.2. The distribution of burn injuries, in Fig. 18.2, indicates that the areas covered by uniforms and personal protective equipment sustained relatively fewer burn injuries. Moreover, burns amongst the wounded soldiers mostly occurred on skin exposed to flash fires.[1] Shafir et al.[16] have analysed the distribution of burns amongst burned tank crewmen and reported that the tank crewmen who used

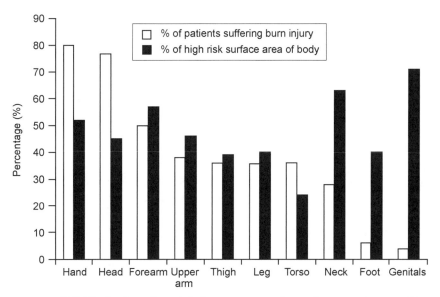

18.2 Distribution of burn injuries amongst combat and non-combat casualties in Iraq and Afghanistan.[15]

fireproof coveralls, suffered burn injuries on most of their forearms. They observed that 2.5% suffered hand burns despite wearing Nomex aramid gloves, whereas 25% sustained hand burn injuries with exposed hands. In general, the high-risk areas include the hand, forearm, neck, head and genitals. Burns to the hands and face are unavoidable unless fire resistant gloves and hoods are used. Such burns can have significant long-term morbidity and functional consequences.[17] The main factors that influence burn injures are:[18]

1. the incident heat flux intensity and the way it varies during exposure;
2. the duration of exposure (including the time it takes for the temperature of the garment to fall below that which causes injury after the source is removed);
3. the level and effectiveness of insulation between heat source and skin;
4. extent of degradation of garment materials during exposure;
5. condensation on the skin of any vapour or pyrolysis products released as the temperature of the fabric rises.

A separate study[17] revealed that most serious burns often involve the ignition of clothing and often result in more serious burn injuries than if no clothing had been worn, particularly when the initial fire is small or of short duration. In the case of clothing being worn next to the skin, ignition of the clothing results in a higher total body surface area (TBSA) burn, which increases the full thickness skin (or 3rd degree) burns thus resulting in a higher mortality rate of almost four times that of the burn victim involved in fire other than garment fires. Contamination of clothing with liquid fuels, oils or flammable liquids is also a major problem since it increases the risk of the garment igniting on exposure to fire or flash fires.

Historically, military burn injury statistics have shown that the extent of the burns suffered by service personnel was determined by the quality of clothing they happened to have on at the time of the attack. For example, the Japanese attack on Pearl Harbour in 1941 resulted in about 60% burn casualties with some of the men suffering as much as 80% of the body surface area. Most of the burns were extensive and superficial and those with the least amount of clothing suffered the most extensive burns. The correlation between the amount of uncovered body surface and the amount of body surface affected was strikingly high. All the doctors who reported on the Pearl Harbour burn cases remarked upon the protection that clothing offered against the so-called flash burns.[19]

Military burn injuries have a broad impact that ranges from individual patients to the overall status of military operations. In a combat situation, a loss of a particular soldier from his unit can jeopardise its mission and with a large number of burn casualty victims, the failure of combat mission is certain. Depending on the anatomical location, even a minor burn can severely handicap the combatant, thus making them non-functional to their units. Highly sophisticated and extensive burn care is required if the military service men/women are to survive and retain combat effectiveness.[17] Moreover, long hospitalisation periods and prolonged absence from the military services can be very taxing on medical facilities and military personnel.

18.4 Clothing protection for military personnel

Textiles and their composites are extensively used in protecting military personnel from fire. Clothing can be a barrier to some types of burns due to inherent fire resistant properties or by the ability to trap insulating air between clothing layers. The insulating properties of textiles and their ability to hold air pockets within the fibre and the fabric construction make it an ideal barrier material to protect skin from intense heat and flames. However, ignition of clothing items may exacerbate an injury. Uncontrolled and unpredictable ignition of fabrics usually takes place when the fabric is subjected to heat of high intensity for short durations. The fabric undergoes an initial pyrolysis with the production of both volatile and non-volatile products. When the decomposition products are oxidised, more heat is generated which, in addition to the heat supplied, is transferred to the skin causing severe burns.

Clothing can be designed to provide very effective thermal protection provided the required properties do not impair military efficiency in various environments. It is, however, difficult to balance comfort, durability and burn protection in the design of military clothing. The thermal protection provided by a clothing ensemble will be determined by several factors such as:

- colour and reflectivity,
- composition,
- intrinsic fire retardant properties,
- thermal conductivity,
- ease of ignition and flammability,
- contamination with combustible materials such as oil and most importantly
- air trapping within or between layers.

From a thermodynamic viewpoint, the protective character of a fabric is a function of the quantity and intensity of heat supplied and generated thereafter, the area over which the heat is effective and the coefficient of heat transfer.[20] However, depending on the coefficient of heat transfer of the fibres, the fibrous nature and the woven construction of the fabric contribute to poor thermal conductivity. To render the garment protective, it should limit the amount of heat supplied to it, secondly the fabric should not support spreading of combustion area and finally for the garment to be fully protective, the fabric should resist the heat transfer to the body or to any flammable material in close vicinity.

Fireproof garments currently used in many armed forces are made of single layer aramid material (e.g., Nomex, DuPont). The material itself does not sustain flame once an ignition source is removed and burns with difficulty under continuous flame exposure; therefore it does not contribute significantly to the severity of thermal injury due to burning of clothing. However, due to the high cost of these and similar materials they are used only for specialised applications such as coveralls for tank crewmen or to protect service personnel involved in military operations with potential for catastrophic fires. The low cost, lower performance flame retardant treatments applied to more conventional fibres as finishes, flame retardant-modified synthetic and regenerated fibres, use of fibre blends and alteration of fabric construction and/or garment configuration are often preferred.

18.5 Existing flame retardant solutions for military uniforms

Flame retardant clothing has been available since early in the Second World War. Flame retardant treatments available during this period were only temporary or non-durable treatments using various combinations of ammonium salts of sulphuric, hydrochloric, or phosphoric acids to impart flame resistance on to cotton fabrics in the main. Synthetic flame resistant fibres were developed from the 1950s onwards and high performance flame retardant fabrics comprising inherently flame resistant fibres such as the meta- and para-aramids, polybenzimidazole (PBI), semi carbon (oxidised acrylic) and phenolic (novaloid) appeared from 1960 onwards. Some of the commercial flame retardant materials used in military clothing are discussed in context with their specific properties and applications.

18.5.1 Durable flame retardant treatments for 100% cotton

Cotton fibre is very popular in apparel fabrics due to its low cost and comfort properties. Two major flame retardant treatments are widely used for cotton fibres. Proban (Rhodia, formerly Albright and Wilson) and Pyrovatex (Ciba) processes are described in details elsewhere.[21] While Proban forms an insoluble polymer in the fibre voids and the interstices of the cotton yarn that is held mechanically in the cellulose fibres and yarns, Pyrovatex process chemically bonds the flame retardant substance to the cellulose fibre. The Proban process uses phosphorus containing tetrakis (hydroxymethyl) phosphonium chloride (THPC) which is reacted with urea. The reaction product is padded onto cotton fabric and dried. The fabric is then reacted with ammonia and finally oxidised with hydrogen peroxide. Pyrovatex process involves application of flame retardant with a crosslinked resin and curing of the fabric at high temperature of 160 °C. Examples of commercially available flame retardant cotton fabrics using aforementioned durable flame retardant treatments are described below.

Proban®

This is the first and principal example of the use of THPC-condensate/ammonia cure process. Proban treated cotton or polycotton fabric is used for normal working clothing for Royal Navy personnel. Owing to the ever-present risk of fires on warships, even during peacetime, the Royal Navy personnel use flame retardant clothing. During military operations, when the risk of weapon flashes, explosions and fires is increased, the Royal Navy personnel wear Action Dress (AD) which consists of Action Coverall made of two-layered Proban treated FR cotton, anti-flash hood and gloves.[22] Proban finished fabrics retain their flame retardant properties for the lifetime of the garment. Fabrics acquire their flame retardant properties from the polymer which is embedded into the fabric which can only be removed by powerful oxidising agents. When exposed to flame, Proban fabrics form an insulating char which stays in place and helps protect the wearer. These fabrics do not smoulder, have no afterglow, do not melt and the flame does not spread outside the charred area.

Indura®

This is also a Proban processed 100% cotton. Such exceptionally flame retardant phosphonium finished cotton fabrics had been used for NASA flight uniforms to be used in enhanced oxygen atmospheres.[23]

Antiflame®

This is an example of Pyrovatex processed 100% cotton. The fabric is treated with dialkylphosphonamide flame retardant followed by heat curing. Chemically bonded flame retardant guarantees flame retardancy for an infinite number of wash cycles as long as the wash and care instructions are followed.

18.5.2 Durable flame retardant cotton blends

In the past, military uniforms in the US were made from the so-called Battledress Uniform (BDU) fabric which was essentially nylon/cotton (50/50) blend twill fabric. Initially the fabric was not flame retarded due to lack of suitable treatments. It was difficult to impart durable flame retardancy to nylon and nylon-containing fabrics due to its low reactivity and poor penetration of a finishing solution into the fibres. Flame retarding nylon fibres during the fibre-forming stage using flame retardant additives such as organophosphorous and halogenated aliphatic/aromatic compounds is not possible owing to the reactivity of the polymer melt towards these additives encountered during the spinning process. Flame retardant finishing of nylon fabrics is another approach but it has not been exploited commercially because of poor durability. A number of patents have been filed on flame retardant treatments of nylon/cotton blends.[24–26] Recent research on flame retardancy of BDU fabrics suggested the use of a commercial melamine-formaldehyde product with a trimethylol-melamine in combination with a commercial product Fyroltex HP (Akzo) based on oligomeric phosphate-phosphinate species which imparts high levels of flame retarding performance and laundering durability.[27,28] For durable flame retardant cotton blends, either cotton must be in the majority (>50%) and so a durable cotton FR (e.g., Proban/Indura) finish is possible or it is in the minority (<50%) with an inherently FR fibres, e.g. modacrylic. Some of the commercial cotton blends with durable flame retardant treatments used in military uniforms are discussed below.

Indura Ultra Soft®

These fabrics are constructed with a 75% cotton/25% nylon warp and a 100% cotton weft thus resulting in an overall blend of 88% cotton and 12% nylon. The fabrics are finished with the ammonia cure process and are designed to provide increased abrasion resistance and guaranteed flame resistance for the life of the garment. The nylon blend is completely absorbed by the majority of cotton fibre when exposed to thermal flux. Thus the nylon does not flow or lead to skin contact.

Cotton: Polyester (65:35)

This provides limited protection from a flame and is used in military uniforms where the wearer is not exposed to a severe fire hazard.

Valzon®

This is a blend of 60% flame retardant acrylic fibre and 40% cotton. The flame retardant acrylic fibre is treated with a flame retardant during the fibre-forming process. The cotton is not treated for flame resistance but derives its self-extinguishing characteristics in the presence of flame retardant acrylic fibre.

18.5.3 Inherently flame retardant synthetic materials

While this group of fibres has been comprehensively reviewed elsewhere,[12] the salient features of more important examples are presented below.

Nomex (Du Pont)

This is an inherently flame-retardant fabric having a meta-aramid chemistry and is predominantly used in military clothing systems to provide protection from intense heat and flame. Similar poly (meta-aramid) fabrics are used in tank crew coveralls, submariners' action dress shirt and trouser, aircrew coverall, aircrew anti-G trouser, aircrew life preserver vest, aircrew immersion coverall and explosive ordnance disposal suits. Nomex has good thermal stability and does not melt with the fibre decomposing between 370–430 °C. When exposed to high heat fluxes, the Nomex fibre consolidates and thickens thus preventing exposure of skin to the incident heat flux and hence second and third degree skin burns. One of the most effective ways to reduce the incidence of such skin burns is to make sure that the barrier of protective clothing between the heat source and skin remains intact during exposure. This is called 'non-break-open protection' or, 'break-open resistance'. For this reason, Nomex is often blended with the para-aramid Kevlar (Du Pont). The higher temperature resistance and strength of para-aramid fibres enables these swollen fabrics to remain intact. The principle commercial blend is Nomex IIIA which is a blend of 93% Nomex, 5% Kevlar, and 2% static dissipative fibre. Garments made of Nomex® IIIA are inherently flame resistant, extremely durable and can be expected to perform well for many years in military applications where its longevity can be utilised. The presence of 5% Kevlar provides added protection both in terms of flame resistance and fragmentation resistance. Inclusion of the proprietary antistatic fibres provides permanent antistatic characteristics. For more cost-effective and comfort characteristics, Nomex can be blended with wool which is a natural flame retardant fibre or a synthetic flame retardant viscose. Nomex blended with FR viscose (35% Nomex/65% FR viscose) is also widely used in military uniforms.

Kermel

This is a polyamide-imide fibre used in heat and fire protective clothing and having comparable properties in many respects to poly (meta-aramid) fibres. 100% Kermel fabric is often used where high resistance to physical damage is required while fire fighting. 100% Kermel® fabric made from long staple fibres is particularly suitable for the outer layers of multi-layered outfits. The fabric does not burn or melt and maintains its chemical characteristics. For these reasons, flame resistant fabric made from Kermel® fibre blends is suitable for general combat uniforms across the globe. Due to its suppleness and intrinsic softness, Kermel fibre allows the manufacture of high-quality fire resistant knits contributing to improved protection against thermal hazards. Kermel® also offers a range of knitted fabric options including underwear. In order to prevent ignition due to sparks, Kermel can be blended with 1% of antistatic fibre.

When blended with flame retardant viscose fibres, Kermel®-FR Viscose blends provide good protection against the hazards associated with electric arcs. The blend is inherently non-flammable, antistatic and durable, providing long-lasting protection. Depending on the end use and specific requirements, the blend ratio can be varied. Examples of such blends are Kermel VMC40 (40% kermel, 60% FR viscose) which the French Army has selected as a part of its Future Soldier Solution Programme.[29] In addition, Kermel V50 fabric made from 50% Kermel and 50% flame retardant viscose is specially intended for manufacturing camouflaged outfits. Improved comfort due to the flexibility and softness of the blend, Kermel V50 is claimed to be ideal for desert combat situations. In addition to the physiological comfort, the fabric also guarantees protection against exposure to heat and fire flashes. For increased fire protection, the Kermel content in the blend can be increased up to 70%. Kermel V70 (70% Kermel: 30% FR viscose) has reduced weight as compared to Kermel V50 and has a better balance between comfort and fire resistance.

One fibre having very high fire and heat resistance and yet very high levels of moisture absorption and comfort is poly (benzimidazole).[12] However, it is expensive and has a metallic brown or bronze colour. When blended with the para-aramid Kevlar, it creates a more cost-effective, high performance PBI Gold branded product. PBI Gold is a 60% Kevlar aramid and 40% poly(benzimidazole) (PBI) (Celanese) blend fabric having inherently flame retardant properties which cannot be degraded by laundering. The fabric does not ignite and does not melt. It has excellent thermal stability and retains fibre integrity and suppleness when exposed to flame. It is used in turnout gear for professional fire fighters and air force flight suits. When tested for personnel protection in simulated aircraft accident fires, flight suits made from PBI Gold fabrics resulted in less than 10% body surface area (BSA) being burned.[30]

A final generic group are the novoloid fibres, typified by Kynol (Kynol, Japan) either alone or in combination with other fibres and materials which are used in protective clothing to provide protection from acute risk of flame or fire, as well as insulating protection from intense heat. When exposed to a flame, Kynol does not melt at any temperature but gradually chars until it is completely carbonised.[31] By virtue of its non-melting property, the charred material retains its integrity as a barrier, thus keeping the heat away from the skin. However, like PBI it has an inherent metallic brown colour which often favours its use in interlinings or as barriers.

18.5.4 Garment design and fabric construction

Despite the use of inherently flame retardant fibres, military clothing still might not pass designated flame/thermal tests. The fabric properties are derived not only by the fibre type but also the fabric construction. For example, researchers at the US Army Natick Soldiers Centre compared Nomex interlock knit fabric (9–10 oz/yd^2) with Nomex Simplex knit. They found that despite interlock knit being more open than the simplex knit, the heat transfer properties determined by measuring thermal resistance of the fabric[32] of the interlock knit were not as good as those of the simplex knit. Furthermore, the interlock knit had different stretch characteristics which did not meet the military requirements.[33]

A single layer of fire-proof garment lacks the characteristics of thermal insulation in terms of maximum air entrapment, and insulation properties are comparable to those of regular cotton clothes. Therefore, when the fire resistant garment is subjected to a fire and the garment is pressed against the skin, localised flame or contact burn injury occurs. Thus single-layer fire resistant garments fail to prevent local thermal injury and so multilayered composites which entrap air are preferred. Most recently designed fire resistant garments have air dispersion systems sewn inside them so that when a direct flame stream hits the multilayered fireproof garment, the heated air expands blowing the garment away from the skin thus reducing the extent of burn injury dramatically.[34] Ideally, a clothing system should provide the necessary insulation only when thermally challenged by flame and heat, and under normal circumstances offer the minimum thermal insulation concordant with comfort requirements.[35] This requirement can be achieved by introducing an expanding air gap between two layers of fabric whenever the challenge occurs. The protective spring can be installed into any double fabric garment, such as in Royal Navy's Action Coverall, by attaching one end of the protective spring to the inner fabric (see Fig. 18.3) and the other end to the outer fabric. Research programme at Defence Clothing and Textiles Agency, Science and Technology Division developed a cone calorimeter test to determine the time at which an individual wearing clothing assembly with and without SMA spring would suffer a second degree burn when exposed to range of heat fluxes. Furthermore, a conductive disc could be attached to the outer end of the spring to increase conduction of thermal energy to actuate shape memory alloy and a second insulating disc attached to the inner end of the spring to improve burn protection. Such a system would prove effective for preventing localised burns.

Increased protection can also be achieved by increasing the number of layers and the air entrapped between them. In a layered configuration, the outer-shell fabric should offer high radiant protection, sufficient strength after heat exposure, high durability in terms of abrasion and should be lightweight. Three-dimensional structures such as in spacer fabrics[36] are ideal materials to be used

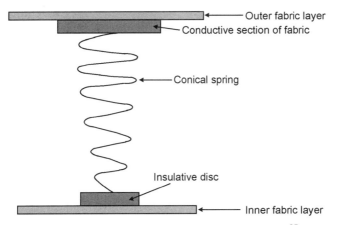

18.3 Protective spring assembly in a double fabric garment.[35]

Table 18.1 Safe exposure limits for military flame resistant clothing using instrumented manikin (ASTM F 1930)[37]

Configuration → Layer ↓	1	2	3	4
One	T-shirt, briefs Cotton	Long underwear	Nomex long underwear	Nomex long underwear
Two	Coverall (Nomex)	Coverall (Nomex)	Coverall (Nomex)	Coverall (Nomex)
Three			Jacket (Nomex)	Bib_overalls (Nomex)
Four				Jacket (Nomex)
% Body burn	Less than 20% At 3 s	Less than 20% At 4 s	Less than 20% At 6 s	Less than 20% At 10 s

in their normal uncompressed state as lining materials. These spacer materials offer high thermal protection when in their uncompressed state, but lower thermal protection when compressed. However, the high cost associated with such fabrics limits their use as lining materials and so battings with a comfort filament face cloth often provide sufficient thermal protection at a more economic rate.

Winterhalter et al.[37] have studied safe exposure limits for different garment configurations and found that those with a higher number of layers gives a higher safe exposure time limit. Safe exposure limits for various military flame retardant clothing are shown in Table 18.1. This garment configuration with four layers gives a safe exposure limit of 10 s with only 20% body surface area suffering from first or second degree burns. However, the increased number of clothing layers increases thermal load on the wearer and heat stress thereby impairing the wearer's ability to perform their tasks effectively and efficiently.[35]

18.5.5 Nonwoven fabrics

Nonwoven composite fabrics are also being considered for incorporation in military uniform, heavy-duty shelter, tentage and equipage applications where enhanced fire resistance properties, high strength, softness, improved abrasion resistance and other related properties are required. The nonwoven fabric composite includes the outer layer of nonwoven fibres, a semi-permeable membrane or additive that imparts flame retardancy to the composite fabric and an inner layer of nonwoven fibres contacting skin which is usually moisture absorbent to give comfort to the wearer. For example, the flame retarded nonwoven apparel fabric developed by The United States Marine Crops

(USMC) is self-extinguishing, non-drip and forms char when subjected to heat/ flame. Fibres explored at the experimental stage include staple polyester with flame retardant nylon 6, nylon 66 and viscose rayon. Nonwoven fabrics may have advantages over woven fabric in that they may comprise finer individual fibres which provide better insulation, filtration and barrier properties as opposed to coarser fibres and twisted yarns present in woven fabrics which contribute towards poor insulation and barrier properties. Durable nonwoven fabrics have an added advantage over woven fabrics of being able to integrate blends of various fibres to impart fire and thermal resistance, water absorbency or repellency and antimicrobial properties in cost effective manner.[38]

18.6 Flame retardant textiles: military applications

18.6.1 Non-clothing applications

With the advancement in technical textiles, a wide range of properties can be assimilated within the textile structures. Owing to their lightweight and low bulk properties, fire resistant textiles have been used extensively in non-clothing items such as tents, shelters, inflatable decoys, covers, nets, load-carriage items and sleeping systems. Shipboard blankets used in navy should be flame resistant and should not melt or drip when exposed to flame. Jeep side-curtains, truck traps and collapsible military storage tanks which can contain up to 800 000 litres of fuel are also examples of applications where fire resistant properties are essential. Fire resistant textile materials find a wide range of applications in the armed forces in the form of woven, knitted, nonwoven, coated, laminated and other composite forms. Almost all branches of the armed forces use architectural fabrics in tension membrane structures to provide dehumidified warehousing space for military equipment as well as military personnel (see Figs 18.4 and 18.5). Vinyl-coated synthetic fabrics which utilise nylon or polyester yarns as the base fabric and PTFE (Teflon) coated fibreglass are the textile materials predominantly used in air-supported and tension membrane structures. Very recently, general service military tentage fabric has been made of polyester/ cotton core-spun base fabric coated with a mixture of polyvinyl chloride resin, antimony oxide for synergising flame inhibition and pentachlorophenyl laurate (PCPL) as rot-proof agent.[39] Owing to the detrimental effects of PCPL on the environment, future tentage material will be made from 100% polyester-coated fabrics which do not require rot proofing, although, of course the use of antimony oxide may still be questioned on environmental grounds.

The primary concern, for the textile materials used in such membrane structures, is the fire performance property of the fabrics employed. For example, in the US the building code agencies have stated four primary standards to evaluate the fire performance characteristics and these have been widely accepted by the military fire protection associations:

18.4 General purpose tent system. (Picture courtesy: Seaman Corporation.)

1. NFPA-701: 'Flame-Resistant Textiles, Films', Including Federal Standard Method 5903, and the California State Fire Marshal's Test;
2. ASTM E 84: 'Standard Test method for Surface Burning Characteristics of Building Materials', involving a horizontal tunnel test;
3. ASTM E 108: 'Standard Test Method of Fire Tests of Roof Coverings';
4. ASTM E 136: 'Standard Test Method for Behaviour of Materials in a Vertical Tube Furnace at 750 °C'.

In view of the fact that textiles burn vigorously in a vertical orientation, the structural fabrics are required to pass the flammability test in a vertical position as would exist for the wall areas of a building. In the NFPA-701 test method, a Bunsen burner flame is applied to 25.4×6.9 cm specimen for 12 s. For the fabric to pass the test, it must self-extinguish within 2 s after removing the flame. The ASTM E 84 tunnel test evaluates the flammability of a membrane fabric on

18.5 NATO dehumidified First Alert warehouse. (Picture courtesy: Seaman Corporation.)

the underside when in the horizontal position, as would exist for the roof area, for flame spread and smoke formation during a 10 min fire exposure. Moreover, military bases are often situated in isolated areas where wind can have significant impact in spreading small fires, the ASTM E 108 test method evaluates the flammability of a membrane fabric and its resistance to penetration by fire on the outer surface. Fabric for membrane structures also needs to be classed as a non-combustible material according to the ASTM E 136 test method which measures flaming, temperature rise and weight loss.

A systematic study[40] on actual fire situations involving fabric membranes have shown that the fabric membranes that meet NFPA 701 do not support combustion nor do they propagate fire. Furthermore, due to their ability to melt away and open up, they vent dangerous smoke and fumes and rapidly dissipate heat away from fire area. This helps to minimise the structural damage and provides quick and easy access to fire fighters.

18.6.2 Clothing applications

After the war between Lebanon and Israel in 1982, army tank crews and aviators of all services in the military are required to wear flame resistant clothing. The flame retardant clothing gives the tank crew and aviators more time to escape with a reduced extent of injury when such vehicles are ablaze. Owing to the changing nature of war, as mentioned earlier, the army infantry also need flame resistance in their uniforms. The US Army tank crew and aviators wear aramid blend protective clothing; however, such materials are too expensive to be issued to infantry soldiers. Use of improvised explosive devices (IED) in the Iraq War has forced the commander of the US Marine Corps Expeditionary Force in Iraq to ban the wearing of polyester and nylon undershirts even while conducting personnel activities outside the military bases since these types of clothing melt when ignited and cause severe burn injuries.[41]

Service personnel on board navy vessels ideally wear flame retardant utility uniforms since they undoubtly face a major hazard of fire on board. Firstly, because the fires in the small compartments of such crafts often burn hotter than other fires[42] and secondly there is nowhere to escape a fire that breaks out on a sea-bound ship. In the UK, the normal Royal Navy clothing, even during peacetime, is made from Proban-treated flame retardant cotton or a polyester/cotton blend fabric. The Royal Navy fire fighters are provided with a newly developed two-ply hood made from 20:80% polybenzimidizole and permanent flame retardant viscose. For specific military operations, a range of flame resistant underwear are used which includes, balaclava, helmet cap, long underwear, turtle-neck shirt, mittens and socks. Some of the clothing systems which require fire resistance are described in the section below.

18.7 Types of military clothing

Nowadays, military personnel are equipped with clothing systems designed to resist a number of hazards including heat and fire. Examples of these are presented below.

Multi-climate protection system[43]

This was developed by the US Naval Air Warfare Centre (NAWC) to enhance the safety, comfort and performance of aviators and aircrew. This modular garment system is made using materials specifically developed to provide flame resistance, moisture management, insulation and wind and water protection. The multi-climate protection system (MCPS) uses seven garments in five layers as shown in Fig. 18.6. The garments are made from flame resistant fabrics specifically engineered for this application. Since this layer is next to skin layer, the fabric must be breathable and flexible. In order to accommodate these properties, the fabric is usually a composite structure. The composite fabric is designed in different ways to achieve desired levels of safety and comfort. For example, the GORE-TEX® fabric is constructed by laminating a highly breathable membrane with flame resistant Nomex fabric. The membrane is made of ePTFE integrated with an oleophobic substance that allows moisture vapour to pass through, but creates a physical barrier to contaminating substances such as oils, cosmetics, insect repellents and food substances. The Polartec® fabrics used for similar application use different approach to achieve durable flame resistance and superior moisture management. Polartec® fabrics comprises a durable

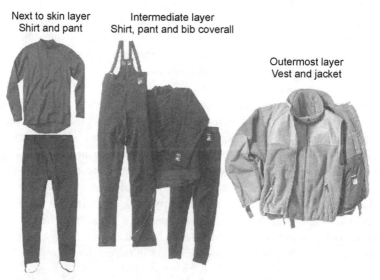

18.6 Different garment layers used in multi-climate protection system (MCPS).

nylon outer layer which offers wind and abrasion resistance. The flame resistant inner layer is made from Nomex fabric which provides flame resistance and prevents melting or dripping next to skin while keeping the wearer dry, warm and comfortable.

The second and third layers in the MCPS are mainly insulating layers. The garments in this configuration are made from fabrics that provide thermal comfort, breathability as well as fire resistance. For construction of such fabrics, textured fibres are used and the weave of the fabric is selected such that numerous air pockets are created within the fabric structure to trap air which can retain body heat. Such fabrics offer superior warmth, are lightweight and are quick to dry.

The fourth and the fifth layer (which is the outermost layer) garments in the MCPS are made from fabrics with flame resistant Nomex fibres. The yarn as well as fabric construction is extremely tight such that it imparts high wind resistance and warmth. The velour structure of the fabric imparts tight knitted construction which imparts high wind resistance and warmth. The lofted fibres on the inner face of the fabric entrap air and retain body heat. Such engineered fabric is highly durable, non-pilling and maintains its insulating properties even after repeated laundering.

Flame Resistant Organisational Gear[44]

The main aim of the Flame Resistant Organisational Gear (FROG) is to produce clothing that will provide protection against flames from improvised explosive devices, heat and flashes of 427 °C to the face, neck, upper torso and extremities for at least 4 s. The intent is to procure a system of clothing consisting of, but not limited to the following items: gloves, face protection, undergarments, and a combat shirt and trouser ensemble. This capability is intended to increase survivability and mobility of marines and sailors supporting combat operations. The long-sleeve t-shirt will increase torso and arm protection and the balaclava will increase face and neck protection. The glove features increased protection of the hand and wrist. The combat shirt and trouser ensemble is expected to provide a higher level of flame resistant performance over the current utility uniform and under garments. The fabrics used in all items are blend of many flame resistant fibres. When exposed to flames for 4 s, such fabrics will self-extinguish and do not melt or drip. Laboratory testing has shown that the outer garment layers provide the first and most effective defence against burn injuries.

Army Combat Uniform

A fire-resistant Army Combat Uniform (ACU) is designed to enhance fire protection without sacrificing mobility and comfort. This includes a fire-resistant Army Combat Shirt (ACS), which can be worn directly under flexible body armour, reducing the need for additional layers, and hence reducing heat stress while adding comfort and protection. The long-sleeved shirt also has a balaclava, which adds fire resistance to the head, face, and neck areas that previously were unprotected. With fire resistant eyewear and pants, soldiers will have full-body protection from burns. The fire resistant clothing is washable and maintains protective properties for the life of the garments.

Nomex Limitedwear for Military[45]

This coverall is designed to protect servicemen and -women from the dangers of flash fires. It is a lightweight, limited use garment that can be comfortably worn over most uniforms and can be vacuum packed for compact storage and mobility. This flexibility makes it ideal for mechanics, infantry or any person performing a task where the risk of flash fire exists. Since the coverall is made primarily with Nomex®, it is claimed to be breathable, does not melt or drip, dries quickly and can be machine-washed without compromising its inherent fire resistance protection. In simulated flash fire tests using DuPont's Thermo-Man, the Battle Dress Uniform (BDU) with NOMEX® Limitedwear coverall worn on top showed reduction in Total Predicated Body Burn by almost 80% when compared to BDU assembly without coverall.

First Attack Fire Suit[46]

This new fire suit, developed by the US Navy Clothing and Textile Research Facility (NCTRF) was driven by reduced manpower requirements for the new generation of Navy ships manned with fewer personnel on board. With reduced manning, about one-third of current ships, the Navy cannot afford to have many dedicated fire fighters and will need more personnel to assist in fighting fires. The First Attack Fire Suit has been developed as a lightweight and comparatively cheaper fire garment which can be easily deployed throughout the ship. Prior to deployment of newly developed fire suits, a range of unique materials were investigated at NCTRF.[47] Aluminised materials, three ends knit fleeces and conventional woven lightweight materials were selected for outershells. Conventional battings with different thicknesses and three-dimensional spacer materials were chosen as thermal liners. These materials were evaluated for their thermal protective performance, burn injury in a thermal environment as well as from moving through the fire using thermal protective performance tester and instrumented manikins in convective oven and dynamic full scale fire chamber respectively. Test results[46] suggested that aluminised material offered high radiant protection but lower convective heat protection and hence was rejected. Despite having high thermal protective performance, knitted material was rejected due to low knit durability. Lightweight woven material was selected as outershell material for First Attack Fire Suit as this material had higher durability and higher residual strength after heat exposure. Although spacer materials offered high thermal protection when in natural uncompressed state, this material was rejected due to high cost associated. Filament, warp-faced woven liner was selected as lining materials. The fire suit is designed as a coverall so that the clothing is less likely to get caught in the confined spaces shipboard, and it restricts hot air and gases that might enter with a two-piece garment. The garment designers are considering the adoption of building reflectivity into the outer-shell, to replace reflective stripes attached along the arms, legs and torso, thereby reducing the bulk and the weight of the fire suit.

The Flame Retardant Environmental Ensemble[48]

The US Army is currently working on a flame retardant environmental ensemble which will provide armored crewmen and aviation crew members with an updated multilayered, versatile, all-climate, flame-retardant system that will be adaptable for use in varying mission requirements and environmental conditions. The Flame Retardant Environmental Ensemble (FREE) system will consist of a base layer, a midweight underlayer, a lightweight outer layer, an intermediate outer layer, an extreme outer layer, cold weather gloves, balaclavas for hot and cold weather, a rigger belt, and wool socks. The outer layers will provide protection from wind and rain as well as fire.

Armoured Fighting Vehicle crewmen's clothing

The NATO Allied Combat Clothing Publication (ACCP)[8] defines characteristics necessary for Armoured Fighting Vehicle (AFV) crewmen's clothing. According to the ACCP, the outer layer should be flame resistant and self-extinguishing on removal of an ignition source with both convective/radiant flames of up to $200\,kJ/m^2$ and purely radiant thermal energy of up to $600\,kJ/m^2$. On self-extinction, the material of the outer layer should still be intact with a residual strength of not less than 25% of the original material strength, and should not have shrunk by more than 10%. The under layers should be substantial in thickness and should fully cover the body and limbs. Textiles in contact with skin should not be thermoplastic or comprise thermoplastic components. Furthermore, the compatibility between clothing items should prevent the exposure or reduced protection of neck, wrist, ankle and lower limbs. In addition to this the clothing system should provide comfort in NW European summer conditions. The NATO Standard Agreement (STANAG) in its guidelines for performance and protective clothing merely states that of the items of combat clothing system, the outer layer in particular should not be easily ignitable nor should propagate the flame readily.

18.8 Testing standards and methodologies

Flammability standards for military textiles are not clearly defined and different commercial suppliers to different armies across the globe use different standards and test methods to specify their products. Research and development centres in the US, UK and other developed nations engaged in the development of improved flame retardant clothing and non-clothing textiles for military applications have their own testing protocols. For example, suppliers to the US Army use ASTM, ISO or NFPA standards for specifying their products whereas in the UK and the European Union, harmonised standards such as BS EN or BS EN ISO are being employed. Since precise military standards for assessing flame resistance of flame retardant textiles are not evidently specified, most of the test methods used for measuring flame resistance are generally ones specified by the flammability standards governing performance requirements of protective clothing for fire fighters. The test methods for protective clothing in general are largely based on assessing the resistance of a fabric when tested in a specific geometry (e.g., horizontal, 45° or vertical) and subjected to a small flame igniting source, which usually is a small gas flame applied to the lower edge or face close to the edge of the sample. Parameters measured include, time-to-ignition, rate of flame spread, afterflame and afterglow following a prescribed ignition time (e.g., 10 s), extent of damaged/char or burnt fabric length or a combination of these.

Some of the standards for protective clothing are briefly summarised in Table 18.2, whereas mandatory tests for NFPA 2112 are described in Table 18.3. Other property evaluations or performance requirements for fabrics used in protective clothing include fabrics weight (ASTM D 3776), tensile strengths (Grab method: ASTM D 5034), tear strength (Elmendorf method: ASTM D 1424), material burst strength (ASTM D 3787), laundering shrinkage (AATCC 135), seam efficiency (ASTM D 1683), etc. In the UK and the EU, ISO standards are used for used for similar testing. For fire fighters' clothing in particular, the tensile strength (ISO 5081) should be $\geq$ 450 N whereas the tear strength (ISO 467/A2) should be $\geq$25 N. The dimensional changes when tested in accordance to ISO 5077 should be $\leq$ 3%. The seam efficiency of the garment is tested using hot plate test method (EN ISO 3146) and the seam must not melt at a temperature less than 260 ± 5 °C. While flammability requirements and test protocols for protective clothing in general have been extensively reviewed and reported[12,49,50] and are beyond the scope of this chapter, test methods for flash fires and high heat fluxes that are more common in an military environment will be discussed below.

Flammability test methods for protective clothing and specialised military uniforms, in particular, represent a defined military environment or fire scenario which are very complex. Combat clothing testing using Allied Combat Clothing Publication (ACCP-2) protocol[8] has been standardised through NATO. It requires combat clothing to be tested in 'worst case scenarios' with given thermal loads defined in the ACCP-2 protocol. An overall fire resistance determined under realistic real fire exposure conditions may be undertaken using an instrumented manikin yielding information regarding a complete clothing system's ability to resist heat flux and protect areas of the torso to first, second or third degree burn injury. Resistance to heat transfer by convective flame, radiant energy or plasma energy sources is quantified in terms of thermal protective index (TPI) often related the time taken for an underlying skin sample with or without an insulating air gap to achieve a minimum temperature or energy condition sufficient to generate a second degree burn. The most important aspect of military clothing testing is the evaluation of burn injury protection and thermal characteristics of clothing systems. Skin burn injury evaluation and subsequent modelling has been studied extensively[51–53] and has been recently reviewed by Song.[54] Of more relevance to this chapter is the thermal hazard and its laboratory simulation.

18.8.1 Bench-scale tests

Prior to full-scale garment testing using a manikin, the materials are tested for flammability properties using a range of bench-scale testing equipment.

Vertical flame tester

The apparatus measures flame resistance of textiles in a vertical orientation when subjected to a propane/butane flame for a specified length of time, usually 10–12 s (see Table 18.3). It also measures after flame, afterglow and char characteristics of materials. Melting and dripping can also be observed. BS EN ISO 15025:2002 (see Table 18.2), ASTM D 6413 and NFPA 2112 (Table 18.3) standards for fire-resistant garments use this type of bench-scale test method for measuring flame spread properties of fabrics.

Thermal performance test apparatus

This apparatus is used to evaluate flame and thermal protective performance at a heat flux simulating battlefield flame and thermal hazards. This portable flammability apparatus is the most advanced and fully automated compact bench-scale system. It can burn fabrics at multiple angles and at specified heat flux levels, the burn time and air gap between fabric and skin are computer controlled. The instrument measures skin temperature and predicts burn injury level almost instantaneously. Test method described (in Table 18.3) in ASTM F 1939 standard requires exposure of sample (127 × 178 mm) to the heat flux of a flash fire (84 kW/m^2) for specific length of time. The performance is measured using skin sensors that can predict temperature changes and burn injuries on the human skin surface resulting from heat penetrating the fabric sample. For fabrics

Table 18.2 Standards and performance requirements for protective clothing[12]

Standard Code	Standard title	Property measured	Performance requirement
ISO 2801:1998	Clothing for protection against heat and flame – general recommendations for selection, care and use of protective clothing	–	–
BS EN ISO 6942:2002 at 40 kW[2]	Protective clothing – protection against heat and fire method of test: Evaluation of materials and material assemblies when exposed to a source of radiant hear	Heat transfer (radiant)	$RHTI^*_{24} \geq 22$ s
ISO 9151:1995 BS EN 367:1992	Protective clothing against heat and flame – Determination of heat transmission on exposure to flame	Heat transfer (radiant)	$RHTI^*_{24} \geq 13$ s
ISO11612:1998 BS EN 469:2005	Clothing for protection against heat and flame – test method and performance requirements for heat-protective clothing	Heat resistance	No melting, dripping ignition
ISO/DIS 12127:1996 (BS EN 702:1995)	Clothing for protection against heat and flame – Determination of contact heat transmission through protective clothing or constituent materials	Contact heat transfer, Manikin (optional)	Defined by manufacturer/customer
ISO 17492:2003 Cor 1:2004	Clothing for protection against heat and flame – Determination of heat transmission on exposure to both flame and radiant heat	–	–
BS EN ISO 15025:2002	Protective clothing – protection against heat ans flame – test method for limited flame spread	Flame spread	No flame extending to top or edge, no hole formation and afterflaming and afterglow times ≤ 2 s

ISO 17493:2000	Clothing and equipment for protection against heat – test method for convective heat resistance using a hot air circulating oven		
ISO 5081 after BS EN ISO 6942 Method A at 10 kW/m^2	The determination of the breaking strength and elongation at break of woven textile fabrics (except woven elastic fabrics)	Residual strength	Tensile strength $\geq$ 450 N
NFPA 2112	Standard on Flame-Resistant Garments for Protection of Industrial Personnel Against Flash Fire	–	–
ASTM F 1930, ISO/DIS 13506	Test method for evaluation of flame resistant clothing for protection against flash fire simulations using an instrumented manikin	Burn injury prediction	–

Note: * RHTI = Radiant heat transfer index and ** HTI = Heat transfer index.

Table 18.3 Test methods and performance requirements for flame resistant garments meeting requirements of NFPA 2112

Property tested	Test method	Application of test method	Performance requirements
Thermal protective performance	152 × 152 mm of specimen is exposed to heat and flame source with heat flux of 84 kW/m^2. The amount of heat transferred through the specimen is measured using a copper calorimeter. The test measures the time taken to transfer amount of heat sufficient to cause second degree burn. This time multiplied by the incident heat flux gives TPP rating.	This test is used to measure the thermal insulation provided by garment materials. The TPP test uses an exposure heat flux that is representative of flash fire environment.	TPP rating of: 3 or more when tested in 'contact', simulating direct contact with skin and 6 or more when measured with 'spaced', simulating an air gap of 6.35 mm between the skin and the garment material.
Flame resistance	76 × 30 mm specimen is held vertically over a small flame for 12 s. Afterflame time, char length and length of tear along the burn line is measured. Melting and dripping is also observed.	This test is used to determine ease of ignition and ease of flame spread.	Afterflame time $\leq$ 2 s Char length $\leq$ 102 mm No melting or dripping
Thermal shrinkage resistance	381 mm^2 fabric specimen is suspended in a forced air-circulating oven at 260 °C for 5 min to determine amount of shrinkage. The specimen is examined for evidence of melting, dripping, separation or ignition.	The test measures resistance to shrinkage of a fabric when exposed to heat since this property is considered important in minimising the effects of a flash fire.	Shrinkage $\leq$ 10%

Heat resistance	Same as above	The specimen should not melt, ignite or separate when exposed to heat.
Manikin testing	Standardised coverall design placed on an instrumented manikin wearing cotton underwear is subjected to an overall heat and flame exposure averaging 84 kW/m² for 3 s. Sensors embedded in the manikin's skin predict occurenec of 2nd or 3rd degree burns. Percentage of body sustaining 2nd or 3rd degree burns is determined using computer program.	This test provides an overall evaluation of how the fabric performs in a standardised coverall design.	Body burn rating $\leq$ 50% Lower body burn ratings indicate greater protection provided by the fabric.
Thread melting resistance	The test involves soaking of the thread used in stitching FR garment in an organic solvent to extract substances that would interfere with the melting of thread. Melting temperature is determined by slowly heating the thread.	Measures the melting temperature of the thread used in flame-resistant garments.	Thread fails the test if melting temperature < 260°C
Label legibility	Sample label containing product information are subjected to 100 wash/dry cycles and then examined for legibility.	This requirement checks for label durability.	Label must remain legible from 0.3 metres.

used in flame resistant garments, a thermal protective performance (TPP) of 3 or more is required when tested in 'contact' condition, simulating direct contact with skin whereas TPP of 6 or more is required when tested in 'spaced' condition, simulating an air gap of 6.35 mm between the skin and the garment material. Similar bench-scale experimental set-up is used for test methods described in EN ISO 6942 and EN 367:1992 (see Table 18.2) for measuring radiant heat transfer index (RHTI) and convective heat transfer index (HTI) respectively. RHTI values of ≥ 22 s and HTI values of ≥ 13 s are required for materials used in protective clothing.

Combustion furnaces

In some research laboratories, thermal processes and combustion by-products are also evaluated using a vapour combustion furnace, drop tube furnace and porosometer.[55] Fire gases from combustion of military fabrics can be analysed to identify any toxic species present using a gas-chromatograph/mass spectrometer and other pieces of analytical equipment.

18.8.2 Full-scale tests

For many years the military used an outdoor fire pit to test and evaluate developmental flame protective clothing systems. However, the outdoor facilities were not very reliable, since their use is dependent on the existing weather conditions and moreover, it is difficult to maintain constant experimental conditions. A full-scale test for flame-resistant garments initially used a fully dressed manikin exposed to open-pit fuel fires. Nowadays the military uses a state-of-the-art instrumented manikin and an environmentally controlled chamber. Propane fire pits are normally used for testing test tents, stationary or walking manikins for the effects of flash fires on clothing systems. Propane is the industry standard fuel and is environmentally safe.

The manikin is equipped with 100–122 individual heat-flux sensors distributed over the surface of the body. The test garment is placed on the manikin at ambient temperature conditions and exposed to flash fire simulation with controlled heat flux, duration, and flame distribution. The sensors measure incident heat flux during and after exposure. The changing temperature of the human tissue at two skin thicknesses, one representing a second degree burn injury point and the other a third degree burn injury point is calculated.[56] Computerised data acquisition system also calculates surface heat fluxes, skin temperature distribution histories and predicts the skin burn damage for each sensor location. In ISO/DIS 13506 (see Table 18.2) standard method for protective clothing a dressed manikin is subjected to a full flame exposure with gas burner flames of about 800 °C and heat flux of about 80 kW/m^2 for 8 s. The manikin test method in ASTM F 1930 requires exposure of fully dressed manikin to the heat flux of $84\,kW/m^2$ for 3 s. According to NFPA 2112 standard (see Table 18.3), the performance requirement for the materials to be used in flame-resistant garments is that the body burn rating should be $\leq 50\%$ when tested in accordance to ASTM F 1930. The ASTM F 1930 test method is being recommended for adoption as the new military standard to evaluate military flame protective materials and clothing systems.

Carbon dioxide laser cell

Military uniform materials can be evaluated for their ability to protect against flame, laser and thermal radiation using a carbon dioxide laser. The energy associated with this type of laser is absorbed by most materials resulting in rapid heating. The ability to concentrate a laser beam into small area with controlled intensity and duration provides a method of localised heating of material.

18.9 Performance standards and durability requirements

The protective materials used for manufacturing combat clothing should be capable of protecting the wearer in the most extreme environmental conditions and more importantly, the military combat clothing must work and perform effectively even after many days and weeks of continuous use. Furthermore, durability to repeated laundering is important in addition to wrinkle resistance. All these requirements add up to significant challenges in fibre type selection, fibre size, dyestuff selection, material technical design and functional finish. Quiet often, different chemicals and processing steps needs to be incorporated during manufacturing military clothing to meet stringent military requirements. This presents a significant challenge when different chemistries and performance characteristics compete with each other.

Performance of flame retardant treatment under the conditions of use is equally important as the initial efficiency of flame retardant when tested indoor. This is particularly so for military textiles which are exposed to varied climatic conditions. For prolonged exposure times the durability of the flame retardant treatments are of major significance. For example, temporary water-soluble flame retardant treatments are suitable for fabrics that are used indoors and do not undergo leaching action by moisture vapour or water solutions. Prolonged exposures to moist air cause migration of flame-retardants and significant loss of flame resistance.

For textiles intended for use in marine uniforms or other marine applications, more complete evaluation of the durability of flame resistant treatment is required. These include measuring resistance of the flame retardant upon exposure to sea water, perspiration and climatic conditions by accelerated weathering or by actual indoor and outdoor storage. Fabrics could also lose their flame retardant treatment by ion exchange when exposed to mineral containing hard water or sea water. Battlefield uniforms require severe and repeated laundering. The uniforms are washed using commercial practices where soda ash is employed with soap solutions which results in serious impairment of any durable flame retardant treatment.

House and Squire[22] studied the effectiveness of Proban treated used clothing and found that used garments provided better protection due to physical changes to the materials during wear. They proposed that when garments are worn, the material layers and the fibres within the fabric structure rub-together to increase the thickness of the layers and hence the amount of air trapped within them. It is expected that material layers have an increased 'loft' due to rubbing action and the air entrapped within the layers provides increased thermal insulation and protection against fire until the garment catches fire.

18.10 Future trends

Flame resistant clothing is still the largest area of concern to military personnel. The fact that the threat from burns is unlikely to diminish and new developments in weapons seek to exploit this vulnerability, research and development is continuously directed to seeking solutions for new military situations. The US military in particular is undertaking major actions to provide the most effective level of fire protection. They are considering flame retardant fabrics for base layer garments through to full battle uniforms. Any new development that has a good chance of becoming standard issue is the one which when exposed to a flame; it will not burn in the flame or self-extinguish once the flame is removed and will also bring into play all the technical qualities such as antimicrobial, moisture wicking and anti-static properties.

For example, at Natick Soldier Research Development and Engineering Centre, there is a specific interest in developing a flame resistant elastomeric fibre that can be woven into an undergarment and wicks away moisture.[57] With the advances in the devolvement and exploitation of bi/tri-component fibres, it is conceivable to produce fibres that combine several of the desired characteristics. For example, a transparent fibre with appropriate additives or secondary components might sense and react to a chemical or physical change by sending an electronic message, advising the user of a hazardous situation. In addition, new cross-sectional geometries could improve the performance of bi/tri-component fibres for specific applications and research effort is needed to explore more fundamental issues such as delamination of the different polymers that could be used in developing multi-functional fibres, the use of compatibilisers to facilitate dispersion of additives in various fibre forming polymers and the use of 'migrating' additives (e.g., nanoclays) to modify desired areas of the fibre.

Nonwoven materials for potential use in combat clothing are also under consideration based on increasing cost-effectiveness and ease of fibre blending.

However, one of the important characteristics that is required for military use is the abrasion resistance, and nonwovens do not perform well in this respect. Furthermore, efforts are under way to develop and improve the performance of textile materials through nanotechnology.

A comparison of current capabilities versus battlefield requirements dictates interest in the following major areas of scientific knowledge and technological capabilities:

1. New, low cost fibres for clothing applications (woven, nonwoven, knit and batting fabric structures) which provide flame and thermal resistance without melt drip characteristics.
2. Improvements to existing fibres, e.g. incorporation of novel flame retardant chemicals, flame suppressors or char formers into conventional low cost fibres.
3. Novel concepts and approaches to integrate protection capabilities into materials and clothing systems. Such concepts should integrate flame and thermal protection with other protective capabilities such as environmental protection, thermal signature management, and electrostatic dissipation without significantly increasing weight.
4. Novel developmental flame resistant treatments, coatings and films that are moisture vapour permeable, lightweight and chemically compatible with a wide variety of substrate materials.
5. Testing methodologies and supporting instrumentation to characterise and evaluate the melt burn potential of thermoplastic fibre-based fabric in layered configurations at either the bench scale or a fully instrumented manikin system.

Clothing and life support equipment are effective only if worn in the manner for which they are designed. The health consequences of any soldier violating clothing rules are expensive in terms of lives and recovery costs from injury. A proactive programme focused on educating commanders, officers and soldiers on the dangers of wearing clothing that will not protect them in the event of fire, is a key to the prevention of thermal injuries. Obviously, a final research goal will be to create flame resistant textiles that require minimal aftercare and so maintain their performance for considerable time during arduous battlefield conditions.

18.11 Sources of further information and advice

Websites

There are several commercial and government websites that provide valuable information on flame retardant fabrics and clothing systems for military usage.

- www.kermel.com
- www.bulwark.com

- www.natick.army.mil
- http://stinet.dtic.mil/
- http://www.dstl.gov.uk/
- www.patentstorm.us
- www.defenselink.mil
- https://www.peosoldier.army.mil/

Organisations active in development of protective clothing and equipment

1. Defence Science and Technology Laboratory (DSTL): www.dstl.gov.uk
2. European Society of Protective Clothing: www.es-pc.org,
3. The U.S. Army Natick Soldier RD&E Centre, www.natick.army.mil
4. Combat Equipment Support Systems: http://www.marcorsyscom.usmc.mil
5. Norwegian Defence Research Establishment (FFI): http://www.mil.no/felles/ffi
6. European materials science and technology research institution (EMPA) http://www.empa.ch
7. Protective Clothing And Equipment Research Facility (PCERF): http://www.hecol.ualberta.ca/PCERF/index.html

18.12 References

1. Dougherty P. J. (1990), 'Armoured Vehicle Crew Causalties', *Military Medicine*, **155**, 471–420.
2. Scott R. A. (2005), *Textiles for protection*, Ed. Scott R. A., Woodhead Publishing, Cambridge.
3. Anon. (2005), Report Summary: 'Developments in Military Clothing', *Technical Textile Markets*, 62, 3rd quarter.
4. Crown E. M. and Capjack L. (2005), 'Flight suits for military aviators', Chapter 24, *in Textiles for Protection*, Ed. Scott R. A., Woodhead Publishing, Cambridge, 678–698.
5. Sochet I. and Gillard P. (2002), 'Flammability of kerosene in civil and military aviation', *Journal of Loss Prevention in the Process Industries*, 1, 335–345.
6. Nilsen T., Kudrik I. and Nikitin A. (1997), *The Russian northern fleet nuclear submarine accidents Bellona*, Report nr. 2:96.
7. Elton S. (1996), 'UK Research into Protection from Flame and Intense Heat for Military Personnel', *Fire and Materials*, **20**, 6, 293–295.
8. NATO (1992, June), *Protection from flame and heat resistance provided by AFV crewman's clothing*. NATO Allied Combat Clothing Publication ACCP- 2 (Edition 1) prepared by Sub-Group B on Materials, Mechanical Parts and General Engineering Working Group 5 on Combat Clothing and Personal Equipment; AC/301-N/1.
9. Kim I. Y. (2000, August), *Battlefield flame/thermal threats or hazards and thermal performance criteria*. Natick Technical Report/TR-00/015L. US Army Soldier and Biological Chemical Command, Soldier Systems Centre: Natick, MA 01760-5020.
10. Foster J. A. and Roberts G. V. (1994), 'An Instrument Package to Measure the Fire Fighting Environment: The Development and Results' FRDG Publication Number 4/

93, Home Office Fire Research & Experimental Unit.
11. McLean A. D. (2001), 'Burns and Military Clothing', *Journal of Royal Army Medical Corps*, 147, 97–106.
12. Horrocks A. R. (2005), 'Thermal (heat and fire) protection', Chapter 15, in *Textiles for protection*, Ed. Scott RA, Woodhead Publishing, Cambridge, 398–440.
13. Kauvar D. S., Wolf S. E., Wade C. E., Cancio L. C., Renz E. M. and Holcomb J. B. (2006), 'Burns sustained in combat explosions in operations Iraqi and enduring freedom (OIF/OEF explosion burns)', *Burns*, 32, 853–857.
14. Pruitt Jr. B A., Goodwin C. W. and Mason Jr. A. D. (2002), 'Epidemiological, demographic and outcome characteristics of burn injury', in *Total Burn Care*. 2nd edn. Ed. Herndon D.N., Philadelphia: WB Saunders, 19.
15. Kauvar D. S., Cancio L. C., Wolf S. E., Wade C. E. and Holcomb J. B. (2006), 'Comparison of combat and non-combat burns from ongoing US military operations', *Journal of Surgical Research*, 132, 195–200.
16. Shafir R., Nili E. and Kedem R. (1984), 'Burn Injury and Prevention in the Lebanon War', *Israeli Journal of Medicine Science*, 20, 311–313.
17. Baycar R.S., Aker F. and Serowski A. (1983), 'Burn casualties in combat: Need for protective garments', *Military Medicine*, 148, 281–282.
18. Rossi R. (2003), 'Fire fighting and its influence on the body', *Ergonomics*, 46, 10, 1017–1033.
19. Anon. (1946), 'Pearl Harbour Navy Medical Activities', *The United States Navy Medical Department at War, 1941–1945*. vol.1, parts 1–2 (Washington: The Bureau: 1–31.) http://www.history.navy.mil/faqs/faq66-5.htm.
20. Coppick S., Church J. M. and Little R. W. (1950), 'Thermal behaviour of fabrics at flaming temperatures', *Industrial and Engineering Chemistry*, 42, 3, 415–418.
21. Hall M. E. (2000), Finishing of technical textiles, in *Handbook of Technical textiles*, Ed. Horrocks A. R. and Anand S. C., Woodhead Publishing, Cambridge, 152–172.
22. House J. R. and Squire J. D. (2004), 'Effectiveness of proban flame retardant in used clothing', *International Journal of Clothing Science and Technology*, 16, 361–367.
23. Dawn F.S. and Morton G.P. (1979), 'Cotton protective apparel for the space shuttle', *Text. Research Journal*, 49 (4), 197–201.
24. Lunsford C. C., Riggins P. H. and Stanhope M. T. (2005), US Patent 6867154.
25. John H. H. (1989), 'Flame-resistant nylon.cotton fabric and process for production thereof'. US Patent 4812144.
26. Fleming G. R. and Green J. R. (1995), 'Long wear life flame-retardant cotton blend fabrics', US Patent 5468545.
27. Holme I. (2000), 'Finishes for protective and military textiles', *Technical Fabrics*, 59, 11.
28. Yang H. and Yang C. Q. (2005), 'Durable flame retardant finishing of the nylon/cotton blend fabric using a hydroxyl-functional organophosphorous oligomer', *Polym. Degrad. Stab.*, 88, 363–370.
29. http://www.worldsecurity-index.com/newsdet.php?id=1554.
30. Schulman, S. S. and Robert M. (1970, November), 'Nonflammable PBI fabrics for prototype air force flight suits', Accession Number: AD0880047.
31. Shishoo R. (2002), 'Recent developments in materials for use in protective clothing', *International Journal of Clothing Science and Technology*, 14, 3, 201–215.
32. Test data for Nomex and Kevlar: http://www2.dupont.com/Personal_Protection/en_HK/tech_info/kevlar_thermaltestdata.html.
33. Gomes C. A. (2006), 'Fabrics in the Military: material demands', *Industrial Fabric Products Review*, 91 (7), 48–53.

34. Salmon A. Y., Breiterman S., Chaouat M., BenBassat H. and Eldad A. (2003), 'Flame burn protection: Assessment of a new, aircooled fireproof garment', *Military Medicine*, **168**, 8, 595–599.
35. Congalton D. (1999), 'Shape memory alloys for use in thermally activated clothing, protection against flame and heat', *Fire and Materials*, **23**, 223–226.
36. Anand, S.C. (2006), Knitted three-dimensional structures for technical textiles applications. Proceedings of the International Conference on Technical Textiles, IIT, New Delhi, November 11–12.
37. Winterhalter C. A., Lomba R. and Tucker D. (2004), 'Novel approach to soldier flame protection', in International Soldier Systems Conference and Exhibition, https://www.dtic.mil.
38. Szczesuil S. (2004), Development of nonwoven fabrics for military applications, http://www.acq.osd.mil/osbp/sbir/.
39. Scott R. A. (2000), Textiles in defence, in *Handbook of Technical Textiles*, Ed. Horrocks A. R. and Anand S. C., Woodhead Publishing, Cambridge, 425–460.
40. Seaman R. N. (1984), 'Fire performance history of flame-retardant membrane structures', *Building Standards*, **53**, 1, 13–17.
41. Anon. (2006), 'The Iraq factor in textile technology', *World Sports Activewear*, **12** (6), 10–14.
42. https://www.natick.army.mil.
43. https://www.polartec.com.
44. https://www.marcorsyscom.usmc.mil.
45. https://www2.dupont.com.
46. https://www.navy-nex.com/command/nctrf/nctrf-index.html.
47. https://www.dtic.mil.
48. https://www.peosoldier.army.mil/highlights/coverall.asp.
49. Bajaj P. (2000), 'Heat and flame protection', Chapter 10 in *Handbook of Technical Textiles*, Eds. Horrocks A. R. and Anand S. C., Woodhead Publishing, Cambridge, 223–263.
50. Haase J. (2005), 'Standards for protective textiles', Chapter 2 in *Textiles for protection*, Ed. Scott R. A., Woodhead Publishing, Cambridge, 31–59.
51. Mehta A. K. and Wong F. (1973), *Measurement of flammability and burn potential of fabrics*, Report from Fuel Research Laboratory, Massachusetts Institute of Technology, Cambridge, MA.
52. Henriques Jr. F. C. (1947), 'Studies of thermal injuries V. The predictability and the significance of thermally induced rate processes leading to irreversible epidermal injury', *Archives of Pathology*, **43**, 703–713.
53. Stoll A. M. and Greene L. C. (1959), 'Relationship between pain and tissue damage due to thermal radiation', *Journal of Applied Pathology*, **14**, 373–382.
54. Song G. (2005) 'Modelling thermal burn injury protection', Chapter 11 in *Textiles for protection*, Ed. Scott R. A., Woodhead Publishing, Cambridge, 261–292.
55. Biberdorf C. (2004, July–August), 'Burn to learn: military thermal, flame testing consolidated in future on-site facility', *The Warrior*, 6–7.
56. Song G. (2007), 'Clothing air gap layers and thermal protective performance in single layer garment', *Journal of Industrial Textiles*, **36** (3), 193–205.
57. Anon. (2007, June), 'Future Warrior Technology Integration', US Army Natick Soldier Research Development and Engineering Centre.

19
Flame retardant materials for maritime and naval applications

U SORATHIA, Naval Surface Warfare Center, USA

Abstract: Naval ships and maritime vessels employ both non-combustible and combustible materials and components. One recent trend is the use of fibre reinforced plastics (FRP). Materials technology, in some limited sense, is driving the next generation of light weight and fast moving naval and maritime vessels. The use of plastics and polymer composites in maritime and naval vessels also raises the issue of the inherent vulnerability of plastics to common fire threats found in maritime operations. This chapter will review potential fire threats, fire safety regulations, relevant fire test methods, and approaches to material flame retardancy and fire protection. This chapter also provides flammability data for several conventional and advanced polymer composite materials including flame spread, smoke generation, heat release rates, fire resistance, and structural integrity.

Key words: fibre reinforced plastics (FRP), flame retardancy, flame spread, smoke generation, fire resistance.

19.1 Introduction

Naval ships and maritime vessels employ both non-combustible metallic (steel, aluminum) and combustible materials and components. Examples of combustible materials include polyethylene and polyvinyl chloride in tubes, pipes, and electrical cables; polyurethane or neoprene foams for mattresses and thermal or acoustic insulation; polyurea or epoxy based non-skid and paint coatings; and textiles in furniture and bedding items, etc. One recent trend is the use of polyester or vinyl ester-based fibre reinforced plastics (FRP), also referred to as polymer composites, which offer advantages over traditional materials with regard to high strength-to-weight ratio, corrosion resistance, low maintenance, and extended service life. Materials technology, in some limited sense, is driving the next generation of light weight, fast moving and manoeuverable naval and maritime vessels.

A significant concern in any application of plastics and polymer composites in confined occupied spaces of maritime and naval vessels is the possibility that an accidental (or deliberate) fire may ignite combustible materials.[1] After

---

The technical views expressed in this paper are the opinions of the contributing authors, and do not represent any official position of the US Navy.

ignition of combustible materials, flames and fire may spread, produce smoke and toxic fire gases, release heat, and suffer significant reduction in structural properties.

In order to ensure an adequate level of fire safety aboard ships, both maritime and naval regulatory bodies have specific fire performance requirements for the use of plastics and FRP materials. This chapter will review potential fire threats, fire safety regulations, relevant fire test methods, and approaches to material flame retardancy and fire protection. This chapter also provides flammability data for several conventional and advanced polymer composite materials including flame spread, smoke generation, heat release rates, fire resistance, and structural integrity. Sources for additional information are provided for the reader to further explore related topics.

19.2 Fire threat

Fires represent a serious hazard for ships, off-shore units, and high speed craft. They rank in third place in terms of costs to insurance companies, after collisions and grounding.[2] A review of the US Navy Safety Center database for naval vessels for a period from 1983–87[3] indicated that overwhelming majority of fires in ships occur in engineering spaces (47%), followed by supply spaces (16%), habitability spaces (11%) and aviation spaces (8%). The data further indicated that majority of fires occurred at shipyards followed by those at sea and in port.

A significant concern in shipboard application of plastics and polymer composites is the possibility that an accidental (or deliberate) fire may ignite the combustible material. This may result in the spread of flame on the plastic surface and also release heat and generate potentially toxic smoke. If this fire is in the compartment containing large exposed surfaces of combustible material such as insulation or cables, the localized incidental fire may become a larger fire involving the insulation or cables, which now becomes the fuel for the growing fire. Hazardous conditions in small pleasure boats and passenger ships can become rapidly untenable in the event of a fire unless controlled in early stages of fire development. Relatively low ceilings with concealed service lines, high fire load density and long and narrow corridors are some of the factors affecting fire growth and the survivability of occupants and crew members. In enclosed and confined spaces, such as in ships, the growing fire can lead to a 'flashover' condition. 'Flashover' is a term that is used to indicate the point during a fire when the internal temperatures in the upper regions of the compartment have increased to the point where the radiant energy from the hot upper layer spontaneously ignites all combustible materials within the compartment. Typically, this is on the order of 600 °C. If the ship structure is constructed from fibre reinforced plastic (polymer composite) and if the affected composite component is part of a primary critical structure, the structure may collapse when exposed to fire.[4,5] In order to ensure adequate level of fire safety aboard

Flame retardant materials for maritime and naval applications 529

ships, both maritime and naval regulatory bodies have specific fire performance requirements for the use of such materials.

19.3 Fire safety regulations

19.3.1 Maritime vessels

Fire performance requirements of maritime vessels in international waters are contained in the International Convention for the Safety of Life at Sea.[6] The Safety of Life at Sea (SOLAS) is a compilation of several International Conventions and contains detailed provisions for the use of materials in fire resistant divisions, internal bulkheads, claddings, and paneling.[7] The International Maritime Organization (www.imo.org) is the United Nation's (UN) maritime safety organization and is the primary regulatory body for all commercial shipping in the international waters.

The 1974 SOLAS Convention has been amended several times by means of resolutions adopted by IMO Maritime Safety Committee (MSC). In recognition of the growth of high-speed craft, IMO by resolution MSC.97(73) adopted the International Code of Safety for High-Speed Craft (HSC 2000).[8] This Code allows for use of non-conventional shipbuilding materials, defined as fire-restricting, in fire resisting divisions. By Resolution MSC.101(73), Maritime Safety Committee added tests for fire restricting materials and fire-resisting divisions of high speed craft. Fire-restricting materials are defined as those materials which have low flame spread characteristics, limited rate of heat release and smoke emissions. All of these requirements were combined into a single consolidated edition of SOLAS in 2004.[6]

Divisions of ships by thermal and structural boundaries and restrictions on the use of combustible materials are important principles which underlie the regulations in the SOLAS. Over the years, a number of fire test procedures and related guidelines have been developed by the Maritime Safety Committee and adopted by the Assembly of the International Maritime Organization (IMO). These fire test procedures were published by the IMO as a part of Assembly resolutions and compiled in its Fire Test Procedures (FTP)[7] for easy reference by ship designers, material manufacturers, surveyors of classification societies and the Administrations. The FTP Code[7] provides international requirements for laboratory testing, type approval and fire test procedures for the:

- **Part 1** – Non combustibility test: when a material is required to be non-combustible, it shall be verified in accordance with the test procedures in the standard ISO 1182.[9]
- **Part 2** – Smoke and toxicity test: where a material is required not to be capable of producing excessive quantities of smoke and toxic products or not to give rise to toxic hazards at elevated temperatures, the material shall comply with this part and be tested in accordance with ISO 5659, Part 2.[10]

- **Part 3** – Test for 'A', 'B', and 'F' class divisions: where products such as decks (overheads), bulkheads, ceilings, linings, windows, fire dampers, pipe penetrations and cable transits are required to be 'A', 'B', or 'F' class divisions, they shall comply with this part and be tested in accordance with resolution A.754(18).[11]
- **Part 4** – Test for fire door control systems: where a control system of fire doors is required to be able to operate in case of fire, the system shall comply with this part and be tested in accordance with the test procedure presented in the appendix to this part.
- **Part 5** – Test for surface flammability: where a product is required to have a surface with low flame spread characteristics, the product shall comply with this part and be tested in accordance with resolution A.653(16).[12]
- **Part 6** – Test for primary deck coverings: where the primary deck coverings are required to be not readily ignitable, they shall comply with this part and be tested in accordance with resolution A.687(17).[13]
- **Part 7** – Test for vertically supported textiles and films: where draperies, curtains and other textile materials are required to have qualities of resistance to the propagation of flame not inferior to those of wool of mass 0.8 kg/m^2, they shall comply with this part and be tested in accordance with resolution A.563(14).[14]
- **Part 8** – Test for upholstered furniture: where upholstered furniture are required to have qualities of resistance to the ignition and propagation of flame, the upholstered furniture shall comply with this part and be tested in accordance with resolution A.652(16).[15]
- **Part 9** – Test for bedding components: where bedding components are required to have qualities of resistance to the ignition and propagation of flame, the bedding components shall comply with this part and be tested in accordance with resolution A.688(17).[16]
- **Part 10** – Test for fire restricting materials for high speed craft: surface materials on bulkheads, wall and ceiling linings including their support structure, furniture, and other structural or interior components required to be fire restricting materials shall be tested and evaluated in accordance with the fire test procedures specified in resolution MSC.40 (64)[17] as amended by resolution MSC.90 (71).[18]
- **Part 11** – Test for fire-resisting divisions of high speed craft: fire resisting divisions of high speed craft shall be tested and evaluated in accordance with the fire test procedures specified in resolution MSC.45 (65).[19] Such constructions include fire-resisting bulkheads, decks, ceilings, linings, and doors. Materials used in fire-resisting divisions shall be non-combustible or fire restricting as verified in accordance with Part 1 or Part 10.

All vessels operating in US waters fall under the cognizance of the US Coast Guard. The requirements of the US flagged vessels are established by Section 46 of the US Code of Federal Regulations. The US Coast Guard (http://www.uscg.mil) has the responsibility of reinforcing the SOLAS and IMO HSC Code requirements. The requirements of the US Coast Guard are published (www.uscg.mil/hq/g-m/mse4) in various so-called Navigation and Vessel Inspection Circulars (NVIC).

19.3.2 Naval vessels

The primary regulatory body for the fire performance of materials in US Navy ships and submarines is the Naval Sea Systems Command (NAVSEA), also referred to as the Naval Technical Authority (NTA). In 2004, the American Bureau of Shipping (ABS) and the United States Navy formalized their established strategic partnership through a Cooperative Agreement. The ABS Guide for Building and Classing Naval Vessels, published in 2004, is now used for most new designs of navy ships.

MIL-STD-1623[20] provides the fire performance requirements and approved specifications for various categories of interior finish materials and furnishings for use on naval surface ships and submarines. This design criterion standard applies to materials for bulkhead sheathing, overhead sheathing, furniture, draperies and curtains, deck coverings, insulation and bedding applications.

19.3.3 Fire performance regulations for composites

Use of polymer composite materials, also referred to as fibre reinforced plastic (FRP), for structural applications in both maritime and naval vessels is a recent and growing trend. The high speed craft code[8] permits the use of fire restricting[18] polymer composites for fire-resisting divisions.[18,19] The US Navy also allows the use of polymer composite materials for topside construction in surface ships when they meet the fire performance requirements of NAVSEA Design Data Sheet DDS-078-1[21] summarized in Table 19.1. This design data sheet provides performance requirements and test methods for surface flammability, smoke generation, fire gas toxicity, fire resistance and structural integrity during fire. The use of polymer composites in submarines is governed by MIL-STD-2031.[22] This military standard, in addition to other requirements, provides acceptance criteria for cone calorimeter heat release rates at four different incident heat flux levels ranging from 25 to 100 kW/m^2.

19.4 Methods of reducing flammability

All organic polymers will burn when exposed to sufficient heat energy for a long enough time. The terms flame retardant, fire retarded, flame resistant, fire resistant, fire safe, fire proof, etc., express degrees of resistance of a material to the ignition source of specific severity.[23] Methods of improving flammability

Table 19.1 Summary of fire performance requirements for composite topside structure (DDS-078-1)[21]

Category	Test method	Criteria
Surface flammability	ASTM E-84, 'Standard Test Method for Surface Burning Characteristics of Building Materials'.	• Interior applications: Flame spread index: 25 max Smoke developed index: 50 max • Exterior applications: Flame spread index: 25 max
Fire growth	ISO 9705 'Fire Tests-Full-scale room test for surface products' Annex A, standard ignition source fire of 100 kW for 10 min and 300 kW for 10 min	• Net peak heat release rate less than 500 kW • Net average heat release rate for test less than 100 kW • Flame spread must not reach 0.5 m above the floor excluding the area 1.2 m from the corner with the ignition source
Smoke production	ISO 9705; 100 kW for 10 min and 300 kW for 10 min	• Peak smoke production rate less than 8.3 m^2/s • Test average smoke production rate less than 1.4 m^2/s
Smoke toxicity	ASTM E662, Specific Optical Density of Smoke Generated by Solid Materials. (Flaming and non flaming mode)	• CO: 600 ppm (max); HCl: 30 ppm (max); HCN: 30 ppm (max) • Fire Gas IDLH Index, $I_{IDLH} < 1$
Fire resistance and structural integrity under fire	• UL 1709 fire exposure for 30 min using Appendix A from MIL-PRF-32161. • Hose stream test applies • Maximum fire test load • Passive fire protection system shall be attached so as to survive the fire and other loads.	• There should be no passage of flame, smoke, or hot gases on the unexposed face • Average temperature rise on the unexposed surface not to exceed 139 °C • Peak temperature rise on the unexposed surface not to exceed 180 °C • Structural integrity under fire (under load): • No collapse or rupture of the structure for 30 min

characteristics of polymeric materials include molecular modification of the chemical structure itself, incorporation of flame retardants and additives, and the use of intumescent coatings or fire insulation as a protective cover.

19.4.1 Modification of chemical structure

A simple linear aliphatic polymer, such as polyethylene, consists of carbon and hydrogen. The chemical bond in this molecule is weak and breaks upon exposure to a small ignition source. This polymer melts and burns with high heat release rates. However, an aromatic-structured phenolic polymer is a product of condensation reaction between a phenol and formaldehyde. The resulting polymer has low flame spread, smoke generation, and heat release rate. The main difference between the two polymers is that polyethylene is aliphatic and linear, while phenolic is aromatic and cross-linked. Thus the chemistry of a base polymer and its morphology to a large extent governs the flammability characteristics of a polymer.

Table 19.2 summarizes the ASTM E 1354 cone calorimeter[24] heat release data from several polymers with different chemical structures at an incident heat flux of 50 kW/m^2. This incident heat flux in the flaming mode simulates a fully developed fire on the exposed surfaces of the test specimen. All the data shown in Table 19.2 is from resin plaques with no fibre reinforcement. Both polyethylene and polyurea have chemical bonds which rapidly degrade with high heat release rates. In case of vinyl ester, modification of chemical structure with bromine reduces heat release rate by 72% but increases specific extinction area (smoke) by 84%. Epoxy and phenolic polymers are extensively used in

Table 19.2 Effect of polymer composition on heat release rates at 50 kW/m^2 [25-29]

Polymer	Thickness (mm)	Time to ignition (s)	PHRR, kW/m^2	AHRR (300s), kW/m^2	ASEA, m^2/kg
Polyethylene (1525)*	25	10	1719	596	421
Polyurea (1487)	16	34	1393	198	284
Vinyl ester	6	82	1197	NA	1015
Brominated vinyl ester	6	42	332	212	1864
Epoxy (1543)	6	101	724	292	804
Phenolic (1555)	6	487	104	81	43
Silicone	6	259	74	48	NA
Phthalonitrile	6	NI	0	0	0
Bisphenol C-based cyanate ester	6	NI	0	0	0
FireQuench PM 1287.2 (POSS)	6	58	117	99	161

PHRR: peak heat release rate; AHRR (300s): average heat release rate at 300 seconds; ASEA: average specific extinction area (smoke); NI: no ignition; NA: not available.
* Numbers in parenthesis represent designation number assigned to the material by the test laboratory.

aircraft applications with epoxy resins mostly utilized in exterior structural applications and phenolics in interior cabin applications. Phenolic resins have low flame spread, smoke generation, heat release rates, and may be the most cost-effective resin for interior non-structural applications.

The resins described above are commercially available and do not necessarily require elevated temperature or pressure for curing. Aircraft applications requiring structural performance at elevated temperatures need resins which have high glass transition temperatures. Such resins, by virtue of their high glass transition temperatures, have high thermal stability and inherently improved fire performance. Bisphenol C-based cyanate ester[25,26] and phthalonitrile[27] resins have a high degree of thermal stability and do not ignite at an incident heat flux of 50 kW/m^2. When phthalonitrile and bisphenol C-based cyanate ester resins cure, they form a triazine network with benzene rings in the backbone structure which is very stable against fire.[28]

In recent years, several studies have been conducted on polyhedral oligomeric silsesquioxanes or (POSS). POSS technology, developed by Hybrid Plastics, Inc. has a unique feature in that it is a hybrid intermediate between that of silica and silicone.[29] Each POSS molecule contains covalently bonded reactive functionalities suitable for polymerization or grafting POSS monomers to polymer chains. As a result, this technology bridges the property space between hydrocarbon-based plastics and ceramics. The organic portion of the POSS molecule provides compatibility with existing resins, thereby enabling their easy incorporation into conventional resins. FireQuench PM 1287 is such a fire retardant hybrid nanocomposite resin based on POSS technology with low heat and smoke release rates as shown in Table 19.2.

19.4.2 Flame retardants

A flame retardant is the substance which when added to a combustible material reduces the flame spread of the resulting material when exposed to fire. The terms fire retarded or flame retarded are used interchangeably in this chapter.

In general, flame retardants reactively function either in the gas or solid (condensed) phases. In the gas phase, flame retardants act as free radical scavengers. They bond with free radicals generated from the fire and thus slow down the combustion chain reaction kinetics. An example of a gas phase flame retardant is brominated vinyl ester. Solid phase fire retardants promote char-formation or cross-linking in polymers to form a thermal/gas diffusion barrier. An example is the phosphorus-based group of flame retardants which generally promotes char formation.[30]

Fire retardants are usually halogen, phosphorus, nitrogen, antimony, or boron compounds. In some specific cases, they can be used in combination with synergistic effects. Chlorine, bromine, and phosphorus compounds are most generally employed in polymer fire retardant applications. Fire retardants may be incorporated in the molecular chain, as is typically the case for brominated vinyl esters, or physically mixed in the bulk polymer as is the case for alumina trihydrate in polyester or acrylic resins.

A recent trend is the use of nanoclays to reduce heat release rates of polymeric materials. Clay platelets are attracting particular interest due to their high performance at low filler loadings, rich intercalation chemistry, high surface area, high strength and stiffness, high aspect ratio of the individual platelets, abundance in nature and low cost.[31–33] The most commonly used clay in polymer nanocomposites is the smectite group of mineral montmorillonites. The layered structure of clay particles allows them to be chemically modified. Surface modification of the clay is an important step in achieving polymer nanocomposite dispersion and performance.

Table 19.3 summarizes the cone calorimeter heat release rate data[24] at incident heat flux of 50 kW/m^2 obtained from different flame retardants studied in conjunction with polyurea and vinyl esters. Vinyl ester resins are the binary resin systems containing a dimethacrylate monomer from which the cured material gains most of its properties, and a reactive monomer such as styrene which acts as a reactive diluent and also takes part in the cross-linking reaction. They are widely used in the applications of fibre reinforced plastics such as for boats, chemical tanks, pipes, etc. Polyureas are synthesized from a di-isocyanate and an amine functional so-called soft polymeric segment that is either a polyether or polyester. Polyureas have an attractive set of properties which include high elongation-to-break, high ductility at high strain rates, toughness, durability, etc. They are extensively used as linings, coatings and tiles. Despite their excellent mechanical properties, both vinyl esters and polyureas are limited in some applications due to their relatively high reaction to fire characteristics such as ease of ignition, flame spread, smoke generation, fire gas toxicity, and heat release rates.

Several flame retardant additives have been studied in conjunction with vinyl ester and polyurea.[34–36] These include nano-clay (e.g. Cloisite15A, Southern Clay Products, USA), phosphorus containing tricresylphosphate (TCP) and resorcinol di-phosphate (RDP), minerals such as alumina trihydrate (ATH), chlorinated phosphonate ester (Antiblaze 78), alkyl aryl phosphate ester (Santicizer 2158), ammonium polyphosphate (APP), bis pentaerythritol diphosphite (Doverphos), a proprietary phosphonate (Fyrol PMP), and pentaerythritol (PER).

The results in Table 19.3 from the study of fire retardants with vinyl esters indicate that brominated vinyl ester exhibited over 60% reduction in peak heat release rate. However, this was achieved with 87% increase in the specific extinction area (smoke). Both RDP and TCP, in synergy with 6% nano-clay, also decreased the heat release rates, but at concentrations which caused the viscosity of the resin to be too high for processing into fibre reinforced plastic with preferred methods such as resin transfer moulding.

Table 19.3 Summary of heat release rate test data (ASTM E 1354) on fire retardant studies with vinyl esters and polyureas at 50 kW/m² [34–36]

Material	Tig (s)	PHRR, kW/m²	ASEA (m²/kg)
Pure PVE	82	1197	1015
Brominated PVE	76	460	1900
PVE+6%15A+5%RDP	68	856	931
PVE+6%15A+10%RDP	74	643	1003
PVE+6%15A+15%RDP	56	512	1044
PVE+6%15A+30%RDP	81	535	1238
PVE+6%15A+15%TCP	44	670	976
PVE+6%15A+30%TCP	29	299	1350
PVE+6%15A+40%TCP	38	397	1721
Pure PU (1520)	15	1652	548
PU + 2% clay (1521)	13	1221	498
PU + 4% clay (1522)	12	719	547
PU + 8% clay (1523)	14	647	527
Pure PU	19 ± 1	2538	426 ± 19
PU + ATH (12%)	17	1479	481
PU + ATH (8%)	18	1519	435
PU + Antiblaze 78 (12%)	20 ± 3	1190	639 ± 11
PU + Antiblaze 78 (8%)	22 ± 4	1169	607 ± 14
PU + Santicizer 2158 (4%)	NA	1470	318
PU + Santicizer 2158 (8%)	NA	1372	550
PU + Santicizer 2158 (10%)	NA	1139	530
PU + APP (20%)	16 ± 1	306 ± 44	519 ± 13
PU + APP (16%)	14 ± 1	344 ± 62	490 ± 77
PU + APP (12%)	16 ± 1	537 ± 78	688 ± 66
PU + APP (8%)	15 ± 1	712 ± 51	543 ± 28
PU + APP (6%)	16 ± 5	746 ± 59	438 ± 18
PU + APP (4%)	15 ± 1	686 ± 28	422 ± 18
PU + Doverphos (20%)	20 ± 2	802 ± 33	694 ± 5
PU + Doverphos (16%)	19 ± 4	1040 ± 36	633 ± 26
PU + Doverphos (12%)	19 ± 2	874 ± 51	730 ± 22
PU + Doverphos (8%)	19 ± 1	1255 ± 172	616 ± 49
PU + Doverphos (6%)	20 ± 4	1342 ± 225	553 ± 80
PU + Doverphos (4%)	17 ± 1	1087 ± 123	403 ± 73
PU + Doverphos (2%)	17 ± 1	1123 ± 81	474 ± 33
PU + Fyrol PMP (12%)	20 ± 0	929 ± 0	428 ± 0
PU + PER (2%) + APP (6%)	18 ± 2	721 ± 54	443 ± 6
PU + PER (3%) + APP (9%)	18 ± 1	524 ± 56	491 ± 12
PU + PER (5%) + APP (15%)	18 ± 1	492 ± 113	467 ± 78

PVE: poly vinylester; PU: poly urea; Tig: time to ignition; PHRR: peak heat release rate; ASEA: average specific extinction area (smoke); NA: not available.

The results from the study of fire retardants with polyurea indicate that ammonium polyphosphate (APP) was most effective in reducing the heat release rate of polyurea. All samples in Table 19.3 with polyurea were tested at the thickness of 5 mm. For APP, there is a decrease in the peak heat release rate from 2538 to about 700 kW/m^2 at low loadings (~4%), falling to 300–350 kW/m^2 at 16–20% APP. There is a parallel change in the mass loss rate and the total heat released, while the time-to-ignition and the smoke produced are about constant.

19.4.3 Reducing flammability by the use of intumescent coatings

Intumescent coatings are formulated to intumesce (i.e. swell) when exposed to direct flame impingement and high temperatures. Intumescence requires a carbonific (char former) such as a polyol, a catalyst, or acid source such as a phosphate, and a spumific (gas generator) such as a nitrogen source.[30] The surface char that is formed by intumescence insulates the substrate from flame, heat, and oxygen. Commercial intumescent coatings contain about 20% of a polymer as a binder.

The porosity of the char produced during intumescence itself can play an important role during the combustion process. Char that is fragile can blow away or fall off from the substrate during large turbulent fires exposing an underneath vulnerable substrate to direct flames.

In recent studies, several intumescent coatings were evaluated over combustible substrates such as polyurea[36] and fibre reinforced plastics.[37] Intumescent coatings were applied by trowel on the exposed surface of 16 mm polyurea flat sheets and were cured for 24 hrs before testing in cone calorimeter in accordance with ASTM E 1354.[24] The test results are shown in Table 19.4. The intumescent coating, 6 mm thick over 16 mm polyurea (#1499), reduced the peak heat release rates from 1393 to 103 kW/m^2 (93% reduction). In some cases, the intumescent coating exhibited a propensity to separate (become loose or fall off) from the substrate.

In another example, six different intumescent coatings were selected and tested on a sandwich polymer composite substrate.[37] The sandwich composite consisted of 6 mm thick glass fibre reinforced, brominated vinyl ester resin face sheets and 76 mm thick balsa wood core. The coatings are identified as A, E, G, I, J, and O and were applied at the manufacturer's recommended thickness. Coatings A, E, G, and I were applied at 1.3 mm thickness, and J was applied at 5 mm thickness. Coated GRP composite samples were tested in the cone calorimeter and Table 19.4 summarizes the test results. Under 50 kW/m^2 heat flux, as defined in ASTM E 1354, only coatings A and I prevented ignition of the polymer composite substrate. In the same study, it was also found that the use of intumescent coatings as a fire protective cover is not reliably effective in providing fire protection for 30 min because current intumescent technologies

Table 19.4 Effect of intumescent coatings on heat release rates at 50 kW/m^2 [36,37]

Material	Thickness of IC, mm	Tig (s)	PHRR, kW/m^2	ASEA, m^2/kg
Polyurea (1487) (16 mm)	0	34	1393	284
PU+IC-White (1505)	1.3	684	828	444
PU + IC-Red (1498)	11	21	124	137
PU + IC-Gray (1499)	6	22	103	142
GRP (88 mm)	0	82	238	794
GRP + IC-A	1.3	NI	13	219
GRP + IC-E	1.3	22	69	59
GRP + IC-G	1.3	17	78	184
GRP + IC-I	1.3	NI	15	1
GRP + IC-J	5	60	135	168
GRP + IC-O	1.3	23	158	460

PU: polyurea; GRP: 88 mm thick glass reinforced plastic (GRP) consisting of 6 mm glass/vinyl ester face sheet with 76 mm thick balsa wood core.
IC: Intumescent coating; Tig: time to ignition; PHRR: peak heat release rate; ASEA, average specific extinction area (smoke); NI: no ignition.

have not been demonstrated to be reliable in the naval combat environment. In some situations, they exhibit propensity to prematurely fall off the substrate.

19.5 Fibre reinforced plastics

A recent trend in the maritime and naval shipbuilding applications is the use of polymer composites, also referred to as fibre reinforced plastics (FRP), or glass fibre reinforced plastic (GRP), or carbon fibre reinforced plastic (CFRP). FRPs offer inherent advantages over conventional materials, such as steel or aluminum, with regard to high strength-to-weight ratio, corrosion resistance, extended service life and signature reduction.[38–41] Materials technology, in some limited sense, is driving the designs of the next generation of light weight, fast moving and manoeuverable naval and maritime vessels.

Introduced in the 1960s, fibre reinforced plastics have evolved as functional substitutes for wood and metals, where weight is an important consideration. Since weight is a premium consideration in aircraft design, the use of continuous carbon and glass fibre reinforced polymer composites for aircraft cabin interiors, as well as other secondary structures, has grown with concurrent developments in process technology.[42]

Fibre reinforced plastics, or polymer composites are engineered materials in which the major component is a high strength fibrous reinforcement (up to 70% by weight) and the minor component is an organic resin binder, often referred to as the matrix resin. Thermoset polymers comprise the majority of composite resins and consist primarily of the chemical families which include polyester

(PE), vinyl ester (VE), epoxy (EP), bismaleimide (BMI), phenolic (PH), cyanate ester (CE), silicones (SI), polyimides (PI), and phthalonitriles (PN), etc. Thermoset polymers are cross-linked and do not exhibit melting or dripping when exposed to high temperatures due to fire. High temperature engineering thermoplastic resins used in military applications include polyetherimide (PEI), polyphenylene sulfide (PPS), polyether sulfone (PES), polyaryl sulfone (PAS), polyether ether ketone (PEEK), and polyether ketone ketone (PEKK), etc.[43,44]

Reinforcing fibres include E and S glass, aramid (Kevlar), carbon, quartz, polyethylene (Spectra), phenylene benzobisoxazole (PBO), boron, etc. These fibres are used either alone or as hybrids in the form of woven rovings, fabrics, unidirectional tapes, bundles (tows), or chopped to various lengths.[45]

Polymer composites may be fabricated in the form of laminates (single skin panel) or sandwich construction. The use of sandwich composites has grown in naval applications. A typical sandwich composite for naval applications consists of a low density thick core, such as balsa wood or poly vinyl chloride (PVC) foam, sandwiched between two thin face sheets which consist of fibre reinforced plastic (polymer composite). It is a type of stressed skin construction in which the facings resist nearly all of the applied edgewise (in-plane) loads and flatwise bending moments. The thin spaced facings provide nearly all of the bending rigidity to the construction. The core spaces between the facings transmit shear between them and provides shear rigidity of the sandwich construction. By proper choice of materials for facings and core, constructions with high ratios of stiffness-to-weight can be achieved. The sandwich composite design is analogous to an I-beam in which the flanges carry direct compression and tension loads, as do the sandwich facings, and the web carries shear loads, as does the sandwich core.[46,47] Examples of large scale usage of sandwich polymer composite in naval applications include the advanced enclosed mast/sensor system (AEM/S) on US Navy amphibious transport dock ship LPD-17,[40,48] deckhouse on US Navy destroyer (www.ddg1000.com), and the Swedish Visby class of stealth corvette.[49]

The topside polymer composite structure on US Navy ships typically comprises a sandwich configuration with balsa core and glass fibre reinforced brominated vinyl ester face sheets. The hull material in Visby Class constructions comprises a PVC core with a carbon fibre and vinyl laminate.[49] Composite components and structures are fabricated by impregnating the fibrous reinforcement with liquid resin using various processes, including infusion of pre-forms in a closed cavity mold, filament winding, pultrusion, lamination of pre-impregnated fibre mats, fabrics, or tapes, etc. A composite manufacturing method known as SCRIMP (Seemann Composites Resin Infusion Moulding Process), shown in Fig. 19.1, was used in the fabrication of the LPD-17 AEM/S structure.

Fibre reinforced polymer composites pose fire safety concerns due to the combustibility of the organic polymer constituents. If the fire is not controlled,

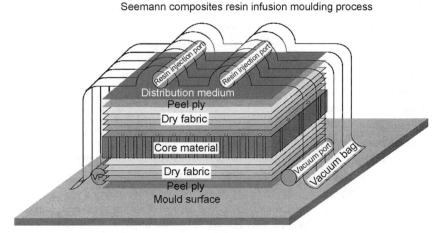

19.1 Sketch of a polymer composite fabrication method by SCRIMP.

polymer composites may lose their structural integrity and collapse under load. A recent example is the fire that took place on board a Norwegian mine sweeper, Orkla, which resulted in the total loss of the vessel (no loss of life).[50]

19.6 Fire performance and test methods for fibre reinforced plastics

Maritime and naval vessels are built with fire safety objectives which include containing, controlling and suppressing a fire or explosion to the compartment of origin and to provide adequate and readily accessible means of egress and escape for passengers and crew. To accomplish these objectives, the ship is typically divided into main horizontal and vertical fire and structural boundaries with fire resistant divisions. To fully utilize polymer composites in shipbuilding, they must be capable of being qualified as fire resistant divisions. Fire resistance is the ability of the construction to withstand fire, retain structural integrity and prevent fire spread to adjoining compartments for a period of time. As applied to shipbuilding elements, such fire resistant divisions are called 'A' Class bulkheads and decks (overheads). As mentioned previously, HSCC[8] permits the use of polymer composites in maritime vessels as fire resisting divisions when they are constructed of fire restricting materials. Fire restricting materials are internationally qualified in accordance with Resolution MSC.40(64)[17] as amended by MSC.90(71).[18] Fire restricting is a measure of the ability of polymer composite structure to withstand a growing fire for a period of 20 minutes without reaching threshold limits of heat release and smoke production rates. Materials qualified as fire restricting are considered to have low flame spread, low smoke, and low heat release rates.

Fire performance characteristics of polymer composite systems may be divided into several categories, including the following here:

1. Flame spread (minimize the ignition and spread of fire within the compartment).
2. Smoke and toxicity (visibility, tenability).
3. Heat release (fire growth, minimize the hazard to personnel escaping the fire or their ability to fight the fire).
4. Fire resistance (contain the fire to designated spaces and or zones).
5. Structural integrity under fire (reduce the risk of structural collapse).

It is the combination and totality of such performance characteristics which ensure the fire safety of composite systems which must perform in most hostile fire threat environments.

19.6.1 Flame spread

The flame spread determines the relative burning behavior of construction materials along the exposed surfaces. The ASTM E-84, Standard Test Method for Surface Burning Characteristics of Building Materials[51] is often the test method of choice in most US codes such as NFPA 301, Code for Safety to Life from Fire on Merchant Vessels.[52] This is also the test method of choice in US Navy MIL-STD-1623, Fire Performance Requirements and Approved Specifications for Interior Finish Materials and Furnishings.[20] Other small-scale test methods for surface flame spread include ASTM E 162, Surface Flammability of Materials Using a Radiant Heat Energy Source,[53] and IMO Resolution A.653 (16), Surface flammability of bulkhead, ceiling, and deck finish materials.[12] Materials which meet the requirements of Resolution A.653 (16) are considered to have low flame spread.

The ASTM E 84[51] test method exposes a nominal 7.32 m (24 ft) long by 508 mm (20 in.) wide specimen in a Steiner Tunnel to a controlled air flow and flaming fire exposure adjusted to spread the flame along the entire length of the select grade red oak specimen in five and a half minutes. Flame spread index and smoke developed index material test data are reported on the basis of calibration procedure which sets each index of cement board and red oak wood at 0 and 100 respectively. This test has been widely adopted for use by the building code authorities to regulate the use of interior finish materials. However, this test has been found to be misleading for materials that do not remain attached to the ceiling of the Steiner tunnel during the fire test. Table 19.5 provides data on flame spread and smoke developed indices of several polymer composite materials when tested to ASTM E 84.[54]

ASTM E 162[53] is a small-scale test for flame spread. It uses a test specimen of nominal 152 × 457 mm (6 × 18 in.) size which is placed in front of a 305 × 457 mm (12 × 18 in.) radiant panel. The radiant panel consists of a porous

Table 19.5 Summary of ASTM E 84 test results[54]

Material	Flame spread index	Smoke developed index
DDS-078-1 Requirement	25	50
Red oak wood	100	100
Glass/vinyl ester/balsa core (unprotected, no paint)	20	750
Glass/vinyl ester/balsa core with F150/silicone alkyd paint)	20	750
Fire insulation, Thermal Ceramics Structogard FB	10	30
Glass/vinylester/balsacore/Structogard FB	10	30
Carbon/vinylester/balsa core	25	630
Glass/vinyl ester solid composite (no balsa core)	15	1100

refractory material and is capable of operating up to 815 °C (1500 °F). A small pilot flame about 50.8 mm (2 in.) long is applied to the top centre of the specimen at the start of the test. The test is completed when the flame front has travelled 381 mm (15 in.) or after an exposure time of 15 minutes. A factor derived from the rate of progress of the flame front and another relating to the rate of heat liberation by the material under test is combined to provide flame spread index (FSI). This fire-test-response standard is used for research and development purposes. It is not intended to for use as a basis of ratings for building code purposes. Table 19.6 provides data on ASTM E 162 flame spread index of several polymer composite materials.[43,44,55]

The High Speed Craft Code[8] uses Resolution A.653 (16)[12] to evaluate the surface flammability characteristics of polymer composite materials. It uses nominal 155 × 800 mm size specimens in a vertical orientation. The specimens are exposed to a graded radiant flux field supplied by a gas fired radiant panel. Experimental results are provided in terms of critical flux at extinguishment, heat for sustained burning, total heat release, and peak heat release rate. The critical flux at extinguishment is a flux level at the specimen surface corresponding to the distance of farthest advance and subsequent extinguishment of the flame on the centre line of the burning specimen. The heat for sustained burning is the product of time from initial specimen exposure until arrival of the flame front and the incident flux level at that same location as measured with a dummy specimen during calibration. The total heat release is given by integration of the positive part of the heat release rate during the test period. The peak heat release rate is the maximum of the heat release rate during the test period. Resolution A.653(16)[12] acceptance criteria and test results for several polymer composites are shown in Table 19.7.[56] Materials giving average values for all of the surface flammability criteria not exceeding those listed in Table 19.7 are considered to meet the HSCC regulations for low flame spread.

Table 19.6 ASTM E 162 flame spread index (FSI) for polymer composites[43,44,55]

Composite	FSI
Red oak wood	100
Carbon/Vinyl ester (1478)	11
Glass (Gl)/Vinyl ester (VE), NFR, 1087	156
Gl/Vinyl ester, FR (1031)	27
Gl/Vinyl ester, FR (1259)	15
Gl/VE Sandwich Composite (1257)	24
Gl/Epoxy (1066)	43
Gl/Epoxy (1067)	12
Gl/Epoxy (1089)	11
Gl/Epoxy (1091)	11
Gl/Epoxy (1092)	23
Graphite (Gr)/Modified bismaleimide (1095)	13
Gl/Bismaleimide (1096)	17
Gr/Bismaleimide (1097)	12
Gr/Bismaleimide (1098)	3
Gl/Phenolic (1099)	1
Gl/Phenolic (1100)	5
Gl/Phenolic (1101)	4
Gl/Phenolic (1014)	4
Gl/Phenolic (1015)	4
Gl/Phenolic (1017)	6
Gl/Phenolic (1018)	4
Gr/Phenolic (1102)	6
Gr/Phenolic (1103)	20
Gr/Phenolic (1104)	3
PE/Phenolic (1073)	48
Aramid/Phenolic (1074)	30
Gl/PMR-15 (1105)	2
Gl/Silicone	2
Gl/J-2 (1077)	13
Gl/Polyphenylene sulfide (PPS) (1069)	7
Gr/ Polyphenylene sulfide (PPS) (1083)	3
Gl/ Polyphenylene sulfide (PPS) (1084)	8
Gr/ Polyphenylene sulfide (PPS) (1085)	3
Gr/Poly aryl sulfone (PAS) (1081)	9
Gr/Poly ether ether ketone (PEEK) (1086)	3

19.6.2 Smoke and combustion gas generation

Fibre reinforced plastics or polymer composite materials give off smoke and fire gases when they burn. Smoke is a combination of airborne solid and liquid particulates and gases that are generated when a material undergoes pyrolysis or combustion. Smoke affects visibility and hinders the ability of the occupants to escape and fire fighters to locate and suppress the fire.

Smoke production of combustible construction materials is typically measured either in small-scale tests, such as ASTM E 662[57] or ISO 5659 Part 2,[10] or full-

Table 19.7 IMO Resolution A. 653(16) surface flammability test results for polymer composite materials[56]

Material	CFE, kW/m²	Q_{sb}, MJ/m²	Q_t, MJ	q_p, kW
IMO requirement for bulkhead, wall and ceiling linings	≥ 20	≥ 1.5	≤ 0.7	≤ 4.0
Glass/Vinyl ester, 12 mm (1209)	16.5	NA	NA	NA
Glass/VE balsa core sandwich composite, 88 mm (1257)	12.4	4.27	6.64	6.77
Sandwich composite with Scrimped Intumescent mat, 88 mm thick (1274)	32	NA	NA	NA
FR Phenolic, USCG ID No. 1	42	NA	NA	NA
Fire restricting material, USCG ID No. 2	69	NA	NA	NA
FR Polyester, USCG ID No. 3	23.16	4.33	1.04	1.83
FR Vinyl ester, USCG ID No. 4	23.16	4.54	1.21	2.47
FR Epoxy, USCG ID No. 5	39.84	8.09	0.04	0.31
Coated FR epoxy, USCG ID No. 6	40	NA	NA	NA
Textile wall covering, USCG ID No. 7	19	NA	NA	NA
Polyester, USCG ID No. 8	11.65	2.10	2.00	6.11
FR modified acrylic, USCG ID No. 9	17.69	5.09	1.18	2.62

CFE: Critical flux at extinguishment; Q_{sb}: Heat for sustained burning; Q_t: Total heat release; q_p: Peak heat release rate; USCG: United States Coast Guard; NA: not available

scale tests, such as ASTM E 84[51] or ISO 9705.[58] The full-scale ASTM E 84 test is widely used for building and construction materials. The small-scale smoke density chamber ASTM E 662 or ISO 5659 tests are widely used for characterization of smoke density of materials as well as combustion products from both flaming (pilot ignition) and non-flaming (smouldering) modes. Both small-scale tests are conducted in a closed chamber of fixed volume and the light attenuation is recorded over a known optical path length. In the ASTM E 662 test, a 75 mm square sample in a vertical orientation is subjected to a radiant heat flux of 25 kW/m² under piloted ignition (flaming) or non-flaming mode. In the ISO 5659 test, a 75 mm square sample in horizontal orientation is subjected to an irradiance of 25 kW/m² in flaming and non-flaming modes and an irradiance of 50 kW/m² in the absence of pilot flame. The light transmission measurements through the smoke in the chamber provide a specific optical density (Ds). Visibility through smoke is inversely related to this specific optical density. In simplified terms, the chamber is calibrated to initial light transmission of 100%, i.e. no smoke. As the sample is heated by a radiant flux of 25 kW/m², either in non-flaming or flaming mode, the amount of light transmitted as a fraction of initial light is used to calculate the specific optical density. The maximum optical density (Dm) over the duration of the test is used to identify materials with relatively high smoke production.

The most common of fire gases generated from decomposing and burning polymer composite include carbon monoxide (CO) and carbon dioxide (CO_2). In addition, nitrogen-containing materials may evolve hydrogen cyanide (HCN) and nitrogen oxides, sulphur-containing materials may evolve sulphur oxides and sulphides, and chlorine-containing materials may evolve hydrogen chloride (HCl). Other gases may also be generated depending upon the chemistry of the polymer resin used in a given composite material. When making toxicity measurements, the fire gases are analyzed and their concentrations measured either by colorimetric tubes or by other analytical means such as a gas chromatograph or a Fourier transform infrared spectrometer (FTIR).[59]

Several regulatory bodies have prescribed gas concentration limits at specific test conditions. The Committee on Fire Toxicology of the National Academy of Science has concluded that as a basis for judging or regulating materials performance in a fire, combustion product toxicity data must be used only within the context of a fire hazard assessment.[60] The committee believes that the required smoke toxicity is currently best obtained with animal exposure methods for the purposes of predicting the fire hazard of different materials.

Table 19.8 provides data on smoke density and the concentrations of fire gases determined via the Draeger colorimetric tube method in flaming mode during the smoke obscuration test (ASTM E 662) for several polymer composite materials.[44] Polymer composites based on brominated vinyl ester exhibit high

Table 19.8 ASTM E 662 (flaming mode) smoke and combustion gas generation test data for polymer composite materials[43,44]

Composite	Ds (300s)	Dm	CO ppm	CO_2 % vol	HCN ppm	HCl ppm
Gl/Vinyl ester (1031)	463	576	230	0.3	ND	ND
Gl/Vinyl ester (1087)	310	325	298	1.5	1	0.5
Gl/Vinyl ester (1167)	103	173	300	2.0	2	ND
Gl/Vinyl ester (1168)	503	593	800	0.5	2	TR
Gl/Vinyl ester (1169)	154	217	200	2.0	TR	ND
Gl/Vinyl ester (1170)	185	197	300	2.0	TR	ND
Gl/Vinyl ester (1259)	NA	191	180	NA	ND	ND
Carbon/Vinyl ester (1478)	NA	214	1041	1.6	ND	ND
Gl/VE/nanoclay (0%) (1279)	NA	736	750	0.5	TR	TR
Gl/VE/nanoclay (1%) (1276)	NA	624	300	0.5	TR	ND
Gl/VE/nanoclay (3%) (1277)	NA	734	600	0.5	TR	TR
Gl/VE/nanoclay (5%) (1278)	NA	699	400	0.5	TR	TR
Gl/VE/balsa core sandwich composite (1257)	550	900	2900	0.6	2	TR
Balsa core (1213)	NA	67	700	1.0	5	ND

Table 19.8 Continued

Composite	Ds (300s)	Dm	CO ppm	CO_2 % vol	HCN ppm	HCl ppm
PVC foam (1214)	NA	410	190	0.3	20	>10
Polyimide foam (1428)	NA	33	97	0.6	TR	TR
Gl/Polyurethane (1130)	NA	615	200	0.5	2	ND
Gr/Polyurethane	NA	471	250	0.5	5	ND
Gl/Modar	11	109	400	0.5	5	ND
Gl/Modar with filler (1176)	NA	191	600	2.0	10	TR
Gl/Epoxy (1089)	56	165	283	1.5	5	ND
Gl/Epoxy (1066)	2	408	200	2.0	5	2
Gl/Epoxy (1067)	16	456	250	1.0	2	ND
Gl/Epoxy (1071)	17	348	80	0.5	3	1
Gl/Epoxy (1090)	96	155	50	0.2	ND	ND
Gr/Epoxy (1091)	75	191	115	0.9	15	TR
Gr/Epoxy (1092)	66	210	313	2.0	1	0.5
Gr/Epoxy (1093)	3	353	160	0.5	2	1.5
Gr/Epoxy (1094)	1	301	300	0.6	2	1
Gl/Cyanate ester (CE) (1046)	4	84	NA	NA	NA	NA
Gr/Bismaleimide (BMI) (1095)	24	158	30	0.3	1	ND
Gr/BMI (1097)	6	171	175	0.8	3	ND
Gr/BMI (1098)	9	117	10	TR	TR	1
Gl/BMI (1096)	34	127	300	0.1	7	TR
Gl/Phenolic (1100)	4	18	300	1.0	1	1
Gl/Phenolic (1099)	4	43	50	TR	TR	ND
Gl/Phenolic (1101)	1	23	300	1.0	1	1
Gl/Phenolic (1014)	1	1	300	1.0	TR	ND
Gl/Phenolic (1015)	1	3	190	1.0	TR	TR
Gl/Phenolic (1017)	1	4	200	1.0	ND	ND
Gl/Phenolic (1018)	1	1	200	1.0	TR	ND
Gr/Phenolic (1102)	1	24	115	0.5	1	1
Gr/Phenolic (1103)	40	138	100	0.1	1	ND
Gr/Phenolic (1104)	1	4	600	1.0	2	ND
PE/Phenolic (1073)	1	241	700	2.0	2	ND
Aramid/Phenolic (1074)	2	62	700	1.5	2	ND
Gl/PMR-15 (1105)	1	16	200	1.0	TR	2
Gl/Phthalonitrile (PN) (1136)	1	5	60	0.5	TR	ND
Gl/Silicone (1116)	1	2	50	0.5	TR	TR
Gl/J-2 (1077)	NA	328	180	1.0	10	ND
Gl/Polyphenylene sulfide (PPS) (1069)	8	87	70	0.5	2	0.5
Gr/PPS (1083)	2	32	100	0.5	1	ND
Gl/PPS (1084)	4	54	100	1.0	TR	TR
Gr/PPS (1085)	1	26	100	1.0	TR	TR
Gr/Poly aryl sulfone (PAS) (1081)	2	3	55	0.1	TR	ND
Gr/Poly ether sulfone (PES) (1078)	1	5	110	1.0	1	1
Gr/Poly ether ether ketone (PEEK) (1086)	1	1	TR	TR	ND	ND
Gl/Poly ether ketone ketone (PEKK) (1079)	1	4	200	1.0	ND	TR

Ds(300s): specific optical density at 300 seconds; Dm: maximum specific optical density; ND: not detected; TR: trace; NA: not available; Gl: Glass; Gr: Graphite; PE: poly ethylene.

Flame retardant materials for maritime and naval applications 547

smoke density values whereas glass or graphite reinforced phenolic composites have low smoke values. This is also true for all advanced thermoplastics which have low maximum smoke density. The thermoplastic panels evaluated in this study expanded or foamed up slightly in the middle during smoke density tests, presumably due to the gases escaping through the softened front face during fire exposure.

19.6.3 Heat release rate

Heat release is defined as the heat generated in a fire due to a polymer composite undergoing combustion by various chemical reactions occurring within a given weight or volume of a material. The major contributors are those reactions where CO and CO_2 are generated and oxygen is consumed.[61] Material burning rates are influenced by the intensity of the ignition source. Different levels of incident heat flux can simulate fire scenarios in which the composite material is itself burning or in which it may be near another burning material. Heat release rate data provides a relative fire hazard assessment for materials.[62] Materials with a low heat release rate per unit weight or volume will do less damage to the surroundings than those with high heat release rates. The rate of heat release, especially the peak amount, is the primary characteristic determining the size, growth, and suppression requirements of a fire environment.[61]

Heat release rates of materials, which are used in large amounts with exposed surfaces such as lining or insulation materials, are typically measured in a full-scale room test, such as ISO 9705.[58] Materials which are used in small components or furniture items are typically characterized in small-scale tests such as ASTM E 1354[24] or ISO 5660[63] oxygen consumption cone calorimeter.

ASTM E 1354[24] and ISO 5660[63] are the two most widely used cone calorimeter test methods which cover the measurement of the response of materials exposed to controlled levels of radiant heating. A specific heat flux of 25, 50, 75 or 100 kW/m² may be used. These thermal flux levels correspond to a small Class A fire (see Section 19.6.5), a large trash can fire, a significant fire and a pool oil fire respectively.[62] A 100 × 100 mm sample is placed beneath the conically-shaped heater that provides a uniform irradiance on the sample surface. A spark igniter 12.5 mm from the sample surface is used to initiate the burning of any combustible gas mixture produced by the sample. The sample mass is constantly monitored using a load cell and the effluent from the sample is collected in the exhaust hood above the heater. In the duct downstream of the hood, the flow rate, smoke obscuration, and O_2, CO_2 and CO concentrations are continuously measured. Once the sample ignites, the burning of the sample causes a reduction in the oxygen concentration within the effluent collected by the hood. This reduction in oxygen concentration has been shown to correlate with the heat release rate of the material, 13.1 MJ/kg of O_2 consumed. This is known as the oxygen consumption principle. Using this principle, the heat release rate per unit area of the sample is determined with time using measurements of mass flow rate and oxygen depletion in the gas flow. Smoke obscuration is measured in the flow system by means of a laser photometer and the results are computed in the form of specific extinction area (SEA).

This test method produces a wealth of data including time-to-ignition, mass loss rates, heat release rates at various time intervals, total heat release, effective heat of combustion, visible smoke development, etc. These values are becoming increasingly important in determining fire growth from burning materials and are needed in the various fire models to predict thermal conditions in compartments. For this reason, the cone calorimeter test method has now become the most widely used small scale test method to determine the flammability characteristics of combustible materials. Table 19.9 provides the time-to-ignition, peak and average heat release rates, total heat release, and specific extinction area for various polymer composite materials at different heat fluxes.[44,64–67]

19.6.4 Full-scale room test (ISO 9705)

The fire growth in a compartment depends on the rate at which the initiating source fire ignites materials and other ignitable items in the compartment and the heat release rate of the ignited items. To effectively fight the fire, the fire must not reach a flashover condition before fire extinguishment procedures are initiated. Flashover is the condition which inevitably ignites all combustible items within the compartment, engulfing the compartment with fire. When flashover is reached, the ability of firefighters to fight the fire will be significantly reduced.

For most interior applications of polymer composites with large exposed surfaces, the potential for fire growth should be the first issue addressed for habitable environments. In small scale tests, fire growth potential is measured in terms of heat release rates in the ASTM E 1354[24] or ISO 5660 methods,[63] using a cone calorimeter, as described above. Under full-scale conditions, fire growth potential is measured by ISO 9705,[58] 'Full-Scale Room Test for Surface Products'.

The High Speed Craft Code[8] requires polymer composite materials used in fire resisting divisions (bulkheads and decks) to be fire restricting. Resolution MSC.40(64),[17] as amended by Resolution MSC.90(71)[18] is used to qualify fire-restricting materials by testing in accordance with ISO 9705.[58] The ISO 9705 room/corner test consists of an 2.44 m (8 ft.) wide, 3.66 m (12 ft.) deep, and 2.44 m (8 ft.) high room constructed of non-combustible material with a 2.0 m (6.5 ft) high and 0.8 m (2.6 ft.) wide door. The enclosure provides an enhanced heat feedback effect, due to accumulating hot smoke which is not present in an open fire exposure. A propane burner is located on the floor in either back corner of the room. The test procedure requires an ignition source of 100 kW heat output for 10 min and thereafter a 300 kW heat output for another 10 min. The product is

Table 19.9 ASTM E 1354 heat release rates of polymer composite materials[44,63–67]

Material system	Irradiance (kW/m²)	WL (%)	Tig (s)	PHRR (kW/m²)	AHRR (kW/m²)	THR (MJ/m²)	ASEA (m²/kg)
Douglas Fir plywood	25	74	306	188	90	69.2	98
	50	82	22	314	98	76.5	75
	75	87	8	335	157	84.8	141
Bass wood (1502)	25	80	64	107	39	65	47
	50	83	14	186	117	124	62
	75	87	5	254	156	119	45
Yellow poplar (1503)	25	83	114	139	82	131	137
	50	84	16	187	117	132	237
	75	84	6	274	177	152	–
Sheet Rock (Gypsum board) (1290)	25	15	NI	13	3	3	39
	50	12	34	77	10	3	4
	75	14	13	96	10	3	2
	100	16	6	149	12	4	12
Vinyl ester resin, brominated, (no fiber) (1483)	25	–	180	176	126	81	1656
	50	–	40	335	206	78	1831
	75	–	21	409	231	77	1997
Glass/Vinyl ester (VE) (1031)	25	14	278	75	29	11	1185
	50	26	74	119	78	25	1721
	75	29	34	139	80	27	1791
	100	28	18	166	–	22	1899
Glass/VE (1087)	25	36	281	377	180	55	1188
	75	34	22	499	220	68	1218
	100	33	11	557	–	64	1466
Glass/VE (1167)	25	–	320	308	180	64	836
	50	–	85	276	184	59	999
	75	–	42	281	190	59	986
Glass/VE (1168)	25	–	214	147	92	29	1341
	50	–	52	152	86	28	1524
	75	–	29	217	108	33	1569

Table 19.9 Continued

Material system	Irradiance (kW/m²)	WL (%)	Tig (s)	PHRR (kW/m²)	AHRR (kW/m²)	THR (MJ/m²)	ASEA (m²/kg)
Glass/VE (1169)	25	–	302	342	211	68	796
	50	–	85	302	198	62	815
	75	–	42	303	203	63	872
Glass/VE (1170)	25	–	259	356	190	58	914
	50	–	75	348	179	55	1027
	75	–	36	432	202	61	1050
Glass/VE/nanoclay, 1%, 1276	25	19	257	122	–	23	1225
	50	24	80	141	–	30	1685
	75	25	36	175	–	27	1798
Glass/VE/nanoclay, 3%, 1277	25	22	262	132	105	32	1324
	50	26	67	143	106	33	1688
	75	26	35	159	–	29	1875
Glass/VE/nanoclay, 5%, 1278	25	21	273	134	104	31	1319
	50	25	73	134	100	31	1664
	75	23	30	153	–	23	1794
Glass/VE/nanoclay, 0%, 1279	25	24	246	133	81	34	1229
	50	28	68	166	–	30	1765
	75	29	33	211	–	28	1895
Carbon/Vinyl ester (1478)	25	–	457	285	140	147	850
	50	–	145	366	145	153	941
	75	–	27	459	172	154	953
Gl/VE/balsa core Sandwich Composite (1257)	25	15	306	121	58	121.1	933
	50	18	70	126	93	126.0	1063
	75	23	28	150	99	149.7	986
Glass/polyurethane (1130)	25	19	179	184	79	97	206
	50	22	29	209	93	98	282
	75	23	13	466	146	123	830
Graphite/polyurethane (1131)	25	29	189	189	92	88	299
	50	35	21	293	134	87	577
	75	39	10	183	156	87	435

Material	%						
Glass/Modar (1161)	25	—	421	149	85	58.2	100
	50	—	119	160	91	64.3	126
	75	—	61	181	105	67.0	161
Glass/epoxy (1089)	25	—	535	39	30	10	470
	50	—	105	178	98	30	580
	75	—	60	217	93	28	728
	100	—	40	232	93	24	541
Glass/epoxy (1066)	25	20	140	231	158	52	1096
	50	23	48	266	154	48	1055
	75	24	14	271	157	48	1169
	100	24	9	489	—	46	1235
Glass/epoxy (1067)	25	18	209	230	120	41	1148
	50	20	63	213	127	39	1061
	75	21	24	300	138	43	1109
	100	20	18	279	—	32	1293
Glass/epoxy (1040)	25	—	—	—	—	—	—
	50	19	18	40	2	29	566
	75	21	13	246	1	38	605
	100	23	9	232	5	47	592
Glass/epoxy (1071)	25	7	128	20	4	1	1356
	50	5	34	93	—	3	1757
	75	23	18	141	99	30	1553
	100	25	10	202	108	34	1310
Glass/epoxy (1006)	25	14	159	81	63	28	2690
	50	28	49	181	108	39	1753
	75	24	23	182	—	35	1917
	100	29	14	229	131	41	1954
Glass/epoxy (1070)	25	23	229	175	95	45	1119
	50	28	63	196	143	49	1539
	75	27	30	262	133	43	1440
	100	30	23	284	—	36	1640
Glass/epoxy (1003)	25	17	198	159	93	36	1162
	50	22	50	294	135	43	1683
	75	22	73	191	121	41	1341
	100	22	19	335	122	37	1535

Table 19.9 Continued

Material system	Irradiance (kW/m²)	WL (%)	Tig (s)	PHRR (kW/m²)	AHRR (kW/m²)	THR (MJ/m²)	ASEA (m²/kg)
Glass/epoxy (1090)	25	19	479	118	67	38	643
	50	28	120	114	90	55	803
	75	34	54	144	115	64	821
	100	34	34	73	150	71	1197
Graphite/epoxy (1091)	25	7	NI	NI	NI	NI	601
	75	25	53	197	90	30	891
	100	38	28	241	–	28	997
Graphite/epoxy (1092)	25	–	275	164	99	32	525
	50	–	76	189	116	37	593
	75	–	32	242	112	37	363
	100	–	23	242	113	71	235
Graphite/epoxy (1093)	25	13	338	105	69	–	–
	50	24	94	171	93	–	–
	75	23	44	244	147	–	–
	100	22	28	202	115	–	–
Glass/Cyanate ester (1046)	25	8	199	121	74	30	794
	50	22	58	130	71	49	898
	75	23	20	196	116	58	1023
	100	24	10	226	141	47	1199
Glass/Cyanate ester (bisphenol C) (1427)	25	NI	NI	–	–	–	–
	50	NI	NI	27	15	7	72
	75	11	254	48	23	8	63
	100	11	140				
Graphite/Modified Bismaleimide (BMI) (1095)	25	19	237	160	103	32	645
	75	24	42	213	115	36	685
	100	26	22	270	124	38	706
Graphite/BMI (1097)	25	5	NI	NI	NI	NI	238
	75	30	66	172	130	45	933
	100	31	37	168	130	41	971

Material							
Graphite/BMI (1098)	25	–	NI	NI	NI	NI	NI
	50	13	110	74	51	14	228
	75	15	32	91	65	17	370
	100	16	27	146	75	22	383
Glass/BMI (1096)	25	17	503	128	105	40	324
	50	25	141	176	161	60	546
	75	30	60	245	199	76	604
	100	30	36	285	219	73	816
Glass/phenolic (1099)	25	–	NI	NI	NI	NI	NI
	50	–	121	66	43	18	4
	75	–	33	102	86	33	85
	100	–	22	122	95	40	–
Glass/phenolic (1100)	25	–	NI	NI	NI	NI	NI
	50	–	125	66	48	17	308
	75	–	20	120	63	21	365
	100	–	40	163	74	21	441
Glass/phenolic (1101)	25	–	NI	NI	NI	NI	NI
	50	–	210	47	38	14	176
	75	–	55	57	40	16	161
	100	–	25	96	70	22	620
Glass/phenolic (1014)	25	–	NI	NI	NI	NI	NI
	50	12	214	81	40	17	83
	75	16	73	97	54	20	246
	100	16	54	133	78	21	378
Glass/phenolic (1015)	25	–	NI	NI	NI	NI	NI
	50	6	238	82	73	15	75
	75	8	113	76	37	7	98
	100	13	59	80	62	12	58
Glass/phenolic (1017)	25	–	NI	NI	NI	NI	NI
	50	10	180	190	139	43	71
	75	14	83	115	84	17	161
	100	18	43	141	73	19	133
Glass/phenolic (1018)	25	–	NI	NI	NI	NI	NI
	50	3	313	132	22	12	143
	75	11	140	56	44	11	74
	100	13	88	68	58	13	66

Table 19.9 Continued

Material system	Irradiance (kW/m²)	WL (%)	Tig (s)	PHRR (kW/m²)	AHRR (kW/m²)	THR (MJ/m²)	ASEA (m²/kg)
Graphite/phenolic (1102)	25	4	NI	NI	NI	NI	NI
	75	28	79	159	80	28	261
	100	–	45	196	–	–	–
Graphite/phenolic (1103)	25	–	NI	NI	NI	NI	NI
	50	28	104	177	112	50	253
	75	27	34	183	132	50	495
	100	29	20	189	142	51	493
Graphite\phenolic (1104)	25	–	NI	NI	NI	NI	NI
	50	9	187	71	41	14	194
	75	11	88	87	–	11	194
	100	11	65	101	–	11	232
Polyethylene\phenolic (1073)	25	30	714	NI	NI	NI	NI
	50	61	129	98	83	107	294
	75	60	28	141	92	104	500
	100	67	10	234	131	96	580
Aramid/phenolic (1074)	25	4	1,110	NI	NI	NI	NI
	50	43	163	51	40	57	156
	75	40	33	93	54	45	240
	100	65	15	104	72	95	333
Glass/polyimide (1105)	25	–	NI	NI	NI	NI	NI
	50	11	175	40	27	21	170
	75	13	75	78	49	22	131
	100	14	55	85	60	20	113
Glass/Phthalonitrile (PN) (1273)	25	–	NI	0	0	0	0
	50	–	437	35	24	10.9	157
	75	–	165	83	49	18.5	75
	100	–	88	109	57	22.3	58
Graphite/PN (1080)	25	–	–	–	–	–	–
	100	13	75	119	36	12	610

Material	%						
Glass/Polypropylene (1082)	25	37	168	187	153	88	702
	50	36	47	361	248	82	959
	75	37	23	484	265	82	1,077
	100	36	13	432	—	82	1,120
Glass/J-2 (nylon) (1077)	25	—	193	67	38	—	803
	50	—	53	96	49	—	911
	75	—	21	116	48	—	866
	100	—	13	135	76	—	1,011
Glass/Poly phenylene sulfide (PPS) (1069)	25	—	NI	NI	NI	NI	NI
	50	12	105	52	25	32	585
	75	12	57	71	56	24	575
	100	14	30	183	106	41	749
Graphite/PPS (1083)	25	—	NI	NI	NI	NI	NI
	50	34	69	81	60	37	431
	75	23	26	141	80	37	752
	100	—	NI	NI	—	—	—
Glass/PPS (1084)	25	13	244	48	28	39	690
	50	15	70	88	67	35	954
	75	16	48	150	94	35	613
	100	—	NI	NI	—	—	—
Graphite/PPS (1085)	25	—	NI	NI	NI	NI	—
	50	16	173	94	70	26	604
	75	17	59	66	50	23	—
	100	26	33	126	88	33	559
Graphite/Poly aryl sulfone (PAS) (1081)	25	—	NI	NI	NI	NI	NI
	50	3	122	24	8	1	79
	75	18	40	47	32	14	211
	100	18	19	60	44	14	173
Graphite/Poly ether sulfone (PES) (1078)	25	—	NI	NI	NI	NI	NI
	50	—	172	11	6	3	145
	75	—	47	41	23	22	88
	100	—	21	65	39	23	189
Graphite/Poly ether ether ketone (PEEK) (1086)	25	—	NI	NI	NI	NI	NI
	50	2	307	14	8	3	69
	75	18	80	54	30	35	134
	100	16	42	85	56	28	252

Table 19.9 Continued

Material system	Irradiance (kW/m²)	WL (%)	Tig (s)	PHRR (kW/m²)	AHRR (kW/m²)	THR (MJ/m²)	ASEA (m²/kg)
Graphite/Poly ether ketone ketone (PEKK) (1079)	25	–	NI	NI	NI	NI	NI
	50	6	223	21	10	15	274
	75	10	92	45	24	20	–
	100	6	53	74	46	24	891
Carbon foam, 25 mm (1444)	25	2	NI	5	2	2	0
	50	6	NI	12	4	12	0
	75	11	NI	31	4	24	170
Carbon foam, 25 mm (1500)	75	25	NI	NI	NI	NI	422
Carbon foam, 25 mm (1501)	75	46	NI	NI	NI	NI	NI
Polymethacrylimide foam, 25 mm (1445)	25	100	63	344	124	37	413
	50	100	8	489	128	38	400
	75	100	3	709	135	41	424
Polymethacrylimide foam, 25 mm (1446)	25	100	57	327	208	64	443
	50	100	12	530	224	66	432
	75	100	4	703	244	73	427
Balsa wood core, D100, 25 mm (1447)	25	36	21	93	30	10	17
	50	74	6	134	51	31	–
	75	87	4	169	74	42	24
PVC foam, 25 mm (1448)	25	85	450	31	3	15	1028
	50	86	13	122	72	22	864
	75	99	5	245	87	27	1100
PVC foam, 25 mm (1449)	25	66	196	143	22	7	775
	50	90	15	177	110	34	1072
	75	94	3	284	68	30	1464
Polyurethane foam, 25 mm (1203)	25	16	43	263	42	14	666
	50	83	10	359	156	209	469
Phenolic foam, 11 mm (1119)	75	22	9	41	22	10	109

WL: weight loss; Tig: time to ignition; PHRR: peak heat release rate; AHRR: average heat release rate at 300 s; THR: total heat release; ASEA: average specific extinction area (smoke); NI: no ignition; –: not available.

Flame retardant materials for maritime and naval applications 557

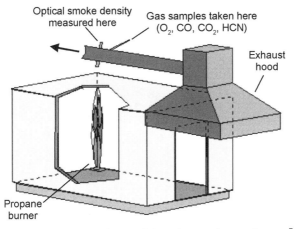

19.2 Schematic of ISO 9705 full-scale room/corner fire test.[54]

mounted both on the walls and the ceiling of the test room. Instrumentation for measuring the heat release rate, smoke production rate and combustion gas concentration is installed in the exhaust duct. The room also contains a heat flux gauge located in the centre of the floor. A schematic of the ISO 9705 test arrangement is shown in Fig. 19.2. A surface material or lining is considered to be 'fire-restricting' if, during a testing time of 20 min according to the standard ISO 9705 room/corner test, the following criteria listed below are fulfilled.

1. The time average of heat release rate (HRR), excluding the HRR from the ignition source, does not exceed 100 kW;
2. The maximum HRR, excluding the HRR from the ignition source, does not exceed 500 kW averaged over any 30 s period of time during the test;
3. The time average of the smoke production rate does not exceed $1.4 \, m^2/s$;
4. The maximum value of the smoke production rate does not exceed $8.3 \, m^2/s$ averaged over any period of 60 s during the test;
5. Flame spread must not reach any further down the walls of the test room than 0.5 m (1.6 ft.) from the floor excluding the area which is within 1.2 m (3.9 ft.) from the corner where the ignition source is located;
6. No flaming droplets or debris of the test sample may reach the floor of the test room outside the area which is within 1.2 m from the corner where the ignition source is located.

Table 19.10 summarizes the ISO 9705 room/corner test results from several polymer composite materials.[56,68–73] The heat release rate test results from the sandwich composite (1257) which contains a balsa core are similar to those of plywood. Whereas most of the polymer composites without a protective covering fail the test criteria, phenolic composites modified with fire retardants passed all the criteria of this test. The sandwich composite with fire insulation also passed all test criteria. For both load-bearing and non-load-bearing polymer

Table 19.10 ISO 9705 room/corner test data for polymer composite materials[56,68–73]

Material	Max HRR kW	AHRR, kW	Peak SPR m^2/s	ASPR m^2/s
Requirement	500	100	8.3	1.4
Plywood*	745	442	NA	NA
Solid Composite, Gl/VE*	287	158	NA	NA
Sandwich Composite* (1257)	631	271	NA	NA
GRP/StructoGard®	57	25	0.500	0.348
GRP/Super Wool®	42	12	0.130	0.058
GRP/intumescent mat	66	16	0.153	0.077
FR Phenolic (1407)	123	59	0.61	0.39
FR Phenolic, USCG ID No. 1	159	62	5.41	1.50
Fire restricting material, USCG ID No. 2	129	31	0.47	0.15
FR Polyester, USCG ID No. 3	677	191	21.7	10.00
FR Vinyl ester, USCG ID No. 4	463	190	32.1	9.08
FR Epoxy, USCG ID No. 5	421	115	26.4	6.39
Coated FR Epoxy, USCG ID No. 6	134	28	3.46	1.45
Textile wall covering, USCG ID No. 7	131	17	0.16	0.10
Polyester, USCG ID No. 8	568	170	4.10	2.28
FR Modified acrylic, USCG ID No. 9	542	109	3.81	0.42

HRR: heat release rate; AHRR: average heat release rate; SPR: smoke production rate; ASPR: average smoke production rate; NA: not available.
* indicates that tests were conducted in open room corner configuration.

composite structures, the US Navy accepts the use of fire insulation as a protective cover over combustible polymer composite structures, such as brominated vinyl esters with a balsa core. The US Navy accepts this practice because it is expected that naval shipboard configuration control will assure the presence of the protective cover over the life of the ship.

19.6.5 Fire resistance and structural integrity

The fire resistance of bulkheads and decks (overheads) is their ability to withstand fire, retain structural integrity and prevent fire spread to an adjoining area for a period of time. They limit the spread of fire by containing it within designated spaces or zones and preventing structural collapse.

Polymer composites consist of fibrous reinforcement with resin binder. The viscoelastic nature of resins used in the fabrication of polymer composites render them inherently prone to degradation in mechanical properties at elevated temperatures due to fire. As shown in the Fig. 19.3, the flexural modulus of glass fibre reinforced vinyl ester composite is significantly reduced during isothermal aging between 121–149 °C (250–300 °F) which corresponds to its glass transition temperature.[74–76] The inherent chemical nature and complexity of polymeric materials do not lend themselves to easy analytical prediction of their structural behaviour when exposed to a high heat flux from a fire source. They burn,

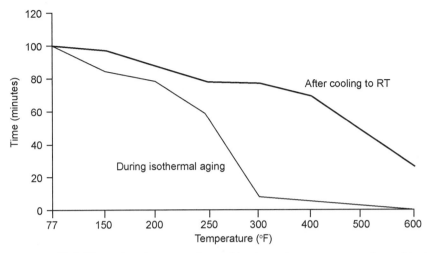

19.3 Effect of ageing at an elevated isothermal temperature on the residual flexural modulus of glass/vinyl ester polymer composite.[74]

spread flame, produce smoke and gases, release heat, chemically degrade, produce char, lose weight, and suffer a reduction in mechanical structural properties. As such, polymer composite-based, fire resisting divisions are typically fire tested with a structural loading to ensure structural integrity under design loads.

International Maritime Organization 'A' Class Divisions

In general, IMO classifies 'A' divisions as those formed by bulkheads and decks (overheads) which comply with the following:

- constructed of steel or other equivalent material;
- they shall be suitably stiffened;
- they shall be so constructed as to be capable of preventing the passage of smoke and flame to the end of the one hour standard fire test;
- they shall be insulated with approved non-combustible materials such that the average temperature of the unexposed side will not rise more than 139 °C above the original temperature, nor will the temperature at any one point, including any joint, rise more than 180 °C above the original temperature, within the time listed below:

- Class A-60   60 min
- Class A-30   30 min
- Class A-15   15 min
- Class A-0   0 min.

The fire resistance of a division is determined by testing in accordance with Resolution A. 754(18), Recommendation on Fire Resistance Tests For 'A', 'B',

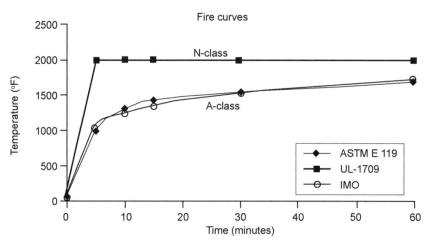

19.4 Comparison of ASTM E 119 (Class A), IMO, and N-Class (UL-1709, Class B) Fire Curves.[81]

And 'F' Class Divisions.[11] It is applicable to merchant vessels with products such as bulkheads, decks, overheads, doors, ceilings, linings, windows, fire dampers, pipe penetrations, and cable transits. It uses the standard IMO time-temperature fire curve in accordance with FTP Code[7] which is similar to the ASTM E 119[77] fire curve typically used in building construction. These fire curves are shown in Fig. 19.4.

High Speed Craft Fire Resisting Divisions

For use in high speed craft constructions, the fire resistance of the polymer composite-based fire resisting divisions is determined in accordance with Resolution MSC.45(65)[19] which, in general, uses IMO A.754(18)[11] test procedures. The test should continue for a minimum of 30 min for fire resisting division 30, or 60 min for fire resisting division 60. In addition, the Resolution MSC.45(65)[19] requires that the load bearing divisions of polymer composites should be tested with the prescribed static load and they should maintain their load-bearing ability within the classification period. It also provides the static loading level for bulkhead and decks (overheads), and also provides the performance criteria for load-bearing ability.

Navy N-Class Divisions

For classification of fire resistant boundaries, the US Navy has introduced N-Class Division system. The Navy N-Class system for classifying fire resistant boundaries is analogous to the commercial IMO system of A-Class divisions which are made of non-combustible materials. The fire resistance test for the A-

Flame retardant materials for maritime and naval applications 561

Class division is conducted in accordance with Resolution A.754(18)[11] which uses the IMO standard time-temperature fire curve. The IMO fire curve is similar to the ASTM E 119[77] fire curve. US Navy N-Class divisions are designed to protect against structural failure and prevent the passage of flame and hot gases when exposed to a UL-1709[78] rapid rise hydrocarbon (Class B) fire curve, after shock testing in accordance with MIL-S-901.[79] The shock test prior to fire resistance test ensures the capability of passive fire protection materials and their attachment methods to survive in a combat environment.

The US Navy previously used fire curve in accordance with the ASTM E 119[77] test method for all fire zone bulkheads. In 1993, Naval Research Laboratory (NRL) performed post-flashover fire tests in ex-USS Shadwell.[80] The results indicated that the UL-1709 fire curve[78] more closely approximated the thermal conditions in the ship compartment during the post flashover fire. The N-Class hydrocarbon pool fire curve[78] uses the more severe temperature and heat flux requirements as shown in Fig. 19.4. Hydrocarbon pool fire exposure provides an average total heat flux of $204 \pm 16 \, \text{kW/m}^2$ (65,000 $\pm$ 5,000 Btu/h-ft^2) and an average temperature of $1,093 \pm 111\,°C$ ($2,000 \pm 200\,°F$).

The minimum fire test duration for N-Class fire resistance test is 30 minutes. In addition to the requirement for no passage of flame to the unexposed face, N-Class Divisions are also designed to prevent an excessive temperature rise similar to IMO 'A' Class divisions. The average unexposed face temperature rise should not be more than 140 °C and the temperature rise recorded by any of the individual unexposed face thermocouple should not be more than 180 °C during the periods given below for each classification:

- Class N-60 60 minutes
- Class N-30 30 minutes
- Class N-0 0 minutes (there is no unexposed face temperature rise requirement).

The fire resistance test for the N-Class division is similar, but with exceptions, to Resolution A.754(18)[11] with the IMO standard time-temperature fire curve replaced with the N-Class fire curve shown in Fig. 19.4. The test method involves mounting the structural component in a load bearing restraint frame. The test specimen is introduced to the furnace and acts as one side of the furnace and may be mounted in a vertical (bulkhead) or horizontal (deck) orientation. The specimen is subjected to the heated furnace environment for the desired duration. If the endpoint criteria are not reached prior to the end of the test period, the assembly is rated as acceptable for the test period, e.g. 30 or 60 minutes. The sample may be tested structurally loaded or unloaded.

Polymer Sandwich Composite Fire Resistant Divisions (N-Class)

Polymer sandwich composites were briefly described in Section 19.5. Table 19.11 provides the summary of material designations for sandwich composites

Table 19.11 Description of test specimen for fire resistance tests of polymer composites[54,81]

Material ID	Composite description
Material A	Bulkhead, carbon/vinyl ester (brominated) face sheets (7.2 mm) with balsa core (50 mm), no insulation, load of 7,120 kg/m (4,780 lb/ft) was applied.
Material B	Bulkhead, carbon/vinyl ester (brominated) face sheets (6 mm) with balsa core (76 mm), 16 mm thick Structogard, load of 5,500 kg/m (3,700 lb/ft) was applied.
Material C	Deck (overhead), carbon/vinyl ester (brominated) face sheets (5.3 mm) with balsa core (76 mm), 16 mm Structogard, load of 1,221 kg/m^2 (250 lb/ft^2) was applied.
Material E	Bulkhead, glass vinyl ester (brominated) face sheets (6 mm) with balsa core (76 mm), 32 mm Structogard, load of 9,000 (kg/m) 6,000 lb/ft was applied.
Material F	Bulkhead, glass vinyl ester (brominated) face sheets (6 mm) with balsa core (76 mm), no insulation; a load of 9,000 kg/m (6,000 lb/ft) was applied.
Material G	Deck (overhead), glass vinyl ester (brominated) face sheets (6 mm) with balsa core (76 mm), 32 mm Structogard, load of 871 kg/m^2 (180 lb/ft^2) was applied.

evaluated in fire resistance tests.[81] The large-scale fire resistance tests were conducted in the SouthWest Research Institute (SWRI) vertical (bulkhead) and horizontal (deck, overhead) furnaces using the UL-1709 (N-Class) fire curve. Instrumentation consisted of thermocouples and deflection gauges placed at various locations in the test specimen. Load was applied to the assembly using hydraulic jacks. After the 30-minute fire test, the hose stream test was conducted as described in Resolution A.754 (18), appendix A.I.

Figure 19.5 shows the thermocouple locations at the face sheet/balsa core interface on the exposed (fire) side, centre of balsa core, and face sheet/balsa core interface on the unexposed (cold) side. Figure 19.6 shows the temperatures at exposed face sheet/balsa interface during large-scale fire tests using UL 1709 (N-Class) fire exposure. The temperatures at this interface are above the glass transition temperature of 149 °C (300 °F) for all composites shown in this figure.

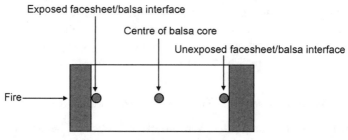

19.5 Locations of thermocouples for sandwich composite thermal profile.

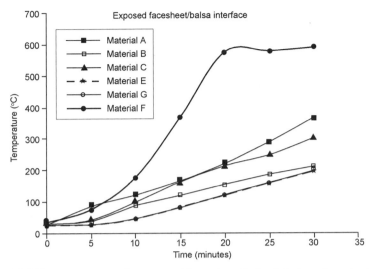

19.6 Temperatures at the exposed face sheet/balsa interface for selected composites.

Figure 19.7 shows the temperatures in the centre of balsa core during large-scale fire tests using UL 1709 (N-Class) fire exposures. All composites, except Material F, are at or below 93 °C (200 °F). The temperature in the centre of balsa core for the Material F (glass reinforced, uninsulated sandwich composite) is significantly higher. It may be that the glass reinforced composite skin, as well as part of the balsa core, has thermally degraded and fallen off and the flames have penetrated into some part of the balsa core.

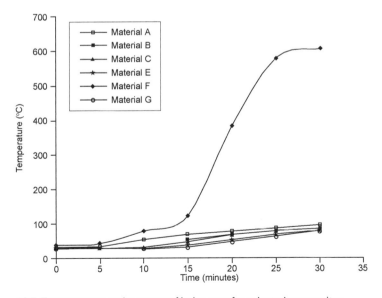

19.7 Temperatures at the centre of balsa core for selected composites.

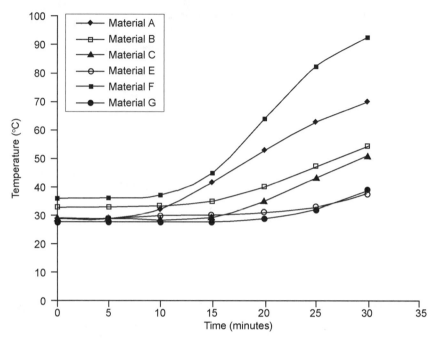

19.8 Temperatures at the unexposed face sheet/balsa interface for selected composites.

Figure 19.8 shows the temperature at the unexposed face sheet/balsa interface during large-scale fire tests using UL 1709 (N-Class) fire exposure. The temperatures at this location are at or below 93 °C (200 °F) for all composites. The highest temperature recorded at the unexposed face sheet/balsa interface was by Material F which may have lost its fire-exposed face sheet during the test. However, since none of the composites collapsed during the fire tests, this may suggest that unexposed face sheet of Material F retains most of its structural load bearing capability and is strong enough by itself to support structural load without buckling or collapse.

Figure 19.9 shows the temperature gradient through the thickness of the sandwich polymer composite test during the fire test. At the end of 30-minute test period, there is a large thermal gradient between the exposed face sheet (400 °C) and unexposed face sheet (35 °C). As such, polymer sandwich composites need to be cooled in order to prevent re-flashing or re-ignition during fire fighting operations.

19.7 Summary

Owing to their inherent advantages of light weight, corrosion resistance, and extended service life, plastics and polymer composites are being utilized in a

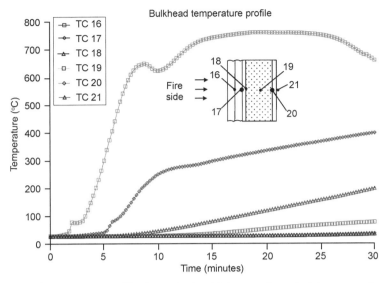

19.9 Temperature profile through the thickness of sandwich composite (bulkhead).

variety of maritime and naval applications traditionally reserved for steel and aluminum. These attributes are also highly desirable in other transportation modes such as commercial aviation, surface mass transit vehicles and infrastructure. Advanced materials technology is driving our next generation of fast moving and manoeuverable naval vessels. The expanded use of plastics and polymer composites in confined spaces of maritime and naval vessels also raises the issue of the inherent vulnerability of plastics to common fire threats found in maritime operations.

Fire and safety requirements of vessels operating in international waters are contained in the International Convention for the Safety of Life at Sea (SOLAS 2004) and in the International Maritime Organization International Code of Safety for High Speed Craft (HSC Code). Relevant test methods and acceptance criteria are published in the Fire Test Procedure Code (FTP Code).[7] The fire test procedures cover fire performance such as non-combustibility (Part 1), smoke and toxicity (Part 2), fire resisting divisions (Part 3), test for fire door control systems (Part 4), surface flammability (Part 5), primary deck coverings (Part 6), textiles and films (Part 7), upholstered furniture (Part 8), bedding components (Part 9), fire restricting materials for high speed craft (Part 10), and fire resisting divisions in high speed craft (Part 11).

There are several approaches to improve the fire retardancy of plastics and polymer composites. These include flame retardants, intumescent additives in the bulk of the polymer, and intumescent coating or fire insulation protective covering. In general, when fire protection is desired against a small ignition source, such as a cigarette, electrical spark, or a small trash can, modification with flame retardants may be satisfactory for short exposure times. When fire protection is desired against a large fire, such as a fully developed fire in a compartment, fire insulation may be required to protect involvement of polymer composite and structural integrity of fire resisting divisions for periods of 30–60 minutes. Any protective covering used to protect a polymer composite against a large fire threat must also be able to remain on the composite system after a shock event and during a fire event in naval vessels. These protective coverings must also be tested for their long term effectiveness and durability.

New polymers with high thermal stability are under development, and some of them are already in the market. However, some of these advanced polymers tend to be cost prohibitive in maritime vessels due to the need for high temperature processing. Such materials are more suitable for use in aircraft industry where structural properties at elevated temperatures are more important, for example in situations such as the leading edge of a fast moving aircraft.

19.8 Sources of further information and advice

- www.astm.org
- www.bfrl.nist.gov
- www.dt.navy.mil
- www.faa.gov
- www.iafss.org
- www.imo.org
- www.nfpa.org
- www.sfpe.org
- www.swri.org
- www.uscg.mil
- www.uscg.mil/hq/g-m/mse4
- www.usfa.fema.gov
- ASTM E603-98, Standard Guide for Room Fire Experiments, ASTM Fire Standards, American Society for Testing and Materials, Conshohocken, PA.
- *Handbook of Fire Protection Engineering*, Second Edition, National Fire Protection Association, Quincy, MA, 1995.
- Federal Railroad Administration, *Guidelines for Selecting Materials to Improve Their Fire Safety Characteristics*, Federal Register, 49:162, 1984.
- Urban Mass Transportation Administration, *Recommended Fire Safety Practices for Rail Transit Materials Selection*, Federal Register, 49:158, 1984.
- W. Hathaway, Fire Safety in Mass Transit Vehicle Materials, *Proc. 36th International SAMPE Symposium, 1900* (1991).
- R.D. Peacock, R.W. Bukowski, W.W. Jones, P.A. Reneke, V. Babrauskas,

and J.E. Brown, *Fire Safety of Passenger Trains*, NIST TN 1406, National Institute of Standards and Technology, 1994.
- National Railroad Passenger Corporation, *Specification for Flammability, Smoke Emissions and Toxicity*, Amtrak Specification No. 352, 1991.
- B.C. Levin, A.J. Fowell, M.M. Birky, M. Paabo, A. Stolte, and D. Malek, Further Development of a Test Method for the Assessment of the Acute Inhalation Toxicity of Combustion Products, NBISR 82-2532, National Bureau of Standards, 15, 1982.
- R.G. Hill, T.I. Eklund, and C.P. Sarkos, Aircraft Interior Panel Test Criteria Derived from Full-Scale Fire Tests, Federal Aviation Administration, DOT/FAA/CT-85/23, 1985.
- C.P. Sarkos and R.G. Hill, Heat Exposure and Burning Behavior of Cabin Materials During an Aircraft Post-crash Fuel Fire, National Materials Advisory Board, National Research Council, NMAB Report 477-2, National Academy Press, 25, 1995.
- D. A. Nollen, Flammability Regulations Affecting Advanced Composite Materials, *J. Fire Sciences*, 8, 227–238 (1990).
- R. E. Lyon, *Fire-Safe Aircraft Cabin Materials, Fire and Polymers II*, G. L. Nelson ed. American Chemical Society Symposium Series 599, 618 (1995).
- National Materials Advisory Board, Fire and Smoke Resistant Interior Materials for Commercial Transport Aircraft, National Research Council, NMAB Report 477-1, National Academy Press, 1995
- V. Wigotsky, *Plastics Engineering*, Volume 44, Number 7, July 1988.
- S. Goodman, *Handbook of Thermoset Plastics*, Noyes Publications, 1986.
- D.B. Miracle and S.L. Donaldson, *ASM Handbook*, Volume 21, Composites, ASM International, Metals Park, Ohio.
- A. Knop, L.A. Pilato, *Phenolic Resins Chemistry, Applications and Performance*, Springer-Verlag, Printed in Germany, 1985, ISBN 3-540-15039-0.
- B. Karlsson, J. G. Quintiere, *Enclosure Fire Dynamics*, CRC Press, ISBN 0-8493-1300-7.
- IMO, MSC Circular 732, Interim Guidelines on the Test Procedure for Demonstrating the Equivalence of Composite Materials to Steel Under the Provisions of the 1974 SOLAS Convention, 28 June 1996.

19.9 References

1. Usman Sorathia, Richard Lyon, Richard G. Gann, and Louis Gritzo, Materials and Fire Threat, *Fire Technology*, 33 (3), 260–275 (1997).
2. T. Herzberg, SP Technical Research Institute of Sweden, *BrandPosten* No. 35, 2007, Pg. 8.
3. Naval Safety Center Database, 1983–87, Norfolk, VA.
4. T. Ohlemiller and T. Cleary. Upward Flame Spread on Composite Materials. Chapter 28 in *Fire and Polymers II – Materials And Tests for Hazard Prevention*, G.

L. Nelson, Editor, American Chemical Society, Washington DC (1995).
5. U. Sorathia, R E. Lyon, T. Ohlemiller, and A. Grenier. A Review of Fire Test Methods and Criteria for Composites. *Society of Advanced Materials and Process Engineering (SAMPE) Journal*, Vol. 33, No. 4, (July/August 1997), 23–31.
6. SOLAS, Consolidated Edition, 2004, consolidated text of the International Convention for the Safety of Life at Sea, 1974, and its Protocol of 1978: articles, annexes and certificates.
7. Fire Test Procedure (FTP) Code, International Code for Application of Fire Test Procedures, Resolution MSC.61 (67), International Maritime Organization, London (1998).
8. 2000 HSC Code, *International Code of Safety for High Speed Craft, 2000*, Resolution MSC.97(73), International Maritime Organization, London (2001).
9. ISO 1182:2002, *Reaction to fire tests for building products – Non-combustibility test*, ISO copyright office, Case postale 56, CH-1211 Geneva 20 (www.iso.ch).
10. ISO 5659-2:1994, *Determination of optical density by a single chamber test*, ISO copyright office, Case postale 56, CH-1211 Geneva 20 (www.iso.ch).
11. Resolution A.754(18), *Recommendation on Fire Resistance Tests for 'A', 'B' and 'F' Class Division, Fire Test Procedures Code*, International Maritime Organization, London (1998), www.imo.org.
12. Resolution A.653(16), Recommendation on Improved Fire Test Procedures for Surface Flammability of Bulkhead, Ceiling and Deck Finish Materials, Fire Test Procedures Code , International Maritime Organization, London (1998), www.imo.org.
13. Resolution A.687, Fire Test Procedures for Ignitability of Primary Deck Coverings, Fire Test Procedures Code, International Maritime Organization, London (1998), www.imo.org.
14. Resolution A.563(14), Recommendation on Test Method for Determining the Resistance to Flame of Vertically Supported Textiles and Films, Fire Test Procedures Code, International Maritime Organization, London (1998), www.imo.org.
15. Resolution A.652(16), Recommendation on Fire Test Procedures for Upholstered Furniture, Fire Test Procedures Code, International Maritime Organization, London (1998), www.imo.org.
16. Resolution A.688(17), Fire Test Procedures for Ignitability of Bedding Components, Fire Test Procedures Code, International Maritime Organization, London (1998), www.imo.org.
17. Resolution MSC.40(64), Standard for Qualifying Marine Materials for High Speed Craft as Fire-Restricting Materials, Fire Test Procedures Code, International Maritime Organization, London (1998), www.imo.org.
18. Resolution MSC.90(71), Amendment to the Standard for Qualifying Marine Materials for High Speed Craft as Fire Restricting Materials (Resolution MSC.40(64)). Fire Test Procedures Code, International Maritime Organization, London (1998), www.imo.org.
19. Resolution MSC.45 (65), Test Procedures for Fire Resisting Divisions of High Speed Craft, Fire Test Procedures Code, International Maritime Organization, London (1998), www.imo.org.
20. MIL-STD-1623, Fire Performance Requirements and Approved Specifications for Interior Finish Materials and Furnishings (2006) (http://assist.daps.dla.mil/quicksearch).
21. NAVSEA Design Data Sheet DDS-078-1, Composite Materials, Surface Ships,

Topside Structural and Other Topside Applications-Fire Performance Requirements, (August 2004), Department of Defense Resource Center (http://www.dodsbir.net).
22. MIL-STD-2031 (SH), Fire and Toxicity Test Methods and Qualification Procedure for Composite Material Systems Used in Hull, Machinery, and Structural Applications Inside Naval Submarines, February (1991).
23. *Fire Safety Aspects of Polymeric Materials, Volume I, Materials: State of the Art*; A report by National Materials Advisory Board, National Academy of Sciences, Technomic Publishing Co., Inc. (1977), ISBN 0-87762-222-1.
24. ASTM E 1354, Standard Test Method for Heat and Visible Smoke Release Rate for Materials and Products Using an Oxygen Consumption Calorimeter. ASTM Fire Standards, American Society for Testing and Materials, West Conshohocken, PA. (www.astm.org)
25. R. E. Lyon, R. Walters, and S. Gandhi. Combustibility of Cyanate Esters. Federal Aviation Administration, Final Report, DOT/FAA/AR-02/44 (June 2002).
26. R. N. Walters. Fire Resistant Cyanate Ester-Epoxy Blends. Federal Aviation Administration, Final Report, DOT/FAA/AR-02/53, (May 2002).
27. T.M. Keller and T.R. Price. Amine-Cured Bisphenol-Linked Phthalonitrile Resins. *J. Macrmol. Sci.-Chem.*, A18, 931 (1982).
28. U. Sorathia and I. Perez. Navy R & D Programs: Materials Technology for Fire Safety of Composite Structures. In *Proceedings of the Society of Advanced Materials and Process Engineering*, Long Beach, CA (May 1–5 2005) (www.sampe.org).
29. J. D. Lichtenhan. Reduced Flammability Vinyl Ester Resin Containing No Halogen for use in Large Composite Ship Surface Structures via Nanocomposite Technology, Hybrid Plastics Inc. (www.hybridplastics.com).
30. A. F. Grand and C. A. Wilkie, editors. *Fire Retardancy of Polymeric Materials*. Marcel Dekker Inc. (2000), ISBN 0-8247-8879-6 (www.dekker.com).
31. A. B. Morgan, J. W. Gilman, T. Kashiwaga, and C. L. Jackson. Flammability of Polymer Clay Nanocomposites. In *Proceedings of the Fire Retardant Chemical Association*, Washington D.C. (March 12–15, 2000).
32. A. El Harrak, G. Carrot, J. Oberdisse, E. Baron and F. Boue, *Macromolecules*, 37, 6376-6384 (2004).
33. A. Shabeer, K. Chandrashekhara, T. Schuman, N. Phan and T. Nguyen, Mechanical Thermal and Flammability Properties of Pultruded Soy-Based Nanocomposites. In *Proceedings of the Society of Advanced Materials and Process Engineering*, Baltimore, Maryland (June 3–7, 2007).
34. M.C. Costache, E.M. Kanugh, U. Sorathia and C. A. Wilkie, Fire Retardancy of Polyureas, *Journal of Fire Sciences*, 24 (6), 433–444 (2006).
35. G. Chigwada and C. A. Wilkie. Enhanced Fire Retardancy of Vinyl Ester Resins by Combinations of Additives. Department of Chemistry, Marquette University, PO Box 1881, Milwaukee, WI 53201.
36. U. Sorathia, T. Gracik, C. Beck. Flame Retardancy of Polyurea for Naval Applications. In *Proceedings of the 18th Annual BCC Conference on Flame Retardancy* (May 21–23 2007), Stamford, CT.
37. U. Sorathia, T. Gracik, J. Ness, A. Durkin, F. Williams, M. Hunstad, and F. Berry. Evaluation of Intumescent Coatings for Shipboard Fire Protection. *J. Fire Sci.*, 21 (6), 423–450 (2003).
38. J. E.Gagorik, J. A. Corrado, and R. W. Kornbau. An Overview of Composite Developments for Naval Surface Combatants. In *Proceedings of the 36th International SAMPE Symposium and Exhibition*, Volume 36 (April 1991).

39. J. E. Beach and J. L. Cavallaro. *Structures and Materials: An Overview*. Naval Surface Warfare Center Carderock Division Technical Digest (December 2001).
40. E. T. Camponeschi Jr. and K. M. Wilson. *The Advanced Enclosed Mast Sensor System: Changing U.S. Navy Ship Topsides for the 21st Century*. Naval Surface Warfare Center Carderock Division Technical Digest (December 2001).
41. *Marine Composites*, Second Edition, Eric Greene Associates, ISBN 0-9673692-0-7, www.marinecomposites.com.
42. National Research Council. *New Materials for Next Generation Commercial Transport*. NMAB-476, National Materials Advisory Board, National Academy Press (1996).
43. U. Sorathia, T. Dapp and C. Beck. Fire Performance of Composites. *Mater. Eng.*, 109 (9), 10–12 (1992).
44. U. Sorathia and C. P. Beck. Fire-Screening Results of Polymers and Composites. In *Proceedings of Improved Fire and Smoke Resistant Materials For Commercial Aircraft Interiors*, National Research Council, Publication NMAB-477-2, National Academy Press, Washington, DC (1995).
45. L. English, editor. Part I, The Basics; Part II, Reinforcements; Part III, Matrix Resins. *Mater. Eng.*, 4 (9) (1987).
46. MIL-HDBK-23A, Structural Sandwich Composites, Department of Defense, Washington, DC (2002).
47. MIL-HDBK-17-3F, *Composite Materials Handbook, Volume 3: Polymer Matrix Composites Materials Usage, Design, and Analysis*, June 2002, Department of Defense, Washington, DC (2002).
48. (http://www.fas.org/man/dod-101/sys/ship/lpd-17.htm).
49. (http://www.naval-technology.com/projects/visby).
50. http://www.mil.no/multimedia/archive/00027/Orkla_report_27673a.pdf, Report from the Technical Expert Group submitted to The Norwegian Defense Logistics organization (25 September 2003).
51. ASTM E 84, *Standard Test Method for Surface Burning Characteristics of Building Materials*. ASTM Fire Standards, American Society for Testing and Materials, West Conshohocken, PA.
52. NFPA 301, *Safety to Life from Fire on Merchant Vessels*. National Fire Protection Association, Quincy, MA (2001).
53. ASTM E 162, Standard Test Method for Surface Flammability of Materials Using a Radiant Heat Energy Source. 1997 *Annual Book of ASTM Standards*, Vol. 04.01: (1997).
54. U. Sorathia, T. Gracik, C. Beck and A. Le. Fire Performance of Glass Reinforced Sandwich Composite for Naval Applications. In *Proceedings of the SAMPE*, Long Beach, CA (April 30–May 4, 2006).
55. U. Sorathia. Fire Performance of Composites in Marine Applications. In *Proceedings of the SAMPE*, Long Beach, CA (May 23-27, 1999).
56. A.T. Grenier, M.L. Janssens and L.Nash. Developing Cone Calorimeter Acceptance Criteria for Materials Used in High Speed Craft. *Fire Mater.*, 24, 29–35 (2000).
57. ASTM E 662-93, *Standard Test Method for Specific Optical Density of Smoke Generated by Solid Materials*. ASTM Fire Standards, American Society for Testing and Materials, West Conshohocken, PA.
58. ISO 9705, *Fire Tests – Full-Scale Room Test for Surface Products*. International Organization for Standards, Geneva, Switzerland, (1993).
59. ISO/DIS 19702, *Toxicity testing of fire effluents – Guide for analysis of gases and vapours in fire effluents using FTIR gas analysis*. International Organization for

Standards, Geneva, Switzerland (2004).
60. *Fire & Smoke, Understanding the Hazards*. Committee on Fire Toxicology, National Research Council, National Academy Press, Washington, DC (1986).
61. A. Tewarson. Generation of Heat and Chemical Compounds in Fires. *Fire Protection Handbook*, Society of the Fire Protection Engineering, J.P. Dinenno, editor (1988).
62. R. A. DeMarco. Composite Applications at Sea: Fire Related Issues. In *Proceedings of the 36th International SAMPE Symposium* (April 15–18, 1991) (www.sampe.org).
63. ISO 5660, *Reaction to fire tests – Heat release, smoke production and mass loss rate, Part 1: Heat release rate (cone calorimeter method)*. ISO copyright office, Case postale 56, CH-1211 Geneva 20 (www.iso.ch).
64. D. P. Macaione and A. Tewarson. *Flammability Characteristics of Composite Materials*. ACS symposium series 425, Washington, DC, 542–565 (1990).
65. J. E. Brown, E. Braun and W. H. Twilley. *Cone Calorimeter Evaluation of the Flammability of Composite Materials*. NBSIR 88-3733, NIST, Gaithersburg, MD (March 1988).
66. J. H. Koo, B. Muskopf, U. Sorathia, P. Van Dine, B. Spencer and S. Venumbaka. Characterization of Fire Safe Polymer Matrix Composites for Marine Applications. In *Proceedings of SAMPE 2001*, Long Beach, CA (May 6–10 2001).
67. R. N. Walters and R. E. Lyon. Flammability of Polymer Composites. In *Proceedings of SAMPE 2007*, Baltimore, MD (June 3–7 2007).
68. M. L. Janssens, A. Garabedian and W. Gray. Establishment of International Standards Organization (ISO) 5660 *Acceptance Criteria for Fire Restricting Materials on High Speed Craft*. United States Coast Guard Report, CG-D-22-98 (1998).
69. Janssens, M. L. and A. L. Orvis, *Fire, Smoke, and Toxicity of Composites on High Speed Craft*, Technical Report to DOT: US Coast Guard R&D Center, Report No. CT CG-D-27-98, Southwest Research Institute (October 1998).
70. C. L. Beyler, S. P. Hunt, B. Y. Lattimer, N. Iqbal, C. Lautenberger, N. Dembsey, J. Barnett, M. Janssens, S. Dillon and A. Grenier. Prediction of ISO 9705 Room/Corner Test Results. USCG Report, CG-D-22-99, Vol I and II (November 1999).
71. B. Y. Lattimer and U. Sorathia. Thermal Characteristics of Fires in a Noncombustible Corner. *Fire Safety J.*, 38, 709–745 (2003).
72. B. Y. Lattimer and U. Sorathia. Thermal Characteristics of Fires in a Combustible Corner. *Fire Safety J.*, 38, 747–770 (2003).
73. B. Y. Lattimer, S. P. Hunt, M. Wright and U. Sorathia. Modeling Fire Growth in a Combustible Corner. *Fire Safety J.*, 38, 771–796 (2003).
74. C. A. Harper, editor-in-chief. *Handbook of Building Materials for Fire Protection*, McGraw-Hill, ISBN 0-07-138891-5, pp. 9–56 (2004).
75. U. Sorathia and T. Dapp. Structural Performance of Glass/Vinyl Ester Composites at Elevated Temperatures. *SAMPE Journal*, 33 (4), pp. 53–58 (July/August 1997).
76. U. Sorathia, C. Beck and T. Dapp. Residual Strength of Composites during and after Fire Exposure. *J. Fire Sci.*, 11 (3), pp. 255–270 (1993).
77. ASTM E119-98, *Standard Test Methods for Fire Tests of Building Construction and Materials*. American Society for Testing and Materials, Vol. 4.07 Building Seals and Sealants; Fire Standards; Dimension Stone, Conshohocken, PA, (1999).
78. UL 1709, *Rapid Rise Fire Tests of Protection Materials for Structural Steel*. Underwriters Laboratories, Inc., Northbrook, IL (1991).
79. MIL-S-901, Military Specification, Shock Tests, H.I. (High Impact) Shipboard Machinery, Equipment and Systems, Requirements for.

80. Darwin, R.L., Leonard, J.T., and Scheffey, J.L., Fire Spread by Heat Transmission Through Steel Bulkheads and Decks, *Proceedings of IMAS 94, Fire Safety on Ships*, London, England, Institute of Marine Engineers, ISBN 0-907206-57-3, Ppr 6, p 71 (May 1994).
81. U. Sorathia. Fire Resistant Divisions in U.S. Naval Ships. In *Proceedings of SAMPE 2007*, Baltimore, MD (June 3–7 2007) (www.sampe.org).

20
Materials with reduced flammability in aerospace and aviation

R E LYON, Federal Aviation Administration, USA

Abstract: This chapter reviews the history of aircraft fire regulations, the types of materials used in transport category aircraft, their location in the fuselage, and the test methods and requirements for their flammability.

Key words: aircraft, fire, flammability, plastics, regulations.

20.1 Introduction

Fire safety is designed into transport aircraft in order to prevent in-flight fires and mitigate post-crash fires, which account for about 20% of the fatalities resulting from airplane accidents.[1,2] Before an airplane can carry passengers, the manufacturer must satisfy airworthiness criteria[3] established by the regulatory authorities that certify the airplane design. In addition to the certification requirements, manufacturers may impose supplementary safety criteria that go beyond the regulatory requirements.[4] The regulatory authorities will issue a type certificate for a particular design (e.g., Boeing 747-100 Series) if the design complies with the regulations, including the fire-worthiness (flammability) requirements for all of the parts used to construct the airframe, power plants, and cabin interior. An airplane having new or novel materials or designs not envisioned when the regulations were promulgated can be certified under Special Conditions if the manufacturer demonstrates a level of safety equivalent to the existing certification requirements. Once the airplane is certified, changes in the construction of cabin materials and components can only be made with the approval of the regulatory authorities and only if the materials/components meet the minimum fire performance standards for the design. The Federal Aviation Administration (FAA) in the United States and its European counterpart, the European Aviation Safety Agency (EASA) are two of the regulatory authorities that certify the airworthiness of aircraft in addition to the individual countries of registry. Manufacturers of transport category (>20 passenger) aircraft in service worldwide include Boeing, Airbus Industries, Embraer, and Bombardier. Bilateral Aviation Safety Agreements are used to facilitate approvals between countries so that aircraft can be certified and sold outside of their country of manufacture.

Once the airplane is manufactured, certificated, and sold to airlines, it is the responsibility of the owner airlines to assure the safe operation of their airplanes in accordance with flight and maintenance standards imposed and monitored by the local regulatory authorities (e.g., FAA, EASA) in cooperation with the International Civil Aviation Organization (ICAO) and the aviation authorities of Canada, Brazil, Israel, China and Russia. These aviation authorities ensure continued airworthiness through working arrangements aimed at harmonizing standards and promoting best practices in aviation safety worldwide. Mandatory changes to materials and components of in-service aircraft may be required if an unsafe condition is identified by the regulatory authorities. In this case, the regulatory authority issues an Airworthiness Directive that mandates a fire safety upgrade or replacement according to a prescribed schedule. Materials and components of in-service aircraft can only be changed voluntarily if the operator or manufacturer demonstrates an equivalent level of safety to the certificated design. As with the manufacturers, the airline operators often impose additional internal safety criteria that go beyond the regulatory requirements.[4]

The following sections describe the history of aircraft fire regulations, the types of materials used in transport category aircraft and their location in the fuselage, as well as the test methods and requirements for the fire-worthiness of aircraft cabin materials.

20.2 History of aircraft fire regulations

Aircraft fire regulations cover many aspects of aircraft as well as many different kinds of aircraft including small airplanes, large airplanes, airplanes that combine passenger and cargo spaces, helicopters, etc. In the United States the rules governing Aeronautics and Space are codified in Title 14 of the Code of Federal Regulations (14 CFR).[3] Parts 1 through 59 of Chapter 1, 14 CFR, define the rules and responsibilities of the US aviation regulatory authority, the Federal Aviation Administration (FAA). The 14 CFR Parts 1–59 are commonly referred to as the Federal Aviation Regulations (FARs). The Airworthiness Standards for Cargo and Commuter Category Airplanes are the subject of 14 CFR Part 23, while the Airworthiness Standards for Transport Category Airplanes are the subject of 14 CFR Part 25. The paragraphs in 14 CFR Part 25 that govern the flammability of materials in transport category aircraft are Paragraphs 853 (Compartment Interiors), 855 (Cargo or Baggage Compartments), 856 (Thermal/ Acoustic Insulation), and 869 (Electrical Wiring), and these will be referred to in the sections that follow as FAR Part 25.853 or Section (§) 25.853, etc. Further definition, test criteria, and guidance for showing compliance with § 25.853, § 25.855, § 25.856, and § 25.869 are given in Appendix F of 14 CFR 25 and the *Materials Fire Test Handbook*,[5] which is updated periodically by the FAA to reflect best practices. These fire regulations are derived from research and development conducted principally by the Federal Aviation Administration (FAA) in anticipation of, or in response to, aircraft incidents and accidents.[2,6] Since the US fire standards are used internationally, and typically adopted by

Materials with reduced flammability in aerospace and aviation 575

reference, the discussion of fire regulations will focus on the history of the FARs as outlined by Peterson.[7]

In 1945 the first regulatory fire performance requirements covering the certification of transport category aircraft design in the United States were introduced as Civil Air Regulation 4b (CAR 4b) by the Civil Aeronautics Agency (CAA). The first jet transports manufactured by Boeing (707, 727) and McDonnell Douglas (DC-8, DC-9) were certified to CAR 4b.

In 1965, the US Department of Transportation (DOT) was created and the CAA was reorganized and made a part of the DOT as the Federal Aviation Administration (FAA). Simultaneously, the existing CARs were reissued without change as parts of the FAR, with CAR 4b becoming FAR Part 25. The FAA has periodically introduced new regulatory requirements through amendments to FAR Part 25 or Special Conditions. Several of these have been issued that have imposed more stringent flammability requirements on interior cabin furnishings. These Part 25 amendments are effective only for new designs and do not directly affect airplanes being manufactured to existing design criteria. The mechanism used by the FAA to mandate changes to existing and/or newly manufactured airplanes built to existing designs is to amend FAR Part 121, which covers what requirements an airplane must meet, in addition to its Part 25 certification basis, before the airlines can operate it in passenger service. Included in the requirements of CAR 4b and FAR Part 25 Original Issue were that for each compartment to be used by the crew or passengers the materials must be at least 'flash-resistant,' which meant that the construction, when tested horizontally using a Bunsen burner with a 15-second ignition time, will have a burn rate less than 508 mm (20 in.) per minute. Large area materials including wall and ceiling linings and the covering of upholstery, floors, and furnishings were held to a higher standard of 'flame-resistant,' which reduced the maximum burn rate to 102 mm (4 in.) per minute in the 15-second horizontal Bunsen burner test. Receptacles for towels, paper or waste were required to be 'fire-resistant,' which meant that the construction, when tested at a 45 degree angle with respect to horizontal using a Bunsen burner with a 30-second ignition time, allows no flame penetration, self-extinguishes in less than 15 seconds, and glows for less than 10 seconds after removal of the Bunsen burner flame.

In 1967 the FAA introduced Amendment 15 to FAR Part 25 (FAR 25-15) to reduce the ignitability of aircraft materials. Amendment 25-15 imposed the more stringent requirement of a vertical Bunsen burner test having a 12-second ignition time and introduced Appendix F, which is a description of most of the test procedures used to demonstrate compliance with the flammability requirements in FAR Part 25. The Amendment 25-15 vertical Bunsen burner test applied to interior wall and ceiling panels, draperies, structural flooring, baggage-racks, partitions, thermal insulation and coated fabric insulation covering. Amendment 25-15 included a char length of 20 cm

(8 in.) for the 12-second vertical Bunsen burner test and modified the acceptance criteria for the horizontal Bunsen burner test to include a maximum char length of 10 cm (4 in.). The FAR Part 121 was not amended to require any changes to existing or newly manufactured aircraft when FAR Part 25-15 was introduced.

In 1972, the FAA introduced Amendment 32 to FAR Part 25 (Amendment 25-32), which was developed by the Aerospace Industries Association (AIA) in their 1966–1967 Crashworthiness Program. FAR Part 25-32 superseded FAR 25-15 for new designs and is still required for certification of new designs (see Section 20.4, Fire Test Methods for Aircraft Materials). Amendment 32 added a requirement for electrical wire insulation that was a 30-second, 60-degree Bunsen burner test and upgraded the requirement for large area interior material (ceiling and wall panels, stowage compartment materials, partitions, galley structure, large cabinet walls, and structural flooring) to a 60-second vertical Bunsen burner test (see 60VBB for performance criteria). The 12-second vertical Bunsen burner test of FAR 25-15, modified to include flame times of sample and dripping, was retained for large area materials not specifically covered by the new 60-second rule, e.g. floor covering, textiles, cargo liners, seat trays, air ducting, insulation blankets, transparencies, thermoformed parts, electrical conduit. All materials not covered by the vertical Bunsen burner test with 12- or 60-second ignition were tested in the 15-second horizontal Bunsen burner test of the 1965 FAR Part 25 Original Issue. The Boeing 747 airplane was certified in 1969 to special conditions equivalent to FAR 25-32, while the McDonnell Douglas DC-10 and Lockheed L-1011 were certified to FAR Part 25-32 in 1972. FAR Part 121 was not amended to require any changes in existing or newly manufactured aircraft when FAR Part 25-32 went into effect.

In the early 1970s the FAA initiated proposed rulemaking activities that involved smoke and toxic gas emissions of burning cabin materials that would amend FAR Parts 25 and 121. Although these proposals were withdrawn in 1978 in favor of a more systematic approach recommended by the Special Aviation Fire and Explosion Reduction (SAFER) committee, aircraft manufacturers had already developed internal acceptance criteria for smoke and toxic gas emission of large area cabin materials. These manufacturers' voluntary criteria were intended to provide guidance for material suppliers in the development of new cabin materials and are still in effect. These internal guidelines vary by manufacturer but go beyond those proposed by the FAA. Boeing criteria varied with the type of airplane. Airbus Industries unilaterally established requirements in 1979[8] that were similar to the Boeing guidelines. The McDonnell Douglas Aircraft Company imposed smoke and toxic gas emission criteria that varied depending on the application of the materials.

In 1984, the FAA introduced Amendment 59 to FAR Part 25 (FAR 25-59) to reduce the extent that seat cushions, usually made of polyurethane foam, would become involved in a cabin fire. Amendment 25-59 added a requirement to FAR

Materials with reduced flammability in aerospace and aviation 577

25 that seat cushions, except those on flight crewmember seats, must pass an oil burner test. FAR Part 121 was amended (FAR 121-184) to require retrofitting of all passenger seats to comply with FAR 25-59 by 1987.

In 1986, the FAA introduced Amendment 60 to FAR Part 25 (FAR 25-60) to contain cargo fires within cargo compartments by imposing an oil burner test for flame penetration of Class C and D compartment liners.

Also in 1986, the FAA introduced Amendment 61 to FAR Part 25 (FAR 25-61) to reduce the flammability of passenger cabin sidewalls, ceilings and large area cabin linings in a post-crash, fuel-fed fire in order to increase the time available for passengers to escape from a burning airplane. Amendment 25-61 added a heat release requirement to FAR 25 for interior ceiling and wall panels, partitions, and the outer surfaces of galleys, large cabinets and stowage compartments. FAR Part 121 was amended (FAR 121-189) simultaneously with Amendment 25-61 and required that all aircraft manufactured after 20 August 1990 comply with Amendment 25-61.

In 1988, the FAA introduced Amendment 66 to FAR Part 25 (Amendment 25-66) that clarified the heat release requirements of FAR 25-61 and added a smoke requirement to those materials that would reduce the smoke emitted in a post-crash fire. FAR Part 121 was amended (FAR 121-198) simultaneously with FAR 25-66 and required that all aircraft manufactured after 20 August 1990 comply with FAR 25-66.

In 1991, the FAA introduced Amendment 202 to FAR Part 121 (FAR 121-202) and Amendment 31 to Part 135 (FAR 135-31) to require an oil burner flame penetration test, as per FAR 25-60, for Class C and D cargo liners that were not constructed of fiber glass or aluminium.

In 2003, the FAA introduced Amendment 111 to FAR Part 25 (FAR 25-111) to add new fire performance requirements for flame spread and fire penetration of thermal/acoustic insulation. FAR Parts 91, 121, 125, and 135 were amended simultaneously to reflect the new requirements. Newly manufactured airplanes and material replacements on in-service airplanes were required to comply with the flame-spread requirement by 2 September 2005. Newly manufactured airplanes are required to meet the fire penetration (burn through) requirement by 2 September 2009. Figure 20.1 is a timeline of FAA regulatory upgrades for materials flammability.

20.3 Materials used in commercial aircraft

Interior cabin furnishings must meet the basic production and functional requirements of the aircraft manufacturers as well as the serviceability and aesthetic requirements of the airline operators.[4] They must also meet the fire safety requirements specified in the airplane certification process.[3] Materials used in transport category aircraft are roughly the same regardless of the type of aircraft involved. The various Federal Aviation Regulations (FARs) refer mostly

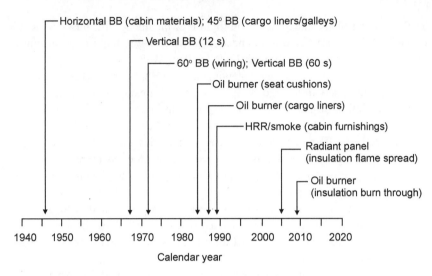

20.1 Timeline of aircraft flammability regulations.

to FAR Part 25 for flammability requirements. For simplicity this discussion will refer only to FAR Part 25. The combustible cabin materials covered by FAR Part 25 weigh about 3000 kilograms for a narrow body aircraft to over 7000 kilograms for a wide body aircraft as shown in Table 20.1.[9] The distribution of weight by major classes of materials for a wide body aircraft is shown in Fig. 20.2. Since the heat of combustion of these materials is roughly similar at about 20 MJ/kg, Fig. 20.2 also represents the distribution of fire load. The total fire load represented by the mass of materials in Table 20.1 is of the order (7600 kg) (20 MJ/kg) = 152 000 MJ, which approximates the heat of combustion of 1000 gallons of jet fuel, or about 2% percent of the fuel capacity of a wide body aircraft.

The construction of major aircraft cabin components below is an update of ref. 5 and the breakdown of materials by application in Table 20.2 is an update of ref. 10.

Table 20.1 Combustible materials in wide body commercial aircraft[9]

Component	Weight (kg)	Component	Weight (kg)
Seats	1500	Linings	500
Acoustic insulation	400	Electrical insulation	200
Decorative panels	1600	Windows	500
Textiles	900	Small parts & rubber	500
Air ducts	500	Safety equipment	1000

Materials with reduced flammability in aerospace and aviation

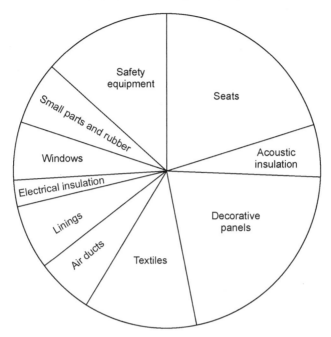

20.2 Distribution of combustible materials in wide body transport category aircraft by weight.

Aircraft seats

Aircraft seats are made from a wide variety of nonmetallic materials. These components can be grouped into five basic areas: foam rubber cushions, upholsteries, fire blocking layer, plastic moldings, and structure. Small nonmetallic seat parts must meet FAR 25.853(b). A pending special condition for large area components is that they also meet the heat release and smoke requirements of FAR 25.853(a) and (a-1) as described in Appendix F Parts IV and V. The cushions, which include the foam rubber, upholstery fabric, and (optionally), a fire blocker, must also meet FAR 25.853(c). Aircraft seats include a wide variety of plastic moldings for items such as decorative closeouts, trim strips, food trays, telecommunication (audio/video/phone) devices, and arm rests. Heat resistant or flame retardant thermoplastics (polycarbonate, polyvinylchloride/acrylic blends, etc.) are commonly used for these purposes. Most seat structures are made of aluminium; however, some manufacturers have introduced carbon composite structures to reduce weight and light metal alloy frames are also being considered.

Thermal-acoustical insulating materials

The entire pressurized section of the aircraft is generally lined with thermal/acoustical insulation, which is by far the largest volume (but not weight) of

Table 20.2 Materials use by application in large aircraft cabins

Application	Material
Floor and floor covering	• Glass/carbon-reinforced epoxy/phenolic over aramid fibre honeycomb with flexible urethane seat track covers and urethane foam edge band • Mylar film over galley and entry floor panels • Wool or nylon carpet attached to floor by double backed tape with optional aramid felt underlay • PVC galley mats
Lower sidewall panel	Glass/carbon-reinforced phenolic with decorative thermoplastic laminated film having PVF outer layer
Upper sidewall panel	Glass/carbon-reinforced phenolic with decorative thermoplastic laminated film having PVF outer layer
Light covers	Polycarbonate
Overhead storage bins	Glass/carbon-reinforced phenolic with decorative thermoplastic laminated surface film having PVF outer layer, urethane foam edge
Gap fillers	Silicone or urethane
Passenger seats	• Wool, wool/nylon, or leather upholstery • Urethane foam cushions • Polybenzimidazole or aramid fibre fire blocking layer • Polyurethane flotation foam • Thermoplastic seat trays and telecommunication equipment
Cabin attendant seats	• Wool, wool/nylon, or leather upholstery • Urethane foam cushions • Polybenzimidazole or aramid fibre fire-blocking layer • Polyethylene flotation foam
Partitions	Glass/carbon phenolic with decorative thermoplastic laminated film with PVF outer layer or wool/aramid fibre textile or leather
Stowage bins	Glass/carbon phenolic surfaced with decorative thermoplastic laminated film with PVF outer layer, wool textile interior liner (infrequent)
Placards	Polyvinylchloride or polyurethane
Insulation	Fibre-glass batting with phenolic binder and PVF or polyimide cover, or foamed polyvinylchloride, nitrile rubber, polyethylene, or polyimide.
Windows	• Outer pane stretched acrylic (polymethylmethacrylate) • Inner pane cast acrylic • Dust cover polycarbonate or acrylic
Passenger service units	Molded heat resistant thermoplastics (PEI, PPSU, PEKK) or aluminium or glass/carbon reinforced phenolic
Hoses	Silicone, nylon, or urethane
Air ducting	• Glass-reinforced phenolic, epoxy, cyanate ester, or polyester over poly(isocyanurate) foam (large ducts) • Fire retarded nylon, glass-reinforced silicone, aramid fibre felt over polyimide foam • Fiber-reinforced polyphenylsulfone and polyetherimide

nonmetallic material in an aircraft. The acoustical requirements for the insulation are more demanding than the thermal requirements, and govern the amount of insulation used. The insulation blanket construction typically consists of a low-density glass-fiber batting surrounded by a protective cover. In some applications, where the insulating material consists of foams or felts, a separate cover is not used. Insulation covers are used to hold the insulation batting in place and to keep out contaminants such as dust and fluids, especially water. Very thin plastic films (0.013 to 0.05 mm) of polyimide, polyvinyl fluoride, or poly(aryletherketones) reinforced with nylon or polyester yarns are used extensively due to their low weight and good tear resistance. In areas that are subject to abuse, lightweight, abrasion-resistant coated fabrics such as vinyl-coated nylon and vinyl-coated fiber-glass are used. Areas subject to higher temperature require the use of silicone-coated fiber-glass, metal-coated fiber-glass, or ceramic covers. Insulation is installed using a variety of attachments, including poly(phenylenesulphide) (PPS) hook and loop fasteners, nylon fasteners, snaps, and splicing tapes.

Interior panels

Although a few monolithic laminate panels have been used, most panels in airplane interiors of newly manufactured airplanes are sandwich structures having high stiffness-to-weight ratio. These panels are typically constructed of fiber-glass or carbon fabric-reinforced thermosetting resin face sheets adhesively bonded to an aramid (e.g., Nomex or Kevlar, Du Pont) paper honeycomb core, and covered with laminated, decorative thermoplastic surface film or paint that must be colourfast and easy to clean. These panels are used for ceilings, galleys, lavatories, sidewalls, baggage racks, floors, partitions, and closets. All panels for these applications must meet FAR 25.853(a) and (a-1) as they are used in the aircraft.

Textiles

Decorative textiles are also used to cover interior panels on surfaces that face the passengers on galleys, lavatories, closets, and partitions. Plush, hand-tufted, predominantly wool tapestries are often used on upper panel surfaces. Lightweight carpeting or a grospoint construction is common on lower panel surfaces. A variety of materials and methods are used to make tapestries. The lower surface textile, wainscoting, is usually fabricated of treated wool with a very lightweight backing or no backing. With the heat release regulation, most of these tapestries and wainscoting could no longer be applied. Tapestry and wainscoting fabrics made from synthetics and wool/synthetic combinations are produced in order to meet the heat release requirements. Draperies are used to close off sections of the aircraft such as galleys and to separate the classes of passenger service. Drapery fabrics are usually wool or polyester fabric that has been treated with a flame retardant. Floor coverings are vinyl or carpets depending on the location in the aircraft. All floor coverings must meet FAR 25.853(b). Carpet covers most of the cabin floor, including the aisle and under the seats. Most aircraft have wool or nylon-face yarns with polyester, polypropylene, cotton, or fiber-glass backing yarns and a fire-retardant back-coating. Wool-faced yarns are treated with a fire retardant. Nylon carpets must have a highly fire-retardant, back-coating for fire resistance. Carpet under-layers of felt are used in some aircraft for noise suppression. Areas where fluid spills are likely, such as galleys and lavatories, use plastic floor coverings typically made of vinyl with a reinforcing fabric backing and a non-slip surface.

Nonmetallic air-ducting

Owing to the relative compactness of an aircraft, much of the conditioned air ducting has to be routed around many different parts. This results in some very complex shapes. Nonmetallic ducting is used to create these complex parts. There are three basic types of nonmetallic duct constructions: fiber-reinforced thermosetting resin, thermoplastic, and rigid thermosetting resin foam. All conditioned air ducting must meet FAR 25.853(b). Fiber-reinforced resins used in ducting consist of woven fiber-glass with polyester, epoxy, cyanate ester, or phenolic thermosetting resin systems. Some aromatic polyamide/epoxy is also used. Ducts made from these materials are usually coated after curing on the outside with a polyester or epoxy resin to seal against leaks. Fiber-glass impregnated with silicone rubber is the industry standard for duct boots because of flexibility, strength, low air permeability, and good fire resistance. Thermoplastic ducting is typically made of vacuum-formed polycarbonate or poly(etherimide). Thermoplastic ducts are not as strong or durable as fiber-reinforced resin; however, thermoplastic ducts are much less costly to fabricate. Polyimide or poly(isocyanurate) foam ducts are used for larger ducts with complex shapes and have the advantage of not requiring additional insulation. Foam ducts are popular for their low weight.

Linings

Sheet products are used to cover areas where strength and flexibility are required in a contoured shape; in addition, linings provide an aesthetically pleasing surface and protect the assemblies behind the liner. Areas such as the exit doors, flight deck, cabin sidewalls, door-frames, and cargo holds utilize liners fabricated of reinforced thermosetting resins or thermoplastics. Decorative sidewall liners made of formed aluminium are used in some aircraft. Depending on the application, the liners must meet FAR 25.853(a), (a-1), or (b) or FAR 25.855(a) or (a-1), or a combination. Linings that are subjected to passenger and food cart

Materials with reduced flammability in aerospace and aviation 583

traffic are typically manufactured from plies of fabric-reinforced resin. Their flexibility, impact resistance, high strength, and low weight make them ideal for lower sidewall kick panels. Cargo liners that are required to meet FAR 25.855(a) and (a-1) are typically fabricated from fiber-glass reinforced resins because of the burn-through and impact resistance. Linings that see less abuse and do not require high strength are fabricated from thermoplastics because less expensive fabrication methods can be used. Flight deck sidewalls, upper door liners, attendant stations, and closeouts are typical applications for vacuum, heat and pressure (thermoformed) plastics such as ABS, polycarbonate, poly(etherimide), poly(phenylsulphone), and polyvinyl chloride/acrylic blends. In many applications, thermoplastics are pigmented and textured and do not require any decorative covering.

Electrical components

Wire and cable insulation consisting of thermosetting resins and thermoplastics comprises a small amount of the nonmetallic material in an aircraft (see Fig. 20.2). For general wire and cable applications inside the pressure shell, the majority of high performance wire jacketing in use today includes radiation cross-linked poly(ethylene-tetrafluoroethylene) (ETFE) and a composite construction of poly(tetrafluoroethylene) (PTFE) wrapped around the conductor, a middle layer wrapped with heat resistant polyimide film, and an outer sheath of PTFE to provide abrasion resistance. In some areas, aromatic polyamide braiding is used to cover power feeder cables for abrasion resistance. For higher temperature and fuel areas, PTFE is used almost exclusively. Where very high temperature or burn-through resistance is a requirement, filled PTFE is typically used. Asbestos had been used as the filler in the past, but has been replaced by proprietary fibers. To withstand the high temperature requirements of fire zones, heavily nickel-plated copper wire is used to ensure continued operation of electrical equipment. All wire insulations in the airplane must meet FAR 25.1713(c) (previously FAR 25.869(a) and FAR 25.1359(d)), and those located in fire zones must also meet FAR 25.1713(b). Conduit and tubing of different types are used for electrical wires and components. Poly(vinyl fluoride) and polyolefin thermoplastic heat shrink tubing, silicone rubber impregnated glass fiber-braid, and extruded and convoluted nylon tubing are industry standards. Most connectors in an aircraft are made of Bakelite with silicone or hardened dielectric material inserts that have no specific FAR burn requirements. Connectors located in firewalls, however, must be fireproof and are made of low-carbon or stainless steel to meet burn-through requirements.

Windows

At present all aircraft windows are fabricated from biaxially-oriented, bulk polymerized poly(methylmethacrylate) (or stretched acrylic) thermoplastic.

Stretched acrylic has the optical clarity, strength, low weight, and solvent resistance necessary to meet the service specifications for pressurized aircraft windows in the cabin. All windows must meet flammability requirements of FAR 25.853(b-2). Transparent dust covers that are vented to the cabin interior are used to protect the windows and are made from tough polycarbonate or acrylic thermoplastic that has been hard-coated for scratch resistance. Transparent dust covers must meet FAR 25.853 (a-1).

Small parts

Except for electrical wire, cable insulation and small parts such as knobs, handles, rollers, fasteners, clips, grommets, rub strips, and pulleys that would not contribute significantly to the propagation of a fire, all parts/materials not identified in FAR 25.853(a), (b), (b-1), or (b-2) must not have a burn rate greater than 10 cm per minute when tested horizontally in accordance with FAR 25.853 (b-3).

Figure 20.3 shows the flammability (as average heat release capacity) of commodity, engineering, and heat resistant polymers compared to the flammability of formulated solid plastic materials used in automotive interiors,[11] consumer electronics,[12] railway car interiors,[13] aircraft overhead spaces (hidden areas), and aircraft cabin interior furnishings as measured for commercial samples in a standardized test.[14] Heat release capacities of railway car materials were estimated from ref. 13 using empirical correlations.[15] It is clear from Fig. 20.3 that the materials used in aircraft cabin interiors are, on average, less

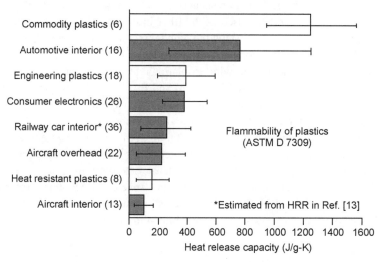

20.3 Average flammability of polymers and plastic products used in various applications. Number of materials tested is in parentheses. Error bars are one standard deviation.

flammable than those used in railway cars, and that aircraft cabin materials are significantly less flammable than the plastics used for consumer electronics and automotive interiors. The flammability of the materials in Fig. 20.3 reflects the severity of the requirements for each application, e.g. a horizontal Bunsen burner test for automotive,[16] a 10-second vertical Bunsen burner for electronics,[17] smoke[18] and flame spread[19] under radiant heat for rail cars.[20] The test methods for aircraft overhead and interior materials[3,5] are discussed in the following section.

20.4 Fire test methods for aircraft materials[3,5]

20.4.1 Vertical Bunsen burner (BB) test for cabin and cargo compartment materials

This test method, shown in Fig. 20.4, is intended for use in determining the resistance of materials to flame when tested according to the 12-second and 60-

20.4 Vertical Bunsen burner test (photo courtesy of the govmark organization).

second vertical flame tests specified in Federal Aviation Regulation (FAR) 25.853 and FAR 25.855. In the test, a 10-mm diameter Bunsen or Tirrill burner having a 38-mm methane or similar combustible gas flame is applied to the bottom edge of a 75-mm wide by 305-mm high specimen clamped to expose 51-mm of width in a draft free cabinet. The sample is positioned in the flame such that the bottom of the sample is 19 mm above the top of the burner at the start of the test and is exposed to the flame for 12 or 60 seconds, depending on the requirement. At least three specimens are tested for each determination. Individual and average values are reported for the ignition time (nominally 12 or 60 seconds), the flame time of the specimen after removal of the burner, the flame time of drips (if any), and the burn length are reported. The requirements are that the average flame time of all of the specimens tested not exceed 5 seconds for either the 12- or 60-second exposure vertical test, and that the average drip time not exceed 3 seconds for the 60-second test, or 5 seconds for the 12-second test. The average burn length for all of the specimens tested shall not exceed 152 mm (6 in.) for the 60-second exposure or 203 mm (8 in.) for the 12-second exposure.

Materials affected (60-second exposure): Interior ceiling panels, wall panels, partitions, galley structure, large cabinet walls, structural flooring and stowage compartments and racks.

Materials affected (12-second exposure): Floor covering, textiles including drapery and upholstery, seat cushions, padding, coated fabrics leather, trays and galley furnishings, electrical conduit, thermal and acoustical insulation and insulation covering, air ducting, joint and edge covering, liners of Class B and E cargo or baggage compartments, floor panels of Class B, C or E cargo or baggage compartments, insulation blankets, cargo covers and transparencies, molded and thermoformed parts, air ducting joints, trim strips.

20.4.2 45-Degree Bunsen burner test (45 BB) for cargo compartment liners and waste stowage compartment materials

This test is shown in Fig. 20.5 and is intended for use in determining the resistance of materials to flame penetration and to flame and glow propagation when tested according to the 30-second, 45-degree Bunsen burner test specified in FAR Part 25. In the test a 10 mm diameter Bunsen or Tirrill burner having a 38 mm methane or similar combustible gas flame is applied for 30 seconds to the center of the bottom surface of a square specimen clamped to expose a 203 mm wide by 203 mm surface oriented at 45 degrees to the horizontal in a draft free cabinet. At least three specimens are tested for each determination. Individual and average values of the flame time of the specimen after removal of the burner, the glow time, and whether or not flame penetration occurs are reported. The requirements are that the average flame time for all specimens tested not

Materials with reduced flammability in aerospace and aviation 587

20.5 45 degree Bunsen burner test (photo courtesy of the govmark organization).

exceed 15 seconds, that the Bunsen burner not penetrate any of the specimens tested, and that the average glow time not exceed 10 seconds.

Materials affected: liners of Class B and E cargo or baggage compartments, floor panels of Class B, C or E cargo or baggage compartments not occupied by crew or passengers.

20.4.3 Horizontal Bunsen burner test for cabin, cargo compartment, and miscellaneous materials

This test method (see Fig. 20.6) is intended for use in determining the resistance of materials to flame when tested according to the 15-second horizontal Bunsen burner tests specified in FAR 25.853. In the test a 10-mm diameter Bunsen or Tirrill burner having a 38-mm methane or similar combustible gas flame is applied for 15 seconds to the bottom edge of rectangular specimen clamped to expose a 51 mm wide by 305 mm long horizontal surface in a draft free cabinet. At least three specimens are tested for each determination. The individual and average linear burn rates (mm/min) in the horizontal direction are reported. The requirement is that the average burn rate for the all of the specimens tested not exceed 6.3 cm/minute (FAR 25.853(b-2) or 10 cm/minute (CFR 14 part 25.853(b-3).

20.6 Horizontal Bunsen burner test (photo courtesy of the govmark organization).

Materials affected: clear plastic windows and signs, parts constructed in whole or in part of elastomeric materials, edge lighted instrument assemblies, seat belts, shoulder harnesses, and cargo and baggage tie-downs including bins, pallets, etc., used in passenger or crew compartments.

20.4.4 60-Degree Bunsen burner test for electrical wire (60 BB)

This test method (see Fig. 20.7) is intended for use in determining the resistance of electrical wire insulation to flame when tested according to the 30-second, 60-degree Bunsen burner test specified in FAR 25.869. In the test a 10-mm diameter Bunsen or Tirrill burner having a 38-mm methane or similar combustible gas flame is applied for 30 seconds to the center section of wire specimen oriented at 60-degrees with respect to the horizontal. One end is clamped to the horizontal surface while the other end passes over an elevated pulley and is attached to a free weight, the mass of which depends on wire gage. At least three specimens are tested for each determination. The individual and average specimen flame time after removal of the burner, flame time of drips (if any), and burn length are reported. The requirement is that the average flame time not exceed 30 seconds, the average flaming drip time not exceed 3 seconds and that the average burn length not exceed 76 mm.

Materials with reduced flammability in aerospace and aviation

20.7 60 (and 45) degree Bunsen burner test (photo courtesy of the govmark organization).

Materials affected: insulation on electrical wire or cable installed in any area of the fuselage.

20.4.5 Heat release rate test for cabin materials

This test method is intended for use in determining heat release rates to show compliance with the requirements of FAR 25.853. Heat release is a measure of the amount of heat energy evolved by a material when burned. It is expressed in terms of energy per unit area (J/m^2 or $kW\,min/m^2$). Heat release rate is a measure of the rate at which heat energy is evolved by a material when burned. It is expressed in terms of power per unit area (kW/m^2). The maximum heat release rate occurs when the material is burning most vigorously. It has been shown that the maximum heat release rate and total heat released by burning materials in an aircraft cabin are predictors of the time available for escape in the event of a post-crash fire. The apparatus, sometimes referred to as the Ohio State University calorimeter, used to measure rate of heat release has a test

20.8 Rate of heat release apparatus (photo courtesy of fire testing technology).

chamber that is approximately 20 cm deep by 40 cm wide by 80 cm high fabricated of 0.6 mm thick stainless steel and covered with 2.5 cm of mineral board insulation (see Fig. 20.8). Ambient temperature compressed air is metered into the test chamber at 0.04 m^3/s. Thermopiles located at the air inlet and exit ports of the test chamber measure the enthalpy rise of the air stream caused by burning of the sample and are used to compute the heat release rate after calibration with a metered methane flame. A 15 cm square test specimen having the thickness used in the aircraft is tested in a vertical orientation under radiant heat exposure of 35 kW/m^2 provided by silicon carbide heating rods. Methane pilot flames are used to force ignition of the sample and burn the gases evolved prior to ignition. Heat release rate is measured for the duration of the test and encompasses the period of ignition and progressive flame involvement of the surface. At least three specimens of each material are tested. The requirement is that the average maximum heat release rate during the five-minute test not exceed 65 kW/m^2 and that the average heat released during the first two minutes of the test not exceed 65 kW-min/m^2 (3.9 MJ/m^2).

Materials with reduced flammability in aerospace and aviation

Materials affected: interior ceiling and wall panels other than lighting lenses and windows, partitions other than transparent panels needed to enhance cabin safety, galley structure, stowage carts, large cabinets and stowage compartments.

20.4.6 Smoke test for cabin materials

This test method is used to determine the smoke generating characteristics of airplane passenger cabin interior materials to demonstrate compliance with the requirements of FAR 25.853. The quantity measured is the specific optical density, which is a dimensionless measure of the amount of smoke produced per unit area by a material when it is burned. The test chamber shown in Fig. 20.9 is a square cornered box with internal dimensions 91 cm wide by 61 cm deep by 91 cm high, containing a sample holder, radiant heat source, and a photometric system capable of detecting light transmittance values down to 1% with an

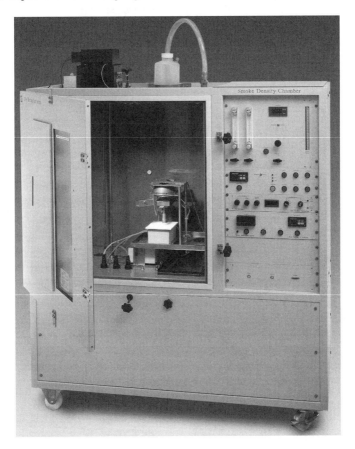

20.9 Smoke test for cabin materials (photo courtesy of fire testing technology).

accuracy of 0.03%. In the test, a 65-mm square sample of the thickness used in the aircraft cabin is exposed to a pilot flame and 25 kW/m² of radiant energy to force ignition and burning. At least three specimens are tested of each material. The requirement is that the average maximum specific optical density measured during the four-minute test not exceed 200, which corresponds to 3% light transmission through the test chamber.

Materials affected: interior ceiling and wall panels other than lighting lenses and windows, partitions other than transparent panels needed to enhance cabin safety, galley structure, stowage carts, large cabinets and stowage compartments.

20.4.7 Oil burner test for seat cushions

This test method evaluates the burn resistance and weight loss characteristics of aircraft seat cushions exposed to a high intensity open flame to show compliance with the requirements of FAR 25.853. In the test shown in Fig. 20.10, a seat back (vertical) and a seat bottom (horizontal) cushion that are representative of the production articles used in aircraft seats with regard to materials and construction are attached with wires to a steel seat frame and exposed for two minutes to a kerosene or fuel oil burner calibrated to produce 120 kW/m² at the seat surface. The weight loss of the cushion set during the test is determined by weighing the cushions before and after the test and the burn lengths are measured along the top and bottom surfaces of the vertical and horizontal cushions. At least three cushion sets are tested for each construction. The requirement is that the average weight loss for the cushion sets not exceed 10%

(a)

(b)

20.10 Oil burner test for seat cushions.

Materials with reduced flammability in aerospace and aviation 593

and that neither the individual nor the average burn lengths of the cushions exceed 43 cm.

Materials affected: passenger cabin seat cushions and some lightweight seat constructions.

20.4.8 Oil burner test for cargo liners

This test method evaluates the flame penetration resistance capabilities of aircraft cargo compartment lining materials utilizing a high intensity open flame to show compliance with the requirements of FAR 25.855. In the test (see Fig. 20.11), a kerosene or fuel oil burner calibrated to a heat flux of 91 kW/m^2 at the sample surface impinges on a specimen of cargo sidewall or ceiling liner that is 61 cm by 41 cm and mounted in a steel frame in the orientation (vertical or horizontal) that it will be used in the aircraft. At least three specimens are tested of each material. The requirement is that none of the specimens allow the flame to penetrate through the thickness or exhibit a temperature at a distance of 10 cm from the rear surface in excess of 204 °C (400 °F).

Materials affected: ceiling and sidewall liners of Class C cargo or baggage compartments.

20.11 Oil burner test for cargo liners.

20.4.9 Radiant panel test for flame spread of thermal/acoustic insulation

This test method shown in Fig. 20.12 evaluates the flammability and flame propagation characteristics of thermal/acoustic insulation when exposed to both a radiant heat source and a flame to show compliance with FAR 25.856(a). In the test a horizontal specimen 31.8 cm wide by 58.4 cm long is placed inside a test chamber that is 1.4 m long by 0.5 m wide by 0.76 m high in a horizontal position beneath a 32.7 cm by 47 cm gas-fired or electrically heated panel inclined at an angle of approximately 20 degrees with respect to the horizontal and calibrated to provide a radiant heat flux gradient. The heat flux decreases from a maximum value of 17 kW/m^2 at the sample leading edge to 16 kW/m^2 at a distance of 10 cm along the sample length. The edge of the specimen is ignited for 15 seconds using a propane burner, after which the burner is removed and the flame spread and time of flaming are measured for a minimum of three replicate tests. The requirement is that the flame cannot spread more than 5.1 cm along the sample length (i.e., the critical heat flux for flame spread is about 17 kW/m^2) and the flame time after removal of the burner not exceed 3 seconds for any specimen.

Materials affected: thermal/acoustic insulation in the fuselage of transport category aircraft.

20.12 Radiant panel test for flame spread of thermal/acoustic insulation (photo courtesy of fire testing technology).

20.4.10 Oil burner test for burn through resistance of thermal/acoustic insulation

This test method evaluates the burn through resistance of thermal/acoustic insulation exposed to a high intensity open flame to show compliance with FAR 25.856(b) Appendix F, Part VII. In the test, shown in Fig. 20.13, two 81 cm by 91 cm samples of thermal/acoustic insulation are mounted side-by-side in a steel frame and exposed to a kerosene or fuel oil burner calibrated to a heat flux of 180 kW/m^2 at the sample surface. The time to flame penetration (if any) and the heat flux behind the sample are measured. At least three specimen sets are tested of each material. The requirement is that none of the specimens allow the flame to penetrate through the thickness for four minutes or register a heat flux of greater than 23 kW/m^2 at a distance of 31 cm behind the sample face.

Materials affected: thermal/acoustic insulation in the lower half of the fuselage of transport category aircraft.

Figure 20.14 is a cross-sectional diagram of the aircraft cabin showing the cabin interior, cargo area and unoccupied areas and the fire test regulations governing the materials in these areas.

20.13 Oil burner test for burn-through resistance of thermal/acoustic insulation.

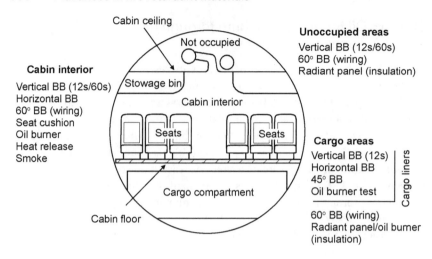

20.14 Fire tests by aircraft cabin location.

20.5 Future trends

The following initiatives are under way at the US Federal Aviation Administration to address or upgrade materials fire safety on aircraft.

20.5.1 Upgraded requirements for materials in unoccupied areas

This program was established to address components installed inside the pressurized hull of the aircraft in areas not visible or readily accessible to passengers and crew. The fire test procedures and acceptance criteria for these 'hidden materials' are being reviewed to determine whether more stringent requirements are needed.

20.5.2 Fireproof aircraft cabin

In 1995, a long-range fire research program was initiated to develop a fireproof aircraft cabin.[21] Ultra fire resistant materials that resist ignition and burning during flight and in a post-crash, fuel-fed fire were identified as a means to achieve this goal. Research is fundamental in nature and seeks to understand the burning process of combustible solids at the molecular level. Applications for ultra fire resistant materials in commercial aviation include thermoplastics for seat parts and transparencies, thermosetting resins for composite wall panels and air ducts, textile fibers for carpets and seat fabrics, and rubber for foam seat cushions, hoses, and sealants.

20.5.3 Special condition for polymer composite fuselage

Industry is currently seeking certification for new design transport category aircraft having a fuselage constructed largely of carbon fiber-reinforced epoxy composites. These composite structures are difficult to ignite and tend to burn slowly[22] because the flammable component is less than 24% of the composite weight after charring of the epoxy is taken into account. There are no fireworthiness requirements for airplane fuselages because they have traditionally been constructed of aluminium, which is noncombustible under normal conditions. Consequently, the composite fuselage aircraft will need to be certified under a Special Condition after demonstrating an equivalent level of safety.

20.6 Acknowledgements

The author is indebted to Jim Peterson of Boeing, and Jeff Gardlin, Gus Sarkos, Pat Cahill and John Reinhardt of the FAA for their help with the manuscript, and to Richard Walters for all of the heat release capacity measurements. Certain commercial equipment, instruments, materials and companies are identified in this paper in order to adequately specify the experimental procedure. This in no way implies endorsement or recommendation by the Federal Aviation Administration.

20.7 References

1. Final Report to President Clinton of the White House Commission on Aviation Safety and Security, February 12, 1997.
2. C.P. Sarkos, An Overview of Twenty Years of R&D to Improve Aircraft Fire Safety, *Fire Protection Engineering*, 5, 4–16 2000.
3. United States Code of Federal Regulations, Title 14: Aeronautics and Space, Volume 1, Chapter I: Federal Aviation Administration, Parts 1–49, 2007.
4. J.M. Peterson and S.E. Campbell, Regulatory and Industry Requirements for Fire Safe Aircraft Interior Materials: Materials and Process Affordability – Keys to the Future, *Proceedings of the 43rd International SAMPE Symposium and Exhibition*, Anaheim, CA, May 31–June 4, pp. 1594–1599, 1998.
5. *Aircraft Material Fire Test Handbook*, Federal Aviation Administration, Report DOT/FAA/AR-00/12, April 2000.
6. W.T. Westfield, The Role of Research and Development on Safety Regulation, Federal Aviation Administration Final Report DOT/FAA/AR-95/84, October 1995.
7. J.M. Peterson, *Flammability Tests for Aircraft, in Flammability Testing of Materials Used in Construction, Transport and Mining*, V.B. Apte, Ed., Woodhead Publishing Ltd., Cambridge, England, 2006, pp. 275–301.
8. J. Troitzsch, *International Plastics Flammability Handbook*, 2nd Edition, Hanser Publishers, Munich, Germany, 1990.
9. *Fire Safety Aspects of Polymeric Material, Volume 6*, Aircraft: Civil and Military, National Materials Advisory Board, Publication NMAB 318-6, Washington, D.C., 1977, p. 68.

10. Fire- and Smoke-Resistant Interior Materials for Commercial Transport Aircraft, National Materials Advisory Board Report, NMAB-477-1, National Academy Press, Washington, DC 1995.
11. R.E. Lyon and R.N. Walters, *Flammability of Automotive Plastics*, Society of Automotive Engineers (SAE) World Congress, Detroit, MI, April 3–6, 2006.
12. A.B. Morgan, Micro Combustion Calorimeter Measurements on Flame Retardant Polymeric Materials, *Proceedings of the Society for the Advancement of Materials and Process Engineering (SAMPE) Technical Conference*, Cincinnati, OH, October 29–November 1, 2007.
13. R.D. Peacock and R.W. Bukowski, Flammability Tests for Railway Passenger Cars, in *Flammability Testing of Materials Used in Construction, Transport and Mining*, V.B. Apte, Ed., Woodhead Publishing Ltd., Cambridge, England, 2006, pp. 336–360.
14. *Standard Test Method for Determining Flammability Characteristics of Plastics and Other Solid Materials Using Microscale Combustion Calorimetry*, ASTM D 7309-07, American Society for Testing and Materials (International), West Conshohocken, PA, 2007.
15. R.E. Lyon, R.N. Walters and S.I. Stoliarov, Screening Flame Retardants for Plastics Using Microscale Combustion Calorimetry, *Polymer Engineering & Science*, 47, 1501–1510 (2007).
16. U.S. Code of Federal Regulations, Title 49: Transportation, Volume 6, Chapter V, National Highway Traffic Safety Administration, Part 571 Federal Motor Vehicle Safety Standards, Standard 302, Flammability of Interior Materials (FMVSS 302), October 1, 2003.
17. *Flammability of Plastic Materials*, UL 94 Section 3 (Vertical: V-0/1/2), Underwriters Laboratories, Northbrook, IL, 1991.
18. *Standard Test Method for Specific Optical Density of Smoke Generated by Solid Materials*, ASTM E 662, American Society for Testing and Materials (International), West Conshohocken, PA, 2006.
19. *Standard Test Method for Surface Flammability of Materials Using a Radiant Heat Energy Source*, ASTM E 162, American Society for Testing and Materials (International), West Conshohocken, PA, 2006.
20. U.S. Code of Federal Regulations, Title 49: Transportation, Volume 4, Chapter II, Federal Railroad Administration, Part 238 Passenger Equipment Safety Standards, Standard 103, Fire Safety (FRA 238.103), October 1, 2006.
21. R.E. Lyon, *Advanced Fire Safe Aircraft Materials Research Program*, Final Report DOT/FAA/AR-95/98, January 1996.
22. J.G. Quintiere, R.N. Walters and S. Crowley, *Flammability Properties of Aircraft Carbon Fibre Structural Composites*, Final Report DOT/FAA/AR-07/57, Federal Aviation Administration, October 2007.

Index

'A' class divisions 559–60
ABS Guide for Building and Classing Naval Vessels 531
absorption of radiation 262
accessory materials, firefighters' 474–5
acetate rayon (cellulose acetate) 24
acoustic insulation 578, 579–81
  fire tests 594, 595
acrylics 56, 163, 164
acrylonitrile 201
active countermeasures 369
  benefits of passive and active measures working together 386–8
additives
  coatings and laminates 165–9
  fire retardants for composites 419–25
  polymeric additive halogen-free flame retardants 79–82
Afghanistan 496, 497
air ducts 578, 579, 580, 582
Airbus Industries 576
  A380 115
aircraft 5, 115, 161, 566, 573–98
  composites 400, 404–5, 582, 597
  fire test methods 280–1, 585–96
  future trends 596–7
    fireproof cabin 596
    polymer composite fuselage 597
    unoccupied areas 596
  history of fire regulations 574–7, 578
  materials used in commercial aircraft 577–85
    electrical components 578, 579, 583, 588–9
    interior panels 578, 579, 581
    linings 578, 579, 582–3, 593
    nonmetallic air ducts 578, 579, 580, 582
    seats 578, 579, 580
    small parts 578, 579, 584–5
    textiles 578, 579, 581–2
    thermal-acoustical insulation 578, 579–81
    windows 578, 579, 580, 583–4
  military fire hazards 494
Airlock 473
airworthiness certification 573, 597
aliphatic polymers 533
alkylation of amino groups 51
Allied Combat Clothing Publication ACCP-2 protocol 514–15
alumina trihydrate (ATH) 114, 133, 150–2, 168–9, 200, 370
aluminized fabrics 473
aminophenyl phosphates 49
aminopropylthoxysilane (APS) 17
ammonium functionalising groups 126–9
ammonium polyphosphate (APP) 18, 42, 52–3
  coatings and laminates 169, 170, 178
  replacement of halogen-containing fire retardants 171–4
  nanocomposites 138–41, 142–5, 147, 150–2
ammonium-substituted clays 99–100
animal-based tests 276–7
anion exchange 100
anti-dripping properties 73–6
Antiflame 501
antimony oxides 191
  risk reduction strategies for decaBDE/antimony oxide 205–6
antimony trioxide (ATO) 167, 192, 201, 370
appliances 310
aramids 403, 499–500
  composite flammability 409–12
  inherently flame retardant 203

see also meta-aramids; para-aramids
area, specimen 242–3
Armoured Fighting Vehicle (AFV) crewmen's clothing 513
armoured vehicles 493
Army Combat Shirt (ACS) 511–12
Army Combat Uniform (ACU) 511–12
aromatic polymers 203, 533
aryl phosphates 50–1
ASTM
  flame spread tests 541–2, 543
  small-flame exposure ignition tests 234, 235, 251
atmospheric pressure plasma technologies 183–4
Attestation of Conformity (AoC) levels 296–7
automotive sector *see* road vehicles
average heat release rate 548, 549–56
azocyclohexane 73

back-coatings 179
  cover fabric fire barriers 199
  halogen replacement 171–4
balance of risk assessment 373–7
barium sulphate nanoparticles 74
barrier fabrics 197–9
basalt fibres 35
Basofil fibre 204
batting
  cotton 200–1
    engineered cotton batting 198
  polyester 201
Battledress Uniform (BDU) fabric 501–2
bedclothes 197–201
benzo(a)pyrene (BaP) 342, 353–4
benzoguanamine-modified phenol biphenylene resin (BG-modified PB resin) 81–2
BGDMSB 85, 86
BHDB-polyphosphonate 82, 83
bi-component fibres 483, 522
bioaccumulation 190
bioavailability 376–7
biopolymers 23–4
bis(benzo-dioxaborolanyl) oxide 85–6
bis(cyclohexylazocyclohexyl) methane 73
bisphenol A bis (diphenyl phosphate) (BDP) 44
bisphenol C-based cyanate ester 534
bis-tetramethyl-dioxaborolanyl oxide 85–6
blends
  fibre 33–4, 37

nanocomposite formation 103
Blumstein, A. 97
Boeing 576
  787 aircraft 115
borates 85–6
boron fibres 35–6
break-open resistance 503
bromine-containing flame retardants 2, 149
  coatings and laminates 166–7
  replacement 171–3
  environmental concerns 191–2
  see also halogen-containing flame retardants
Budit 3118 products 54
building *see* construction/building
bulk polymers 95
  potential application areas 133–6
burn injuries 493, 496–9
burn through resistance 595

cabin, aircraft
  fire test methods 585–6, 587–8, 589–93
  fireproof cabin 596
cables/wires 275
  Fire-LCA model 343–9
    results 345–8
    statistical fire model 344–5
  sheathings 114, 133–4
  60-degree Bunsen burner test 588–9
caged bicyclic flame retardants 75–6
calorific potential 303
candidate designs 454, 455
carbon dioxide 340, 545–6
carbon dioxide laser cell 521
carbon fibres 34–5, 402–3
  composite flammability 409–12
carbon monoxide 260, 276, 545–6
carbon nanofibres 101–2
  mechanism of flammability reduction 109–11
carbon nanotubes (CNT) 29–30, 101–2, 429–30
  mechanism of flammability reduction 109–11, 112
carbonaceous protective layer 169–71
carboxyethylmethylphosphinic acid (CEMP) 83, 84
carboxyethylphenylphosphinic acid (CEPP) 83, 84
cargo compartment, aircraft 585–8, 593, 596
CASICO cable 343–9

Index 601

catalysis 70–3
catalytic cracking 217
cause-related fire incidence 367–8
CBRN (chemical, biological, radiological and nuclear) protection 470–1
CE marking 294, 295, 318
   AoC levels 296–7
cellulose acetate (acetate rayon) 24
cellulosic/viscose rayon (regenerated cellulose) 201–2
ceramic coatings 416–17
ceramic fibres 35–6
ceramic matrix composites (CMCs) 399
certification
   aircraft 573, 597
   electrical engineering and electronics 317–18, 328
Certified Body (CB) Scheme 317
char formation 262, 408–9
Char-Guard 327 54
char promotion 138, 179
chemical analysis 277
chemical recycling 215, 216–17, 218–19
   flame retarded polymers 222–3
chemically cured systems 160
China Compulsory Certification (CCC) Mark 328
chlorine-containing flame retardants
   environmental concerns 191–2
   phosphorus and chlorine-containing flame retardants 167–8
   see also halogen-containing flame retardants
cigarettes 207
Civil Aeronautics Agency (CAA) 575
Civil Air Regulations (CARs) 575
Classification certification 317–18
clay nanocomposites 27, 30–1, 115–16, 427–9, 430, 535
   functionalising group type and stability 126–30
   layered silicates 98, 99–100, 109, 110, 111
Cloisite clays 126, 127, 128, 140, 141, 142, 144–5, 147
closed box tests 279
clothing 275
   environmentally friendly flame resistant cotton 193–6
   ignition of and burn injuries 498, 499
   protective *see* firefighters' protective clothing; military fabrics
Clothing Area Network (CLAN) 484–5
coatings 4, 132, 159–87

additives 165–9
coating levels vs nanofilm challenges 179–80
composites 416–18
cotton 18, 171–3
halogen replacement in back-coated textile formulations 171–4
intumescent 416–17
   formation of protective layer 169–71
   military and naval applications 537–8
main types of fire retardant coatings 161–71
novel/smart approaches 179–80
plasma-initiated 181–4
potential applications of nanocomposites 148–9
role of nanoparticles 174–5
volatile phosphorus-containing species 175–8
cobalt chelate 72–3
co-cured resins 427
Code of Federal Regulations (CFR) 574
codes, fire safety 249–50
cold plasma technologies 19, 21
colemanite 74
collisions 458, 461
combinations of nanoparticles 116
combustion furnaces 520
compatibilisers 131–2
composites 4, 95, 398–442
   aircraft 400, 404–5
      air ducting 582
      fuselage 597
   core materials 403–4, 416
   fire retardant solutions 416–30
      additive and reactive fire retardants 419–25
      co-curing of different resins 427
      hybridisation of reinforcing fibres 418–19, 420
      nanoparticle inclusion 427–30
      resin modification 425–7
      surface coating and thermally insulative layers 416–18
   flammability of composites and their constituents 407–16
   future trends 435
   key issues and performance requirements by sector 404–7
   maritime and naval applications 400, 406–7, 527, 538–64
      fire performance regulations 531, 532

fire performance and test methods 540–64, 565
mechanical properties 430–4, 435
degradation during and after fire 430–3
post-fire mechanical performance of fire retardant composites 433–4
potential applications of nanocomposites 149–52
reinforcing fibres see reinforcing fibres
resins see resins
and their constituents 400–4
thickness 414, 415
types of 398–400
conductive heat transfer 468
measurement of protection against for firefighters' protective clothing 479–80
cone calorimetry 106–8, 240, 241–2, 249, 321, 323
composites 547–8
fire toxicity testing 278–9
ignitability 265
measurement of heat release 269–71, 272
peak heat release rates of car components vs random plastics 446, 447
'confined to building' fires 350–1
'confined to item' fires 350–1
'confined to room' fires 350–1
construction/building 292, 293–309, 326–7
Attestation of Conformity 296–7
CE marking 294, 295
EU requirements and tests 293–4
Euroclasses 294, 298–301, 302
fire testing of construction products 301–9
harmonised product standards 297–8
heat release measurement for regulations and control of construction products 274–5
notified bodies 297
Construction Products Directive (CPD) 294, 326–7
essential requirements (ERs) 294
consumer electronics 310, 318–19
Consumer Product Safety Commission (CPSC) 326
consumer products 2, 363–97
active and passive countermeasures 369
balance of risk 373–7

benefits of flame retardants in 377–81
differentiating the effects of smoking and smoke alarm trends 381–91
fire hazards and risks in residential fires 364–7
future issues 391–4
competing and contradictory regulations 393
controlling and eliminating exposure risks 392–3
maintaining knowledge of fire impacts 391
targets for improvement in fire risk reduction 393–4
understanding exposure to flame retardants 391–2
primary risks of flame retardants in 369–73
product and cause-related fire incidence 367–8
role of flame retardants in reducing fire hazards and risks 368–9
contact heat 468
measuring protection against for firefighters' protective clothing 479–80
convective heat 468
measuring protection against for firefighter's protective clothing 478–9
copolymerisation 82–6
core materials 403–4
and composite flammability 416
core spun yarns 196–7, 198–9
cost 69
cost-benefit analysis (CBA) 357
cotton 15–19
back-coated and halogen replacement 171–3
cotton/nylon blends 33–4
environmentally friendly flame resistant cotton 193–7
potential applications 193–7
research needs 193
military fabrics 196–7, 500–2
flame retardant cotton blends 501–2
flame retardant treatments for 100% cotton 500–1
cotton batting 200–1
engineered 198
cover fabrics 197–9
cracking 216–17
critical flux at extinguishment 542, 544
critical heat flux for ignition 265

Index 603

curing agents 426
cyclic phenoxy-phosphazenes 45
cyclic phosphonates 46, 171–2, 178

DDG-1000 Zumwalt class destroyer 115
death, risk of 374–5
decabromodiphenyl ether (decaBDE) 55, 167, 191–2, 199, 206, 220, 222
  replacement of 171–3
  risk reduction strategies 205–6
  toxicology 370–1
  VECAP risk management 204–5
decontamination 337
decorative textiles, aircraft 578, 579, 581–2
deflagration 495
delaminated nanocomposites 103–4, 105
Department of Transportation (DOT) (US) 575
depolymerisation 216, 262
design criteria 454, 455
detonation 495
developing fire 257
DGDPS 85, 86
DGMPS 85, 86
dialkylphosphinate salts 45
diammonium hydrogen phosphate (DAHP) 78
dihydric phenol product 48–9
dihydrodihydroxycarbonylpropylphospha-phenanthrene oxide (DDP) 83–4
dihydro-oxaphosphaphenanthrene oxide (DOPO) 46, 48
diluents 261–2
dimethyloldihydroxyethyleneurea (DMDHEU) 15–17
dimethylpropylene spirocyclic pentaerythritol biphosphonate (DPSPB) 74–5, 76
dioxins 2
  Fire-LCA model 341–2, 347–8, 351–4
dipentaerythritol 150–2
dispersion 132–3
disposal *see* recycling and disposal
divisions, fire resisting 540, 558–64, 565
dopotriol 84, 85
dose-response assessment 191
draperies 581–2
dripping 262
ducts
  aircraft 578, 579, 580, 582
  fires penetrating through in cars 459
durability requirements, military fabrics 521–2

ease of extinction tests 266–9
economic costs and benefits 378–81, 382
edge effects 245–6
efficiency, flame-retardant 68
  halogen replacement and improving 69–73
electrical components, aircraft 578, 579, 583
  tests on wire insulation 588–9
electrical conductivity 102
electrical engineering and electronics (E&E) equipment 292, 309–19, 327–8
  CE marking 318
  certification 317–18
  external ignition sources 318–19
  fire safety requirements 310
  fire safety standardisation 311
  fire tests 311–16
  German VDE approval procedures 318
Elementis Specialities clays 126, 127, 144–5
emergency firefighting conditions 469
encapsulation 77–9, 207
endocrine disruptors 208
energy recovery 215, 217–20
  recycled polymers 223
energy use in Fire-LCA model 339–40, 345, 346
engine compartment fires 452, 457, 458–60
engine covers 444, 459, 460
engineered cotton batting products 198
engineering design and analysis 250, 463
environmental impact 2
environmentally friendly flame resistant textiles 2, 188–212
  desirable properties of flame retardant chemicals 190–3
  future trends 206–8
  inherently flame retardant fibres 201–4
  key environmental/ecotoxicological issues 189–90
  potential applications for cotton-based textiles 193–7
  potential applications for mattresses, bedclothes and upholstered furniture 197–201
  risk management programmes 204–5
  risk reduction strategies 205–6
epoxy resins 131, 401–2, 412–14
  epoxy/amine hybrid resins 84
  epoxy/LDH nanocomposite 113

604 Index

phosphorus-based flame retardants 48–50, 113
EpSi 85, 86
equivalence ratio 259, 260
essential requirements (ERs) 294
ethylene vinyl acetate (EVA) 114, 133–4
ethylenediamine phosphate (EDAP) 53–4
Euroclasses 294, 298–301, 302
European Aviation Safety Agency (EASA) 573
European Chemicals Agency 205
European Committee for Electrotechnical Standardization (CENELEC) 311
European Union (EU)
  Construction Products Directive (CPD) 294, 326–7
  decaBDE 192
  Registration, Evaluation and Authorisation of Chemicals (REACH) 205, 373
  regulations 292
    building/construction 293–309, 326–7
    fire testing of construction products 301–9
    rail vehicles 320–1, 322, 323, 328
  residential fire statistics 366–7
  Restriction of Certain Hazardous Substances (RoHS) Directive 220
  standards for firefighters' protective clothing 476
  Waste Electrical and Electronic Equipment (WEEE) Directive 220–1
exfoliated nanocomposites 103–4, 105
Exolit OP products 45, 49
explosion 495
exposure assessment 191
exposure to flame retardants 189
  controlling and eliminating exposure risks 392–3
  hazards and risks 372–3, 376–7
  need to better understand 391–2
exposure pathway 189
exposure route 189
extended-chain polyethylene (ECPE) 413–14
external ignition/heat sources
  electrical engineering and electronics equipment 318–19, 328
  road vehicles 452, 461
extinguishment in Fire-LCA model 336–7

fabric construction 504–6

Federal Aviation Administration (FAA) 573, 574, 575–7, 596–7
Federal Aviation Regulations (FARs) 161, 404–5, 574–7
federal countries 293
Federal Motor Vehicle Safety Standard FMVSS 302 test 4, 13–14, 160–1, 237–8, 239–40, 445, 446, 462, 463–4
feedstock recycling 215, 216–17, 218–19
  flame retarded polymers 222–3
fibre-reinforced composites *see* composites
fibres
  inorganic man-made fibres 34–6
  natural fibres 14–23, 36
  potential applications of nanocomposites 136–47
  reinforcement *see* reinforcing fibres
  synthetic fibres 23–34, 36–7
field trials 475–6
FIGRA index 307
fillers 419–20
filling materials for furniture 200–1
films, nanocomposite 136–47
Finland CLAN smart fire suit 484–5
Fire-CBA 357
fire classification 302–3
fire curves 560, 561
fire growth models 286
Fire-LCA model 332, 334–56, 357
  cable 343–9
  furniture 349–55
  television 337–43
fire propagation apparatus 271–3, 285
Fire Protection Research Foundation of NFPA (FPRF) 448–50
fire resistance 575
fire resistance testing 255, 256, 263
fire resistant divisions 540, 558–64, 565
fire restricting materials 540
  room/corner test 548–58
fire risks/hazards
  maritime and naval 528–9
  military 494–5
  residential fires 364–7
fire-safe cigarettes 207
fire safety codes 249–50
fire safety regulations *see* regulations
fire scenarios (road vehicles) 452, 454, 455, 457–61
fire stages 256–7
  defining 258–61
fire standards, EU (fEN) 298

Index

fire statistics 1, 9–10, 11, 335, 337
  importance of collecting 391
  residential fires 365–8
  road vehicle fires 443, 444, 447–8, 449
Fire Test Procedures (FTP) Code 292, 321–4, 328–9, 407, 529–30
fire-test-response characteristics 251
fire testing 3–4, 255–90
  aircraft materials 585–96
  areas of fire testing 255–6
  chemical and physical processes 257–8
  composites for maritime and naval applications 540–64, 565
  conditions of each fire stage 256–7
  construction products 301–9
  defining the stages of a fire 258–61
  electrical engineering and electronics equipment 311–16
  factors affecting ignitability and fire development 261–2
  fire resistance tests 263
  fire toxicity testing 255–6, 275–86
  firefighters' protective clothing 475–83
  flammability tests *see* flammability tests
  measurement of heat release 269–75
  nanocomposites 106–8, 114
  prediction of fire behaviour from material properties and models of fire growth 286
  principles and problems 262–3
  reaction to fire tests 255–6, 263–9
  road vehicles 445, 446, 455–7
  standards and methodologies for military fabrics 514–21
fire toxicity testing 255–6, 275–86
  animal exposure methods 276–7
  chemical analysis methods 277
  flow-through tests 282–6
  general requirements 277–9
  need to quantify toxic gases 275–6
  NIST radiant furnace method 281–2
  tests based on NBS smoke chamber 252, 280–1
  toxic potency data 276
firefighters' protective clothing 4, 467–91, 514
  future trends 483–6
    firefighters' work 483
    new materials 483–4
    test method development 486
    wearable electronic technology 484–5
  heat exposure in different types of firefighting conditions 469–71

materials 472–5
  accessory materials 474–5
  moisture barrier 474
  outer shell 472–3
  thermal liner 473
  underwear 475
measuring flame and thermal performance 475–83
requirements 468–71
types and design of clothing 471–2
Firegard Brand Products 198–9
fireproof aircraft cabin 596
firewalls 459
First Attack Fire Suit 512–13
flame propagation rate 12
flame resistance 575
Flame Resistant Organisational Gear (FROG) 511
flame retardancy 4, 9–40
  developments in assessing 10–14
  future trends 36–7
  inorganic man-made fibres 34–6
  natural fibres 14–23
  synthetic fibres 23–34
Flame Retardant Environmental Ensemble (FREE) System 513
flame spread index (FSI) 542, 543
flame spread test 13–14, 476–7, 594
  composites for maritime and naval applications 541–3, 544
flaming combustion 188
  fire toxicity tests 278
flammability
  aircraft materials compared with other applications 584–5
  defining 233–4
  polymer nanocomposites 106–8
  mechanism of flammability reduction 108–12
flammability tests 11–14, 233–54, 255–6
  challenges in assessing material flammability 249
  electrical engineering and electronics 311–12, 313, 314, 327
  firefighters' protective clothing 476–7
  future trends in material flammability assessment 251
  limitations 250–1
  nanocomposites 106–8
  type I tests 234–40, 251
  type II.A tests 240–6, 251–2
  type II.B tests 246–9, 251–2
  uses of 249–50
flash resistance 575

flashover 495, 528, 548
flashover time 213
flexural modulus 558, 559
flexural strength 431
floor coverings
  aircraft 580, 582
  EN ISO 9239-1 test 308–9
  Euroclasses 298, 300–1, 302
flow-through tests 282–6
fluorescent materials 474
fluorocarbons 163, 164
foams
  nanocomposite applications 152
  polymer 403–4
  polyurethane 50–1, 200
formaldehyde-based resins 163, 164
45-degree Bunsen burner test 194, 586–7
FQ-POSS 28
FR-370 51–2
France
  furniture regulations 326
  rail vehicle regulations 320
FS-2 54
fuel production 261
full-scale tests 462–3
  military fabrics 520–1
  room/corner test for composites with maritime and naval applications 548–58
fully developed fire 257
fumed silica 174, 175
functional unit 333
functionalising groups 99–100
  type and stability 126–30
furans 341–2, 351–4
furniture
  Fire-LCA model 349–55
    results 351–4
    statistical fire model 349–51
  fire safety requirements and tests 324–6, 329
  UK 1988 regulations *see* United Kingdom
  upholstered furniture and environmentally friendly flame retardant applications 197–201
fuselage, composite 597
fusible polymers 163–4
Fyrol PMP 49–50
Fyrol PNX 50
Fyrolflex BDP 44
Fyrolflex RDP 43–4
Fyroltex HP 196

gasification (partial oxidation) 216
gelcoats 417
Germany
  Chemikalienverbotsverordnung 221
  rail vehicle regulations 320
  VDE approval procedures 318
glass fibres 35, 402
  composite flammability 409–12
glow-wire tests 312–16, 327
glowing/smouldering combustion 188
goal and scope definition 333
GORE-TEX 510
grafted-PP 131–2
graphite 99

Haloclean process technology 223
halogen-containing flame retardants 2, 55, 420–1
  challenges posed by replacing 67–9
  coatings and laminates 166–8
  see also bromine-containing flame retardants; chlorine-containing flame retardants
halogen-free flame retardants 2, 67–94
  attempts at successful halogen replacement 69–73
  back-coated textile formulations 171–4
  challenges posed by replacing halogen-containing flame retardants 67–9
  flame retardancy and anti-dripping properties 73–6
  flame retardancy and mechanical properties 76–86
  future trends 86–7
hardeners 426
harmonised product standards (hEN) 297–8
hazardous firefighting conditions 469
hazardous materials firefighting 470
hazards 190
  human exposure hazards and flame retardants 372–3
  identification 190–1
  and risks in residential fires 364–5
  role of flame retardants in reducing 368–9
  see also fire risks/hazards
heat exposure, firefighters and 469–71
heat flux 414
  military fire hazards 494–5
heat reflective finishes 180
heat release 14
  measurement 269–75
    bench-scale measurement 269–73

large-scale measurement 274–5
microscale measurement 273
reduction and road vehicles 464
total (THR) 542, 544, 548, 549–56
heat release rate (HRR)
 aircraft cabin materials 589–91
 charring materials 244
 composites
  fibre type and 409–12
  maritime and naval applications 547–8, 549–56
 peak *see* peak heat release rate (PHRR)
heat source (ignition source) 257
 characteristics 264
 effects and flammability tests 238, 241–2, 248–9
heat for sustained burning 542, 544
heat transfer 12–13, 468–9
 measuring protection against 477–80
heat transfer index (HTI) 479, 520
heavy-duty electrical enclosure applications 134
hexabromocyclododecane (HBCD) 192, 199, 206, 372
hexakis(methoxymethyl) melamine (HMMM) 84, 85
high performance fibres 23, 31–3
high speed craft
 fire resisting divisions 560
 HSC Code 407, 529
Hofmann elimination reaction 100
honeycomb cores 403
horizontal flame test 266–7, 311, 312, 313
 aircraft 576, 587–8
hot melt processes 160, 163–4
human subject tests 475, 481–2
hybridisation of reinforcing fibres 418–19, 420
hydrocarbon emissions 340, 342, 345–7
hydrocracking 217
hydrogen chloride 545–6
hydrogen cyanide 260, 545–6
Hydroweave 484
hydroxide functional organophosphorus oligomer (HFPO) 15–17
hydroxy double salts (HDS) 99, 100

ignition 257
 factors affecting ignitability 264–5
 factors and vehicle fires 448, 449
 reaction to fire tests 264–6
 time to *see* time to ignition
ignition pilot 242

ignition sources *see* heat sources (ignition sources)
ignition temperature 263, 264, 265
imidazolium derivatives 100, 116, 129–30
immiscible nanocomposites 103
impact assessment 333–4
improvised explosive devices 509
incapacitation concentration to 50% of exposed population (IC50) 276
induction period 256–7
Indura 501
Indura Ultra Soft 502
information technology 310, 318–19
inherent flame retardancy 82, 83, 201–4, 502–4
inhibitors 261–2
inorganic flame retardants 168–9
 see also alumina trihydrate (ATH); magnesium hydroxide
inorganic man-made fibres 34–6
inorganic phosphorus compounds 56
insulation
 thermal-acoustical in aircraft 578, 579–81
  fire testing 594, 595
 thermally insulating layers in composites 416–18
intercalated nanocomposites 103–4, 105
interfacial polymer 95–6
interior fabric fire blocking barriers 198–9
interior panels, aircraft 578, 579, 581
interlayer fire blockers 23, 25
International Civil Aviation Organization (ICAO) 574
International Code of Safety for High-Speed Craft 407, 529
International Convention for the Safety of Life at Sea (SOLAS) 247, 292, 529
International Electrotechnical Commission (IEC) 292–3, 311
International Electrotechnical Committee for Conformity Testing to Standards for Electrical Equipment (IECEE) 317
International Forum of Fire Research Directors 285
International Maritime Organization (IMO) 247, 321–4, 328–9, 407, 529
 'A' class divisions 559–60
 FTP Code 292, 321–4, 328–9, 407, 529–30

HSC Code 407, 529
lateral ignition and flame spread test (LIFT) 265–6
International Organization for Standardization (ISO) 293, 311, 332
fire stages 258, 259
intumescent flame retardants 17–18, 20–1, 74
coatings for composites 416–17
coatings and laminates 169–71
combined with nanoparticles 136, 138–41
intumescent/fire retardant fibre interactive fire retardant systems 423–5
maritime and naval applications 537–8
phosphone-based 52–4
inventory analysis 333
Iraq 496, 497, 509
isobutyl bis(hydroxypropyl) phosphine oxide (IHPOGly) 84, 85

K-131 submarine 494
Kabelwerk Eupen AG 114, 133
kaolin 72
Kermel 503–4
Kevlar 203
blended with Nomex for military applications 503
knitted fabrics 29–30
Kynol 32, 504

label legibility 519
laminates 4, 159–87
main types of fire retardant 161–71
additives 165–9
intumescent systems 169–71
novel or smart approaches 179–80
role of nanoparticles 174–5
volatile phosphorus-containing species 175–8
lateral ignition and flame spread test (LIFT) 265–6
laundering 522
layered double hydroxide (LDH) 98–9, 100
layered nanocomposites 98–100
layered products 249
layered silicates 98, 99–100
mechanism of flammability reduction 109, 110, 111
layers, clothing 505–6

Lenzing FR 202
lethal concentration to 50% of exposed population (LC_{50}) 276
life cycle assessment (LCA) 4, 331–59
cable case study 343–9
furniture case study 349–55
future trends 356–7
method 332–7
Fire-LCA system description 334–7
risk assessment approach 334
television case study 337–43
life cycle inventory (LCI) 332–3
limiting oxygen index (LOI) 12, 106, 108, 268–9
linings
aircraft 578, 579, 582–3
test for cargo liners 593
Euroclasses 298, 301, 302
Listing certification 317–18
long-term health effects 376
low-temperature heating elements 241–2
low velocity impact (LVI) response 432
Lyocell 24

macromolecular chelates 72–3
magnesium hydroxide 68, 77, 168–9
manikin tests 12–13
firefighters' protective clothing 475, 481
military fabrics 519, 520–1
maritime and naval applications 5, 115, 527–72
composites 400, 406–7, 527, 538–64
fire performance and test methods 540–64, 565
performance regulations 531, 532
fire safety regulations 529–31
maritime vessels 321–4, 328–9, 529–31
naval vessels 531
fire threat 528–9
methods of reducing flammability 531–8
flame retardants 534–7
intumescent coatings 537–8
modification of chemical structure 533–4
smoke chamber test 280–1
mass loss cone fire test 462, 463
mass loss rate reduction 108–12
mattresses 197–201
maximum optical density 545–6
McDonnell Douglas 576
measurement uncertainty 252

Index   609

mechanical properties
  composites 430–4, 435
    degradation during and after fire 430–3
    post-fire mechanical performance of fire retardant composites 433–4
  halogen-free flame retardants and 68, 76–86
    additive type and intrinsically flame-retardant polymers 79–82, 83
    encapsulation and microencapsulation 77–9
    reactive type flame retardants and copolymerisation 82–6
mechanical recycling 214–16
  flame retarded polymers 222, 223
melamine 200, 371
melamine cyanurate (MCA) 45, 78, 113
melamine phosphates 42, 53, 150–2, 178
  coatings and laminates 169, 170
Melapur 200 53
melt dripping 69
  halogen-free flame retardants and anti-dripping properties 73–6
melt viscosity 130–1
melting 249
membrane structures, military 507–9
meta-aramids 32–3, 203
  blends with PBO 472–3
metal chelates 71–2
metal hydroxides 135–6, 419–20
metal matrix composites (MMCs) 399
methylol dimethylphosphonopropionamide (MDPA) 17
microcapsulated red phosphorus (MRP) 78
microcomposites 103
microencapsulation 77–9
military fabrics 4, 492–526
  applications of flame retardant textiles 507–9
    clothing 509
    structural 507–9
  burn injuries 493, 496–9
  clothing protection 499–500
  existing flame retardant solutions 500–7
    cotton 500–1
    cotton blends 196–7, 501–2
    garment design and fabric construction 504–6
    inherently flame resistant synthetics 502–4

    nonwoven fabrics 506–7
  fire hazards 494–6
  future trends 522–3
  performance standards and durability requirements 521–2
  testing standards and methodologies 514–21
    bench-scale tests 515–20
    full-scale tests 520–1
  types of military clothing 510–13
modacrylics 201
moisture
  barrier 474
  effect on firefighters' protective clothing 482
molecular modification of chemical structure 533–4
monoguanidine dihydrogenphosphate (MGHP) 17
monohydric alcohols 50
montmorillonite (MMT) 19, 98
mounting, specimen 239–40, 245–6
MPC 138–41, 142
multi-climate protection system (MCPS) 510–11
multivariate analysis 356
multi-walled carbon nanotubes (MWNT) 29–30, 101–2
municipal solid waste (MSW) 217

N-Class Divisions 560–1
  polymer sandwich composites 561–4, 565
nanocomposites 3, 19, 20, 37, 95–158, 179–80, 483
  applications for flame retardant nanocomposites 114–15
  applications and processing challenges 125–33
    dispersion 132–3
    effect on rheology 130–2
    functionalising group type and stability 126–30
  clay-based see clay nanocomposites
  combined with conventional flame retardants 87, 112–13, 136, 138–46
  determination of morphology 104–6
  environmentally friendly flame resistant textiles 207
  flammability 106–8
  formation 103–6
  future trends 115–17, 153
  halogen-free flame retardants 70, 87

history in flame retardancy applications 97
mechanism of flammability reduction 108–12
one-dimensional 98–100
plasma-coated polyamide 6 films 182–3
potential application areas 133–52
  bulk polymeric components 133–6
  coatings 148–9
  composites 149–52
  films, fibres and textiles 136–47
  foams 152
role in composites 427–30, 435
three-dimensional 102
two-dimensional 101–2
Nanocor Inc. 126, 132–3, 134
nanofilms 179–80
nanoparticles 96
inclusion in resin and fibre components of composites 427–30, 435
for nanocomposite use 98–102
role in coatings and laminates 174–5
NAS report 192–3
National Fire Protection Association (NFPA) (US)
FPRF white paper 448–50
hazard and risk committee (HAR) 450–2
NFPA 556 draft of 2007 451–63, 464
  approaches to key fire scenarios 457–61
  background information for fire hazard 453, 454
  draft recommendations 462–3
  methodology to decrease fire hazard 453–7
NFPA 2112 testing standards 514, 518–19
room/corner test 247–8
National Highway Traffic Safety Administration (NHTSA) (US) 445, 446
natural fibres 14–23, 36
  see also cotton
natural rubber 162, 164
naval applications see maritime and naval applications
Naval Sea Systems Command (NAVSEA or Naval Technical Authority) 531
NBS smoke chamber 252, 280–1
needle-flame test 316
needle-stick injuries 409

Netherlands, the 484
new materials 483–4
NFX 70-100 test 282–3
nickel formate 70–1
NIST radiant furnace method 281–2
nitrogen-containing flame retardants 87
Noflan 21
Nomex 32–3, 203, 502–3
  Limitedwear military coverall 512
non-break-open protection 503
non-combustibility test 303, 304
non-flaming fire stage 258–61
nonwovens 22–3, 506–7, 522–3
notified bodies 297
novoloid fibres 504
nylon see polyamides

octabromodiphenyl ether (octaBDE) 191
office electronics 310, 318–19
Ohio State University (OSU) calorimeter 14, 589–90
oil burner test
  burn through resistance of thermal/acoustic insulation 595
  cargo liners 593
  seat cushions 592–3
oligomeric phosphate-phosphonate 178
oligomeric phosphates 42
oligosalicylaldehyde complexes 71
one-dimensional nanocomposites 98–100
open tests for fire toxicity 278–9
organically modified clays 99–100
orientation, specimen 239, 244–5, 262
Orkla mine sweeper 540
outer shell protective clothing 472–3
oxide fibres 35, 36
oxidized PAN fibres 32
oxygen consumption principle 547–8

packaging derived fuel (PDF) 220
para-aramids 32, 203
partial oxidation (gasification) 216
passenger compartment fires, in cars 452, 457, 458
passive countermeasures 369
  benefits of passive and active measures working together 386–8
PBI Gold 504
PBT 47
peak heat release rate (PHRR) 96, 262–3
  maritime and naval composites 542, 544, 548, 549–56
  nanocomposites 107–8
Pearl Harbour 498

penetration of fire retardant upholstered furniture 389–91
pentabromodiphenyl ether (pentaBDE) 55, 191
pentabromodiphenyl oxide 50
pentaerythritol (PER) 78–9, 138–41, 142
performance-based codes and regulations 250
performance standards for military fabrics 521–2
persistence 190
phenolic resins 48, 402, 412–14, 533–4
Phoslite series 55
phosphate-phosphonate ester flame retardant 196
phosphine oxides 49
phosphinic acid esters 42
phosphonic acid esters 42–3
phosphonium-substituted clays 99–100
phosphoramides 45
phosphoric acid esters 42
phosphorus-containing regenerated cellulose 202
phosphorus flame retardants 2, 41–66, 87, 206, 421
　for cellulosics 195–6
　coatings and laminates 165–6
　　halogen replacement 172–3
　cover fabrics 199
　epoxy resins 48–50, 113
　future trends 55–7
　main types of 42–3
　nanocomposites 150–2
　phosphorus and chlorine-containing flame retardants 167–8
　polycarbonate and its blends 43–5
　polyesters and nylons 45–8
　polyolefins 51–5
　polyurethane foams 50–1
　vapour-phase active systems 57, 166, 173, 175–6
　volatile phosphorus-containing species 166, 173, 175–8
phosphorus-nitrogen containing intumescent flame retardant (P-N IFR) 75–6
phosphorus oxynitride $(PON)_m$ 47
phthalonitrile resin 534
pilot flames 242
PIPD (M5) 32
plasma graft polymerisation 19, 21, 181–4
plasma treatment 37, 483
Plastics Recognized Component Directory 318

PNX 50
Polartec fabrics 510–11
polyacrylonitrile nanocomposites 147
polyamides 26–7, 113
　nanocomposite films 138–41, 182–3
　nanocomposites 27, 137–8
　phosphorus flame retardants 45–8
polyaromatic hydrocarbon (PAH) emissions 340, 342, 352, 353–4
polybenzimidazole (PBI) 472, 504
　PBI Gold 504
polybenzoxazole (PBO) 32, 472–3
polybrominated biphenyls (PBBs) 220
polybromodibenzodioxins and furans (PBDD/F) 221, 222
polybromodiphenyl ethers (PBDEs) 191–2, 220
polycarbonates 43–5
polycarboxylic acids 196
polychlorodibenzodioxines and furans (PCDD/F) 223
polydimethylsiloxane (PDMS) star polymers 82
polyester 24–6
　batting 201
　cotton/polyester blends 33, 502
　inherently flame retardant 202
　phosphorus-based flame retardants 45–8
polyhedral oligomeric silsesquioxanes (POSS) 85, 102, 534
　mechanism of flammability reduction 111–12
polyhydroxy propylene spirocyclic pentaerythritol biphosphonate (PPPBP) 25–6, 75
polylactic acid (PLA) 24, 26
polymelamine fibre 204
polymer composites *see* composites
polymer foams 403–4
polymer fuel 220
polymer matrix composites (PMCs) 399–400
　see also composites
polymeric additive-type halogen-free flame retardants 79–82
polymerisation 103
polymethyl methacrylate (PMMA) nanocomposites 107, 108
polyolefins 51–5
PolyOne Maxxam FR products 114
polypropylene 27–31, 202
　nanocomposites 141–6
polystyrene nanocomposites 107, 134–5

polysulphonyl diphenylene thiophenylphosphonate (PSTPP) 80
polysulphonyl phenylenephosphonate (PSPPP) 79–80
polytetrafluoroethylene (PTFE) 583
polyurea 535–7
polyurethanes 163, 164
   foams 50–1, 200
polyvinyl alcohols (PVAs) 163, 164
polyvinyl chloride (PVC) 133–4, 162, 164
   cable and LCA 343–9
pool fires 452, 457, 461
precondensate/ammonia process 195
primary fires 336
primary mechanical recycling 214, 215
Proban 500–1
   Proban CC 138–41, 142
product-related fire incidence 367–8
production control 250
protection goals 258
protective barriers 108–12
protective clothing
   firefighters' *see* firefighters' protective clothing
   military *see* military fabrics
   standards and performance requirements 514, 516–17
protective layer, carbonaceous/vitreous 169–71
protective spring assembly 506
PsiNII 81, 82
Purser furnace 283–5, 286
pyrolysis 216–17
pyrolysis-combustion flow calorimetry (PCFC) 273
Pyrovatex 195–6, 500, 501

radiant furnace toxicity test 281–2
radiant heat 468
   heat source effects in flammability tests 241–2
   measuring protection against for firefighters' protective clothing 477–8
radiant heat transfer index (RHTI) 477, 478, 520
radiant panel test 308–9, 321, 322, 594
railway vehicles 280–1
   composites 405–6
   regulations 320–1, 322, 323, 328
reaction to fire tests 255–6, 263–9
   cone calorimeter 265

ease of extinction tests 266–9
ignitability and flame spread 264–5
LIFT/spread of flame apparatus 265–6
see also fire toxicity testing; flammability tests
reactive flame retardants 56
   composites 419–25
   halogen-free 82–6
   phosphorus-based 195–6
Recognition certification 317–18
recycling and disposal 3, 213–30
   directives on 220–1
   processes 214–20
   recyclability of flame retarded polymers 221–3
red phosphorus 47–8, 55
   encapsulation 77–8
reduced ignition propensity (RIP) cigarettes 207
refuse derived fuel (RDF) 217
regenerated cellulose 201–2
Registration, Evaluation and Authorisation of Chemicals (REACH) 205, 373
regulations 3–4, 249–50, 291–330
   building/construction 292, 293–301, 326–7
   competing and contradictory 393
   engineering and electronics equipment 292, 309–19, 327–8
   fire testing of construction products 274–5, 301–9
   furniture 324–6, 329
      UK 1988 regulations *see* United Kingdom
   future trends 326–9
   heat release tests for regulations and control 274–5
   history of aircraft regulations 574–7, 578
   maritime and naval 321–4, 328–9, 529–31
   overview 291–3
   transportation 320–4, 328–9
      rail vehicles 320–1, 322, 323, 328
      road vehicles 445–7
reinforcing fibres 95, 402–3, 539
   fibre type and flammability of composites 409–12
   resin/fibre combinations 412–14
   hybridisation 418–19, 420
   inclusion of nanoparticles 427–30, 435
Reogard products 52, 54
rescue tasks, firefighters' 470, 472

residential fires 363
  fire hazards and risks 364–5
  in the UK, US and Europe 365–7
  targets for improvement in risk reduction 393–4
  see also consumer products; smoke alarms
resins 401–2, 538–9
  co-curing of different resins 427
  nanoparticle inclusion 427–30, 435
  resin/fibre combinations 412–14
  resin modification 425–7, 533–5
  type and flammability of composites 407–9
resorcinol bisdiphenyl phosphate (RDP) 43–4
resorcinol bisdixylyl phosphate (RXP) 44
Restriction of Certain Hazardous Substances (RoHS) Directive 220
retainer frame 245–6
retro-reflective materials 474
rheology
  of decomposing polymer 262
  of nanocomposites 130–2
risk 190
  primary risks of flame retardants in consumer products 369–73
  role of flame retardants in reducing 368–9
  see also fire risks/hazards
risk assessment 334
risk-benefit analysis 363–97
  balance of risk 373–7
  benefits of flame retardants in consumer products 377–81
  differentiating effects of smoking and smoke alarm trends 381–91
  importance of understanding risks and benefits 364–9
risk management programme 204–5, 371, 373
risk reduction strategies 205–6
road vehicles 4–5, 160–1, 443–66
  composites 405–6
  FPRF white paper 448–50
  improving survivability 463–4
  NFPA 556 guide 2007 draft 451–63, 464
  NFPA hazard and risk committee 450–2
  regulatory requirements 445–7
  US statistics 443, 444, 447–8, 449
room/corner test 246–8
  composites with maritime and naval applications 548–58

construction products testing in the EU 304–6
routine firefighting conditions 469
Royal Navy 501, 509
rubbers 162, 164

Safety of Life at Sea (SOLAS) Convention 247, 292, 529
SaFRon 5371 52
sagging 249
sandwich composites 539
  fire resistant divisions 561–4, 565
scaffolding effect 413
scenarios, fire (road vehicles) 452, 454, 455, 457–61
scorch 50, 51
screening tests 250, 462, 463
SCRIMP composite fabrication method 539, 540
seat cushions, aircraft 576–7
  oil burner test 592–3
seats, aircraft 578, 579, 580
secondary fires 336
secondary mechanical recycling 215–16, 222
self-extinguishing epoxy-LDH nanocomposite 113
ships *see* maritime and naval applications
short-term injury, risk of 375–6
shredding technologies 215–16
shrinkage 480, 518
silica, fumed 174, 175
silica-containing regenerated cellulose 201–2
silicon carbide fibres 35
silicon-containing flame retardant epoxy resins 85, 86
silicone-based aliphatic polyamides 82, 83
silicone-based aromatic polyesters and polyamides 82, 83
silicones 163, 164
silk 22
simulated chair 325
single burning item (SBI) test 274–5, 304, 306–7
single European market 293–4
single-wall carbon nanotubes (SWNT) 101–2
60-degree Bunsen burner test 588–9
sleepwear 193–6
small flame tests 234–40, 307–8
small parts, aircraft 578, 579, 584–5
smart fire suit 484–5

smectite clays 98, 99–100
SMOGRA index 307
smoke 262
  aircraft 576
    cabin materials testing 591–2
  composites for maritime and naval applications 543–7
smoke alarms 381–91
  trends 384–6
smoke chamber tests 252, 280–1, 321, 323
smoking 381–91
  trends 381–4
smouldering/glowing combustion 188
socio-economic factors 388–9
solvent-based systems 160
sorting technologies 216
Southern Clays Inc. 126, 127
spacer fabrics 505–6
specialized firefighting 469, 471
specific extinction area (SEA) 548, 549–56
specific optical density 544–6, 591, 592
specifications, technical 250
specimen effects
  area 242–3
  mounting 239–40, 245–6
  orientation 239, 244–5, 262
  thickness 239, 243–4, 245
  type I flammability tests 239–40
  type II.A flammability tests 242–6
  type II.B flammability tests 248–9
sphere/colloid-based nanocomposites 102
spirocyclic flame retardants 74–5
spread of flame apparatus 265–6
spring systems 505
sprinkler systems 394
standard substrates 303
standards 188–9
  electrical engineering and electronics equipment 311
  firefighters' protective clothing 476, 489–91
  harmonised product standards 297–8
  military protective clothing performance standards 521–2
  testing standards and methodologies 514–21
static smoke chamber methods 251–2
statistical fire model 336–7
  cable 344–5
  furniture 349–51
  television 338–9

steady state tube furnace methods 283–5, 286
steam 482
stiffness of composites 430, 432–3, 434
storage compartment fires, in cars 452, 457, 460–1
stored heat energy 483
strength of composites 430–1, 433, 434
structural fabrics, military 507–9
structural firefighting 469–70, 471, 483
structural integrity of composites 558–64, 565
submarines 406–7, 494
substrate 244
  standard substrates 303
Sud-Chemie AG 126, 127
surfactants 97, 100
survivability, improving in road vehicles 463–4
sustainable development 356–7
synthetic fibres 23–34, 36–7
  inorganic 34–6
  military fabrics 502–4
synthetic rubbers 162, 164

tapestries, wool 581
Tecnofire ceramic webs 417, 418
televisions 363
  Fire-CBA 357
  Fire-LCA 337–43
    results 339–42
    statistical fire model 338–9
temperature
  gradient in sandwich polymer composites 564, 565
  ignition temperature 263, 264, 265
tenability criteria 453, 454
tensile strength 431
tension membrane structures 507–9
tents, military 507–9
terrorism, war against 493
tetrabromobisphenol A (TBBPA) 55, 371
tetrabromodibenzo(p)dioxin-equivalents (TBDD-equivalents) 341–2, 351–4
tetrachlorodibenzo(p)dioxin-equivalents (TCDD-equivalents) 341, 342, 347–8, 351–4
tetrakis(hydroxymethyl) phosphonium chloride (THPC) 15, 43, 195, 500
TGPS 85, 86
thermal-acoustical insulation 578, 579–81
  fire tests 594, 595
thermal cracking (pyrolysis) 216–17
thermal inertia 258

Index 615

thermal insulation layers, in composites 416–18
thermal liner 473
thermal protective index (TPI) 515
thermal protective performance 515–20
thermal resistance, measuring 480
 thread heat resistance 480–1
thermogravimetric analysis (TGA) curves 409, 410
thermomechanical models 435
thermoplastic polyurethane (TPU) 133–4
thermoset polymers 538–9
thickness
 coating 180
 composites 414, 415
 specimen 239, 243–4, 245
 thermal thickness 137, 180, 243–4
thinness, thermal 137, 180, 243–4
thread heat/melting resistance 480–1, 519
three-dimensional nanocomposites 102
time to ignition 263, 264–5
 composites 409–11, 548, 549–56
 nanocomposites 96–7
total heat release (THR) 542, 544
 composites 548, 549–56
toxic gas products 364–5
 aircraft 576
 composites with maritime and naval applications 543–7
 fire stages and 260–1
 generation 277–9
 quantification 275–6
toxicity
 characterisation of 190–1
 fire toxicity testing *see* fire toxicity testing
traffic accidents 470, 472
transition metal salts 173–4
transmission electron microscopy (TEM) 104–5
transportation 4–5
 regulations 292, 320–4, 328–9
  rail vehicles 320–1, 322, 323, 328
  ships 321–4, 328–9
 see also aircraft; road vehicles
Trevira CS 25, 202
triaryl phosphates 50–1, 54
triazine polymers 74
tributyl phosphate (TBP) 176–8
tri-component fibres 522
trimethylolmelamine (TMM) 17
triphenyl phosphate (TPP) 44, 176–8
triphenylphosphine oxide (TPPO) 176
tris-chloropropyl-phosphate (TCPP) 372

trunk, fires in 452, 457, 460–1
tube furnace fire toxicity tests 282–5, 286
tube/rod-based nanocomposites 101–2
 see also carbon nanotubes (CNT)
tungsten filament lamps 242
two-dimensional nanocomposites 101–2
type I flammability tests 234–40, 251
 FMVSS 302 test 4, 13–14, 160–1, 237–8, 239–40, 445, 446, 462, 463–4
 general concepts 234
 heat source effects 238
 specimen effects 239–40
 standard ASTM small-flame exposure ignition tests 234, 235
 UL 94 vertical burning test 108, 235–7, 238, 239, 251, 266–8
type II.A. flammability tests 240–6, 251–2
 general concepts 240
 heat source effects 241–2
 observations and measurements 246
 specimen effects 242–6
type II.B flammability tests 246–9, 251–2
 ignition source and specimen configuration effects 248–9
 room/corner test 246–8, 304–6, 548–58

Ukanol FR50/1 46
underwear 475
Underwriters' Laboratories (UL)
 certifications 317–18
 UL-94 tests 310
  classifications 237, 267
  horizontal burn method 266–7
  UL-94 V-rated plastics 114
  vertical burn test 108, 235–7, 238, 239, 251, 266–8
under-ventilated flaming 258–61
unit risk factor (URF) model 342, 353–4
United Kingdom
 building regulations 293
 differentiating the effects of smoking and smoke alarm trends 381–91
 life cycle assessment of furniture 349–55
 1988 legislation on furniture 324–5, 363
 benefits of 378–81, 382
 DTI study on effect of 355
 future prospective trends and benefits 389–91
 railway vehicle regulations 320
 residential fire statistics 365–8

616 Index

Royal Navy 501, 509
United States (US)
   Army 513
      Natick Soldier Center (NSC) 492, 522
   Coast Guard 530–1
   DOT 575
   FARs 161, 404–5, 574–7
   firefighters' protective clothing 470–1
   FMVSS 302 test 4, 13–14, 160–1, 237–8, 239–40, 445, 446, 462, 463–4
   Marine Corps (USMC) 506–7
   Navy 512, 558
      N-Class Divisions 560–4, 565
      polymer composite structures 539
   PBDEs 191–2
   regulations 292, 293
      furniture 326, 329
   residential fires 365–6
   road vehicle fires 443, 444, 447–8, 449
   smoking and smoke alarm trends 382–4, 385
unoccupied areas of aircraft 596
unsaturated polyester resins 401
upholstered furniture 197–201

Valzon 502
vapour-grown carbon nanofibre (VGCNF) 101–2
vapour-phase-active phosphorus systems 57, 166, 173, 175–6
vertical flame test 194–5, 311, 312, 313
   aircraft 575–6, 585–6
   military fabrics 515, 518
   UL-94 108, 235–7, 238, 239, 251, 266–8
vinyl acetate copolymers 163, 164
vinyl ester resins 401, 413–14, 535, 536
vinylidene chloride 201

vinylphosphates 56
viscose-Kermel blends 503–4
viscosity 130–1
visibility of clothing 474
VISIL 201–2
vitreous protective layer 169–71
volatile phosphorus-containing species 166, 173, 175–8
Voluntary Emissions Control Action Program (VECAP) 204–5, 371, 373

wainscoting 581
warehouse structures, military 507–9
Waste Electrical and Electronic Equipment (WEEE) Directive 220–1
waste stowage compartment 586–7
wear, effect of 522
wearable electronic technology 484–5
well-ventilated flaming 258–61
wholly aromatic thermotropic copolyesters (PLCP) 80–1
wildland firefighting 470, 471
windows, aircraft 578, 579, 580, 583–4
windshields, car 459–60
wires *see* cables/wires
wood crib tests 325
wool 19–21
   wool/PPTA blends 34
World Fire Statistics Centre 391
wrinkle-resist resin 196

X-ray diffraction 104

zinc borate 169
zinc hydroxyl stannate 169
zinc stannate 169
zirconium compounds 206
Zirpro process 20